D1713782

Ball Grid Array Technology

Electronic Packaging and Interconnection Series
Charles M. Harper, Series Advisor

CLASSON • *Surface Mount Technology for Concurrent Engineering and Manufacturing*

GINSBERG AND SCHNORR • *Multichip Modules and Related Technologies*

HARPER • *Electronic Packaging and Interconnection Handbook*

HARPER AND MILLER • *Electronic Packaging, Microelectronics, and Interconnection Dictionary*

HARPER AND SAMPSON • *Electronic Materials and Processes Handbook, 2/e*

LICARI • *Multichip Module Design, Fabrication, and Testing*

Related Books of Interest

BOSWELL • *Subcontracting Electronics*

BOSWELL AND WICKAM • *Surface Mount Guidelines for Process Control, Quality, and Reliability*

BYERS • *Printed Circuit Board Design with Microcomputers*

CAPILLO • *Surface Mount Technology*

CHEN • *Computer Engineering Handbook*

COOMBS • *Printed Circuits Handbook, 3/e*

DI GIACOMO • *Digital Bus Handbook*

DI GIACOMO • *VLSI Handbook*

FINK AND CHRISTIANSEN • *Electronics Engineers' Handbook, 3/e*

GINSBERG • *Printed Circuits Design*

JURAN AND GRYNA • *Juran's Quality Control Handbook*

MANKO • *Solders and Soldering, 3/e*

RAO • *Multilevel Interconnect Technology*

SZE • *VLSI Technology*

VAN ZANT • *Microchip Fabrication*

To my daughter Judy (who helped to design the book cover) and my wife Teresa for their love, consideration, and patience in allowing me to work on many weekends for this private project. Their simple belief that I am making my small contribution to humanity was a strong motivation for me, and to them I dedicate my efforts on this book.

Library of Congress Cataloging-in-Publication Data

Ball grid array technology / John H. Lau, editor.
p. cm.—(Electronic packaging and interconnection series)
Includes bibliographical references and index.
ISBN 0-07-036608-X (acid-free paper)
1. Ball grid array technology. I. Lau, John H. II. Series: Electronic packaging and interconnection series.
TK7870.15.B35 1995
621.381'046—dc20 94-38302
CIP

1 2 3 4 5 6 7 8 9 0 DOC/DOC 9 0 9 8 7 6 5 4

ISBN 0-07-036608-X

The sponsoring editor for this book was Stephen S. Chapman and the production supervisor was Donald F. Schmidt. It was set in Century Schoolbook by North Market Street Graphics.

Printed and bound by R. R. Donnelley & Sons Company.

This book is printed on acid-free paper.

Ball Grid Array Technology

John H. Lau
Editor

McGraw-Hill, Inc.
New York San Francisco Washington, D.C. Auckland Bogotá
Caracas Lisbon London Madrid Mexico City Milan
Montreal New Delhi San Juan Singapore
Sydney Tokyo Toronto

Contents

Preface

Ball Grid Array (BGA) is not just an emerging technology—it is quickly becoming the package of choice. The last few years have witnessed an explosive growth in the research and development efforts devoted to Ceramic BGAs (CBGA), Plastic BGAs (PBGA), and Tape Automated Bonding (TAB) BGAs (TBGA) as a direct result of the limitations of Fine-Pitch Surface Mount Technology (FPT) and Pin Grid Array (PGA) technology. The BGA technology offers many advantages over FPT and PGA. The most prominently stated benefits of BGA are: reduced coplanarity problems (no leads); reduced placement problems (self-centering); reduced paste printing problems (bridging); reduced handling issues (no damaged leads); lower profile (smaller size); better electrical performance; better thermal performance; better package yield; better board assembly yield; higher interconnect density; cavity-up or -down options; multilayer interconnect options; higher I/Os for a given footprint; shorter wire bonds; easier to extend to multichip modules; and faster design-to-production cycle time.

However, as with any new technology, BGA is not without its problems. In the development of BGA, the following issues must be noted and understood in order to obtain its full benefits: the infrastructure of FPT and PGA is well established; BGA expertise is not commonly available; rework methods are more difficult; solder joint reliability is more critical; BGA availability is limited; BGA testability is not well established; BGA standardization is just starting; BGA assembly inspection is more difficult; BGA cost is higher at low volume; and PBGA cracking has been reported during reflow.

We are now beginning to obtain useful insight and understanding of the economic, design, material, process, equipment, manufacturing, quality, and reliability issues of BGAs. The important BGA parameters such as the Integrated Circuits (ICs), substrates, packaging, routing capabilities, thermal and electrical management, assembly processes, reliability, inspection, rework, test, burn-in, and infrastructure have been studied by many experts. Their results have already been disclosed in diverse journals or, more incidentally, in the proceedings of many conferences, symposia, and workshops whose primary emphasis is material science or electronic packaging and interconnection. Consequently, there is no single source of information devoted to the state of the art of BGA technology. This book aims to remedy this deficiency and to present, in one volume, a timely summary of progress in all aspects of this fascinating field.

This book covers the three most popular BGA technologies (CBGA, PBGA, and TBGA), and the sequence of this book is basically following the BGA processes. A brief introduction to BGA technologies is presented in Chapter 1.

The book's next two chapters present BGA substrate technologies. In Chapter 2, Richard Sigliano describes ceramic technology and its applications to CBGAs. The merits as well as demerits of plastic technology and its applications to PBGAs are discussed by Osamu Fujikawa and Motoji Kato (Chapter 3).

The advantages in packaging and assembly with BGAs can be offset by the difficulties created in Printed Circuit Board (PCB) routing. The BGA package needs to be carefully chosen with PCB routing in mind. In Chapter 4, Patrick Hession examines the impact of PCB layout on the use of BGAs.

The next four chapters of this book present the CBGA technology. In Chapter 5, Thomas Caulfield, Marie Cole, Frank Cappo, Jeff Zitz, and Joseph Benenati present the design, material, and process of CBGA package carriers. Key considerations of materials, processes, and equipment for assembling CBGAs on PCB are discussed in Chapter 6 by Donald Banks, Karl Hoebener, and Puligandla Viswanadham. In Chapter 7, Yung-Cheng Lee, Jay Liu, Robert Tsai, and Jeffrey Zitz present some specific thermal and electrical characteristics of CBGA assemblies. Chapter 8, by Yifan Guo and John Corbin, provides the mechanical behavior and the reliability of the CBGA assemblies.

The next five chapters present the PBGA technology. In Chapter 9, Robert Marrs presents the design, material, and process of PBGA package carriers. Key considerations of placement, process compliance, yields, factory handling, moisture, and repair for assembling PBGAs on PCB are discussed in Chapter 10 by Bill Mullen, Allen Hertz, Barry Miles, and Robert Darveaux. The electrical and thermal issues of PBGA assemblies are discussed by Phil Rogren (from the vendors' perspective) in Chapter 11, and by David Walshak and Hassan Hashemi (from the users' perspective) in Chapter 12. Extensive data, analytical techniques, and a generalized methodology for PBGA solder joint reliability assessment are provided by Robert Darveaux, Kingshuk Banerji, Andrew Mawer, and Glenn Dody (Chapter 13).

By combining most of the advantages of C4 (Controlled Collapse Chip Connection) technology and TAB technology, IBM was the first to develop the TBGA technology. In Chapter 14, the design, material, and process of TBGA package carriers, and the electrical and thermal issues of TBGA assemblies are discussed by Chin-Ching Huang and Ahmad Hamzehdoost.

The next three chapters present the inspection and rework of BGA assemblies, and the test-socket of BGAs. In Chapter 15, John Adams discusses the nondestructive and automated process test of BGA PCB assemblies with x-ray techniques. Key considerations of the BGA PCB assembly rework process and equipment are presented by Tom Chung and Paul Mescher (Chapter 16). In Chapter 17, Christopher Schmolze presents the material, hardware, and contact issues of BGA burn-in sockets.

The last two chapters present two very important subjects for BGA technology readers. In Chapter 18, E. Jan Vardaman describes the current BGA industry infrastructure (i.e., the relationship and interaction between users and suppliers of the package). In Chapter 19, Ronald Gedney presents, arguably, the most complete glossary in print for electronics packaging.

Some duplication of material between chapters is necessary if each chapter is to offer the reader all the information essential for understanding the subject matter. An attempt has been made to provide a degree of uniformity in perspectives, but diverse views on certain aspects of BGAs are a reality. I hope that their inclusion here is seen as an unvarnished reflection of the state of the art and a useful feature of the book.

For whom is this book intended? Undoubtedly it will be of interest to three groups of specialists: (1) those who are active or intend to become active in research and development of BGAs; (2) those who have encountered practical BGA problems and wish to understand and learn more methods of solving such problems; and (3) those who have to choose a high-performance and cost-effective packaging technique for their interconnect system.

I hope this book will serve as a valuable source of reference to all those faced with the challenging problems created by the ever-more-expanding use of BGAs in electronics packaging and interconnection. I also hope that it will aid in stimulating further research and development on ICs, substrates, solder bumping, wire bonding, TAB, fine-line-and-space high-density low-cost PCB, chip attachment, encapsulant and underfill epoxy, flux, printing, placement, mass reflow, cleaning, inspection, rework, testing, electrical and thermal managements, reliability, equipment, material, process, and design, and more sound use of BGA technologies in either single-chip or multichip packaging applications.

The organizations that learn how to design and manufacture BGAs in their interconnect systems have the potential to make major advances in electronics packaging and to gain great benefits in cost, performance, quality, size, and weight. It is our hope that the information presented in this book may assist in removing roadblocks, avoiding unnecessary false starts, and accelerating design, material, and process development of these technologies. The BGA technologies are limited only by the business constraints, ingenuity, and imagination of engineers, visionless managements, and infrastructures.

John H. Lau, Ph.D., PE
Hewlett-Packard Company

Acknowledgments

Development and preparation of *Ball Grid Array Technology* was facilitated by the efforts of a number of dedicated people at McGraw-Hill and North Market Street Graphics. I would like to thank them all, with special mention to Mr. Steve Chapman of McGraw-Hill and Ms. Christine Furry of North Market Street Graphics for their unswerving support and solving many problems that arose during the book's preparation. It has been a great pleasure and fruitful experience to work with them.

The material in this book has clearly been derived from many sources—including individuals, companies and organizations—and the various contributing authors have attempted to acknowledge, in the appropriate parts of the book, the assistance that they have been given. It would be quite impossible for them to express their thanks to everyone concerned for their cooperation in producing this book, but on their behalf, I would like to extend due gratitude, especially to the following: the American Society of Mechanical Engineers (ASME) Electronic Packaging & Interconnection (EP&I) Conferences, Proceedings, and Transactions (e.g., *Journal of Electronic Packaging*); the Institute of Electrical and Electronic Engineers (IEEE) EP&I Conferences, Proceedings, and Transactions (e.g., *Hybrids, Packaging, and Manufacturing Technology*); the International Society of Hybrid Microelectronics (ISHM) and the International Electronic Packaging Society (IEPS) EP&I Conferences, Proceedings, and Transactions (e.g., *Microcircuits & Electronic Packaging*); and American Society of Metals (ASM) EP&I Conferences, Proceedings, and books (e.g., *Electronic Materials Handbook, Volume 1, Packaging*).

I express my deep appreciation to the 37 contributing authors, all experts in their respective fields, for their many helpful suggestions and cooperation in responding to requests for revisions. Their depth of knowledge, dedication, and patience have been demonstrated throughout the process of preparing this book. Working with them has been an adventure and a privilege, and I learned a lot about Ball Grid Array and other electronic packaging technologies from them. Their brief technical biographies are presented at the end of this book.

Last, I want to thank my employer, Hewlett-Packard, for providing an excellent environment in which completing this book was possible. I also want to thank my manager, Anita Danford, for her trust, respect, and support of my real work at HP. Finally, I want to thank my colleagues (both in HP and the industry) for their stimulating discussions in electronic packaging and interconnections. I learned a lot from them.

John H. Lau, Ph.D., PE
Palo Alto, California

Chapter

1

A Brief Introduction to Ball Grid Array Technologies

John H. Lau

1.1 Introduction

The electronics industry is one of the most fascinating, dynamic, and important industries. It has literally transformed the world and provides many products that affect our daily lives—for example, telephones, television, high-definition television, electronic organizers, Personal Computers (PCs), notebook PCs, subnotebook PCs, laptop PCs, palmtop PCs, PCs with built-in portable phones, workstations, midrange, mainframe, and supercomputers, PCMCIA (Personal Computer Memory Card International Association) PC cards, cellular phones, wireless phones, pagers, portable electronics products, video camcorders, audiovisual products, multimedium products, etc. One of the key technologies that is helping to make these products possible is electronics packaging and assembly technology,[1–157] which is the focus of this book.

Figure 1.1 schematically shows the hierarchy of an electronic package.[45] It should be noted that the wafer is not in the packaging hierarchy. It is included in Fig. 1.1 to show where the Integrated Circuit (IC) chip originates. Packaging focuses on how a chip (or many chips) is packaged efficiently and reliably.

The chip is not an isolated island. It must communicate with other chips in a circuit through an Input/Output (I/O) system of interconnects. Furthermore, the chip and its embedded circuitry are delicate, requiring the package to both carry and protect it. Consequently, the major functions of electronic package are[1–157]: (1) to provide a path for the electrical current that powers the circuits on the chip, (2) to distribute the signals on to and off of the silicon chip, (3) to remove the heat generated by the circuit, and (4) to support and protect the chip from hostile environments.

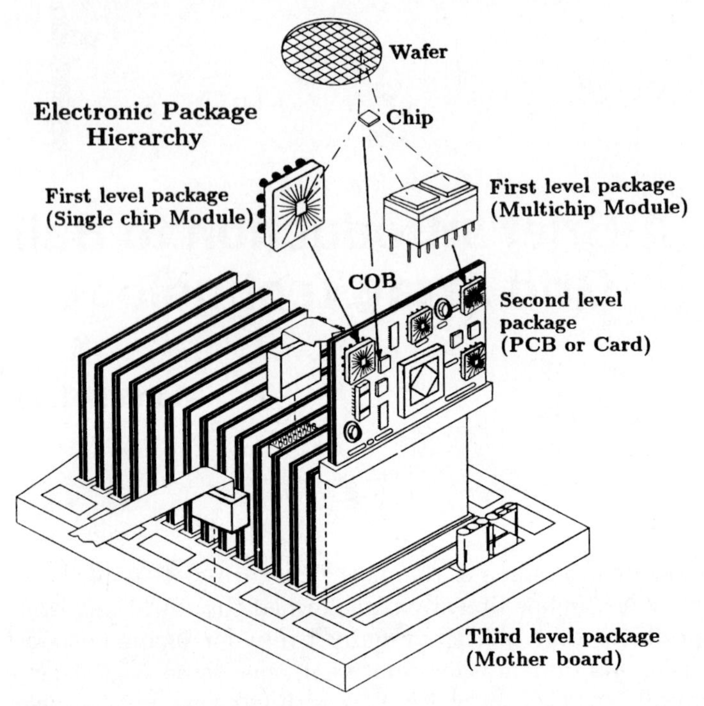

Figure 1.1 Shown here are the first three levels of electronics packaging. While not part of the packaging hierarchy, the wafer shows the origin of the IC chip.

Packaging is an art based on the science of establishing interconnections ranging from zero-level packages (chip-level connections), first-level packages (either single-chip or multichip modules), second-level packages (e.g., Printed Circuit Boards (PCBs)), and third-level packages (e.g., mother boards). (See Fig. 1.1.) Because of the recent trend in wire bonding, tape automated bonding, and flip bare Chips On printed circuit Board (COB) technology, the distinction between the first and second levels of packages is blurred. Usually, COB is called the one-and-one-half (or 1.5-) level package.[16] Figure 1.2 schematically shows the three most common COBs.

The most common methods of chip level interconnects are wire bonding (Fig. 1.3),[12,13,16,38] Tape Automated Bonding (TAB) (Fig. 1.4),[14,16,91–95] and solder bumping (Fig. 1.5).[1,2,15,16,17,60,69] Among these three technologies, solder bumped flip chip provides the highest packaging density with less packaging delay.[1,2,12,14,16]

There are many forms of first-level packages. Some examples are: thin Tape Carrier Package (TCP),[12,75,98] Plastic Quad Flat Pack (PQFP),[12,78–84]

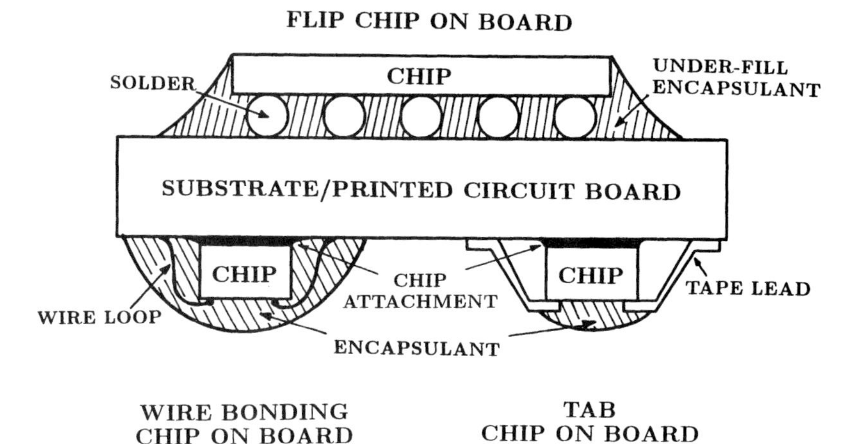

Figure 1.2 Three most common Chip On Board (COB) technologies.

Thin Plastic Quad Flat Pack (TQFP),[12] Square Quad Flat Pack (SQFP),[12] Rectangular Quad Flat Pack (RQFP),[12] Very Small Outline Package (VSOP),[12] Very Small Quad Flat Pack (VSQF),[12] Small Outline Integrated Circuit (SOIC),[12,85] Thin Small Outline Package (TSOP),[12,72–74] Plastic Leaded Chip Carrier (PLCC),[12,88] SOIC with J-Leads (SOJ),[12] Dual In-line

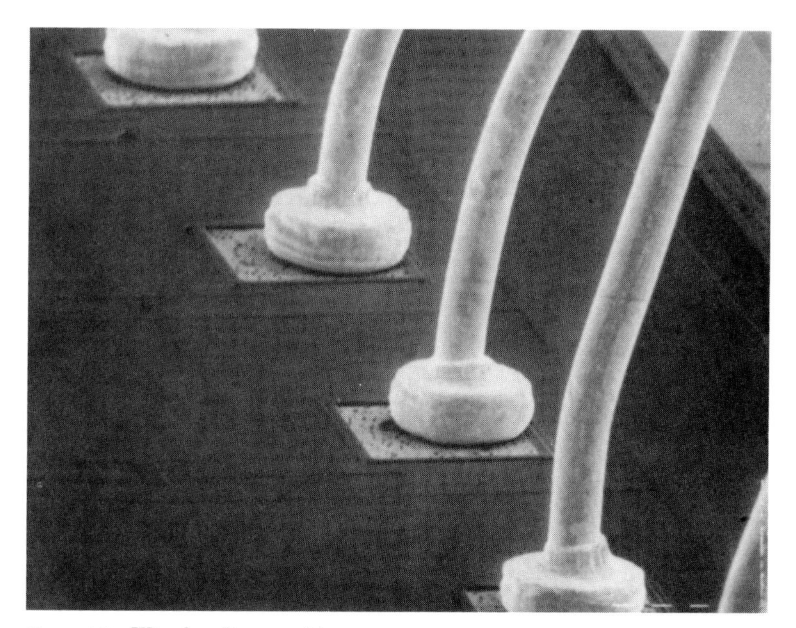

Figure 1.3 Wire bonding on chip.

Figure 1.4 Tape automated bonding on chip.

Package (DIP),[12] Ceramic Quad Flat Pack (CQFP),[97] Leadless ceramics Chip Carrier (LCCC),[13] ceramic Pin Grid Arry (PGA),[12,86,87] Plastic Pin Grid Arry (PPGA), Surface mount Pin Grid Arry (SPGA), Plastic Ball Grid Array (PBGA),[56,58,76,77] Ceramic Ball Grid Arrary (CBGA),[54,55,57] Area Tape Automated Bonding Ball Grid Array (TBGA).[99,100] The trends in single-chip (or multichips with standard single chip) carriers are finer pitch, higher pin count, larger horizontal body size, thinner vertical body size, and area array.[12]

The focus of this book is on the first- and second-level packages. The first-level packages include CBGA, PBGA, and TBGA. The substrates of the CBGA, PBGA, and TBGA packages are discussed in Chaps. 2, 3, and 14, respectively. The material, design, and assembly process of the CBGA, PBGA, and TBGA packages (either single-chip or multichip) are presented in Chaps. 5, 9, and 14, respectively. Also, burn-in sockets of PBGA, CBGA, and TBGA packages are discussed in Chap. 17.

In this book, the second-level packages include PCB assembly of CBGA packages (Chap. 6), PCB assembly of PBGA packages (Chap. 10), and PCB assembly of TBGA packages (Chap. 14). Also, the PCB routing considerations for CBGA, PBGA, and TBGA packages are pre-

Figure 1.5 Solder bumped on chip.

sented in Chap. 4. The thermal and electrical management of CBGA, PBGA, and TBGA assemblies are examined in Chaps. 7, 11 and 12, and 14, respectively. X-ray inspection and rework of PBGA, CBGA, and TBGA PCB assemblies are discussed in Chaps. 15 and 16, respectively. Figure 1.6 shows the flow chart of this book. It can be seen that the infrastructure of BGAs is discussed in Chap. 18 and a detailed packaging glossary is presented in Chap. 19.

In this chapter, a brief introduction to the IC trends (e.g., process technology, density, chip size, feature size, performance, operating voltage, and design-cycle) will be presented. Also, various electronics packaging technologies such as the conventional packages (e.g., PLCC, PQFP, CQFP, and TSOP), and the advanced/emerging packages (e.g., TCP, PGA, CBGA, PBGA, and TBGA) will be discussed. A comparison between the BGA and PQFP packages based on their I/O count, density (area), electrical performance, manufacturing yield, and popcorn effect will be given. Furthermore, some of the next generation BGAs are mentioned. Finally, multichips with CBGA and PBGA technologies will be briefly discussed.

1.2 Integrated Circuit (IC) Trends

Figure 1.7 shows the 1993 total worldwide merchant semiconductor usage ($81.9B),[101] where 14 percent ($11.7B) were for discretes and 86 percent ($70.2B) were for ICs. More than 58 percent ($48.12B) of the total merchant semiconductors were supplied by Intel, NEC, Toshiba,

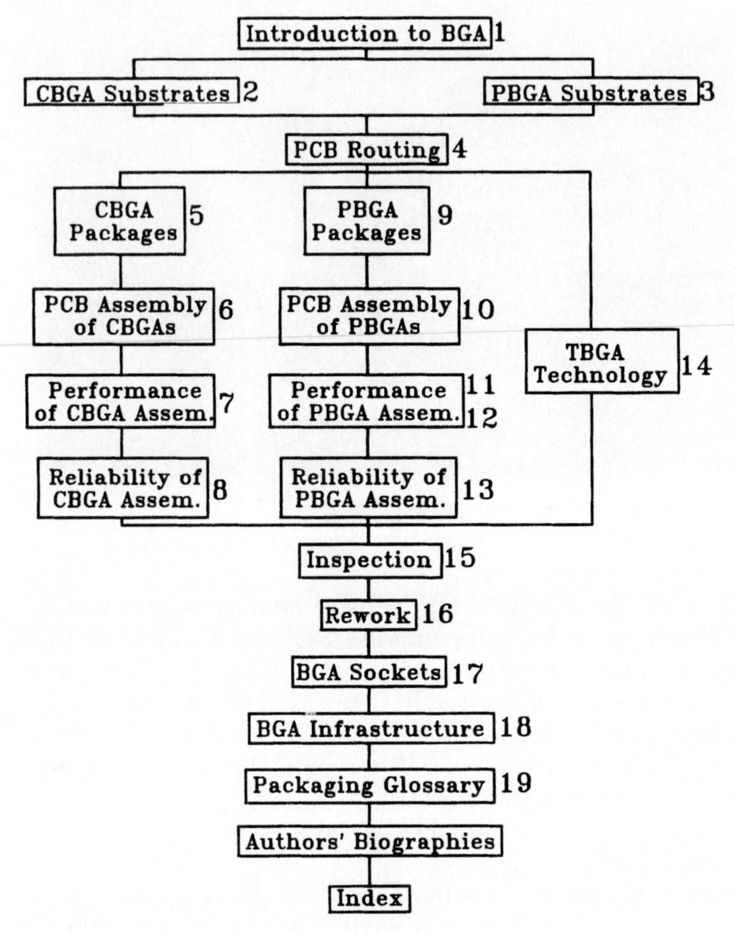

Figure 1.6 Flowchart of the book (numbers are for book chapters).

Motorola, Hitachi, Texas Instruments, Fujitsu, Mitsubishi, Samsung, and Matsushita. The semiconductor market was about 12.8 percent of thc total electronics marketplace ($641B).

1.2.1 IC process technology trends

Figure 1.8 shows the IC process technology market-share trends from 1982 through 1998. It shows that the Complementary Metal-Oxide-Semiconductor (CMOS) has become the primary technology for fabricating ICs. In the past 23 years, no other technology has dominated the IC marketplace like CMOS does now, increasing from a small percentage (12 percent) of all ICs manufactured in 1982 to 82 percent forecast by 1998. Some of the advantages of CMOS are[101]: low power density,

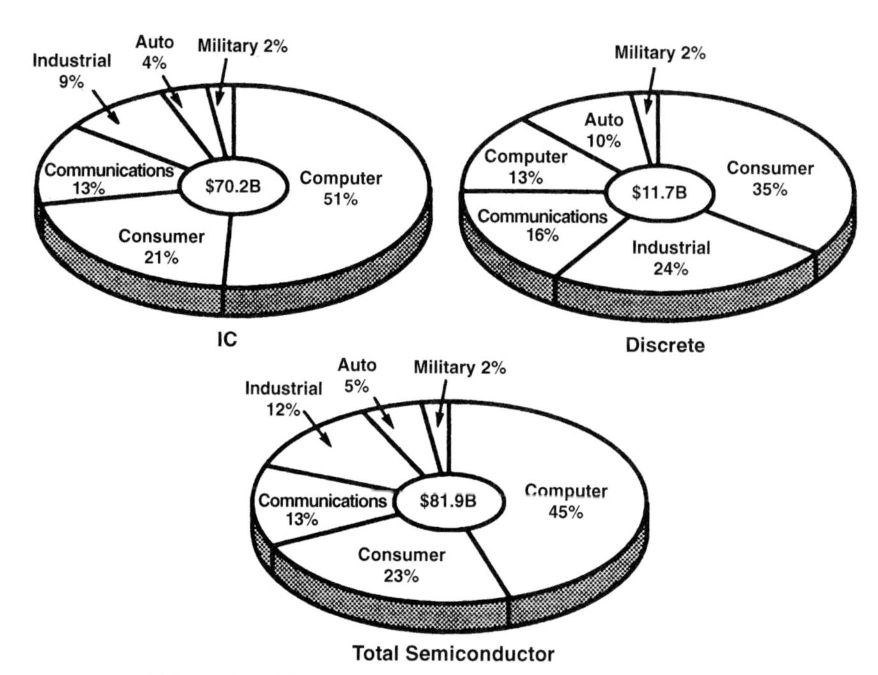

Figure 1.7 1993 total worldwide merchant semiconductor usage. (*Source: ICE Corp. 1993 Status.*)

relatively good noise immunity and soft error protection, low threshold bias sensitivity, design simplicity and relatively easy layout (especially for ASICs), and capability for lower power analog and digital circuitry on the same chip. The Toshiba's CMOS scaling trends are shown in Table 1.1. It can be seen that the gate oxide thickness and gate length (n-channel/p-channel) are decreasing from 11 nm and 0.5/0.6 μm in 1993–1994 to 6–7 nm and 0.25/0.25 μm forecast by 1999–2000, respectively (n = nano = 10^{-9}).

On the other hand, bipolar technology for fabricating ICs in the past 23 years has been rapidly shrinking. It declined from 45 percent in 1982 to less than 11 percent forecast by 1998. Even though Emitter Coupled Logic (ECL) ICs are the fastest silicon-based devices available, the high power dissipation and high cost of ECL circuitry has limited it to only a niche process technology.

1.2.2 IC density trends

Figure 1.9 shows the increase in IC density trends for Dynamic Random Access Memory (DRAM) (1 Kbit to 256 Mbits) and microprocessor/logic (4-bit to 32-bit and the emerging 64-bit).[101] It can be seen that the number of transistors per chip has grown continually by about 50

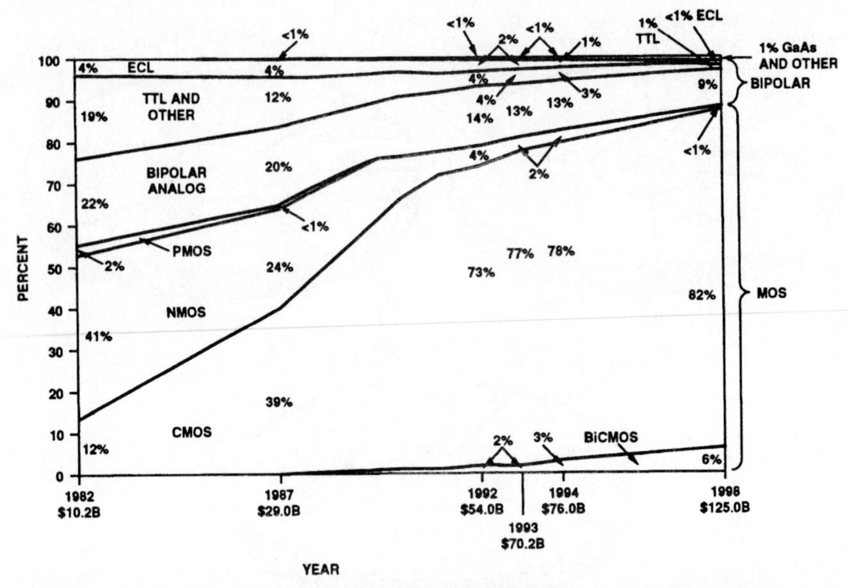

Figure 1.8 IC process technology marketshare trends. (*Source: ICE Corp. 1993 Status.*)

percent per year for the DRAM and 35 percent per year for the microprocessor/logic. The IC density trends for Static Random Access Memory (SRAM) can be found from the Semiconductor Industry Association (SIA) IC technology roadmap, Table 1.2. It can be seen that the 16 and 64M SRAMs are forecast to be available in 1995 and 1998, respectively. Also, by 1998, it is forecast to have 2 million gates on a chip.

TABLE 1.1 Toshiba's Complementary Metal-Oxide-Semiconductor (CMOS) Scaling Trend

	1993–1994	1996–1997	1999–2000
Line widths	0.5 μm	0.35 μm	0.25 μm
Power supply	3.3 V	3.3 V	2.0–2.5 V
Gate oxide thickness	11 nm	9 nm	6–7 nm
Gate length (n-channel/p-channel)	0.5/0.6 μm	0.35/0.35 μm	0.25/0.25 μm
Switching speed (fan-out = 1)	95 ps	68 ps	45 ps
Drain structure			
NMOS	FOLD*	Graded diffused drain	Conventional
PMOS	Conventional	Lightly doped drain	Conventional
Gate structure	n+ Polysilicon	n+ Polysilicon NMOS p+ Polysilicon PMOS	n+ Polysilicon NMOS p+ Polysilicon PMOS

* FOLD: Fully Overlapped Lightly Doped Drain.

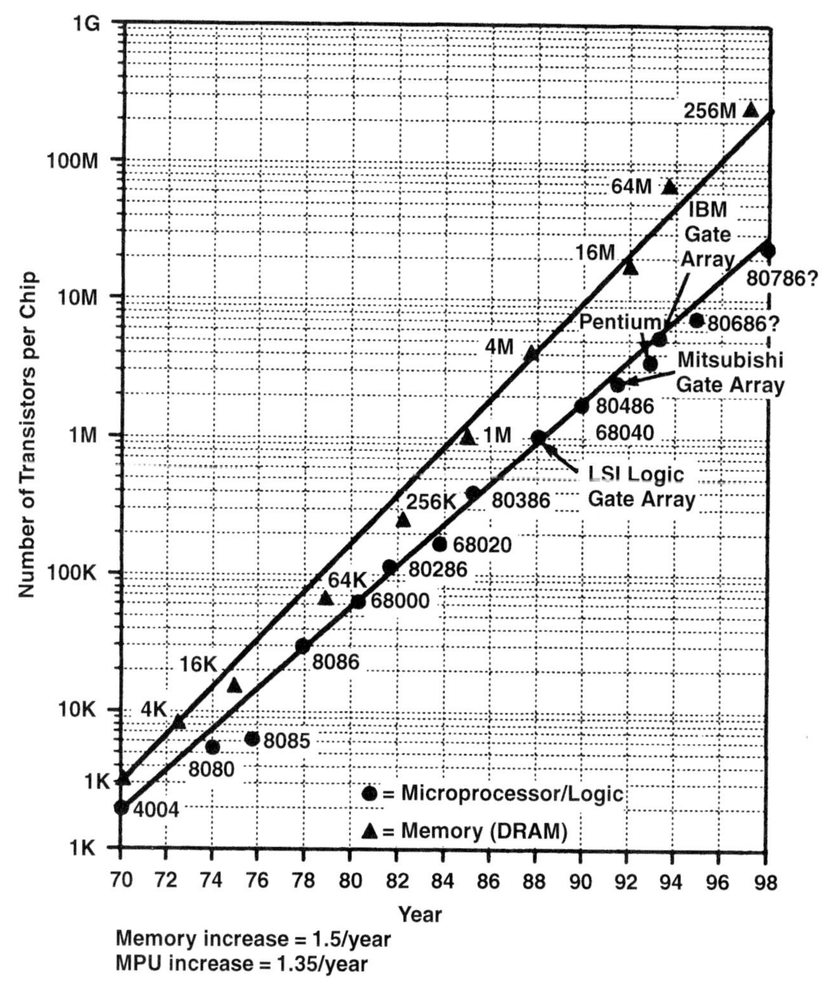

Figure 1.9 IC density trends for DRAMs and Microprocessor/Logic. (*Source: ICE Corp. 1993 Status.*)

1.2.3 IC chip size trends

Figure 1.10 shows the IC chip size trends for the DRAMs and microprocessor/logic. It can be seen that the chip area of the DRAM and microprocessor/logic has grown continually about 13 percent per year. By 1996, the microprocessor/logic chip area is forecast to be 0.9 in^2 (581 mm^2) and the DRAM chip area is forecast to be 0.42 in^2 (271 mm^2). Thus, the dimensions of most of the microprocessor/logic chip would be 0.95 in (24.1 mm) square. Since most of the DRAMs are rectangular, the longer side of the largest DRAM chip would be greater than 1 in (25.4 mm).

TABLE 1.2 Semiconductor Industry Association (SIA)'s IC Technology Roadmap

		1995	1998	2001	2004	2007
Feature size (μm)		0.35	0.25	0.18	0.12	0.10
Gates/chip		800K	2M	5M	10M	20M
Bits/chip	•DRAM	64M	256M	1G	4G	16G
	•SRAM	16M	64M	256M	1G	4G
Chip size (mm^2)	•Logic/microprocessor	400	600	800	1,000	1,250
	•DRAM	200	320	500	700	1,000
Wafer diameter (mm)		200	200–400	200–400	200–400	200–400
Number of interconnect levels (logic)		4–5	5	5–6	6	6–7
Max. power (W/Die)	•High-performance	15	30	40	40–120	40–200
Power supply (V)	•Portable	2.2	2.2	1.5	1.5	1.5
Number of I/Os		750	1,500	2,000	3,500	5,000
Performance (MHz)	•Off-chip	100	175	250	350	500
	•On-chip	200	350	500	700	1,000

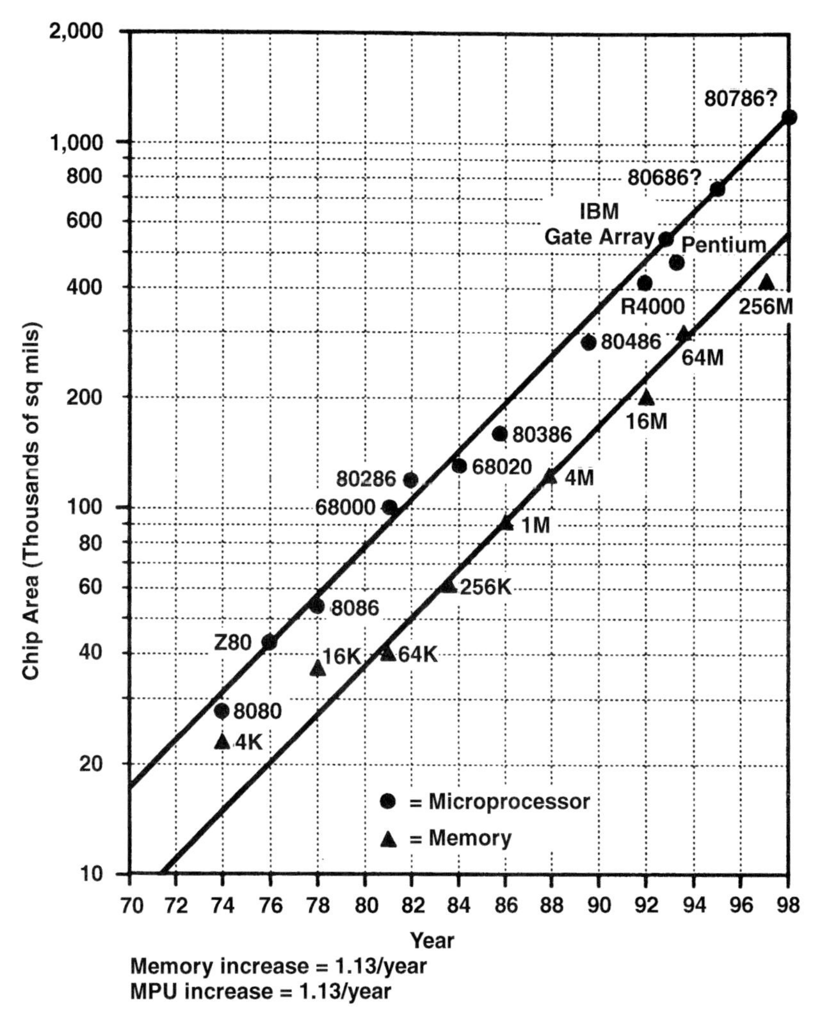

Figure 1.10 IC chip size trends for DRAMs and microprocessor/logic. (*Source: ICE Corp. 1993 Status.*)

1.2.4 IC feature size trends

Figure 1.11 shows the reduction in IC feature sizes for production as well as experimental circuits. It can be seen that for a "tight production resolution" device, the feature sizes have decreased from about 3 μm in 1980 to about 0.35 μm in 1994, which represents approximately 15 percent decrease every year. Before the turn of this century, the 0.2 μm technology is forecast to be in production (Tables 1.1 and 1.2) and the 256M DRAMs would be available, but with a very high price.

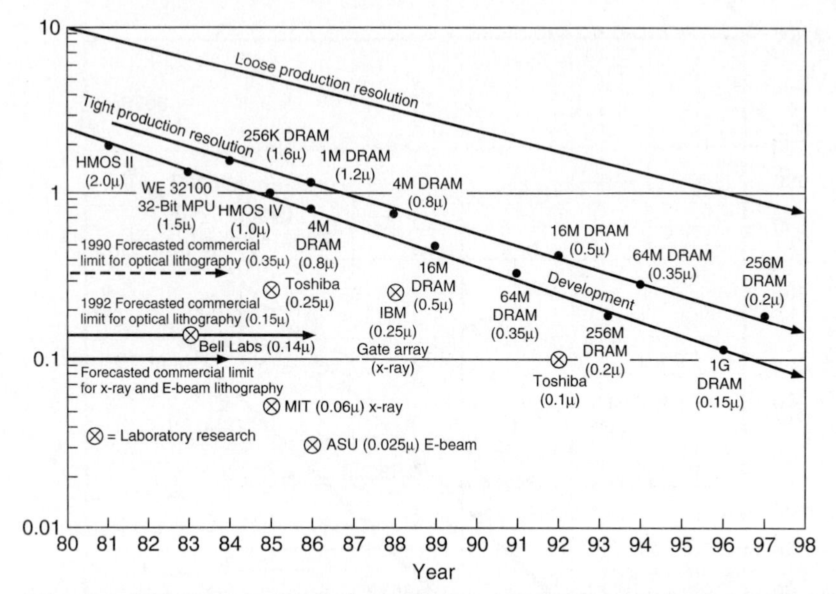

Figure 1.11 IC feature size trends for production and experimental circuits. (*Source: ICE Corp. 1993 Status.*)

1.2.5 IC performance trends

Figure 1.12 and Tables 1.1 and 1.2 show the (average) rapid growth of Very Large Scale Integration (VLSI) and Ultra Large Scale Integration (ULSI) IC operating frequencies. This is the result of reducing feature sizes (Fig. 1.11, Tables 1.1 and 1.2)) and gate delay (Fig. 1.13). By 1995, the average IC device clock frequency is expected to run faster than 200 MHz. Thus, proper transmission line termination, cross-talk protection, and grounding pose technical challenges.

1.2.6 IC operating voltage trends

Figure 1.14 and Tables 1.1 and 1.2 show that the IC power supply (operating voltage) is reducing from 5 V to either 3 or 3.3 V now, and eventually to 2.5, then 1.5 V. The reasons are: (1) the power consumption is proportional to the square array of the operating voltage—lowering the voltage from 5 to 3.3 V would reduce the power by 70 percent (Fig. 1.15), (2) the reduction in feature sizes (0.5 μm, Fig. 1.11 and Tables 1.1 and 1.2) and gate oxide thicknesses (11 nm, Table 1.1), today, a 5-V power supply is not practical (Fig. 1.16), (3) the physical limitations of IC materials such as the dielectric and metallization thickness issues and transistor breakdown characteristics, and (4) the explosive growth in portable electronics products, which demand more battery life by reducing power dissipation (Fig. 1.14).

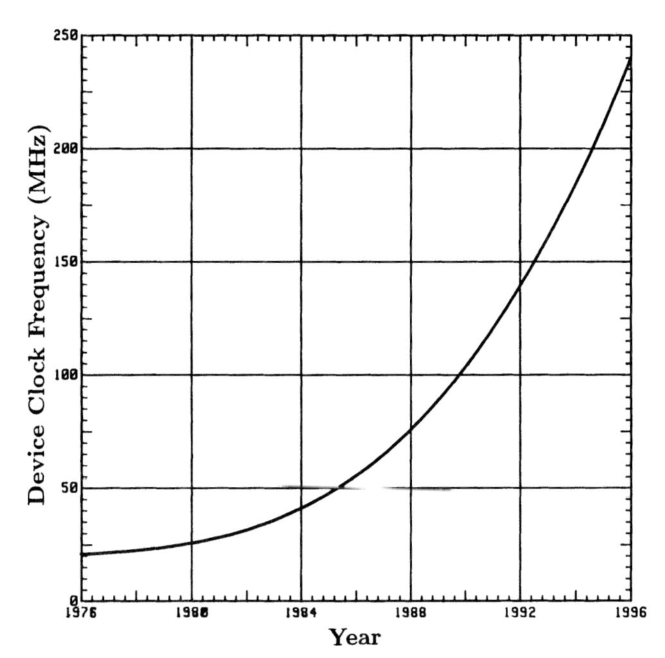

Figure 1.12 Average IC performance trends.

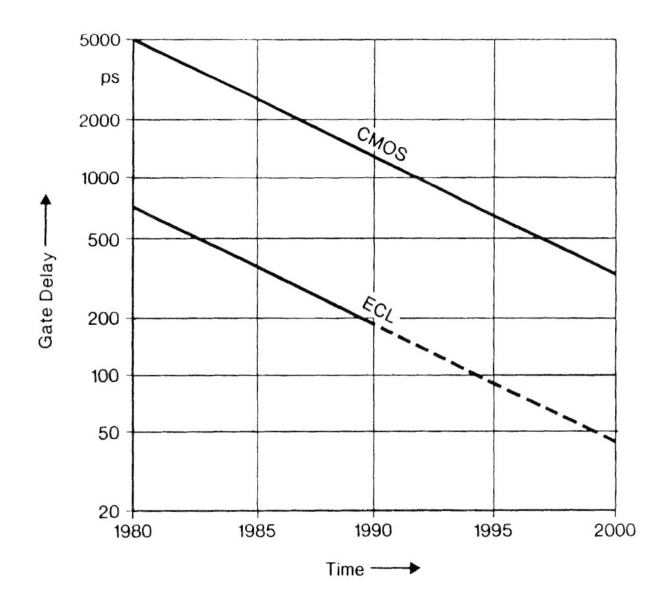

Figure 1.13 Gate delay for CMOS and ECL. (*Source: Wessely, et al.*[47])

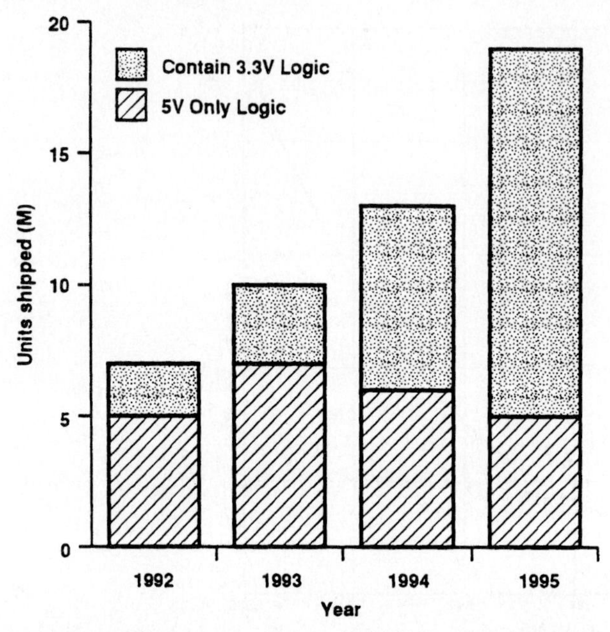

Figure 1.14 IC Power supply for portable PCs. (*Source: AT&T.*)

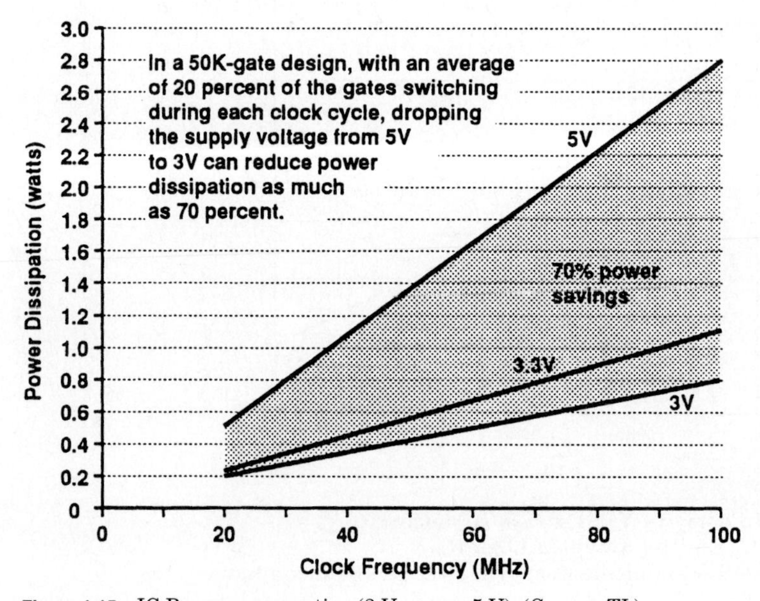

Figure 1.15 IC Power consumption (3 V versus 5 V). (*Source: TI.*)

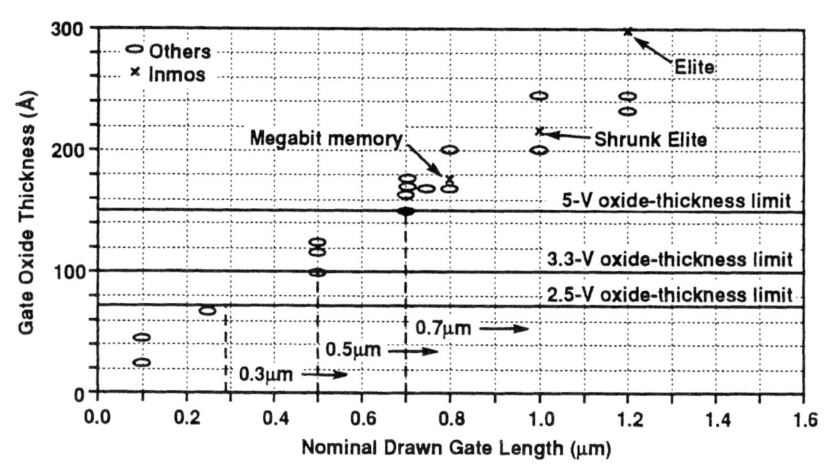

Figure 1.16 Gate oxide thickness versus gate length versus supply voltage. (*Source: SGS-Thomson.*)

The operating speed is a function of power supply, though: the lower the supply voltage the lower the speed (clock frequency). Also, for ICs with 3.3 V or less, the devices are more sensitive to Electrostatic Discharge (ESD) and smaller noise margins. Thus, handling lower-voltage ICs poses a manufacturing challenge.

1.2.7 IC design-cycle trends

Figure 1.17 shows the number of designs versus the average number of productions for a design. It can be seen that, in 1975 there were only 200 designs, each with an average of 500,000 devices manufactured. In 1991 there were more than 100,000 designs. For each design, however, less than 300 devices (on average) were manufactured. As a matter of fact, many of these designs never get into production. Thus, the trend of IC design is to try many different new concepts and put only the most promising ones into production. Also, most were manufactured in small quantities.

In the past decade, the IC design cycles have been reduced because of the rapid decrease in product life cycles (both ICs and equipment). In general, the life cycle of Application-Specific Integrated Circuits (ASIC) is one year, memories is 4 years, microprocessors is 3 to 6 years, and glue logic ICs is less than 6 years. Thus, the time to market of a new product becomes very critical in competing for market share, and reducing design cycles is vital for survival. The penalty for late introduction of new products is to reduce the gross profit, Fig. 1.18. Figure 1.19 shows Intel's Complex Instruction-Set Computing (CISC)-based microprocessors family. It can be seen that Intel *quickens* its Pentium (P5), Sexium (P6), and Septium (P7) product introductions.

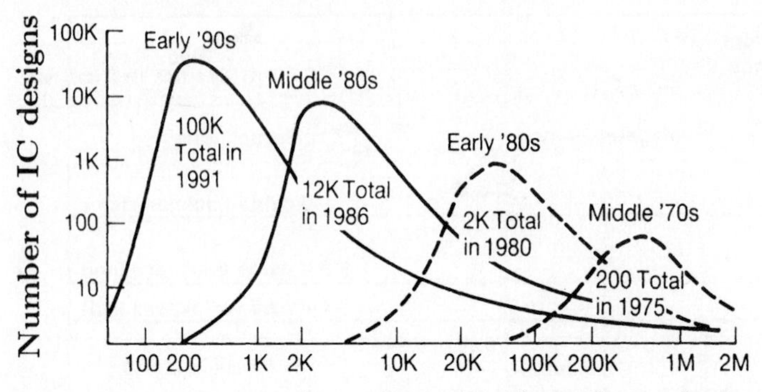

Figure 1.17 Number of IC designs versus the average number of productions for a design.

In the past decade, in order to reduce cost, shorten development time, and reduce risk of IC designs, many (two or more) companies formed a partnership, defined a set of common objectives, took advantage of other companies' strong points, and shared the cost and workloads. Some successful cases have been reported, e.g., IBM-Motorola-Apple's Reduced Instruction-Set Computing (RISC)-based PowerPC microprocessors, Fig. 1.20. Apple Computer is now enjoying the benefit from its Macintosh computers which are based on the PowerPC 601 RISC microprocessors, Table 1.3.

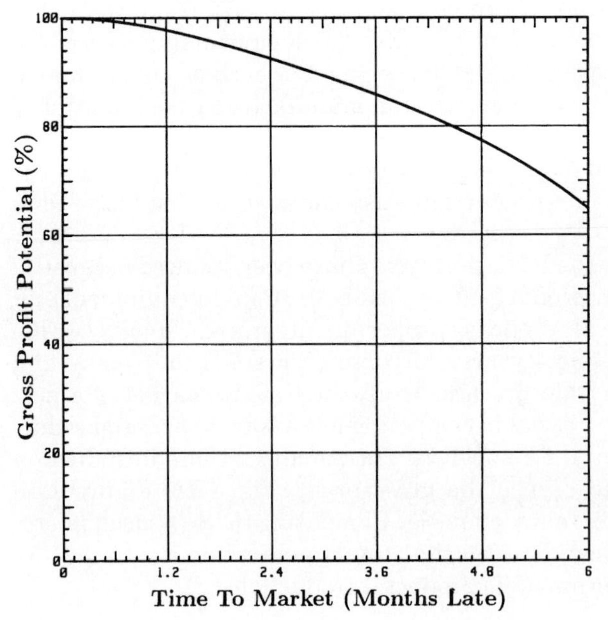

Figure 1.18 Effect of time to market "late" on profit.

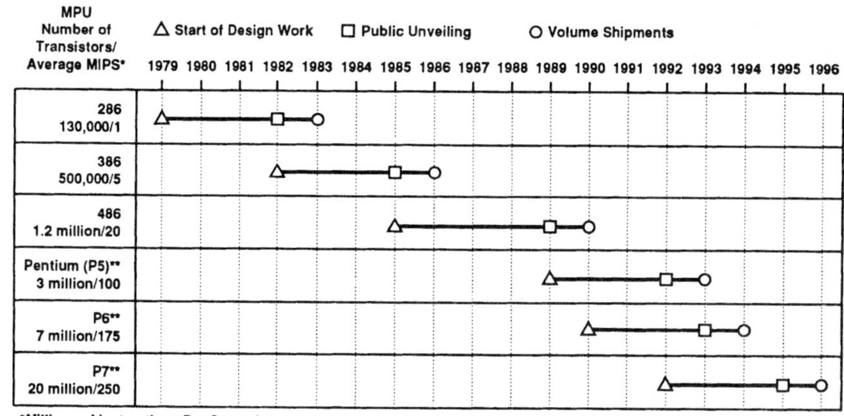

Figure 1.19 Intel quickens P5, P6, and P7 product introductions. (*Source: Business Week.*)

1.3 Packaging Technology Update: BGAs

Figure 1.21 shows the factors affecting the performance, Million Instructions per Second (MIPS), of a computer system. It can be seen that beside IC semiconductor and others, packaging technology also plays a key role in performance. As discussed in Sect. 1.2, the on-chip delay in IC semiconductor devices has been reduced rapidly in the past few years. However, the signal delay in packaged ICs was not able to be

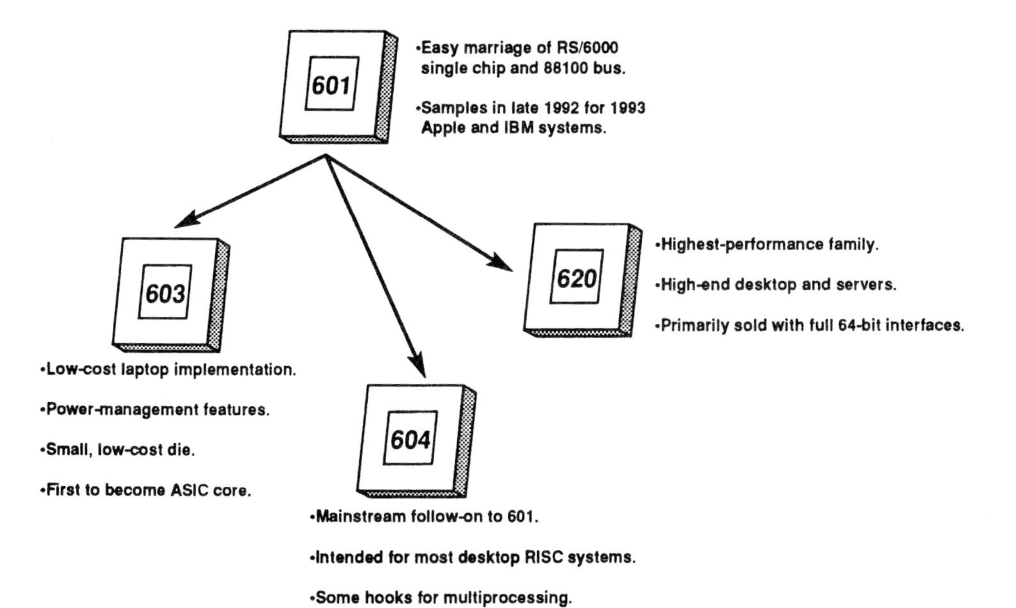

Figure 1.20 IBM/Motorola/Apple's PowerPC evolution. (*Source: EE Times.*)

TABLE 1.3 Macintosh PowerPC 601 Processor Family

Macintosh	Clock frequency (MHz)	Cache memory	Standard RAM (MB)	DRAM expansion (MB)	SIMM (slots)	VRAM video and expansion (MB)
6100/60	60	optional	8	72	2	NA
7100/66	66	optional	8	136	4	3
8100/80	80	256 Kbits	8	264	8	6

NOTE: Networking Ethernet and DRAM video are standard for all models.

reduced as much as that of the on-chip ICs. Thus, packages have become a great *loss function* (or bottleneck)[45] of a computer system. Also, it can be seen from Fig. 1.22 that a 50 percent improvement in logic performance for a computing system results in less than 23 percent overall gain if the packaging does not change. In order to keep pace with the advance of IC semiconductor and logic implementation technologies, packaging engineers have to minimize the loss function by increasing the packaging density and reducing the packaging delay.

Figure 1.23 shows various packaging alternatives (QFPs, TQFPs, TSOPs, PBGAs, TCPs, PGAs, TBGAs, CBGAs, and flip chip) on top of the key categories (e.g., processor, gate array, cache memory, and main memory) of the next-generation high-speed devices (e.g., microprocessors, ASICs, SRAMs, and DRAMs). As discussed in Sec. 1.2, both RISC-based and CISC-based microprocessors are expected to require

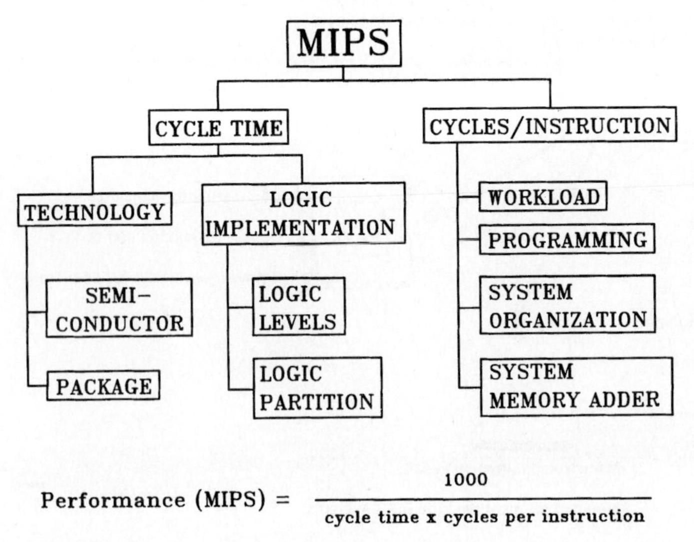

$$\text{Performance (MIPS)} = \frac{1000}{\text{cycle time x cycles per instruction}}$$

Figure 1.21 Factors affecting the performance of a computing system.

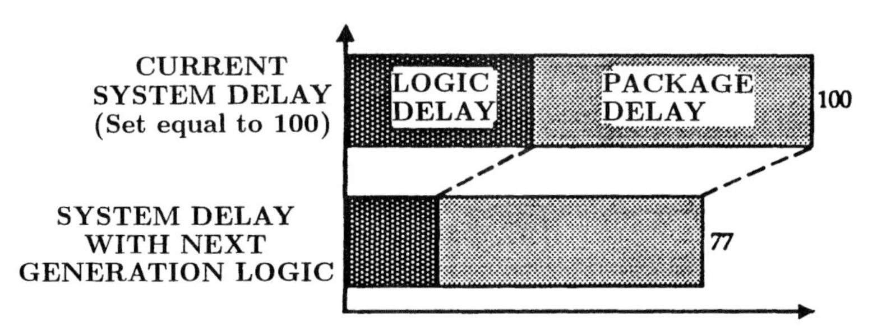

Figure 1.22 Logic and package delay.

over 1000 package pin counts and to perform over 400 MHz on-chip clock frequency. Also, the SRAMs for cache memories are expected to perform at similarly high speeds to prevent system data bottlenecks. Even the gate-array ASICs are expected to run faster than 100 MHz on-chip clock frequency and expand their package pin counts to 900. Packaging technology will be hard-pressed to meet all these future requirements. In the following sections several packaging technologies and their applications will be briefly discussed.

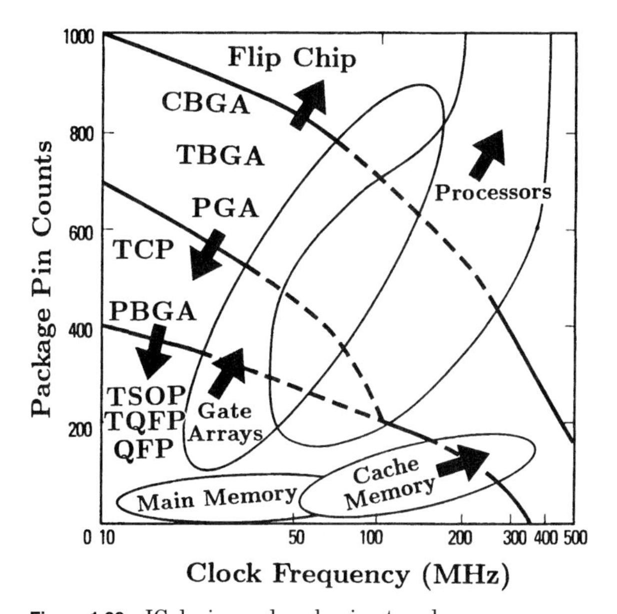

Figure 1.23 IC device and packaging trends.

1.3.1 Conventional packages: PLCC, PQFP, CQFP, and TSOP

PLCC. Figure 1.24 shows a 68-pin PLCC with "J" leads.[12,88] The lead pitch (spacing from center to center) is 0.05 in (1.27 mm). The dimensions of this PLCC are 0.96 × 0.96 × 0.093 inches (24 × 24 × 2.4 mm), and the average thickness of the J-Lead is 0.008 in (0.2 mm). The lead-frame material is usually copper and the molding material is plastic. The next level of this package family is the 84-pin PLCC. This package family is low-cost and usually used for ASIC applications.

PQFP. Figure 1.25 shows a PQFP with "gull-wing" leads.[12, 78–84] The bumpers are for lead protection purpose and are optional. When the lead spacing of a QFP is equal to or less than 0.65 mm, it is commonly referred to as a fine pitch Surface Mount Component (SMC). Any Surface Mount Technology (SMT) assembly with fine pitch SMCs is called fine pitch SMT, or simply Fine Pitch Technology (FPT).[12] The lead-frame material of PQFPs is usually copper and the molding material is plastic. QFP has a very large family ranging from 5 × 5 mm to 40 × 40 mm square.[12] The most commonly used are the 28 × 28 mm body size with 0.65-mm pitch (160-pin), 0.5-mm pitch (208-pin), and 0.4-mm pitch (256-pin). Today, 0.3-mm pitch PQFPs are already in production and 0.2-mm pitch PQFPs are in prototyping. This package family is low-cost and usually used for ASIC applications.

CQFP. Figure 1.26 shows a CQFP with clip lead frames and gull-wing leads. The chip is solder bumped on a Metallized Ceramic (MC) substrate (with Cr/Cu/Cr lines) which serves as the distribution layer

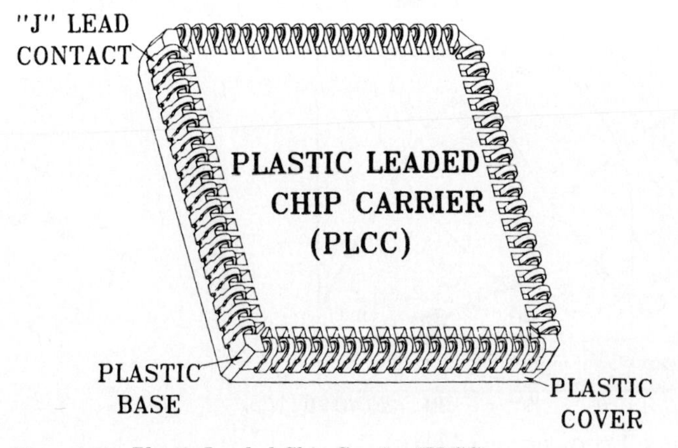

Figure 1.24 Plastic Leaded Chip Carrier (PLCC).

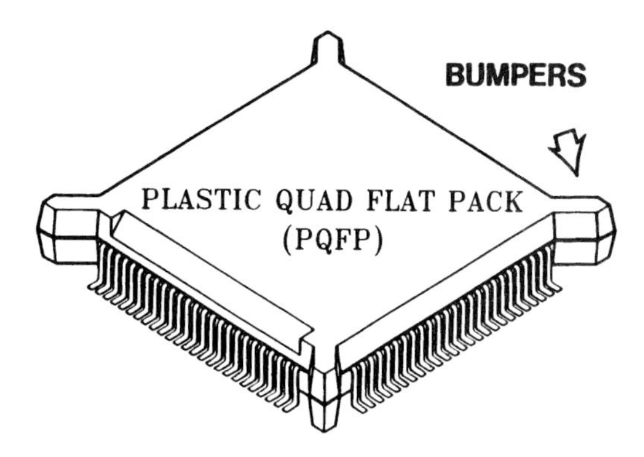

Figure 1.25 Plastic Quad Plat Pack (PQFP).

between the chip and the PCB (not shown). Encapsulant is used to "cement"[16] the chip to the MC substrate in order to enhance the solder joint reliability. Final package encapsulation options include a ceramic cap or urethane coating which offer protection to the ceramic circuitry from the hostile environment. This package is available in 0.4- and 0.5-mm lead pitches, and is used for IBM's PowerPC 601 microprocessor.[97]

PLCCs, PQFPs, and CPFQs are peripheral packages, in which the leads are located around the edges of the carrier. Figure 1.27 shows the possible number of I/Os (pads) for these packages with various lead pitches (0.3, 0.4, 0.5, and 0.65 mm). The dimensions of the package are x and y, with $x = \theta y$, $\theta \geq 1$. For square package, $\theta = 1$.

TSOP. Figure 1.28 shows a TSOP with gull-wing leads. It is a very short (low-profile) package (i.e., very small overall compliance). The silicon-to-package ratio usually is very large (i.e., small thermal coefficient of expansion). Most of the TSOP lead-frame materials are alloy 42, even though copper has been used to enhance the solder joint reliability.[12,72–74] This package is low-cost and used for memory applications.

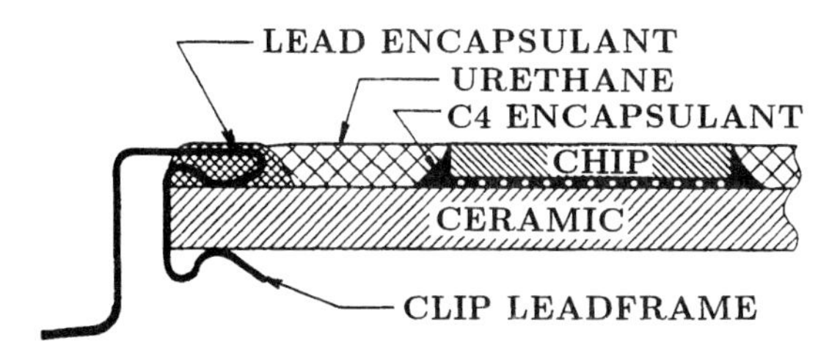

Figure 1.26 Ceramic Quad Plat Pack (CQFP). (Source: IBM.[97])

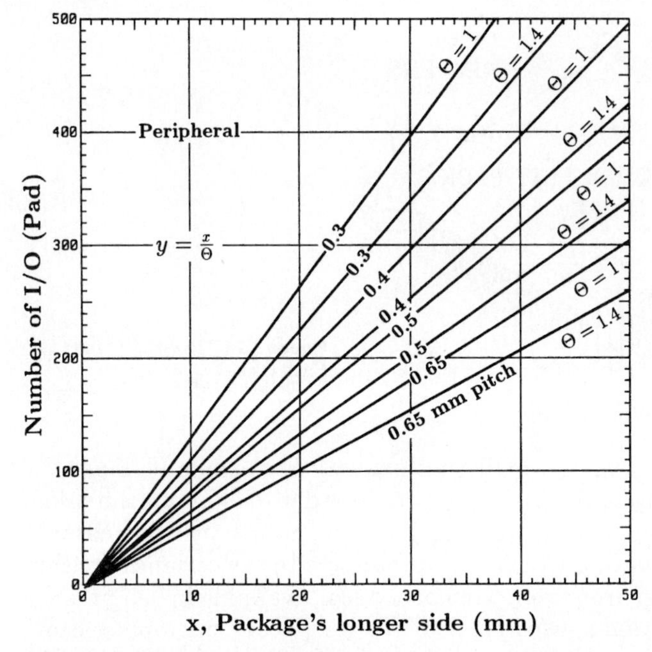

Figure 1.27 Number of package pin counts versus package size.

PLCCs, PQFPs, CPFQs, and TSOPs are SMT-compatible and low-cost solutions for medium speed requirements at package pin counts to about 392-pin (0.3-mm pitch and 32 × 32 mm body size). However, because of lead coplanarity, package cracking, I/O limitation, fine pitch limitation, relatively long lead lengths with inherent inductances, capacitances, and resistances, and manufacturing problems in achieving high yield for very fine pitches, these traditional packages are nearing their practical limits for pin count and performance. Some of their market share has already been taken away by the following advanced/emerging packaging technologies.

1.3.2 Advanced Packages: TCP, PGA, CBGA, PBGA, and TBGA

TCP. Figure 1.29 shows a cross section of an ultra TCP.[75] The thickness of the Si chip and package are 0.2 and 0.4 mm, respectively. In order to achieve this thin package, TAB technology is used. The upper and side surfaces of the chip are encapsulated with resin, however, the bottom surface of the chip is exposed. This package offers smaller pitches, thinner package profiles, smaller footprints on the PCB, and manufacturable handling media—without compromising performance.

Figure 1.30 shows another example of the TCP.[98] The lead pitch and lead dimensions of this TCP are, respectively, 0.2 and 0.08 mm (wide)

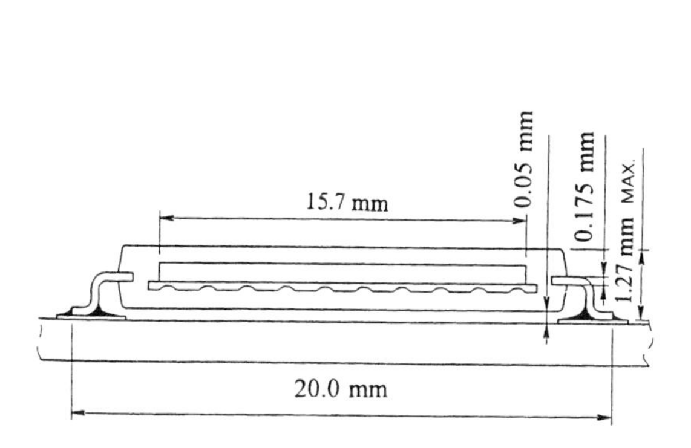

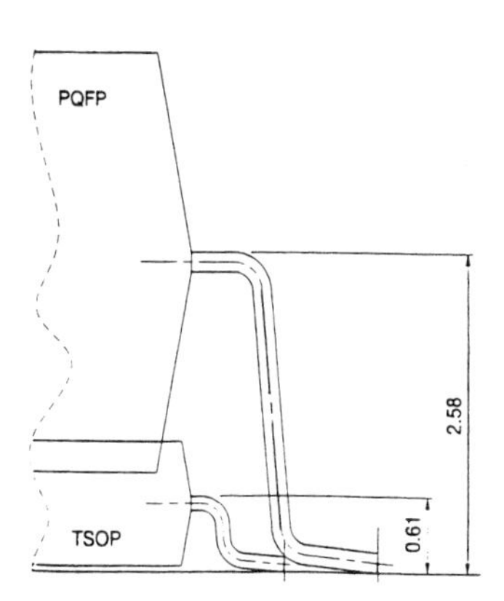

Figure 1.28 Thin Small Outline Package (TSOP).

by 0.035 mm (thick). The package thickness is 0.53 mm and the body size is 20 mm square. The tape used for this TCP is a 48-mm tape format with polyimide down for pick and place handling and slide carrier handling. Shipped flat in slide carrier, the leads are formed into gull-wing shape to increase compliance and are bonded on to the PCB by hot bar reflow. Most of the heat from the chip can be transferred through the thermal vias to the heatsink, then to the air, Fig. 1.31.

TCPs can provide moderate performance solutions for applications (ASICs and microprocessors) with up to about 700 package pin counts. However, unless it is a very high volume production, it may not be cost-effective due to very high development cost of this custom-design package. Also, TCPs suffer the same drawbacks (peripheral chip carrier, long lead length, handling, board level manufacturing yield loss) as the QFPs.

PGA. Figure 1.32*a* schematically shows a cross section of a PGA.[12,86,87] It has had a predominant role in high-density packaging for many years and is commonly used (for ASICs and microprocessors) in high-pin-count, high-power, and high-performance computers. PGAs have the internal power and ground planes needed for excellent electrical performance (controlled impedance), and their I/O pins are spread over the area of substrate making it possible to have a very large number of I/Os.

The drawbacks of PGAs are high cost (e.g., alloy 42 or copper pins), through-holes on the PCB for mounting (not SMT-compatible), 100-mil (2.54-mm) pin pitch, handling, and pin bending and insertion.

CBGA. Since solder balls are cheaper than alloy 42 or copper pins and easier to attach to the bottom of the chip carrier (SMT-compatible), Motorola, IBM, Hitachi, NEC, and others replaced the pins of the ceramic PGA with solder balls and called CBGA, Fig. 1.32*b*.

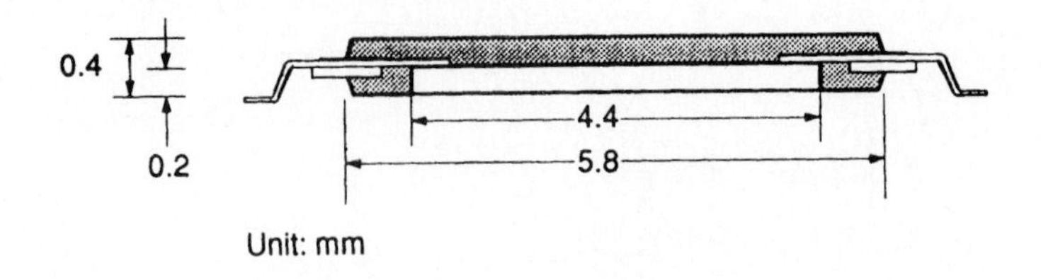

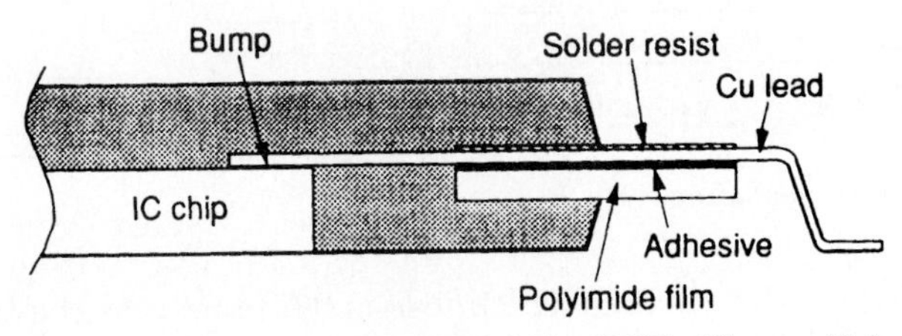

Figure 1.29 Ultrathin Tape Carrier Package (TCP). (*Source: Nakamura, et al.*[75])

Figure 1.33*a* schematically shows a cross section of Motorola's 50-mil (1.27-mm) pitch solder bumped Pad Array Carrier (PAC) and its footprint.[122] It consists of a square alumina base, 0.85 in (21.59 mm) on a side and a thickness of 0.025 in (0.64 mm). The 14 × 14 matrix of 0.03-in (0.76-mm) diameter solder balls was spread on a square area with 0.65 in (16.51 mm) on a side. The total I/O count was 196. The silicon chip was wire-bonded to the alumina substrate and was protected with a cap cover. By a combination of different solder joint and PCB geometries and materials it was demonstrated that the large cofired alumina solder bumped PAC can withstand 1000 cycles of –55 to 125°C air-to-air temperature cycle with a 15-minute dwell at extreme temperatures and a 15-minute transition between extreme temperatures without failure. Also, on an FR-4 epoxy PCB, it was determined that an epoxy underfill (Fig. 1.33*b*) improves the thermal cycle life of the PAC solder joint by factor of 5.

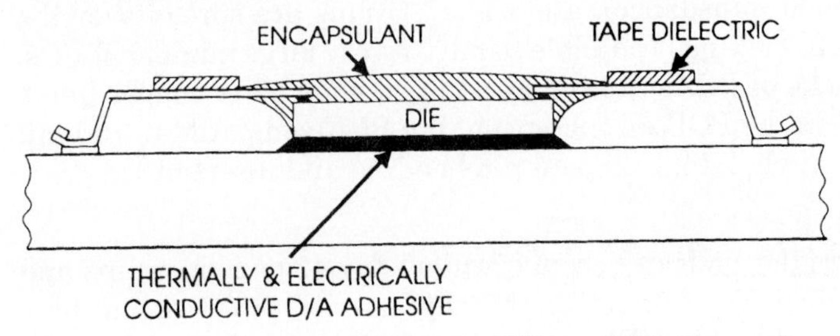

Figure 1.30 Thin-Tape Carrier Package (TCP) on PCB. (*Source: Pope and Do.*[98])

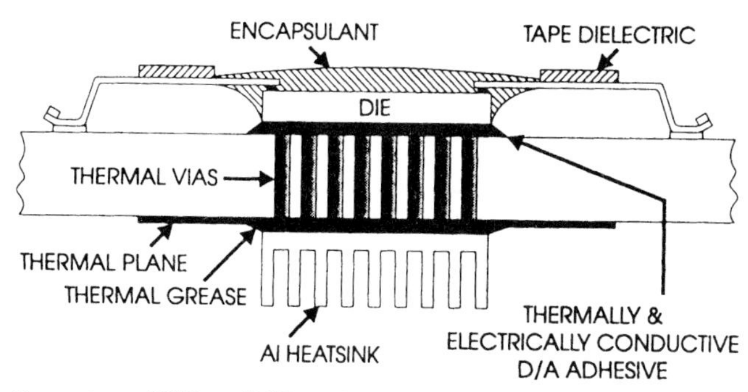

Figure 1.31 TCP on PCB with vias and heatsink. (*Source: Pope and Do.*[98])

Figure 1.34 shows IBM's 50-mil (1.27-mm)-pitch nonhermetically sealed CBGAs. Depending on the size of the (9–20-layer cofired) Multi-Layer Ceramic (MLC) carriers, two different solder interconnects were used. For 28 × 28 mm and smaller carriers, 90%wtPb/10%wtSn solder ball (35 mils or 0.9 mm in diameter) were used and are called Solder Ball Carrier/Connection or SBC. Today, IBM calls it Ceramic Ball Grid Array (CBGA). On the other hand, for 32 × 32 mm and larger carriers, 90%wtPb/10%wtSn solder columns (87 mils or 2.2 mm tall and 20 mils or 0.5 mm in diameter) were used and are called Solder Column Carrier/Connection or SCC. Today, IBM calls it Ceramic Column Grid

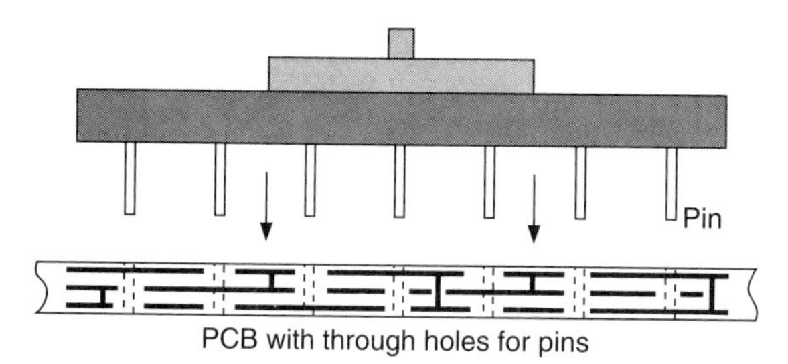

Figure 1.32*a* Ceramic Pin Grid Array (PGA).

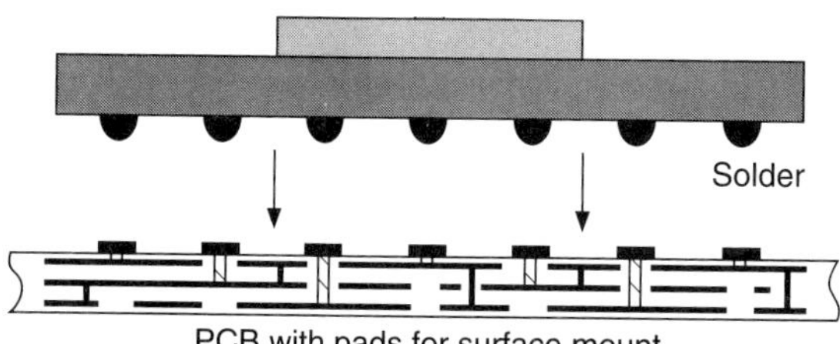

Figure 1.32*b* Ceramic Ball Grid Array (CBGA).

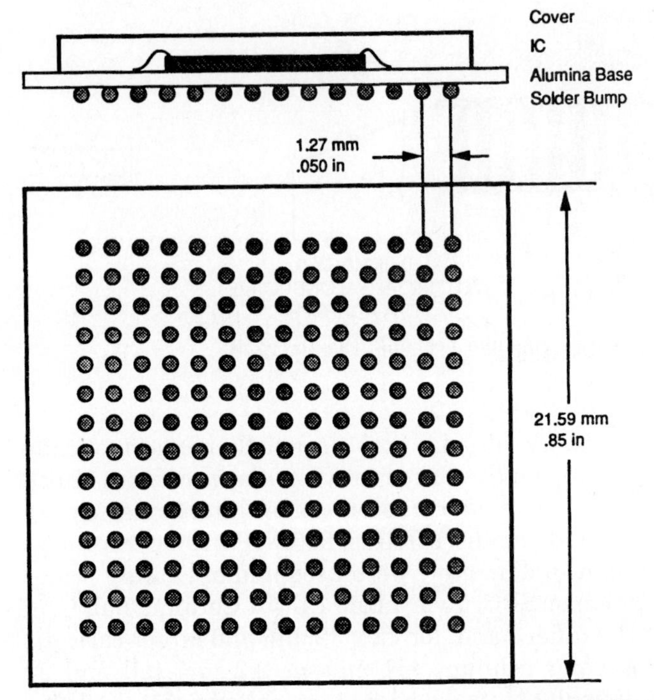

Figure 1.33*a* Motorola's solder bumped Pad Array Carrier (PAC).

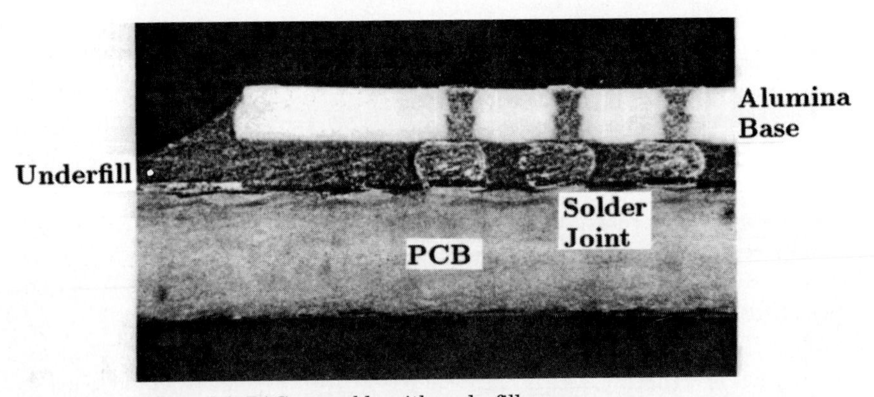

Figure 1.33*b* Motorola's PAC assembly with underfill epoxy.

Array (CCGA). The reason to use the taller solder column for the larger carriers is to increase the compliance and reduce the shear strain due to thermal expansion mismatch between the ceramic carrier (6×10^{-6} m/m–°C) and the FR-4 epoxy PCB (18.5×10^{-6} m/m–°C).

There are two ways to attach the solder columns to the bottom of the MLC substrate. One is by reflowing the 63%wtSn/37%wtPb solder

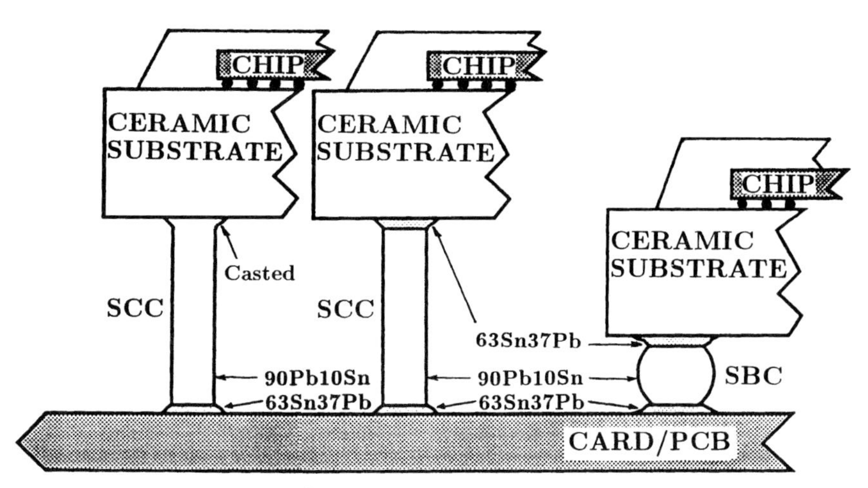

Figure 1.34 IBM's Ceramic Ball Grid Array (CBGA).

(SBC also used this method). In order to increase the stability of the carrier during solder reflows, the other method is to cast the high Pb solder column in place on the carrier. For both SBC and SCC, the 10-mil (0.25-mm) pitch 97%wtPb/3%wtSn C4 (controlled collapse chip connection) chip[63–70] is soldered to the MLC substrate. IBM's CBGAs can be used to package the ASICs and microprocessors. Solder joint reliability data are presented in Chaps. 5 through 8.

Figure 1.35*a* and *b* shows Hitachi's new Micro Carriers for LSI Chip (MCC) for their M-880 processor in supercomputing.[104] The size of MCC is ranging from 10 to 12 mm square and has a total of 528 0.25-mm pitch, area array high-temperature solder bumps (252 for signal, 99 for terminating resistor, 153 for power and ground, and 24 for monitor). In order to match the bumps on the LSI chip with the thick-film pattern on the mullite ceramic base substrate correctly, thin-film layers (five wiring layers and one resistor layer) are deposited on the substrate. (The thin-film conductor material is aluminum and interlayer insulators are a polyimide and a photosensitive polyimide.) The substrate has seven conductor layers, Fig. 1.35*b*. The bottom layer I/O pads are electroplated by Ni/Au for the eutectic solder balls, Fig. 1.35*a*.

A high thermal conductivity aluminum nitride (AlN) cap is soldered to the back of the LSI chip. The AlN cap and the MCC substrate are hermetically sealed with solder, Fig. 1.35*a*. These MCCs (up to 41) are mounted on a 44-layer mullite ceramic substrate (106 × 106 mm) to form a multichip module for high-speed supercomputer application.

Figure 1.36*a* and *b* shows the Nippon Electric Company (NEC)'s CBGA for their Flip TAB Carrier (FTC) in their SX-3/SX-X supercomputers.[105] Their design budgeted roughly 50 percent of the critical path

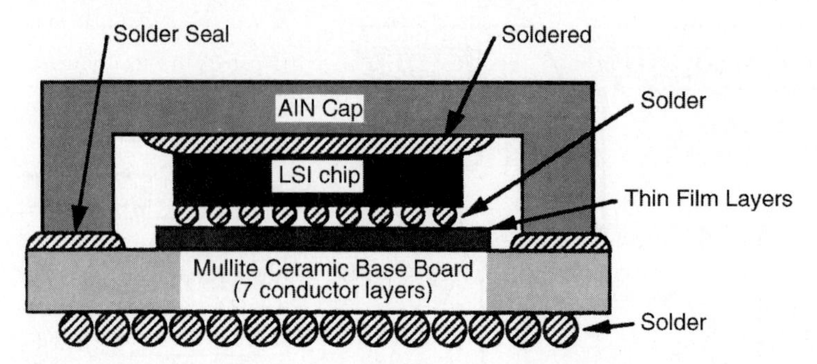

Figure 1.35*a* Hitachi's Micro Carriers for LSI Chip (MCC) for their M-880 processor.

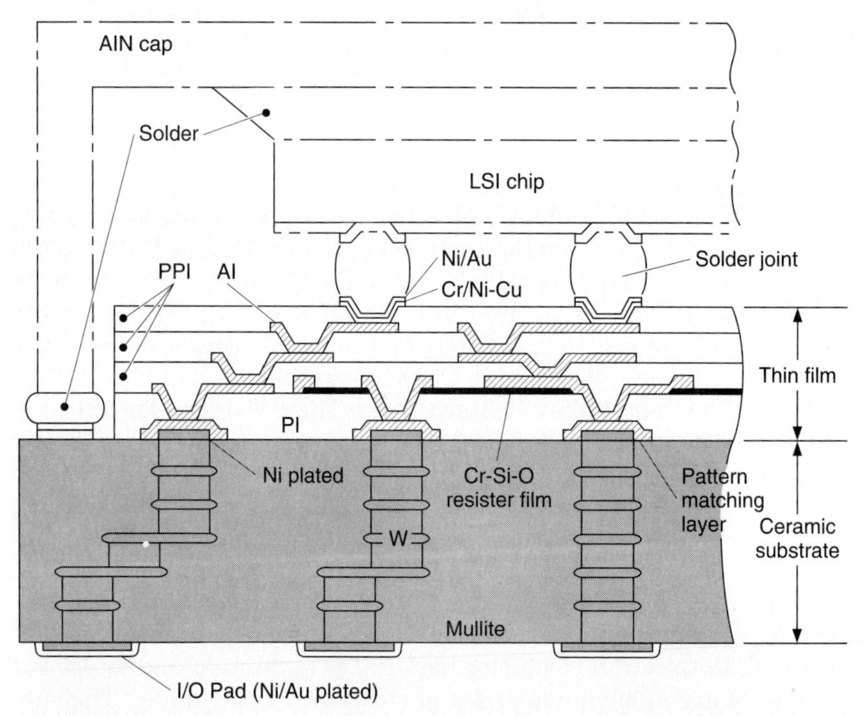

Figure 1.35*b* Hitachi's Micro Carriers for LSI Chip (MCC) for their M-880 processor.

delay for logic and 50 percent for their high-density packaging. The 14-mm square LSI chip is facedown TAB (with 485 Au-plated Cu leads at 0.7 mm pitch) on an 18.5 × 18.5 mm alumina substrate. The outer lead pitch is 0.11 mm. The bottom of the substrate has 604 I/O solder bumped pads at 0.7 mm pitch for the logic FTC and has 588 I/O solder bumped pads at 0.7 mm pitch for the memory FTC. The back of the LSI chip is attached to an AlN cap to transfer the heat from the chip. This FTC is then connected through these solder bumps to the 100-mm square multilayer substrate with polyimide thin-film signal layers with ground planes, Fig. 1.36*b*. The minimum feature size of NEC's thin-film conductors is 25 μm. There is a maximum of 36 FTCs on a single multichip module.

PBGA. Since bismaleimide triazine (BT) resin is cheaper than ceramic and has a lower dielectric constant (i.e., better electrical performance), Motorola replaced the ceramic substrate of the CBGA by BT resin and teamed up with Citizen to develop a plastic molding process for the chip and substrate.[106] They have called it Over Molded Plastic Pad Array Carrier (OMPAC). Today, it is called PBGA. Since 1989, Motorola has been successfully assembling PBGAs in their portable electronic products. Also, since 1993 Compaq Computers[108–109] has been using PBGA packages in their personal computers. In both applications, the pin counts are equal to or less than 225.

Figure 1.37 schematically shows a cross section of a PBGA. It consists of a silicon IC, gold wires and bonds, die attach and pad, molding compound, BT epoxy substrate, solder balls, vias for ground/signal/thermal, plated copper conductors, and a solder mask.

There are at least two forms of molding, one-side molding and overmold.[76] One-side molding (Fig. 1.37) is a transfer molding method used for most of the PBGAs that are mass produced by Citizen and others. Overmold for PBGAs is a new design developed by Citizen (Fig. 1.38). The fabrication of prototypes has been completed. This method is mainly for PBGAs with very high pin counts and for multilayer PBGAs. The advantage of this overmold is to reduce moisture absorption from the edges of the substrate. However, overmold is difficult to apply to thin substrates. (The minimum substrate thickness for the

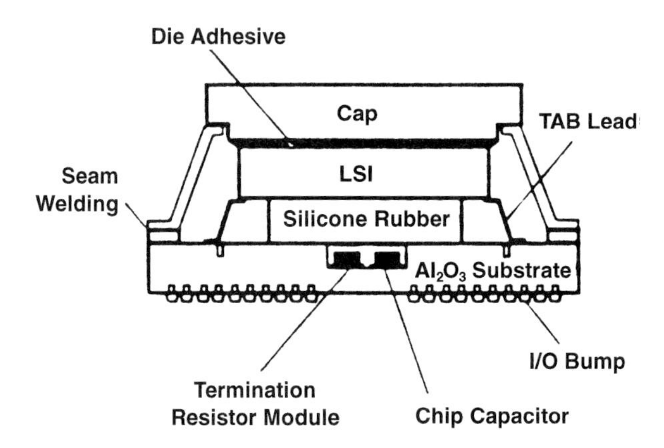

FTC cross section

Figure 1.36*a* NEC's Flip TAB Carrier (FTC) for their SX-3/SX-X supercomputers.

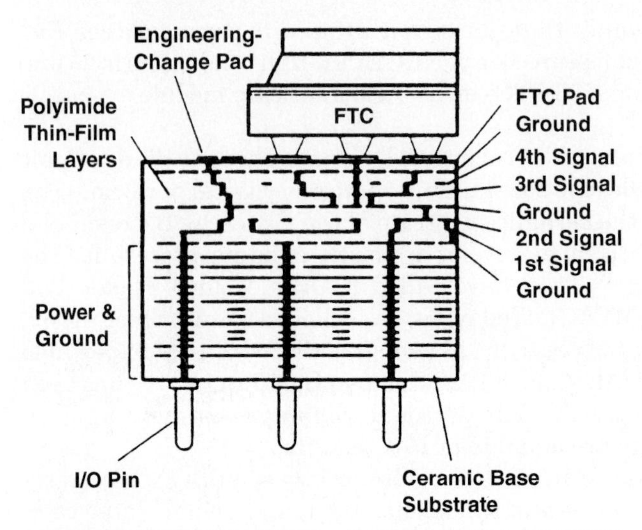

Figure 1.36b NEC's Flip TAB Carrier (FTC) for their SX-3/SX-X supercomputers.

overmold is 0.4 mm.) In general, for both one-side and overmolding, the maximum molding thickness for the wire loop is 0.3 mm and the minimum molding thickness above the wire loop is 0.2 mm.

Most of the PBGA substrate material used is BT resin (CCL-HL 832) manufactured by Mitsubishi Gas Chemical Company. For this material, the glass transition temperature is 170 to 215°C, the thermal expansion coefficient is 15×10^{-6}/°C in the in-plane direction and is 52×10^{-6}/°C in the out-of-plane direction, the peel strength (35 μm) is 1.6 kgf/cm, the flexural strength is 52 kgf/mm^2, and the water absorption is 0.06 percent.

The hole diameters of the BT substrate are dependent on the substrate thickness. For 0.1–0.6-mm-thick substrate the normal hole diameter is 0.4 mm, and for 0.7–1.0-mm-thick substrate the normal hole diameter is 0.5 mm.

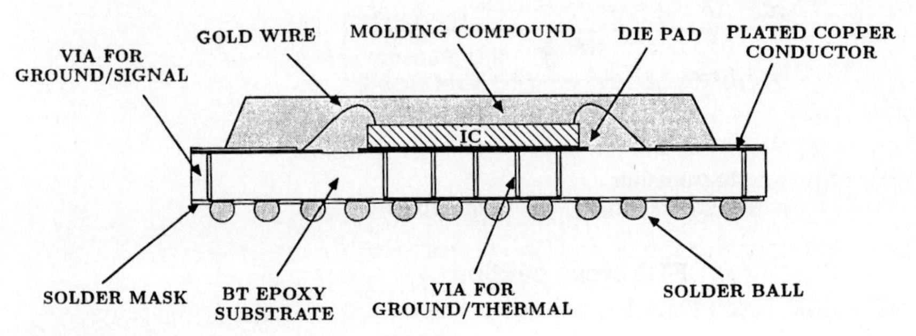

Figure 1.37 Plastic Ball Grid Array (PBGA).

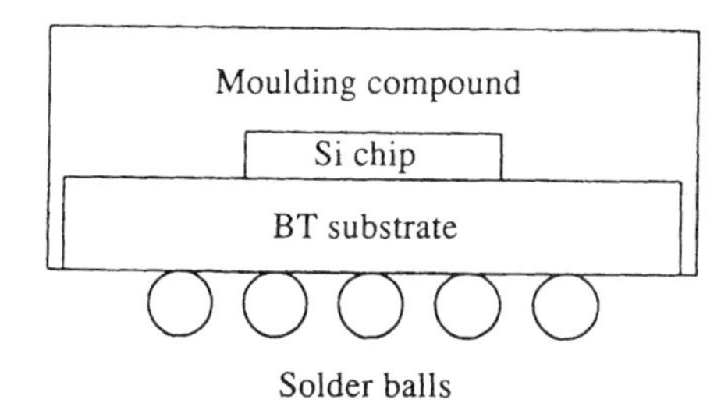

Figure 1.38 Schematic of an overmold Plastic Ball Grid Array (PBGA).[76]

Figure 1.39 shows a solder ball on the bottom surface of the PBGAs. It can be seen that the thickness of copper pattern is 38 μm, the diameter of copper pattern is 889 μm, the thickness of dry film is 75 μm, the opening diameter of copper pattern is 635 μm, the diameter of solder ball is about 760 μm, and the height of solder ball is about 625 μm. The solder composition is 63wt%Sn/37wt%Pb. The Young's modulus of the solder is 1.5×10^6 psi (10,000 MN/m^2) and the Poisson's ratio is 0.4. The thermal coefficient of linear expansion of the solder is 21×10^{-6}/°C.

Most of the PCBs for PBGA applications is made of FR-4 epoxy/glass. One of the detailed pad/mask/trace dimensions is shown in Fig. 1.40. It can be seen that the diameter of the copper pad is 0.635 mm and the diameter of the solder mask opening is 0.889 mm. The thickness of the the copper pad is 0.043 mm and the width of the copper trace is 0.25 mm. The solder mask height is 0.025 mm and is 0.068 mm when it is covering the copper trace. Two common surface finishings have been used, one is organic-coated Entec Cu56 and the other is hot-air-leveled with about 0.017 mm thick of Sn/Pb solder.

Figure 1.41*a* schematically shows a cross section of SGS-Thomson's high-performance PBGA. It is a cavity-down package with gold wires. A metal heatsink plate is attached at the back to the chip and the heat path is through the Ag epoxy adhesive. The thermal resistance of this package is very low.

Figure 1.41*b* shows a cross section of AMKOR's PBGA with solder bumped flip chip (instead of wire bonding) on the BT substrate. The very large thermal expansion mismatch between the silicon chip (2.5×10^{-6}/°C) and the BT substrate (15×10^{-6}/°C) is reduced by the underfill epoxy encapsulant. It reduces the stresses and strains in the flip chip solder joints and redistributes the stresses and strains over the entire chip area that would otherwise be increasingly concentrated near the corner solder joints of the chip.

Other advantages of encapsulant are to protect the chip from moisture, ionic contaminants, radiation, and hostile operating environments such as mechanical shock and vibration. The most desirable epoxy encapsulants should have a high glass transition temperature (>150°C) and low thermal coefficient of linear expansion ($<27 \times 10^{-6}$/°C).

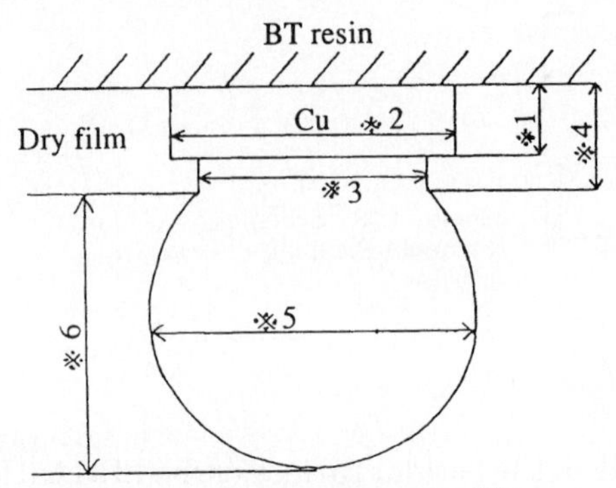

※ 1 = thickness of copper pattern -- 38 μm
※ 2 = diameter of copper pattern -- ϕ 889 μm
※ 3 = opening diameter of copper pattern -- ϕ 635 μm
※ 4 = thickness of dry film -- 75 μm
※ 5 = diameter of solder ball -- ϕ 760 μm
※ 6 = height of solder ball -- 600 μm ± 50 μm

Figure 1.39 Schematic of a solder Ball of the Plastic Ball Grid Array (PBGA).[76]

PBGAs can be used for SRAMs with high speed applications. For ASICs and microprocessors with less than 600 I/Os and below seventy-five-MHz clock frequencies, PBGAs could be the low cost solutions.

TBGA. Figure 1.42 schematically shows a cross section of IBM's TBGA, which uses area array TAB technology. Two-metal tape for ground and signal was used for the substrate. The chip is 95%wtPb/5%wtSn solder bumped by C4 technology. The inner lead bonding (i.e., bonding the solder bumped chip on the Cu-Ni-Au tape) can be done by either pulse-

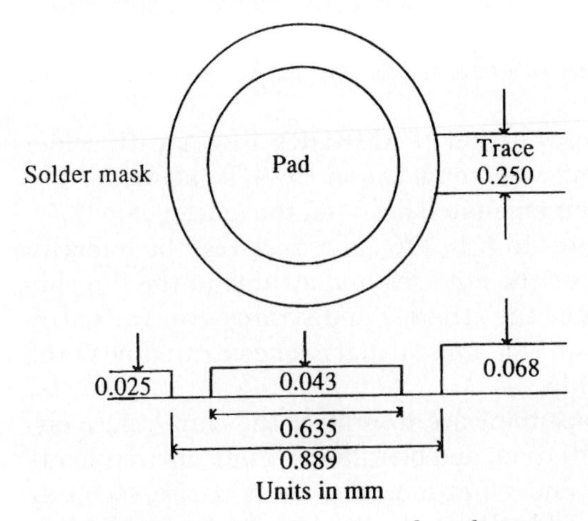

Figure 1.40 Details of pad, trace, and mask geometry on a PBGA PCB.[76]

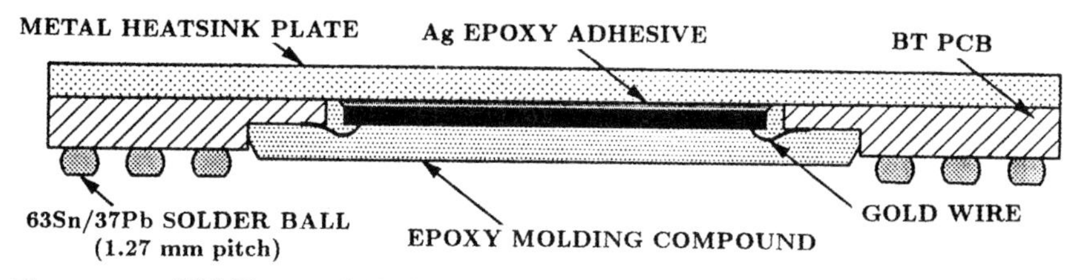

Figure 1.41a SGS-Thomson's Ball Grid Array.

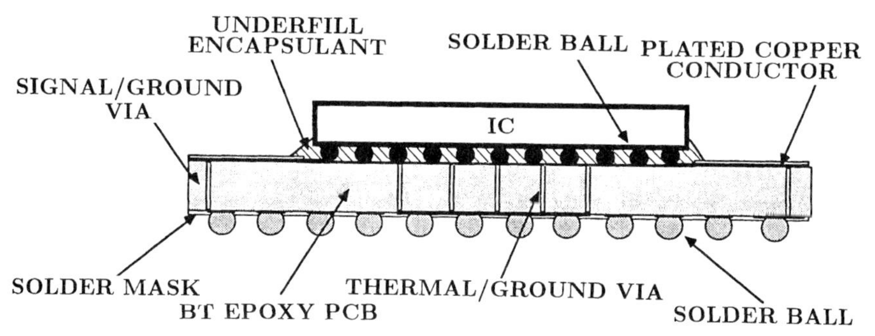

Figure 1.41b Plastic Ball Grid Array (PBGA) with solder bumped flip chip. (*Source: AMKOR.*)

thermode or hot-air-thermode reflow soldering. Again, underfill encapsulant is needed to reduce the thermal expansion mismatch between the chip and the copper ($17.5 \times 10^{-6}/°C$) tape. A copper plate with center opening (for the chip and the underfill incapsulant) is attached (by a stiffener adhesive) to the two-metal tape to provide planarity and rigidity (as a stiffener) to the package. The bottom side of the substrate is attached with area array 50-mil (1.27-mm) pitch 90%wtPb/10%wtSn solder balls (25 mils or 0.65 mm in diameter). The back of the chip is attached to a heatsink to improve package thermal performance. TBGA can be used for ASIC and microprocessor applications.

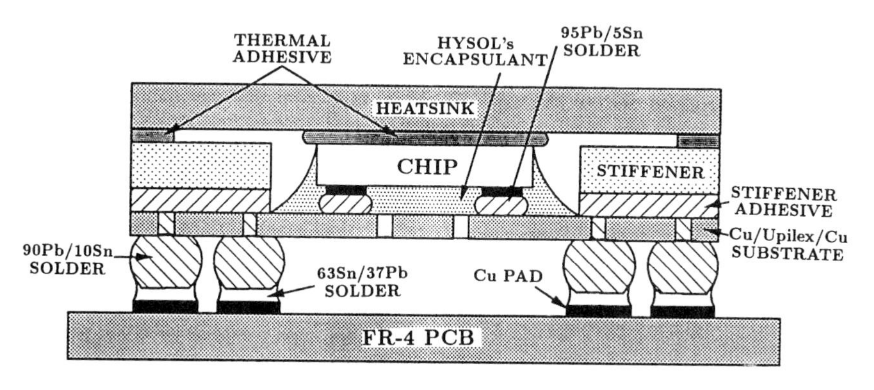

Figure 1.42 IBM's area array TAB Ball Grid Array (TBGA).

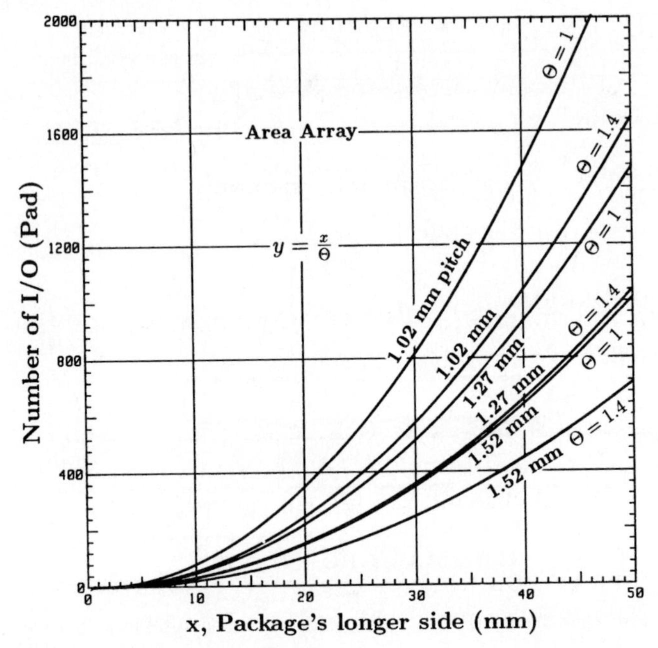

Figure 1.43 Number of package pin counts versus package size.

1.3.3 BGA and PQFP comparison: package pin count

PGAs, CBGAs, PBGAs, and TBGAs are area array packages in which the pins or solder joints are spaced over the bottom surface of the carrier. Figure 1.43 shows the possible number of I/Os (pads) for various package sizes and pitches. The dimensions of the package are x and y, with $x = \theta y$, $\theta \geq 1$. For square package, $\theta = 1$.

Figure 1.44 shows the possible number of I/Os between the peripheral and area array square packages for various sizes and pitches. It can be seen that for a 32×32 mm package size, the possible number of I/Os of a peripheral package (e.g., PQFP) is about 184-pin (0.65 mm pitch). However, for the same package size, the possible number of I/Os of an area array package (e.g., BGA) is about 600-pin (even double the pitch to 1.27 mm). Figure 1.45 shows the possible number of I/Os between the peripheral and area array rectangular packages (the longer side is 1.4 times the shorter side) with various sizes and pitches.

1.3.4 BGA and PQFP comparison: package area

Figure 1.46 shows the package density (area) comparison between the peripheral, area array, and flip chip packaging technologies for various packaging pin counts. It can be seen that for a given package pin count, the peripheral PQFP packages require the largest package area (i.e., occupy the largest PCB real estate). On the other hand, the area array

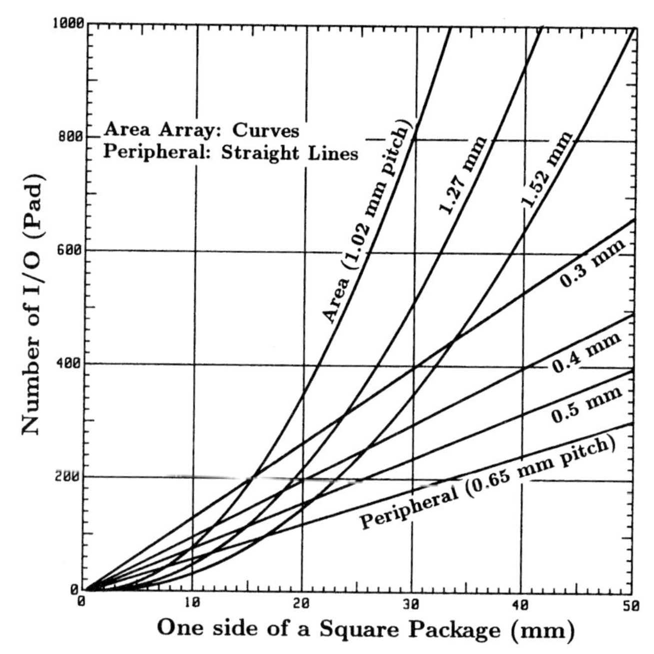

Figure 1.44 BGA and PQFP comparison: package pin counts (square).

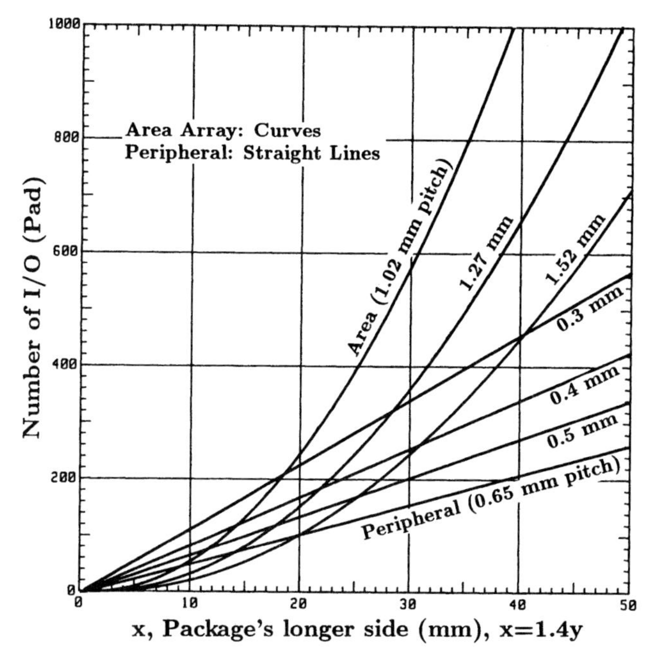

Figure 1.45 BGA and PQFP comparison: package pin counts (rectangular).

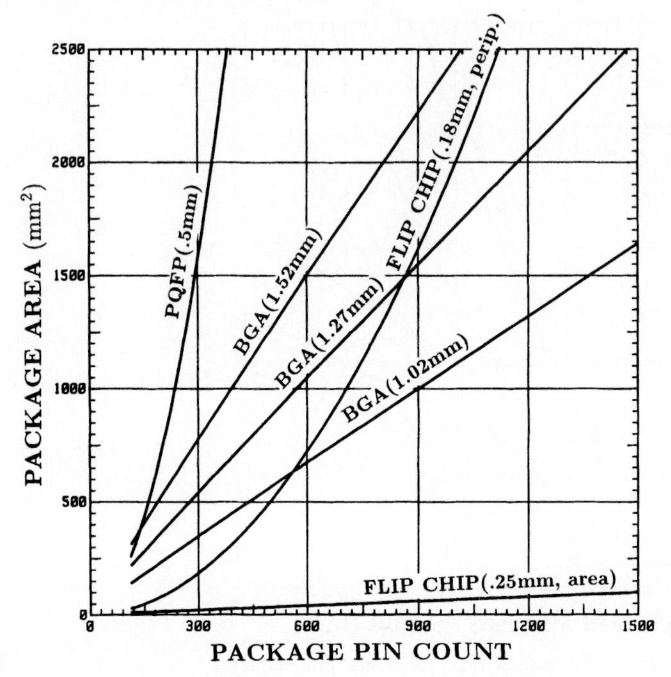

Figure 1.46 BGA and PQFP comparison: package density.

flip chip on board technology takes up the smallest real estate. The required package area for BGA technologies is less than that for PQFPs and more than that for area array flip chip technology. There is a crossover point between the BGA technologies and the peripheral flip chip technology.

1.3.5 BGA and PQFP comparison: package performance

Figure 1.47 shows the inductance (nH) comparison between the 0.5-mm pitch PQFPs and the 1.27-mm-pitch wire-bonding PBGAs. It can be seen that for a given package pin count (>200), the inductance of PBGA is smaller than that of PQFP. This is because of the short runs between the chip and the solder bumps on the bottom of the PBGA packages. Also, the reflections and noise levels can be reduced, respectively, by matching the trace length with the output impedance and by dissipating heat with power and ground planes.

1.3.6 BGA and PQFP comparison: manufacturing yield

It has been shown[76] that PCB assembly of BGAs is very forgiving due to the self-aligning characteristic of the molten eutectic Sn/Pb solder

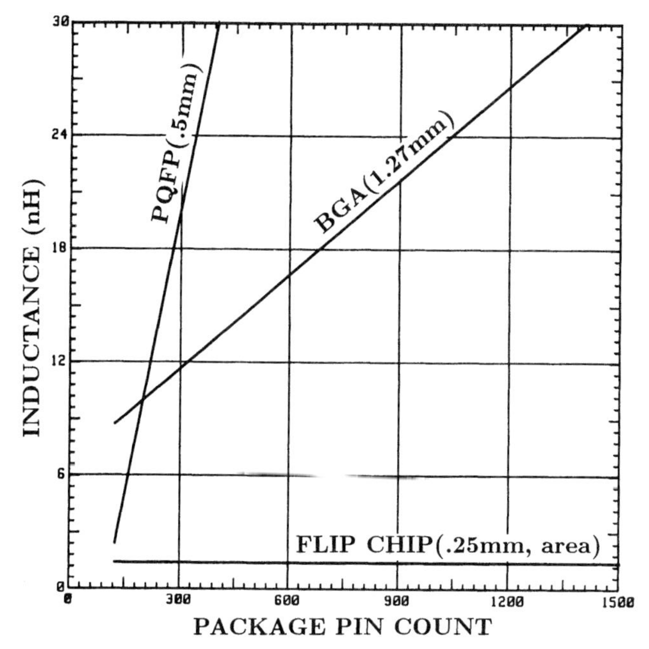

Figure 1.47 BGA and PQFP comparison: performance (inductance).

during reflow, and that BGAs on PCB is a very high yield assembly process. On the other hand, PCB assembly of PQFPs is suffering from poor manufacturing yield due to the limitations of equipment and personnel for very fine pitches.

1.3.7 BGA and PQFP comparison: moisture and popcorn effects

PBGAs are not hermetically sealed packages and, because of the material constructions of these packages, they will adsorb the moisture while they are in storage. During the rapid heating in a reflow oven, moisture in the package vaporizes and creates stresses, which can crack the package ("popcorn" effect).

Figure 1.48 shows the moisture absorption rates (at 85°C/85%RH) of a 119-pin PBGA.[96] (The moisture adsorption rates were obtained by dividing the weight increase from the initial weight after a given storage period by the initial weight.) It can be seen that the saturation of moisture absorption in the PBGA occurred at 0.35 percent after 48 hours of storage at 85°C/85%RH environment. This is much higher than that (0.2 percent) of PQFPs at the same condition, and is due to the PBGA's BT substrate which absorbs and contains more water than epoxy molding compounds.

Table 1.4 (reported as defective of samples inspected) and Fig. 1.49*a* and *b* show that baking (125°C for 24 hours) of the PBGAs can reduce the moisture effect.[76] Figure 1.49*a* and *b* shows the Scanning Acoustic Microscopy (SAM) photos of a stressed (reflowed) PBGA, respectively, under the baked and 85/85 conditions. It can be seen that the 85°C/85%RH for 168 hours preconditioned packages showed significant internal delaminations (Fig. 1.49*b*). The packages that were baked before reflow had no internal delaminations or cracks (Fig. 1.49*a*).

Since the "baked" packages suffered no damage after the reflow stress and the moisture-saturated packages were highly damaged, the thermal mismatch contribution (alone) to interfacial delaminations appears insignificant. The delaminations resulted from the moisture adsorbed by the package. These packages should be considered *moisture-sensitive.* Further evaluation will determine the acceptable level of moisture, or allowable time out of dry pack before reflow.

Figure 1.50*a–d* shows the cracking mechanism due to moisture and eventually popcorn effect.[96] The crack started from the chip adhesive layer and propagated along the chip pad. From that point, there were two types of cracking mechanisms. Type-I was the cracking pattern that the crack propagated through the molded body, parallel to the interface between the BT substrate and molding compound. For Type-II, the crack began to propagate into the dielectric layer of the BT substrate and then propagated through the substrate, parallel to the interface between the molding compound and substrate. PBGA package cracking due to popcorn effect can be avoided (even without baking) if there are some temporary open via holes (e.g., use the thermal via holes as the vent holes) under the chip pad to allow the water vapor to escape during solder reflow.[96]

1.3.8 Next-generation BGAs

Figure 1.51 shows a next-generation BGA package, Slightly Larger than IC Carrier (SLICC) proposed, developed, and tested by Motorola.[110] Unlike the PBGAs, SLICC has a smaller solder ball pitch (35 mils or 0.9 mm) and diameter (20 mils or 0.5 mm). The 97%wtPb/3%wtSn or 95%wtPb/5%wtSn solder bumped C4 chips are available in waffle packs and are picked up, fluxed, and placed on the organic substrate.

One of the reasons the SLICC is so much more area-efficient than wire-bonded packages like PBGAs is because of the increased area that is available to route the runners under the chip. Some of the typical features of SLICC substrate are listed in Table 1.5. As the PCB technologies push themselves to smaller plated vias, better tolerances, and finer lines and spaces, this next-generation BGA can result in carriers that can theoretically be the same size as the chip in the horizontal

TABLE 1.4 PBGA Moisture Study Results[76]

No. of leads	Preconditioning stress	Die surface delamination	BT substrate to moulding compound delamination	BT substrate to solder mask delamination	Solder mask to moulding compound delamination
225	Bake	0/3	0/3	0/3	0/3
324	Bake	0/3	0/3	0/3	0/3
396	Bake	0/3	0/3	0/3	0/3
225	85/85	5/5	5/5	5/5	5/5—least delam.
324	85/85	5/5	5/5	5/5	5/5
396	85/85	3/5	5/5	5/5	5/5—most delam.

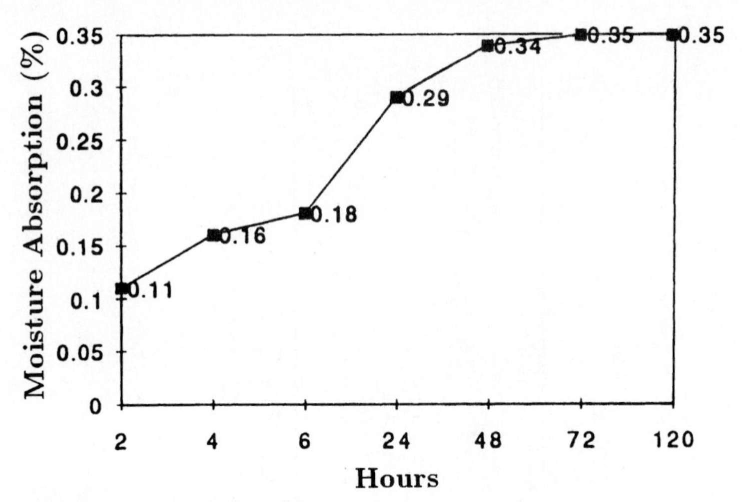

Figure 1.48 PBGA moisture absorption rates at 85/85 conditions.

dimensions. Figs. 1.52, 1.53, and 1.54 show the possible number of I/Os (pads) for the next-generation (square and rectangular) BGAs with various pitches (0.9, 0.75, 0.65, 0.5, 0.38, and 0.25 mm) and sizes.

By combining the best features of BGA and flip chip technologies, Sandia National Laboratories developed a minimally packaged die technology called Mini Ball Grid Array (mBGA), Fig. 1.55*a* and *b*.[123] It is similar to BGA technology in that it protects the active surface of the die and redistributes the peripheral die pads to an area array of bumped pads on relatively large pitch (0.5 to 0.75 mm), Fig. 1.55*a* and *b*. It is similar to flip chip technology in that the mBGA package is no larger than the die itself.

A typical mBGA redistribution fabrication process flow is shown in Fig. 1.56. Because of the large pad pitch, the mBGA is easily placed in nonbonding die carriers for test and burn-in to ensure a known good die (KGD), and assembled on the next-level substrate. The mBGA uses off-the-shelf chips.

By combining most of the advantages of tape automated bonding (TAB), wire bonding, and flip chip technologies, Tessera developed a small, thin, high-performance single-chip package called Micro Ball Grid Array (μBGA), Fig. 1.57.[124–126] It allows a die pitch as small as 50 μm (which is competitive with advanced TAB pad pitch), and the flexible gold beam lead is assembled to the die pad with conventional thermosonic wire-bonding equipment. The leads fan inward from the die pads to an Ni/Au bump array (Fig. 1.58) that serves to make connection to the next-level package. The Ni/Au bump height is about 0.085 mm, and the bump pitch on the tape can be as small as 0.5 to 0.3 mm.

A high level of reliability can be achieved with the elastomer-compliant layer (Fig. 1.58), which also provides the opportunity for full test

Figure 1.49*a* C-mode Scanning Acoustic Microscope (SAM) of baked PBGA.[76]

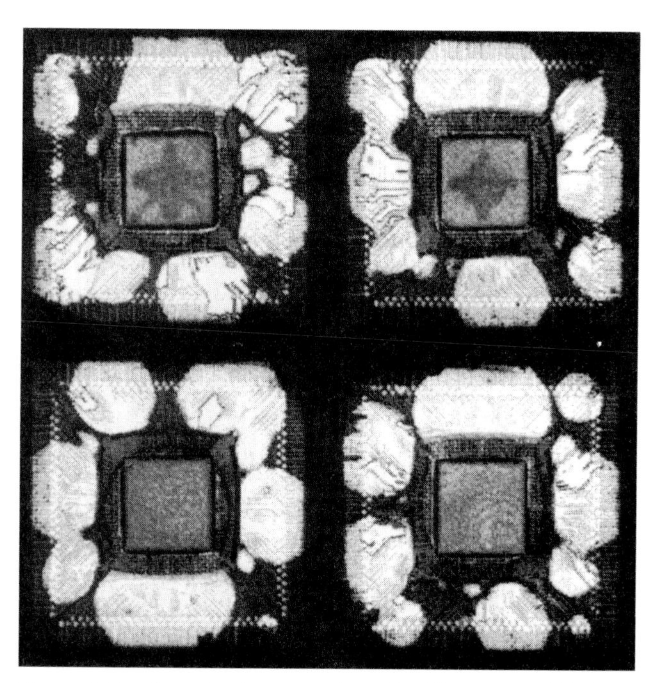

Figure 1.49*b* C-mode Scanning Acoustic Microscope (SAM) of 85/85 PBGA.[76]

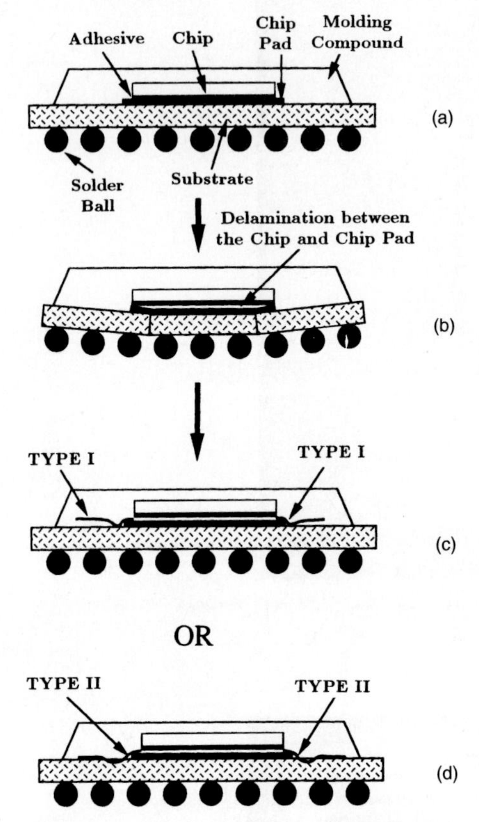

Figure 1.50 PBGA moisture effect, popcorn cracking mechanism. Source: Ahn, et al.[96]

and burn-in for KGD prior to final assembly. Figure 1.59*a* (with package case) and *b* (without package case) show a couple of the cross sections of the μBGA. Because of the low resistance, capacitance, and inductance of the short beam leads, electrical performance is very good. For very high performance applications, an optional ground plane further reduces the self- and mutual inductance as well as the crosstalk. The thermal management of μBGA can be enhanced with a heatsink attaching to the package case or the back of the die with thermal grease.[127] It should be noted that μBGA uses an off-the-shelf die and does not require die postprocessing.

By combining the best features of flip chip and bare chip technologies, Mitsubishi Electric Corporation developed a very high density, performance, and I/Os, and very small and thin package called Chip Scale Package (CSP).[128] It is very easy to handle, test, standardize, and solder. A typical family of the CSP is shown in Fig. 1.60 and the cutout

TABLE 1.5 Some Slightly Larger than IC Carrier (SLICC) Features[110]

Size of Carrier	0.060″–0.100″ larger than IC edge
Thickness of Substrate	0.008″–0.012″
Core Materials	FR4, BT epoxy
Topside Solder Mask	0.0008″–0.001″ thickness
Topside Pad Finish	0.0008″–0.0015″ high 60/40 Sn/Pb
Plated Through Hole Metal	0.0005″ thick Cu (typ), plugged
Plated Thru Hole Dim.	0.008″ drilled, 0.016″ dia. pad
Line widths/Spacings	0.004″/0.004″ minimum
Bottomside Pad Finish	Au flash over Cu-Ni
Bottomside Pad Pitch	0.035″ center to center
Bottomside Pad Diameter	0.020″

of a CSP is shown in Fig. 1.61. It can be seen that the package consists of the LSI chip, electrode pads, wiring conductor pattern, thin resin encapsulation, and external electrode bumps.

Figure 1.62 shows a cross-sectional view of the CSP package. It can be seen that the electrode pad was "transferred" (redistributed) to a new location (external electrode bump) on the LSI chip. Electrical connections between the pads on the chip and the external electrode bumps are done by wiring conductor patterns on the chip; this allows the internal/external pad locations to be adjusted. Thus, there are no additional restrictions on the chip designers. The bump fabrication processes flow is shown in Fig. 1.63. The CSPs can be fully tested and burned in for KGDs.

One of AMKOR's next-generation PBGAs is the Super Ball Grid Array (SuperBGA). It is designed for very light weight (1.9 to 2.9 gram), very low profile (1.2 to 1.45 mm), very good electrical characteristics (total capacitance: 0.93–1.86 pF, resistance: 110–270 mΩ, self-inductance: 2.11–4.24 nH, and impedance: 50–70Ω); and excellent thermal management (up to 10 W for a 27 × 27 mm SuperBGA).[129]

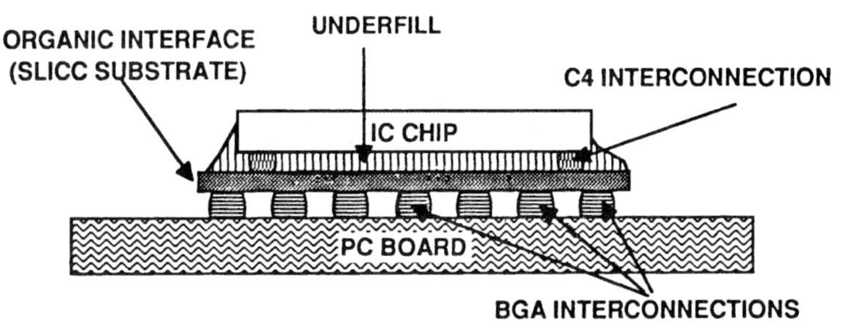

Figure 1.51 Slightly Larger than IC Carrier (SLICC). (*Source: Banerji of Motorola.*[110])

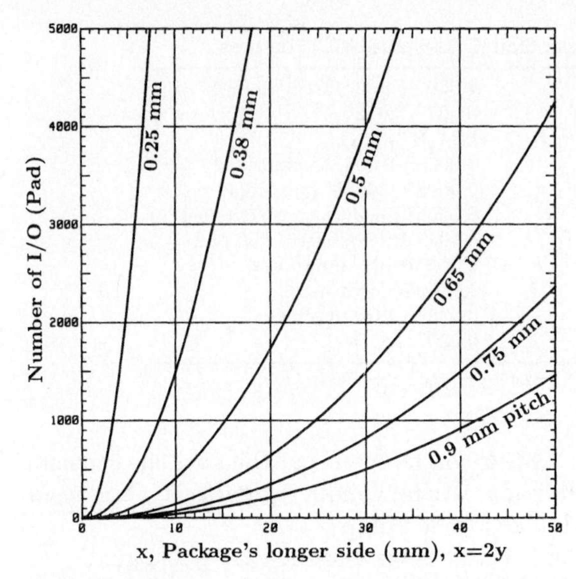

Figure 1.52 Number of I/Os vs BGA package size (long side = 2 short side).

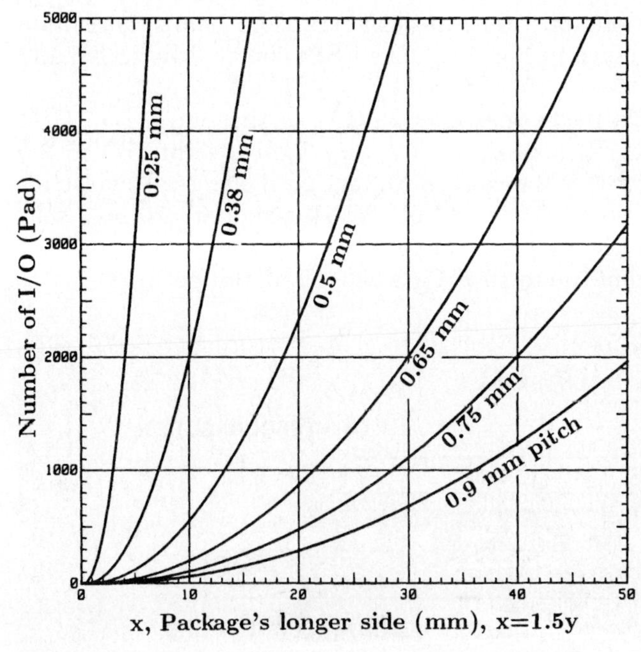

Figure 1.53 Number of I/Os vs BGA package size (long side = 1.5 short side).

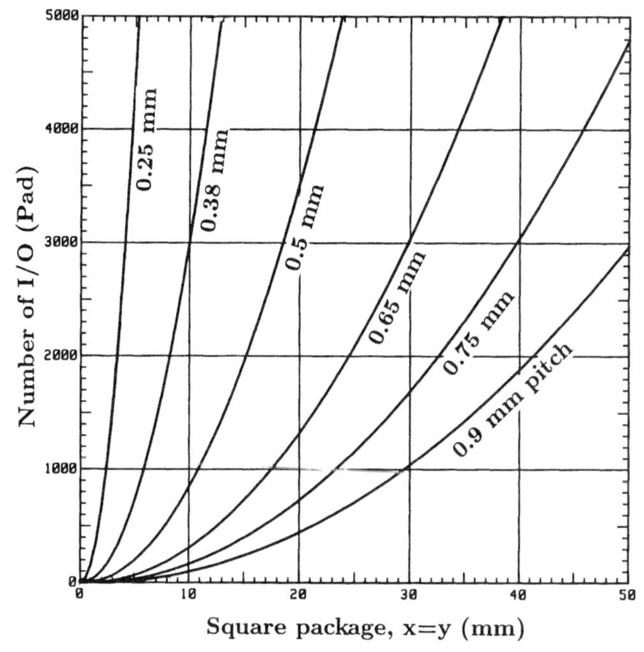

Figure 1.54 Number of I/Os vs BGA package size (square).

Figure 1.64 schematically shows a cross section of AMKOR's Super-BGA. It can be seen that it is a cavity-down package with the chip face placed downward in the cavity with gold wires. A thin copper heatsink plate is attached with silver epoxy to the back of the chip, covering the top surface of the package. This copper plate acts not only like a heatsink but also a stiffener and ground plane of the package. Super-BGA has very good physical attributes, e.g., 0.075–0.1-mm line widths, 0.075-mm line spacing, 0.018-mm line thickness, 0.3-mm typical via

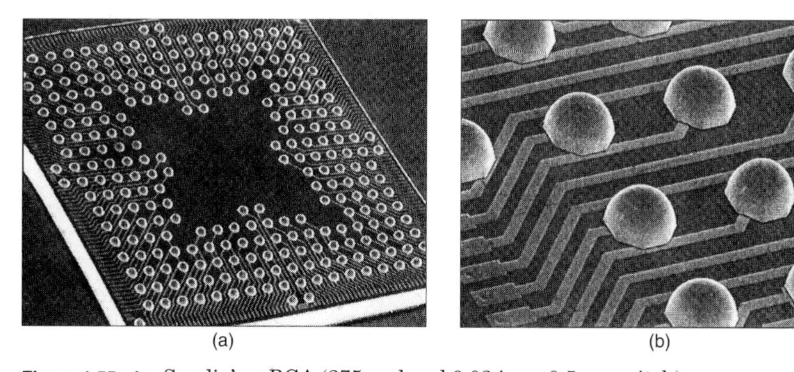

Figure 1.55*a,b* Sandia's mBGA (275-pad and 0.02 in or 0.5 mm pitch).

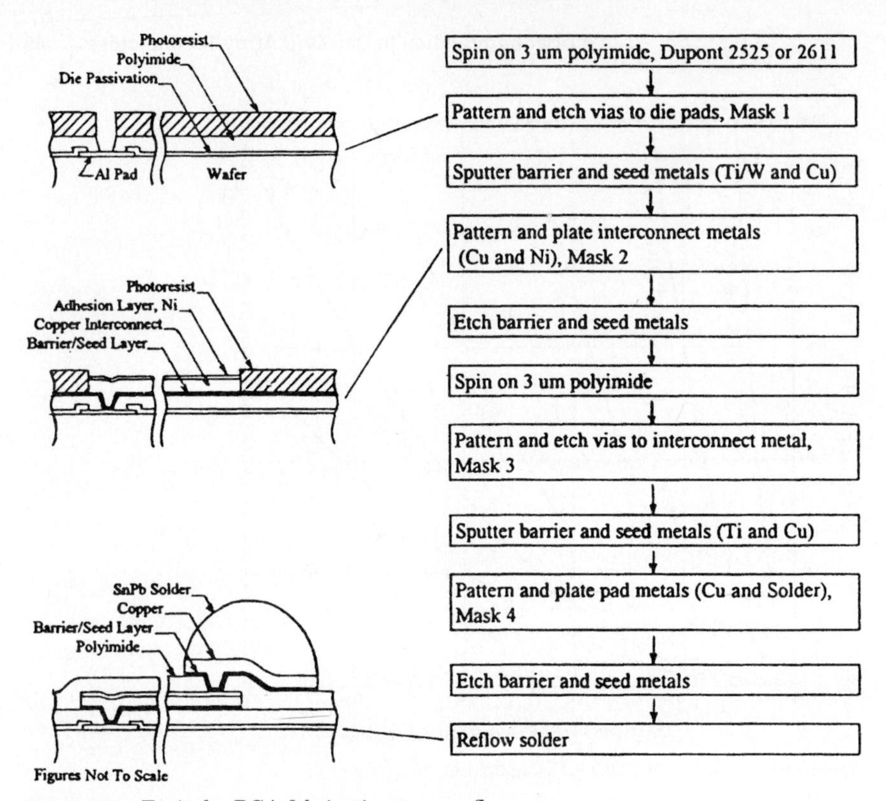

Figure 1.56 Typical mBGA fabrication process flow.

Figure 1.57 Tessera's μBGA (188-pad and 12.7 mm on a side).

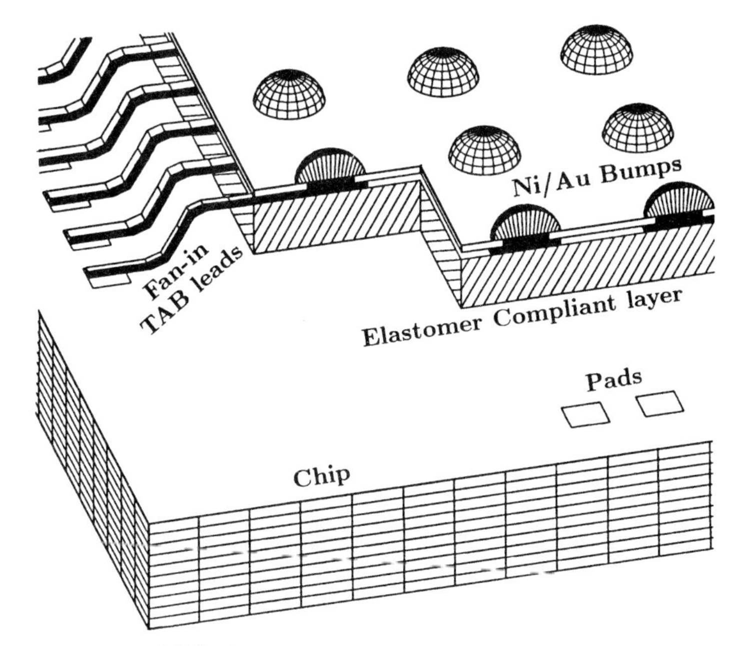

Figure 1.58 μBGA's elastomer-compliant layer.

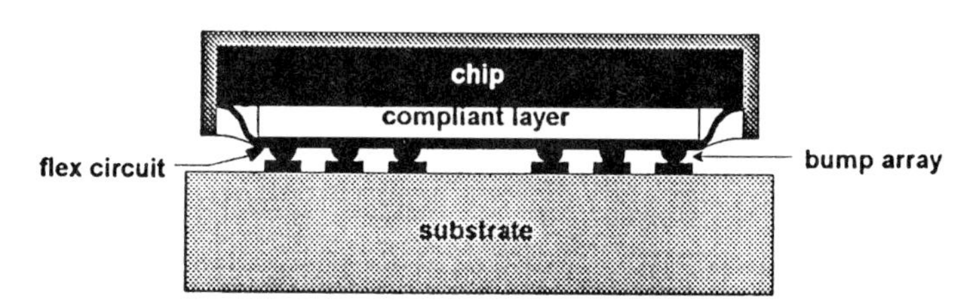

Figure 1.59*a* Standard μBGA with package case.

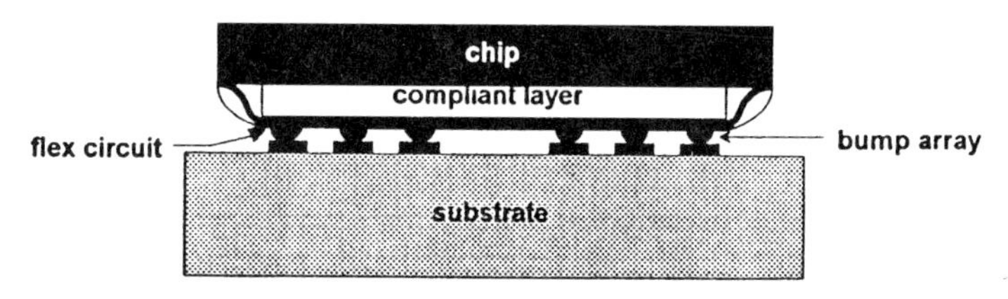

Figure 1.59*b* Chip-size (or less) μBGA for high-performance applications.

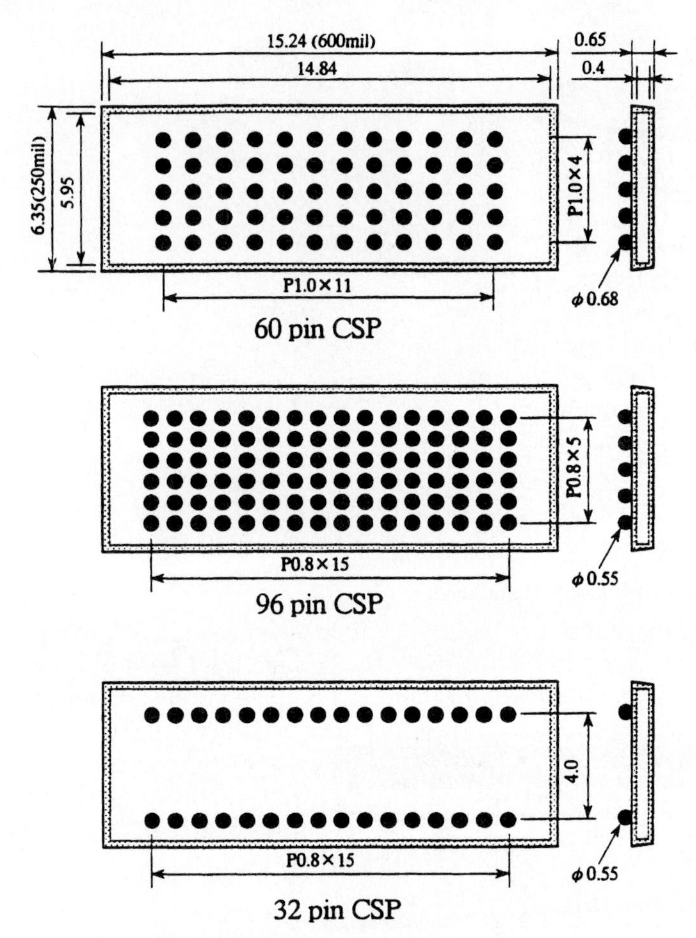

Figure 1.60 Mitsubishi's Chip Scale Package (CSP) family.

diameter, and two to five metal layers. The SuperBGA standard square package family (JEDEC MO-151) ranges from 49 to 2601 mm^2 in size, and from 20 solder balls (one row) to 680 solder balls (five rows). Super-BGA is capable of performing at very high speeds (1.5 to 2 GHz).

By combining the best advantages of PBGA and peripheral leaded PQFP, MCC and Motorola developed a unique, low-cost mixed MCM package to house four Motorola M56002 digital signal processor chips and five capacitors,[130] Fig. 1.65. There is a total of 270 wire-bonded connections on the four-layer laminate module (thickness = 0.5 mm; and dimensions = 30 × 30 mm), 121 off-module peripheral connections, and 196 solder balls (0.89 mm in diameter) on the bottom layer. The package can operate up to 60 MHz with nanosecond edge rates present on the board and may generate up to 4 W of heat. The eutectic solder balls

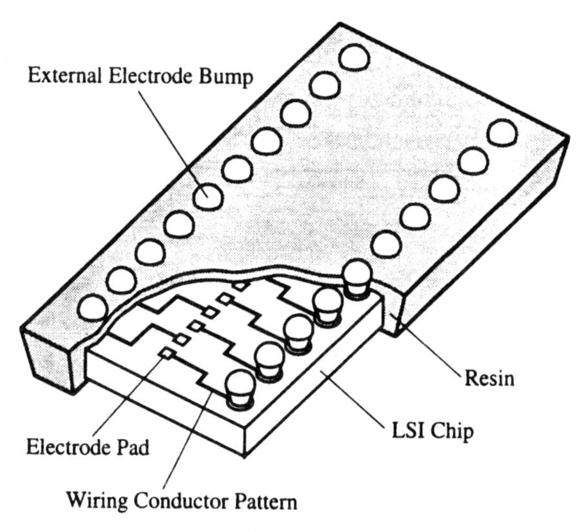

Figure 1.61 Cutout of Mitsubishi's Chip Scale Package (CSP).

are multifunctional, 60 of which are used for power and ground connections to the next level of package (PCB), 2 for clock lines, and the rest for low-resistance thermal paths of the thermal vias located in both the package and the PCB. The peripheral copper leads are available for the functional I/O signals, and are attached to the peripheral

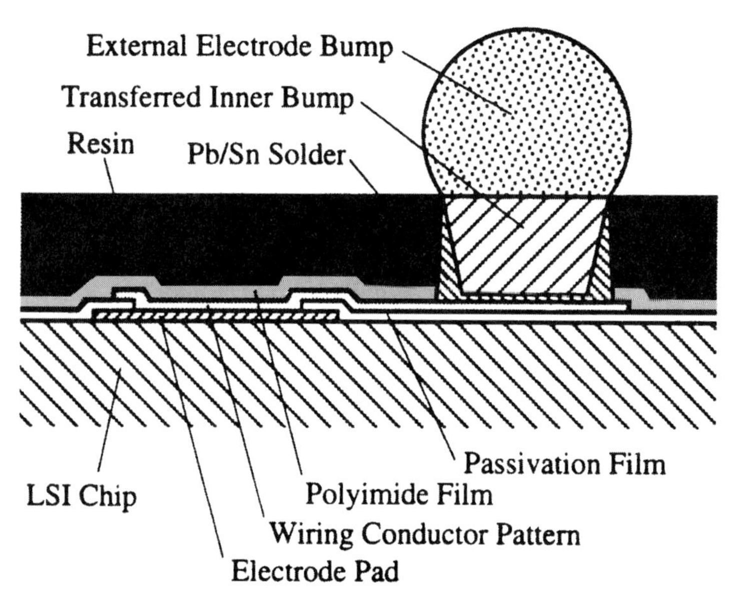

Figure 1.62 Redistribution of Mitsubishi's Chip Scale Package (CSP).

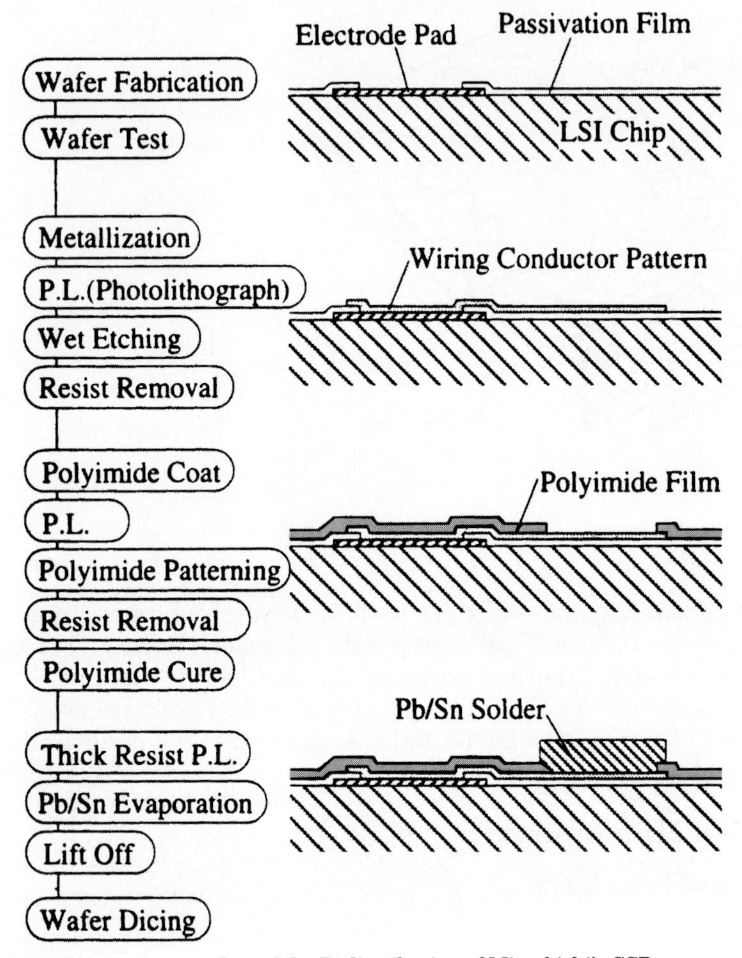

Figure 1.63 Process flow of the Redistribution of Mitsubishi's CSP.

pads on the substrate using a hot-bar, high-temperature solder (95wt%Pb/5wt%Sn) reflow process. The mechanical integrity is more robust with the combination of interconnects.

By combining most of the advantages of BGA and flip chip technologies and a silicon transposer, Aptix developed a very high density (1024-pin) PBGA for their field programmable interconnect device (FPIC), the AX1024R, Fig. 1.66.[131] Instead of directly flip-chipping the FPIC die to a multilayer PCB substrate, they used a silicon transposer to fan out the 12-mil (0.3-mm) pitch high-temperature (95wt%Pb/5wt%Sn) solder bump array to a wire-bondable perimeter (staggered 6.8 mil pitch or 3.4 mil effective). The silicon transposer is 0.96 in (24.38 mm) on a side.

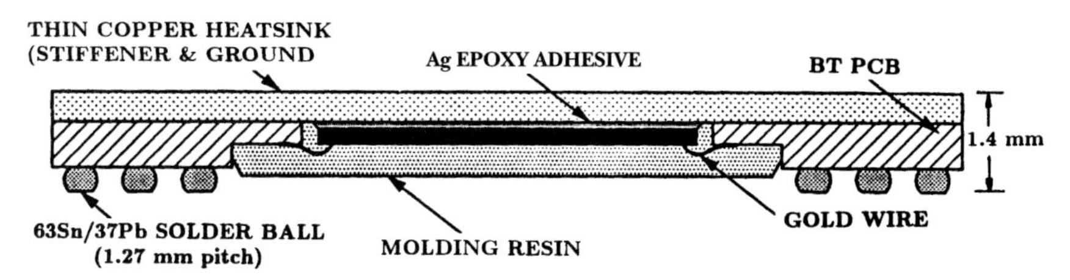

Figure 1.64 AMKOR/ANAM's cavity-down SuperBGA.

Figure 1.65 MCC and Motorola's mixed (lead and solder) BGA technology.

After the FPIC die is flip-chip-mounted to the transposer, the transposer is attached to the multilayer PCB substrate using a low-stress die attach film. Next, the transposer is wire-bonded to the multilayer PCB substrate. The package can then be capped. Finally, eutectic solder balls at 40 mil (1 mm) pitch are attached to the bottom of the substrate.

1.3.9 BGA multichip modules

The simplest definition of multichip module (MCM) is that the package (or chip carrier) has more than one chip in it. The past few years have witnessed an explosive growth in the research and development efforts devoted to MCM as a direct result of the density and performance limitations of single-chip modules.

MCMs combine several to more than a hundred high-performance silicon ICs with a custom-designed substrate structure which takes full advantage of the IC performance. This complex substrate structure is the *heart* of the MCM technology. It can be fabricated on multilayer ceramics, polymers, silicons, metals, glass ceramics, PCB, etc., using thin films, thick films, cofired, and layered methods.

A formal definition of MCM has been given by IPC (Institute for Interconnecting and Packaging Electronic Circuits), which defined three main categories of MCMs:

1. MCM-Cs are multichip modules which use thick-film technology such as fireable metals to form the conductive patterns, and are constructed entirely from ceramic or glass-ceramic materials, or possibly other materials having a dielectric constant above 5. In short, MCM-Cs are constructed on ceramic (C) or glass ceramic substrates.

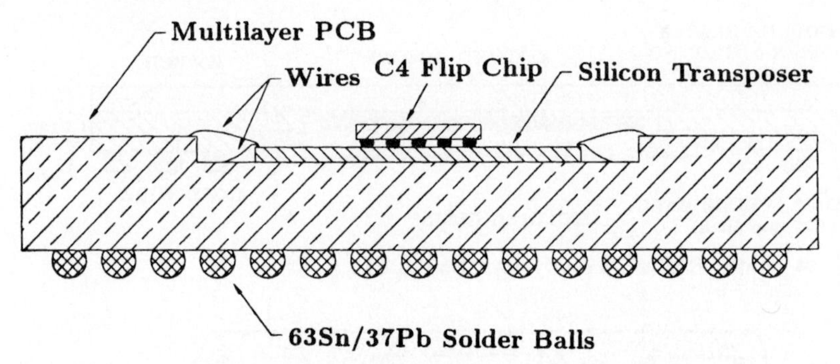

Figure 1.66 Aptix's field programmable interconnect device (FPIC).

2. MCM-Ls are multichip modules which use laminate structures and employ printed circuit board (PCB) technology to form predominantly copper conductors and vias. These structures may sometimes contain thermal-expansion-controlling metal layers. In short, MCM-Ls utilize PCB technology of reinforced plastic laminates (L).
3. MCM-Ds are multichip modules on which the multilayered signal conductors are formed by the deposition of thin-film metals on unreinforced dielectric materials with a dielectric constant below 5 over a support structure of silicon, ceramic, or metal. In short, MCM-Ds use deposited (D) metals and unreinforced dielectrics on a variety of rigid bases.

Table 1.6 shows a more detailed definition of MCMs given by the International Society for Hybrid Microelectronics. It can be seen that the line width and separation for MCM-D are less than 15 and 30 μm, respectively. Also, the line density per layer can be as high as 400 mm/mm. Figure 1.67 shows the potential applications of MCM-L, MCM-C, and MCM-D on top of the key categories (e.g., processor, gate array, cache memory, and main memory) of the next-generation high-speed devices (e.g., microprocessors, ASICs, SRAMs, and DRAMs). It can be seen that, due to its power distribution and wiring density capabilities, the range of application of MCM-C for the future computer requirements is expanding. In the meantime, however, because of availability, reasonable cost, and low risk, MCM-L could be the most used.

Know Good Die (KGD) is a critical issue for MCM applications.[16,48–53] Figures 1.68 and 1.69 show, respectively, the MCM yield versus number of chips in the module and the resultant shipped MCM yield versus test fault coverage (the ability of the tester to identify defects). It can be seen that the chip yield plays a very important role. With these figures, the trade-off between the number of chips, chip yield, MCM yield, MCM rework, and resultant shipped MCM yield can be made.

TABLE 1.6 International Society for Hybrid Microelectronics (ISHM)'s MCM definition

Properties	Multichip module identifier types (substrates)				
	MCM-C(ceramic)		MCM-D(deposited thin films)		MCM-L(layered)
Interconnect substrate	Low-ε ceramics	Cofired ceramics	Semiconductor-on-semiconductor	Low-ε polymers	Printed wiring boards
Possible materials	porous ceramic, SiN, BeO	Al_2O_3	SiO2	polyimide, BCB, PPQ	composites of fiber and epoxy
Dielectric constant (ε)	2.7–6.9	8.9–10.0	3.8	2–10	2–7.2
Subsubstrate materials	AlN	Al_2O_3	silicon, GaAs, diamond, SiC	ceramics, Si, metals, diamond	FR4, etc.
Power dissipation	high-to-very high	high	very high	very high	poor
Line width (μm)	125	125	10	15	750
Line separation (μm)	125–375	125–375	10–30	25	2250
Line density per layer ($mm\text{-}mm^{-1}$)	20	40	400	200	30
Pinout (mm^{-2})	1500–6000			800–3000	1500–3000
Typical size (mm)	100–150	150		100	660
Substrate costs	high	moderate		moderate	low

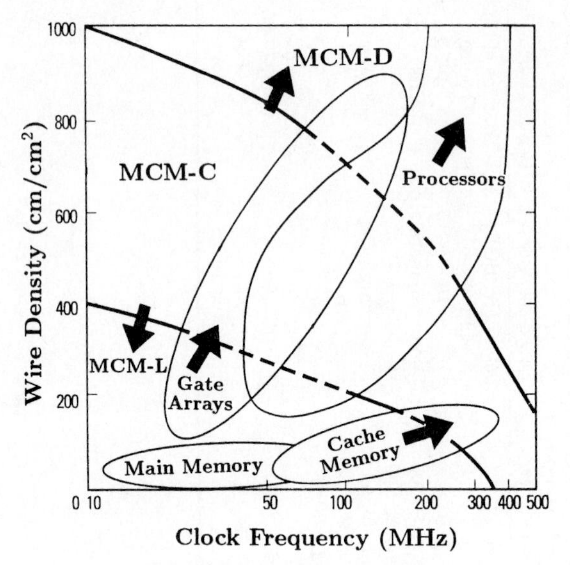

Figure 1.67 IC device and MCM packaging trends.

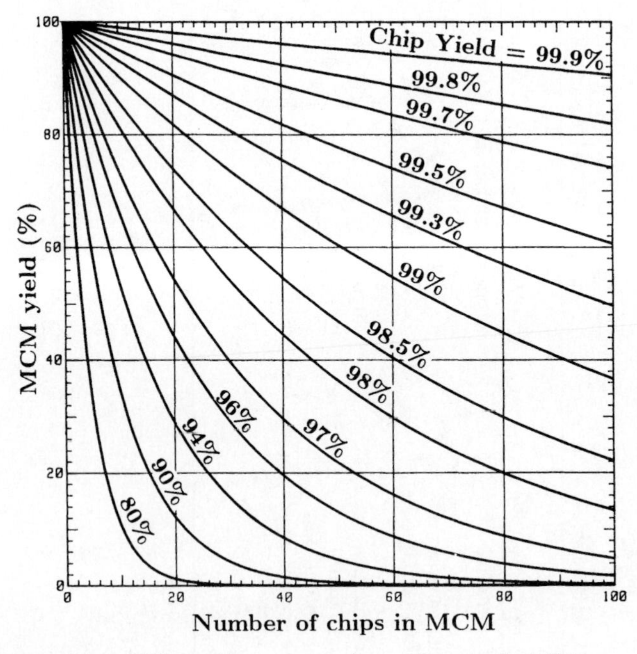

Figure 1.68 MCM yield versus number of chips in the MCM.

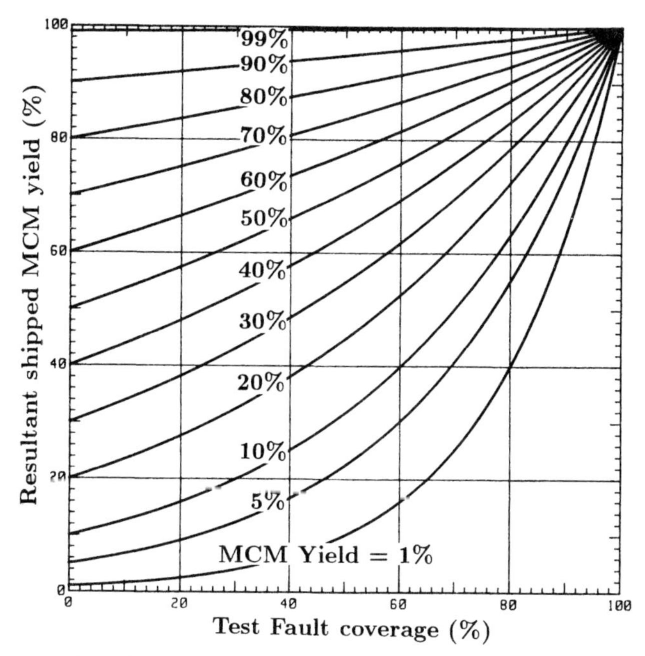

Figure 1.69 Resultant shipped MCM yield versus test fault coverage.

For example, if a 10-chip MCM has a chip yield of 90 percent and test fault coverage of 95 percent, then the MCM yield is 35 percent (Fig. 1.68) and the resultant shipped MCM yield is 95 percent (Fig. 1.69). In this case, 65 percent of the MCM will require at least one rework.

On the other hand, if the chip yield is increased from 90 to 99 percent and the test fault coverage remains the same (95 percent), then the MCM yield is 90 percent and the resultant shipped MCM yield is 99.5 percent, and only 10 percent of the MCM will require rework. This shows the importance of known good dies and the logic that the more bad chips (lower chip yield) there are, the higher the chance that they will escape into MCM assembly.

If a MCM consists of more than one chip type, then the MCM yield is given by

$$Y_m = (Y_A^{N_A})\,(Y_B^{N_B})\,(Y_C^{N_C})...(Y_I^{N_I})$$

where Y_m is the MCM yield, Y_i is the chip yield for chip Type-i, and N_i is the number of Type-i chips on the substrate. For example, there is a 10-chip MCM, where six are Type-A chips ($N_A = 6$) with chip yield $Y_A = 99\%$ and four are Type-B chips ($N_B = 4$) with chip yield $Y_B = 95\%$. Then the MCM yield $Y_m = (.99)^6(.95)^4 = 77\%$.

One of the KGD solutions to system applications is to integrate a few chips with some discretes on a small common substrate (e.g., silicon, ceramic, FR-4 epoxy, BT), which is then assembled into a standard single-chip module (e.g., PBGA, CBGA) and test it. For example, Fig. 1.70 schematically shows a cross section of IBM's CBGA package with more than one chip and Fig. 1.71 shows Fujitsu's PBGA with four chips. This is a welcome package configuration for board-level manufacturers because it is SMT compatible and they have the know-how to assemble the standard single-chip module on the PCB with high yield and low cost.

1.4 Summary

A brief introduction to IC trends (e.g., process technology, density, chip size, feature size, performance, operating voltage, and design cycle) has been presented. Also, various electronics packaging technologies such as the conventional packages (e.g., PLCC, PQFP, CQFP, and TSOP) and the advanced/emerging packages (e.g., TCP, PGA, CBGA, PBGA, and TBGA) have been updated. A comparison between the BGA and PQFP packages has been made based on their I/O count, density (area), electrical performance, and manufacturing yield and popcorn effect. Furthermore, the next-generation BGAs such as the mBGA, μBGA, SuperBGA, Mixed BGA, SLICC, and CSP have been presented. Finally, MCMs with CBGA and PBGA technologies have been briefly discussed.

As the trend toward higher speeds and I/Os for computer applications continues, it seems clear that PBGAs (for SRAMs, and ASICs and microprocessors with up to 600 I/Os and 75-MHz clock frequencies), CBGAs (for high-speed microprocessors and SRAMs, and ASICs with

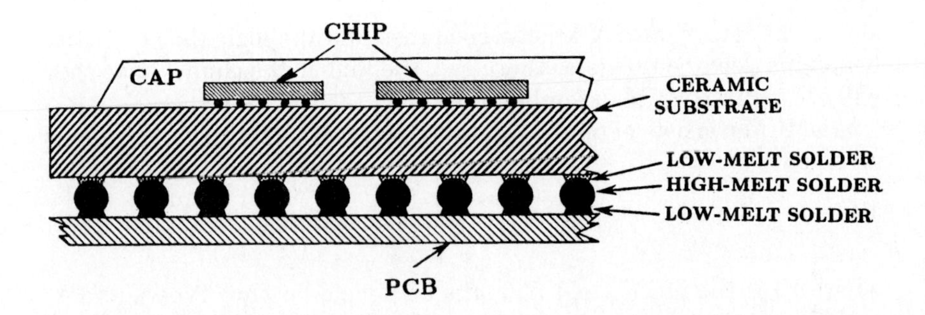

Figure 1.70 IBM's CBGA MCM.

Figure 1.71 Fujitsu's PBGA MCM.

more than 600 I/Os), and MCM BGAs (microprocessor ASIC, with a few cache SRAMs) are excellent electronic packages. BGAs will fill the gap between the conventional packaging technology and the area array flip chip technology. It is our hope that the information presented in this book may assist in removing roadblocks, avoiding unnecessary false starts, and accelerating design, material, and process development of BGA technologies.

1.5 Acknowledgments

The author would like to thank J. Miremadi, J. Gleason, R. Haven, S. Ottoboni, S. Mimura, and A. Danford for their contributions.[76] It has been a great pleasure and fruitful experience to work with them.

1.6 References

1. Tummala, R. R., and Rymaszewski, E., *Microelectronics Packaging Handbook,* Van Nostrand Reinhold, New York, N.Y., 1989.
2. Seraphim, D. P., Lasky, R., and Li, C. Y., *Principles of Electronic Packaging,* McGraw-Hill Book Company, New York, N.Y., 1989.
3. Vardaman, J., *Surface Mount Technology, Recent Japanese Developments,* IEEE Press, New York, N.Y., 1992.
4. Johnson, R. W., Teng, R. K., and Balde, J. W., *Multichip Modules: System Advantages, Major Construction, and Materials Technologies,* IEEE Press, New York, N.Y., 1991.

5. Senthinathan, R., and Prince, J. L., *Simultaneous Switching Noise of CMOS Devices and Systems,* Kluwer Academic Publishers, New York, N.Y., 1994.
6. Sandborn, P. A., and Moreno, H., *Conceptual Design of Multichip Modules and Systems,* Kluwer Academic Publishers, New York, N.Y., 1994.
7. Nash, F. R., *Estimating Device Reliability: Assessment of Credibility,* Kluwer Academic Publishers, New York, N.Y., 1993.
8. Gyvez, J. P., *Integrated Circuit Defect-Sensitivity: Theory and Computational Models,* Kluwer Academic Publishers, New York, N.Y., 1993.
9. Doane, D. A., and Franzon, P. D., *Multichip Module Technologies and Alternatives,* Van Nostrand Reinhold, New York, N.Y., 1992.
10. Messuer, G., Turlik, I., Balde, J., and Garrou, P., *Thin Film Multichip Modules,* Internation Society for Hybrid Microelectrices, Silver Spring, Md., 1992.
11. Wong, C. P., *Polymers for Electronic and Photonic Applications,* Academic Press, San Diego, Calif., 1993.
12. Lau, J. H., *Handbook of Fine Pitch Surface Mount Technology,* Van Nostrand Reinhold, New York, N.Y., 1993.
13. Lau, J. H., *Thermal Stress and Strain in Microelectronics Packaging,* Van Nostrand Reinhold, New York, 1993.
14. Lau, J. H., *Handbook of Tape Automated Bonding,* Van Nostrand Reinhold, New York, N.Y., 1992.
15. Frear, D., H. Morgan, S. Burchett, and J. H. Lau, *The Mechanics of Solder Alloy Interconnects,* Van Nostrand Reinhold, New York, 1993.
16. Lau, J. H., *Chip On Board Technologies for Multichip Modules,* Van Nostrand Reinhold, New York, 1994.
17. Lau, J. H., *Solder Joint Reliability: Theory and Applications,* Van Nostrand Reinhold, New York, 1991.
18. Manzione, L. T., *Plastic Packaging of Microelectronic Devices,* Van Nostrand Reinhold, New York, N.Y., 1990.
19. Hwang, J. S., *Solder Paste in Electronics Packaging,* Van Nostrand Reinhold, New York, N.Y., 1989.
20. Hymes, L., *Cleaning Printed Wiring Assemblies in Today's Environment,* Van Nostrand Reinhold, New York, N.Y., 1991.
21. Gilleo, K., *Handbook of Flexible Circuits,* Van Nostrand Reinhold, New York, N.Y., 1991.
22. Engel, P. A., *Structural Analysis of Printed Circuit Board Systems,* Springer-Verlag, New York, N.Y., 1993.
23. Suhir, E., *Structural Analysis in Microelectronic and Fiber Optics Systems,* Van Nostrand Reinhold, New York, N.Y., 1991.
24. Matisoff, B. S., *Handbook of Electronic Packaging Design and Engineering,* Van Nostrand Reinhold, New York, N.Y., 1989.
25. Prasad, R. P., *Surface Mount Technology,* Van Nostrand Reinhold, New York, N.Y., 1989.
26. Manko, H. H., *Soldering Handbook for Printed Circuits and Surface Mounting,* Van Nostrand Reinhold, New York, N.Y., 1986.
27. Morris, J. E., *Electronics Packaging Forum,* vol. 1, Van Nostrand Reinhold, New York, N.Y., 1990.
28. Morris, J. E., *Electronics Packaging Forum,* vol. 2, Van Nostrand Reinhold, New York, N.Y., 1991.
29. ASM International, *Electronic Materials Handbook, vol. 1, Packaging,* Materials Park, Ohio, 1989.
30. Hollomon, J. K., Jr., *Surface-Mount Technology,* Howard W. Sams & Company, Indianapolis, Ind., 1989.
31. Solberg, V., *Design Guidelines for SMT,* TAB Professional and Reference Books, New York, N.Y., 1990.
32. Hutchins, C., *SMT: How to Get Started,* Hutchins and Associates, Raleigh, N.C., 1990.
33. Bar-Cohen, A., and Kraus, A. D., *Advances in Thermal Modeling of Electronic Components and Systems,* vol. 1, Hemisphere Publishing Corp., New York, N.Y., 1988.

34. Bar-Cohen, A., and Kraus, A. D., *Advances in Thermal Modeling of Electronic Components and Systems,* vol. 2, ASME PRESS, New York, N.Y., 1990.
35. Kraus, A. D., and Bar-Cohen, A., *Thermal Analysis and Control of Electronic Equipment,* Hemisphere Publishing Corp., New York, N.Y., 1983.
36. Harper, C. A., *Handbook of Microelectronics Packaging,* McGraw-Hill Book Company, New York, N.Y., 1991.
37. Pecht, M., *Handbook of Electronic Package Design,* Marcel Dekker, New York, N.Y., 1991.
38. Harman, G., *Wire Bonding in Microelectronics,* International Society for Hybrid Microelectronics, Reston, Va., 1989.
39. Lea, C., *A Scientific Guide to Surface Mount Technology,* Electrochemical Publications, Scotland, 1988.
40. Lea, C., *After CFCs? Options for Cleaning Electronics Assemblies,* Electrochemical Publications Ltd., IOM, British isles, 1992.
41. Wassink, R. J. K., *Soldering in Electronics,* Electrochemical Publications, Scotland, 1989.
42. Pawling, J. F., *Surface Mounted Assemblies,* Electrochemical Publications, Scotland, 1987.
43. Ellis, B. N., *Cleaning and Contamination of Electronics Components and Assemblies,* Electrochemical Publications, Scotland, 1986.
44. Sinnadurai, F. N., *Handbook of Microelectronics Packaging and Interconnection Technologies,* Electrochemical Publications, Scotland, 1985.
45. Lau, J. H., and S. Erasmus, "Review of Packaging Methods to Complement IC Performance," *Electronic Packaging & Production,* June 1993, pp. 51–56.
46. Moresco, L., "Electronic System Packaging: The Search for Manufacturing the Optimum in a Sea of Constraints," *IEEE Transactions on Components, Hybrids, and Manufacturing Technology,* vol. 13, no. 3, September 1990, pp. 494–508.
47. Wessely, H., Fritz, O., Klimke, P., Koschnick, W., and Schmidt, K. H., "Electronic Packaging in the 90s—A Perspective from Europe," *Proceedings of the 40th IEEE Electronic Components and Technology Conference,* May 1990, pp. 16–33.
48. Radke, C. E., Su, L. S., Ting, Y. M., and Vanhorn, J., "Known Good Die and Its Evolution—Bipolar and CMOS," *Proceedings of the 2nd International Conference and Exhibition on Multichip Modules,* April 1993, pp. 152–159.
49. Bracken, R. C., Kraemer, B. P., Paradiso, R., and Jensen, A., "Multichip Modules, Die and MCM Test Strategy: The Key to MCM Manufacturability," *Proceedings of the 1st International Conference and Exhibition on Multichip Modules,* April 1992, pp. 456–460.
50. Trent, J. R., "Test Philosophy for Multichip Modules," *Proceedings of the 1st International Conference and Exhibition on Multichip Modules,* April 1992, pp. 444–452.
51. Smitherman, C. D., and Rates, J., "Methods for Processing Known Good Die," *Proceedings of the 1st International Conference and Exhibition on Multichip Modules,* April 1992, pp. 436–443.
52. Martin, S., Gage, D., Powell, T., and Slay, B., "A Practical Approach to Producing Known-Good Die," *Proceedings of the 2nd International Conference and Exhibition on Multichip Modules,* April 1993, pp. 139–151.
53. Corbett, T., "A Process Qualification Plan for KGD," *Proceedings of the 2nd International Conference and Exhibition on Multichip Modules,* April 1993, pp. 166–171.
54. Cappo, F., Milliken, J., and Mosley, J., "Highly Manufacturable Multi-Layered Ceramic Surface Mounted Package," *Proceedings of the 11th IEEE IEMTS,* September 1991, pp. 424–428.
55. Acocella, J., Benenati, J., Caulfield, T., and Puttlitz, K., *Proceedings of the 2nd International Conference and Exhibition on Multichip Modules,* April 1993, pp. 358–365.
56. Marrs, R. C., Freyman, B., Martin, J., "High Density BGA Technology," *Proceedings of the 2nd International Conference and Exhibition on Multichip Modules,* April 1993, pp. 326–329.
57. Caulfield, T., Benenati, J., and Acocella, J., "Surface Mount Array Interconnections for High I/O MCM-C to Card Assembles," *Proceedings of the 2nd International Conference and Exhibition on Multichip Modules,* April 1993, pp. 320–325.

58. Marrs, R. C., "Recent Developments in Low Cost Plastic MCM's," *Proceedings of the 2nd International Conference and Exhibition on Multichip Modules,* April 1993, pp. 220–229.
59. Chung, T., Chang, J., and Emamjomeh, A., "Tape Automated Bonded Chip on MCM-D," *Proceedings of the 15th IEEE International Manufacturing Technology Symposium,* October 1993, pp. 282–293.
60. Lau, J. H., and Rice, D., "Thermal Fatigue Life Prediction of Flip Chip Solder Joints by Fracture Mechanics Method," *ASME/JSME Proceedings of Advances in Electronic Packaging,* April 1992, pp. 385–392.
61. Lau, J. H., "Thermomechanical Characterization of Flip Chip Solder Bumps for Multichip Module Applications," *Proceedings of the 13th IEEE International Electronics Manufacturing Technology Symposium,* September 1992, pp. 293–299.
62. Lau, J. H., "Thermal Fatigue Life Prediction of Flip Chip Solder Joints by Fracture Mechanics Method," *International Journal of Engineering Fracture Mechanics,* vol. 45, no. 5, July 1993, pp. 643–654.
63. Totta, P. A., and Sopher, R. P., "SLT Device Metallurgy and Its Monolithic Extension," *IBM Journal of Research and Development,* May 1969, pp. 226–238.
64. Goldmann, L. S., "Geometric Optimization of Controlled Collapse Interconnections," *IBM Journal of Research and Development,* May 1969, pp. 251–265.
65. Seraphim, D. P., and Feinberg, J., "Electronic Packaging Evolution," *IBM Journal of Research and Development,* May 1981, pp. 617–629.
66. Tummala, R., and Clark, B., "Multichip Packaging Technologies in IBM for Desktop to Mainframe Computers," *Proceedings of the 42nd IEEE Electronic Components and Technology Conference,* May 1992, pp. 1–9.
67. Tsukada, Y., Mashimoto, Y., and Watanuki, N., "A Novel Chip Replacement Method for Encapsulated Flip Chip Bonding," *Proceedings of the 43rd IEEE/EIA Electronic Components & Technology Conference,* June 1993, pp. 199–204.
68. Tsukada, Y., Maeda, Y., and Yamanaka, K., "A Novel Solution for MCM-L Utilizing Surface Laminar Circuit and Flip Chip Attach Technology," *Proceedings of the 2nd International Conference and Exhibition on Multichip Modules,* April 1993, pp. 252–259.
69. Tsukada, Y., Tsuchida, S., and Mashimoto, Y., "Surface Laminar Circuit Packaging," *Proceedings of the 42nd IEEE Electronic Components and Technology Conference,* May 1992, pp. 22–27.
70. Lau, J. H., Krulevitch, T., Schar, W., Heydinger, M., Erasmus, S., and Gleason, J., "Experimental and Analytical Studies of Encapsulated Flip Chip Solder Bumps on Surface Laminar Circuit Boards," *Circuit World,* vol. 19, no. 3, March 1993, pp. 18–24.
71. Yamada, H., Kondoh, Y. and Saito, M., "A Fine Pitch and High Aspect Ratio Bump Array for Flip-Chip Interconnection," *Proceedings of the IEEE International Electronic Manufacturing Technology Symposium,* September 1992, pp. 288–292.
72. Lau, J. H., Golwalkar, S., Boysan, P., Surratt, R., Rice, D., Forhringer, R., and Erasmus, S., "Solder Joint Reliability of a Thin Small Outline Package (TSOP)," *Circuit World,* vol. 20, no. 1, November 1993, pp. 12–19.
73. Lau, J. H., Golwalkar, S., Rice, D., Erasmus, S., and Foehringer, R., "Experimental and Analytical Studies of 28-Pin Thin Small Outline Package Solder-Joint Reliability," *J. of Electronic Packaging, Trans. of ASME,* vol. 114, June 1992, pp. 169–176.
74. Lau, J. H., S. Golwalkar, and S. Erasmus, "Advantages and Disadvantages of Thin Small Outline Packages (TSOP) with Copper Gull-Wing Leads," *Proceedings of the ASME International Electronics Packaging Conference,* Binghamton, N.Y., September 1993, pp. 1119–1126.
75. Nakamura, Y., Ohta, M., Nishioka, T., Tanaka, A., and Ohizumi, S., "Effects of Mechanical and Flow Properties of Encapsulating Resin on the Performance of Ultra Thin Tape Carrier Package," *Proceedings of the 43rd IEEE Electronic Components and Technology Conference,* June 1993, pp. 419–424.
76. Lau, J. H., Miremadi, J., Gleason, J., Haven, R., Ottoboni, S., and S. Mimura, "No Clean Mass Reflow of Large Over Molded Plastic Pad Array Carriers (OMPAC)," *Proceedings of the IEEE International Electronic Manufacturing Technology Symposium,* October 1993, pp. 63–75.

77. Lau, J. H., K. Gratalo, T. Baker, E. Schneider, T. Marcotte, and S. Mimura, "Reliability of Ball Grid Array Solder Joints Under Bending, Twisting, and Vibration Conditions," to be published in *Circuit World,* 1995.
78. Lau, J. H., Pao, Y., Larner, C., Twerefour, S., Govila, R., Gilbert, D., Erasmus, S., and Dolot, S., "Reliability of 0.4 mm Pitch, 256-Pin Plastic Quad Flat Pack NO-Clean and Water-Clean Solder Joints," *Soldering & Surface Mount Technology,* no. 16, February 1994.
79. Lau, J. H., Govila, R., Larner, C., Pao, Y., Erasmus, S., Dolot, S., Jalilian, M., and Lancaster, M., "No-Clean and Solvent-Clean Mass Reflow Processes of 0.4 mm Pitch, 256-Pin Fine Pitch Quad Flat Packs (QFP)," *Circuit World,* vol. 19, no. 1, October 1992, pp. 19–26.
80. Lau, J. H., Dody, G., Chen, W., McShane, M., Rice, D., Erasmus, S., and Adamjee, W., "Experimental and Analytical Studies of 208-Pin Fine Pitch Quad Flat Pack Solder-Joint Reliability," *Circuit World,* vol. 18, no. 2, January 1992, pp. 13–19.
81. Lau, J. H., "Thermal Stress Analysis of SMT PQFP Packages and Interconnections," *J. of Electronic Packaging, Trans. of ASME,* vol. 111, March 1989, pp. 2–8.
82. Lau, J. H., Powers, L., Baker, J., Rice, D., and Shaw, W., "Solder Joint Reliability of Fine Pitch Surface Mount Technology Assemblies," *IEEE Trans. on CHMT,* vol. 13, September 1990, pp. 534–544.
83. Lau, J. H., and Keely, C. A., "Dynamic Characterization of Surface Mount Component Leads for Solder Joint Inspection," *IEEE Trans. on CHMT,* vol. 12, no. 4, December 1989, pp. 594–602.
84. Lau, J. H., and Harkins, G., "Stiffness of 'Gull-Wing' Leads and Solder Joints for a Plastic Quad Flat Pack," *IEEE Trans. on CHMT,* vol. 13, no. 1, March 1990, pp. 124–130.
85. Lau, J. H., and Harkins, G., "Thermal Stress Analysis of SOIC Packages and Interconnections." *IEEE Trans. on CHMT,* vol. CHMT-11, no. 4, December 1988, pp. 380–389.
86. Lau, J. H., Subrahmanyant, R., Rice, D., Erasmus, S., and Li, C., "Fatigue Analysis of a Ceramic Pin Grid Array Soldered to An Orthotropic Epoxy Substrate," *J. of Electronic Packaging, Trans. of ASME,* vol. 113, June 1991, pp. 138–148.
87. Lau, J. H., Leung, S., Subrahmanyant, R., Rice, D., Erasmus, S., and Li, C., "Effects of Rework on the Solder Joint Reliability of Pin Grid Array Interconnects," *J. of the Institute of Interconnection Technology (Circuit World),* July 1991.
88. Lau, J. H., Harkins, G., Rice, D., Kral, J., and Wells, B., "Experimental and Statistical Analyses of Surface-Mount Technology PLCC Solder-Joint Reliability," *IEEE Trans. on Reliability,* vol. 37, no. 5, December 1988, pp. 524–530.
89. Tummala, R. R., "Multichip Technologies from Personal Computers to Mainframes and Supercomputers," *Proceedings of NEPCPN West,* 1993, pp. 637–643.
90. Lau, J. H., Erasmus, S. J., and Rice, D. W., "An Introduction to Tape Automated Bonding Technology," in *Electronics Packaging Forum,* ed. by Morris, J. E., Van Nostran Reinhold, New York, 1991, pp. 1–83.
91. Kang, S. K., "Gold-to-Aluminum Bonding for TAB Applications," *Proceedings of the 42nd IEEE Electronic Components and Technology Conference,* May 1992, pp. 870–875.
92. Lau, J. H., Erasmus, S. J., and Rice, D. W., "Overview of Tape Automated Bonding Technology," *Circuit World,* Vol. 16, No. 2, 1990, pp. 5–24.
93. Lau, J. H., Erasmus, S. J., and Rice, D. W., "Overview of Tape Automated bonding Technology," *Electronic Materials Handbook,* Vol. 1: *Packaging, ASM International,* Nov. 1989.
94. Chen, W. T., Raski, J., Young, J., and Jung, D., "A Fundamental Study of Tape Automated Bonding Process," *ASME Transactions, Journal of Electronic Packaging,* September 1991, pp. 216–225.
95. Lau, J. H., Rice, W. D., and Harkins, G., "Thermal Stress Analysis of TAB Packages and Interconnections," *IEEE Trans. on CHMT,* vol. 13, no. 1, March 1990, pp. 183–188.
96. Ahn, S., Y. Kwon, and K. Shin, "Popcorn Phenomena in a Ball Grid Array Package," *Proceedings of the 44th IEEE Electronic Components and Technology Conference,* Washington, D.C., May 1994, pp. 1101–1107.

97. Carden, T., J. Clementi, and S. Engle, "Epoxy Encapsulation on Ceramic Quad Flat Packs," *Proceedings of the 44th IEEE Electronic Components and Technology Conference,* Washington, D.C., May 1994, pp. 585–589.
98. Pope, D., and H. Do, "Thermal Characterization of a Tape Carrier Package," *Proceedings of the 44th IEEE Electronic Components and Technology Conference,* Washington, D.C., May 1994, pp. 532–538.
99. Groover, R., C. Huang, and A. Hamzehdoost, "BGA—Is It Really the Answer?" *Proceedings of the 1st International Symposium on Flip Chip Technology,* Feb. 1994, pp. 57–64.
100. Kang, S., W. Chan, R. Hammer, and F. Andros, "Chip Level Interconnect: Wafer Bumping and Inner Lead Bonding," *Chip On Board Technologies for Multichip Modules,* ed. by J. H. Lau, Van Nostrand Reinhold, New York, 1994, pp. 186–227.
101. ICE Corporation, *1994 STATUS Report.*
102. Weldon, R., "3-D Inspection of Solder Paste for High-Quality BGA Assembly," *Proceedings of the 1st International Symposium on Flip Chip Technology,* Feb. 1994, pp. 52–56.
103. Moore, K., Machuga, S., Bosserman, S., and Stafford, J., "Solder Joint Reliability of Fine Pitch Solder Bumped Pad Array Carriers," *Proceedings of NEPCON West,* Feb. 1990, pp. 264–274.
104. Kobayashi, F., et al., "Hardware Technology for HITACHI M-880 Processor Group," *Proceedings of the 41st IEEE Electronic Components and Technology Conference,* May 1991, pp. 693–703.
105. Akihiro, D., Toshihiko, W., and Hideki, N., "Packaging Technology for the NEC SX-3/SX-X Supercomputer," *Proceedings of the 40th IEEE Electronic Components & Technology Conference,* Las Vegas, May 1990, pp. 525–533.
106. Freyman, B., and Pennisi, R., "Overmolded Plastic Pad Array Carriers (OMPAC): A Low Cost, High Interconnect Density IC Packaging Solution for Consumer and Industrial Electronics," *Proceedings of the 41st IEEE Electronic Components and Technology Conference,* May 1991, pp. 176–182.
107. Vardaman, E. J., "Ball Grid Array Packaging," consulting report, January 1993.
108. Johnson, R., Mawer, A., McGuiggan, T., Nelson, B., Petrucci, M., and Rosckes, D., "A Feasibility Study of Ball Grid Array Packaging," *Proceedings of NEPCON East,* June 1993, pp. 413–422.
109. Johnson, R., and Cawthon, D., "Thermal Characterization of 140 and 225 Pin Ball Grid Array Packages," *Proceedings of NEPCON East,* June 1993, pp. 423–430.
110. Banerji, K., "Development of the Slightly Larger than IC Carrier (SLICC)," *Proceedings of the NEPCPN West,* March 1994, pp. 1249–1256.
111. Adams, J. A., "Using Cross-Sectional X-Ray Techniques for Testing Ball Grid Array Connections and Improving Process Quality," *Proceedings of the NEPCPN West,* March 1994, pp. 1257–1265.
112. Walshak, D. B., and H. Hashemi, "Thermal Modeling of a Multichip BGA Package," *Proceedings of the NEPCPN West,* March 1994, pp. 1266–1276.
113. Rogren, P. E., "MCM-L Built on Ball Grid Array Formats," *Proceedings of the NEPCPN West,* March 1994, pp. 1277–1282.
114. Puttlitz, K. J., and W. F. Shutler, "C-4/BGA Comparison with Other MLC Single Chip Package Alternatives," *Proceedings of the 44th IEEE Electronic Components & Technology Conference,* Washington, D.C., May 1944, pp. 16–21.
115. Kromann, G., D. Gerke, and W. Huang, "A Hi-Density C4/CBGA Interconnect Technology for a CMOS Microprocessor," *Proceedings of the 44th IEEE Electronic Components & Technology Conference,* Washington, D.C., May 1944, pp. 22–28.
116. Panicker, M., N. Greenman, J. Forster, and P. Johnston, "Low-Cost Ceramic Thin-Film Ball Grid Arrays," *Proceedings of the 44th IEEE Electronic Components & Technology Conference,* Washington, D.C., May 1944, pp. 29–31.
117. Switky, A., V. Sajja, J. Darnauer, and W. Dai, "A 1024-Pin Plastic Ball Grid Array for Flip Chip Die," *Proceedings of the 44th IEEE Electronic Components & Technology Conference,* Washington, D.C., May 1944, pp. 32–38.
118. Michalka, T., and P. Frank, "A Hige Performance HDI Based Pin Grid Array Package," *Proceedings of the 44th IEEE Electronic Components & Technology Conference,* Washington, D.C., May 1944, pp. 39–42.

119. Huang, W., and J. Casto, "CBGA Package Design for C4 PowerPC Microprocessor Chips: Trade-off Between Substrate Rountability and Performance," *Proceedings of the 44th IEEE Electronic Components & Technology Conference,* Washington, D.C., May 1944, pp. 88–93.
120. Kromann, G., "Thermal Modeling and Experimental Characterization of the C4/Surface-Mount-Array Interconnect Technologies," *Proceedings of the 44th IEEE Electronic Components & Technology Conference,* Washington, D.C., May 1944, pp. 395–402.
121. Panicker, M., and C. Mattei, "A Surface Mount, Thin-Film, Ceramic Package for High-Frequency Wireless Application," *Proceedings of the 44th IEEE Electronic Components & Technology Conference,* Washington, D.C., May 1944, pp. 609–611.
122. Moore, K., S. Machuga, S. Bosserman, and J. Stafford, "Solder Joint Reliability of Fine Pitch Solder Bumped Pad Array Carriers," *Proceedings of NEPCON West,* Feb. 1990, pp. 264–274.
123. Treece, R. K., "mBGA Technology Overview," *Proceedings of ELECTRECON,* May 1994, pp. 2-1–2-7.
124. Matthew, L. C., and T. H. DiStefano, "Future Directions in TAB: The TCC/MCM Interconnect," *Proceedings of ITAP and Flip-Chip Symposium,* Feb. 1994, pp. 228–231
125. Martinez, M., D. Gibson, L. Matthew, T. DiStefano, and J. Cofield, "The TCC/MCM. μBGA on a Laminated Substrate," *Proceedings of the International Conference and Exhibition on Multichip Modules,* April 1994, pp. 161–166.
126. Matthew, L., Z. Kovac, G. Karavakis, and T. DiStefano, "Beyond the Barriers of Known Good Die," *Proceedings of NEPCON East,* June 1994, pp. 153–156.
127. Loo, M., and K. B. Gilleo, "Area Array Chip Carrier: SMT Package for Known Good Die," *Proceedings of ISHM,* November 1993, pp. 318–323.
128. Yasunaga, M., S. Baba, M. Matsuo, H. Matsushima, S. Nakao, and T. Tachikawa, "Chip Scale Package (CSP), A Lightly Dressed LSI Chip," *Proceedings of the 16th IEEE IEMTS,* September 1994, pp. 169–176.
129. AMKOR/ANAM's "SuperBGA Specification Data Sheet," June 1994.
130. Hashemi, H., M. Olla, D. Cobb, P. Sandborn, M. McShane, G. Hawkins, and P. Lin, "A Mixed Solder Grid Array and Peripheral Leaded MCM Package," *Proceedings of the 43rd IEEE/EIA Electronic Components & Technology Conference,* May 1993, pp. 951–956.
131. Switky, A., V. Sajja, J. Darnauer, and W. Dai, "A 1024-Pin Plastic Ball Grid Array for Flip Chip Die," *Proceedings of the 44th IEEE/EIA Electronic Components & Technology Conference,* May 1994, pp. 32–38.
132. Walshak, D., and H. Hashemi, "BGA Technology: Current and Future Direction for Plastic, Ceramic and Tape BGAs," *Proceedings of Surface Mount International Conference & Exposition,* August 1994, pp. 157–163.
133. Mescher, P., and G. Phelan, "A Practical Comparison of Surface Mount Assembly for Ball Grid Array Components," op. cit., pp. 164–168.
134. Fauser, S., C. Ramirez, and L. Hollinger, "High Pin Count PBGA Assembly: Solder Defect Failure Modes and Root Cause Analysis," op. cit., pp. 169–174.
135. Zimerman, M. A., "High Performance BGA Molded Packages for MCM Application," op. cit., pp. 175–180.
136. Bernier, W. E., B. Ma, P. Mescher, A. Trivedi, and E. Vytlacil, "BGA vs QFP: a Summary of Tradeoffs for Selection of High I/O Components," op. cit., pp. 181–185.
137. Anderson, L., "Solder Attachment Analysis of Plastic BGA Modules," op. cit., pp. 189–194.
138. Rooks, S., "X-Ray Inspection of Flip Chip Attach Using Digital Tomosynthesis," op. cit., pp. 195–202.
139. Hart, C., "Vias in Pads for Coarse and Fine Pitch Ball Grid Arrays," op. cit., pp. 203–207.
140. Johnson, R., Moore, D., and Wright, T., "Thermal Characterization of 313 Pin BGA Package," op. cit., pp. 208–211.
141. DiStefano, T., Karavakis, K., Fjelstad, J., and Kovac, Z., "μBGA for High Performance Applications," op. cit., pp. 212–215.

142. Spadafora, J., "A Rework Process for Ball Grid Array Packages Containing Flip-Chip Silicon-on-Silicon Multi-Chip Modules," op. cit., pp. 219–224.
143. Miles, B., and R. Darveaux, "Plastic Ball Grid Array Repair Procedures," op. cit., pp. 232–235.
144. Mawer, A., S. Bolton, and E. Mammo, "Plastic BGA Solder Joint Reliability Considerations," op. cit., pp. 239–251.
145. Attarwala, A., and R. Stierman, "Failure Mode Analysis of a 540 Pin Plastic Ball Grid Array," op. cit., pp. 252–257.
146. Ramirez, C., and S. Fauser, "Fatigue Life Comparison of The Perimeter and Full Plastic Ball Grid Array," op. cit., pp. 258–266.
147. Quinones, H., M. Shatzkes, and J. Jaspal, "Reliability of Solder Ball Connection," op. cit., pp. 267–270.
148. Banks, D., "Assembly and Reliability of Ceramic Column Grid Array," op. cit., pp. 271–276.
149. Cole, M., and T. Caulfield, "Ball Grid Array Packaging," op. cit., pp. 147–153.
150. Miyazaki, T., K. Terashima, "The Improvement of PBGA Solder Ball Strength Under High Temperature Storage," *Proceedings of the 16th IEEE/IEMTS,* September 1994, pp. 333–339.
151. Dobers, M., M. Seyffert, F. Hauschild, and C. Czaya, "Low Cost Multi Chip Modules," op. cit., pp. 340–343.
152. Mitchell, C., "Assemble and Reliability Study for the Micro-Ball Grid Array," op. cit., pp. 344–346.
153. Rao, S., "Ball Grid Array Assembly Issues in Manufacturing," op. cit., pp. 347–348.
154. Towne, D., S. Kaul, B. Verstegen, and E. Selna, "Considerations for Use of High Pin Count, Multilayer PBGAs in Engineering Workstations," op. cit., pp. 349–354.
155. Chidambaram, N., and M. Papageorge, "Stress and Coplanarity Analysis of 225 BGA Package," op. cit., pp. 355–358.
156. Degani, Y., and T. Dudderar, "AT&T Lead-Free Solder Paste for the Cost effective Manufacturer of Flip-Chip Silicon-on Silicon MCMs," op. cit., pp. 20–24.
157. Kelly, M., and J. H. Lau, "Low Cost Solder Bumped Flip Chip MCM-L Demonstration," op. cit., pp. 147–153.

Chapter

2

Ceramic Substrates for Ball Grid Array Packages

Richard Sigliano

2.1 Introduction

Ceramic materials for integrated circuit packaging have been used for the past couple of decades. The first ceramic packages were designed for Transistor Outline (TO) devices manufactured with as many as 14 external I/Os. In the 1960s, integration increased the need for more leads. This led to the development of dual-in-line packages (DIP), introduced by Fairchild in 1963. At about the same time, IBM introduced a ceramic package called Solid Logic Technology (SLD). Although mainly used for IBM's internal consumption, was a forerunner to the modern pin grid array package (PGA). These packages measured 12 × 12 mm in outer dimensions and were made of 96 percent dry pressed aluminum. Subsequently, conductor and resistor metallization was screened and fired at about 800° centigrade. Leads were attached by the pin swaging method, whereby copper alloy pins are mechanically compressed in holes pressed into the ceramic substrate. The entire assembly was then submerged in a solder bath to form electrical continuity between the copper pins and conductor metallization. The devices were soldered in place and the lid sealed by epoxy sealant material. These packages can be considered to be the first area array ceramic enclosures. Even though they were through-hole mount in design, they were the precursor to the modern surface-mount ball grid array package (BGA).

Today, ceramic packages play an extremely important role in the information and data processing era. The demands for integrated circuit packaging differ from the original concept of protecting the device from the harmful effects of the environment. Current packaging demands include electrically tuned circuitry with the device, thermal management, and reduction of package noise parasitics.

Figure 2.1 illustrates the demands and trends for electronic packaging. A paradoxical trend in the industry is that as device size increases, package size is decreasing. The only way to meet this demand is through area array packaging, or more specifically, BGA packaging. Figure 2.2 depicts total electronic packaging technologies that affect the different system levels from chip interconnection through first- and second-level packaging. Chips and packages are commonly connected by these primary methods:

1. Wire bonding through either ultrasonic, thermosonic, or thermocompression methods
2. Tape automated bonding (TAB) utilizing either thermocompression bonding or soldering methods
3. Controlled collapse chip connection (C4-IBM) using solder; commonly referred in the industry as "flip chip" bonding

All of these bonding technologies, however, are approaching their intrinsic limits. Papers presented have stated that wire bonding with pitches of 40 microns and lead counts in excess of 1000 chip connections have been achieved. Yet, in the case of most semiconductor device assemblies, pitches of less than 85 to 100 microns are not in their future developments. One advantage of wire bonding is it can connect a chip to a package on more than one plane or tier. Today's VLSI devices wire-bonded to a package accomplish high I/O counts through at least two tiers, commonly on three tiers, and rarely on four tiers. By utilizing multiple tiers, the semiconductor assemblies can keep package size down with common and trusted interconnect attachment methods.

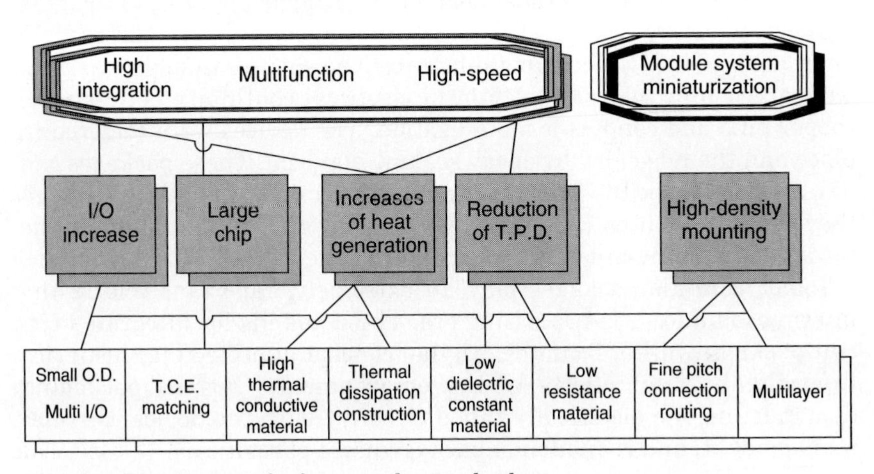

Figure 2.1 Requirements for future package technology.

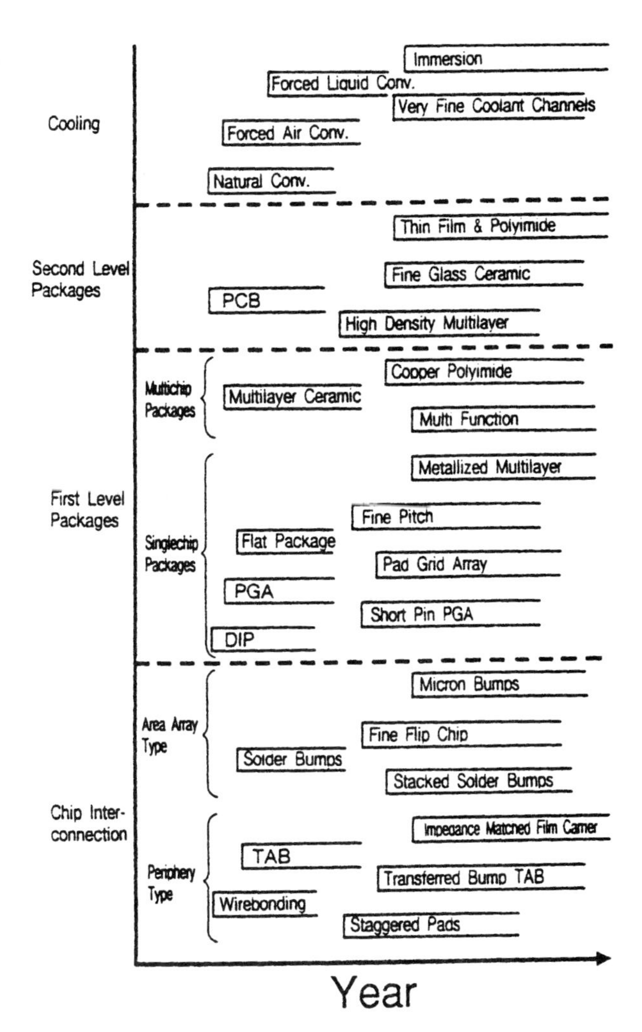

Figure 2.2 Present status and future trends in electronic packaging technologies.

TABs in excess of 600 interconnects have been achieved. However, this technology is limited to a single plane. High lead counts can be accomplished through multiple rows of interconnect pads, but TAB is still considered a peripheral bonding technology. A large and costly concern for TAB interconnects, especially for finer pitches (<100 microns), is the exact registration of TAB pads on the package level. This suggests post-metallization conductor technology must be employed. A second concern for TAB interconnects is the "alluvial fan" effect where TAB pads spread out to meet via hold pitches, usually around 250 microns at best. This results in a cost of real estate causing the package size to increase.

The most promising interconnect technology for high-lead-count single-chip devices, as well as for packages, is area array or "flip chip" (C4) bonding. IBM has used this technology for the past three decades with tremendous success. Flip chip bonding has been performed with solder connections in excess of 2000 I/Os.

As integration and functionality of devices increase, so does their physical size. This causes greater mechanical stress due to the thermal coefficient of expansion (TCE) mismatch between the chip and package-bonding interface. To eliminate this problem on larger chips, flip chip soldering can be enhanced by the following methods:

1. Stretched or stacked solder columns
2. Solder with more forgiving fatigue life
3. Epoxy filled solder columns
4. Matched TCE substrates

The bonding method specified will determine the type of BGA package and mechanical configurations available to the packaging and system engineer.

2.2 Ceramic Material Processing

Ideally, a ceramic raw material would have to be hard and strong enough for industrial applications, possess excellent electrical insulation properties, have high resistance to moisture and corrosive gases, and could be mass-produced economically. In a search for such a raw material that meets the desired industrial application, with material properties found naturally, we would find that sapphire (and ruby, which is made of the same chemical structure) has the needed characteristics. In the characteristics of sapphire, we find the following:

1. *Chemical composition:* Aluminum oxide (alumina)
2. *Crystal type:* Hexagonal crystal system
3. *Hardness:* MOHS Scale 9
4. *Electrical insulation:* Excellent insulation against high frequency and temperature
5. *Chemical stability:* Noncorrosive, not oxidized by air, water insoluble
6. *Strength:* Extremely tough

7. *Thermal characteristics:* Melting temperature exceeds 2050°C; heat conductivity is high, equivalent to Fe-Ni-Co alloy (Kovar)

To use such a raw material would be ideal. However, natural sapphire and ruby are expensive and in scarce supply, making them inappropriate for an industrial ceramic raw material.

The chemical composition of sapphire is aluminum oxide or alumina. A high percentage of alumina is contained in the naturally occurring mineral bauxite. If bauxite is dissolved with soda (NaOH), the impurities are removed, leaving high-purity aluminum hydroxide. With the removal of water, the aluminum hydroxide becomes aluminum oxide. Further removal of oxygen yields the metal aluminum. Hence, aluminum oxide is a by-product of the aluminum industry, is low-priced, and can be manufactured in large volumes.

If alumina is purified and sintered at an extremely high temperature it acquires the same crystalline structure as sapphire. This alumina is of the hexagonal crystal system. If alumina is formulated and sintered for a specific application, then it is referred to as a *alumina ceramic.* Therefore, alumina ceramic has the same material properties as those of sapphire or ruby.

To solidify properly, high-purity alumina must be sintered at a very high temperatures. But for cost-effective manufacturing, alumina ceramic must be fired at a lower temperature. For this reason, a material with a lower melting point is added to the alumina.

Normally, inorganic oxides have very high melting points. But many inorganic oxides, when mixed with another inorganic oxide, melt at temperatures lower than their own individual melting points. If three or more oxides are mixed together, the process becomes complicated, but the combined mixture will give a lower melting point. For example, the melting point of silica is 1710°C; for magnesia (magnesium oxide) it is 2800°C. If these two oxides are mixed with a ratio of 66 percent silica and 34 percent magnesia, the melting point of the mixture will drop to 1557°C. Then, if calcium oxide or calcia is added, the melting point is lowered even further.

Ratio Change of Melting Temperature per Mixing

Silcia (%)	Magnesia (%)	Calcia (%)	Melting temperature (°C)
100	—	—	1,710
—	100	—	2,800
—	—	100	2,572
66	34	—	1,557
55.4	10.6	34	1,357

Therefore, the ceramic manufacturer will add small amounts of other inorganic raw materials to the principal raw material alumina and mix. During sintering, the additive with the lower melting point will melt. Upon cooling, it turns into glass positioned in between the alumina crystals and acts as a binding force holding the alumina crystals.

To make the hardest alumina ceramic, the manufacturer will minimize the amount of additives and prevent the additives from becoming a part of the glassy phase. When alumina crystals are fired at high temperature, the glass affects the surface of the alumina ceramic. It acts to limit the growth of the crystal size and helps alumina crystals to bond together more easily.

2.2.1 Multilayer cofiring method

Alumina is conventionally used as the material of choice for BGA packages today. Ceramic BGA substrates can be fabricated one of two ways: using either ceramic green sheets (multilayer cofired process) or by the dry press method. Figure 2.3 shows the standard multilayer cofiring manufacturing process utilizing ceramic green sheets for the interconnection layers. A traditional ceramic BGA would consist of three layers: a seal ring layer for eutectic solder seal or epoxy seal, a chip interconnect layer for wire bonding to interface the device I/Os with the BGA external outputs by wire, stitch or TAB bonding, and a die attach layer

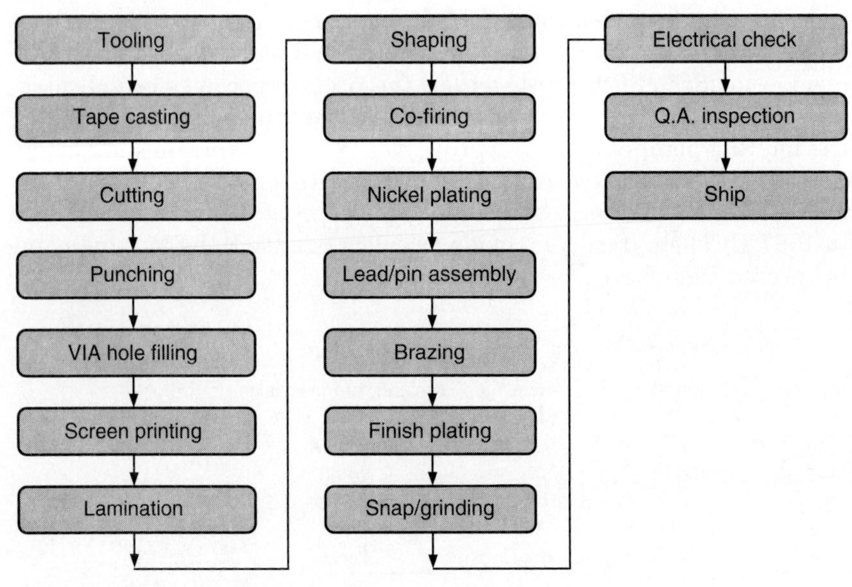

Figure 2.3 Multilayer ceramic technology process flow.

to mount and support the device mechanically. Each layer is laminated together and fired at about 1500 to 1700°C for several hours. Pictured in Fig. 2.4 is a typical alumina multilayer cofired process. The steps of this process for multilayer BGA packages include the following.

Green sheets. Green sheets are created by mixing and combining alumina powder (roughly from 3 to 10 microns in diameter) with various glasses, inorganic darkeners, and sintering additives. A slurry is prepared by adding organic solvents and binders to the system. Common binders used include polymethyl methacrylate (PMMA), polyalphamethyl styrene (PIMS), polyisobutylene (PIB), nitrocellulose, acetate butyryl, cellulose acetate, and polyvinyl chloride acetate. Probably the most popular binder for slurry preparation is polyvinyl butyryl (PVB) because of its favorable thermoplastic properties and bond strength between layers. The polyvinyl butyryl chemical formula is as follows:

$$\left[-CH_2-\underset{\displaystyle O}{\overset{|}{CH}}\overset{\displaystyle CH_2}{}\underset{\displaystyle O}{\overset{|}{CH}}- \quad \underset{\displaystyle CH_2CH_2CH_3}{CH} \right]_N$$

These mixtures are subsequently ball milled and deagglomered to attain a uniform slurry dispersion.

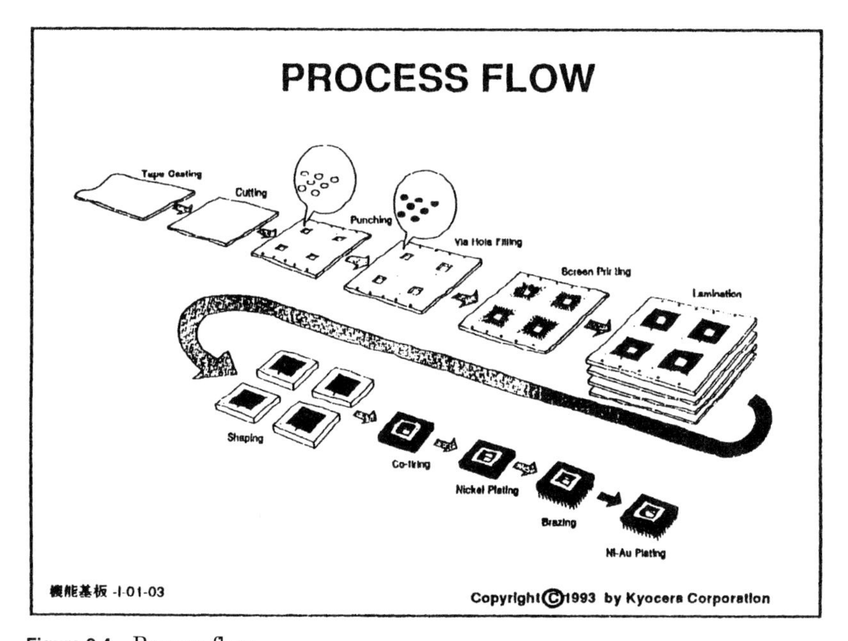

Figure 2.4 Process flow.

Tape casting. Slurry is fed into a continuous hopper and deposited on a waxed paper or plastic sheet which constantly moves into a dryer oven. The hopper is normal-fitted with a constant level liquid reservoir and is doctor-bladed on the sheets. The doctor blade is adjusted to varying heights depending on the desirable green sheet thickness. The deposited slurry is routed through a series of humidity- and temperature-controlled drying ovens. This drying cycle can be as long as 60 meters. The dried green tape is then separated from the waxed paper and rolled onto spools for later use. Each ceramic tape spool is continuously inspected for tape layer thickness and for any particulate and void anomalies. When ready for use, each spool is cut into square or rectangular sheets based upon the manufacturer's tooling. A representative sample is usually laminated together and fired to check for *X, Y,* and *Z* directional shrinkage properties, specific gravity, ceramic density, lamination, and flexural strength.

Punching. Via holes, cavities, and other internal punching is performed at this time. Punching can be accomplished in numerous ways; numerical-controlled mechanical punching or drilling, or by utilizing hard tungsten carbide punches. Green sheets are typically punched with keying holes to accent the alignment of punched layers. Depending on the size of the green sheet, the size of the BGA package will determine the number of punch heads per tool. Numerical-control equipment (sometimes referred to as *soft tooling*) can have a number of punch or drill heads to improve the efficiency and lower the cost of the individual BGA packages.

Via hole filling. Punched via holes in the ceramic green tape layer are either filled or coated with refractory tungsten or molybdenum (for high-temperature-cofired ceramics), or with gold, silver, silver-palladium, or copper thick-film metallization (for low-temperature-cofired ceramics). This metallization is especially controlled for viscosity and flow characteristics to ensure via hole coverage.

Screen printing. Conductor lines, planes, bond pads, backside BGA pads, and markings are printed with screening paste tailored to designed patterns (Figs. 2.5 and 2.6). Typically, the same paste composition used for via hole filling is used for the patterns; the only difference is that paste viscosity is adjusted to control paste flow characteristics. As with alumina milling for green sheets, the ball milling process is used for conductor paste systems. This step is extremely important to control metal powder dispersion in the organic binder materials. Equally important is maximizing screen printing by regulating the screen mesh size and material composition, environmental conditions, and type of printing (contact/noncontact). Another method of screen printing is accomplished by extruding a paste through a nozzle as the nozzle traverses a metal mask in contact with a green sheet. The vias are simultaneously filled as the pattern is defined.

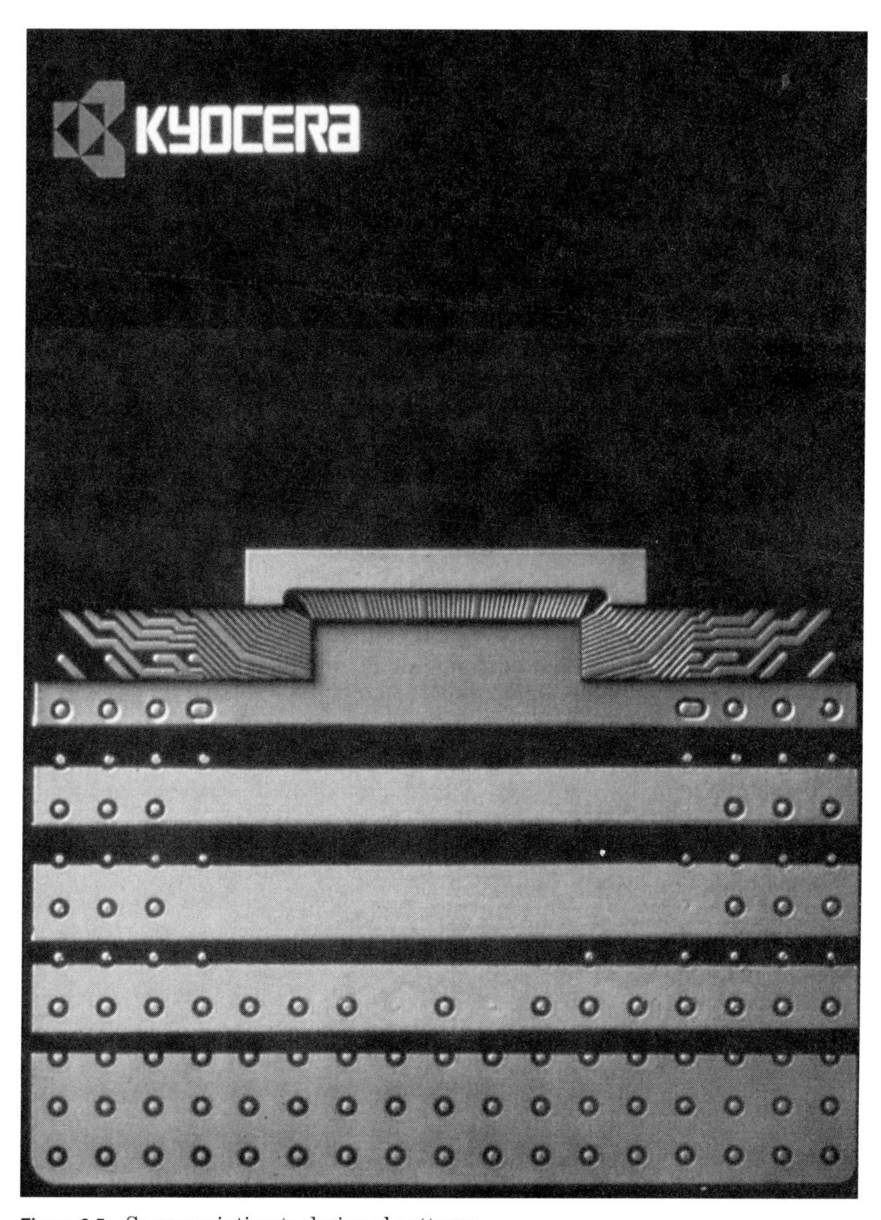

Figure 2.5 Screen printing to designed patterns.

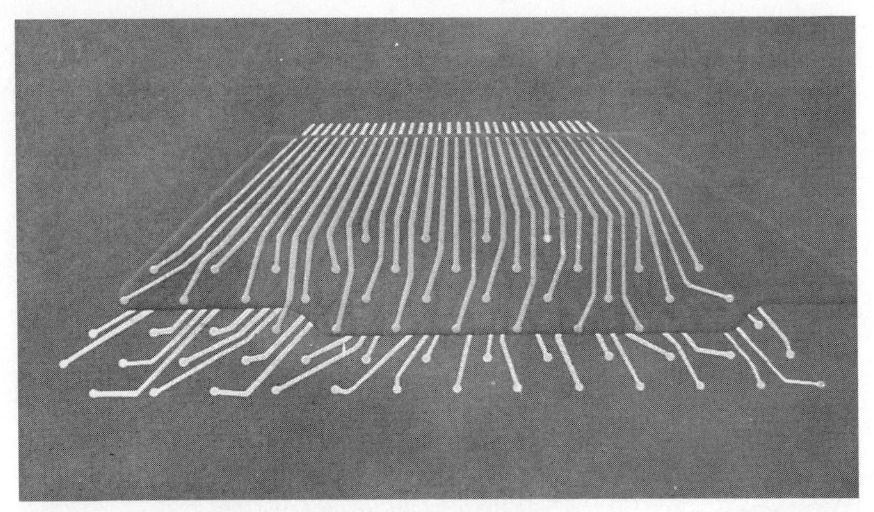

Figure 2.6 Screen printing tailored to design patterns.

The screen-printed green sheets are dried in a forced air circulated oven or air dried. One big advantage of the multilayer cofired process is the ability to visually inspect each layer before committing to the next step. Inspection can be done by either machine vision with a pattern recognition system or inspected manually.

Lamination. Each green sheet is screened with an adhesive slurry material and/or stacked in the designed sequence and bonded together with pressure and heat. It is important to control the rate and pressure of compression in order to optimize cohesion and alignment of green layers.

Shipping and scribing. Since alumina has almost the same hardness as diamonds, it is usually at this stage that laminated green sheets are cut, punched, or grooved before sintering. BGA packages can either be separated from the master array or scored on both sides for future snapping from the master array. The dimensional characteristics and tolerances determine the method of BGA separation. If tighter tolerances are required, they can be achieved by diamond saw or laser-cutting the sintered master array.

Cofiring. Cofiring is the simultaneous sintering of ceramic and metallization. High-temperature cofired ceramics are fired between 1500 to 1700°C, while low-temperature cofired ceramics are fired at about 800 to 1000°C.

Controlling the sintering profile is probably the most complex and important aspect of producing multilayer BGA packages. The sintering stage is a two-step process. The first stage is oxidizing for organic removal of the solvent and plasticizer followed by the pyrolysis of the binder. In the second stage, performed in a slightly reduced atmosphere, the product is ramped up to peak temperature to sinter the multilayer ceramic reactions that cause crystallization. The conductor paste similarly goes through the organic oxidation and crystallization process. For tungsten refractory metallization systems, reducing atmospheres for the sintering stage is necessary to guarantee nonoxidation of tungsten. The reducing atmosphere is generated by using cracked ammonia or fired in a hydrogen/nitrogen mixture containing less than 0.25 percent oxygen.

If the sintering process is not optimally controlled, incomplete organic burn-out and sintering can lead to delamination, camber, cracking, flexural strength, and metallization adhesion issues.

Plating. For functional and environmental protection, all exposed metal and metallization surfaces are either electrolytic or electroless-plated with desired metals (usually gold with nickel underplating). For BGA packages it is important to control the amount of gold plating on the BGA attachment pads. Since the BGA pads are screened or dipped with solder, the amount of gold can contaminate the solder and cause embrittlement of the solder joint. This issue causes extreme concern for package manufacturers. Wire- and tab-bonded packages require a higher thicknesses of gold plating than BGA pads, and these differing gold thicknesses have to be controlled. This is usually accomplished by a series of plating and masking steps to ensure BGA package solder reliability. This issue is alleviated, however, as integration increases and the need for flip chip devices becomes prevalent. These BGA packages will require the same gold-plating thickness on both planes of the package. Motorola has termed these flip chip BGA packages as *C5 packages* (believed to be an acronym for "controlled collapse chip connection carrier").

2.2.2 Dry pressing method

Worth noting is another ceramic firing process for simply designed single-layer BGA ceramic substrates. Although this process is not popular today for BGA packages, it could be an alternative lower-cost manufacturing process for the future.

Generally, dry pressing uses a mechanical or hydraulic press, charging the forming die with the ceramic raw material, and applying pressure from above and below or from one direction only. This method is suitable for a plate form for high-volume production at a lower cost. However, a high degree of technology is required to charge the raw material into forming dies uniformly in order to achieve homogeneous density for each individual cross section over the whole formed structure. There are many methods for dry pressing parts, including rubber press method, mechanical press method, and hot press method.

The popular method for pressing cost-efficient substrates is the mechanical pressing method. The mechanical pressing method, using upward-downward pressure, is not suitable for forming thick parts which require uniform pressure transmission. Forming vias with this method has its advantages. One is that densely packed particles are produced in the ceramic body; another is that shrinkage caused by sintering becomes uniform. A disadvantage of this method, in many cases, is that a part formed in this manner may differ from the mechanical press steel die formed part. These parts cannot be sintered as is and must be machined to the desired shape before firing.

With nonoxide compound materials such as aluminum nitride and silicon carbide, ceramic materials cannot be sintered in the conventional method, but they can be sintered with the hot press method. With the hot press method, the raw ceramic material is put into a graphite mold and the part is sintered while applying pressure and high heat at the same time. This is called *forming and sintering;* the part is sintered perfectly at a temperature normally lower than the general sintering temperature. For this reason, sintering causes less crystal grain growth, and the fired ceramic body is of high specific density, usually with fine grain microstructure. The manufacturer can get a fired substrate which is superior in flexural strength and hardness. The drawback of the hot press method that graphite material is the only material which can withstand the high heat and pressure. Complex and precise formed shapes can be accomplished with steel dies, but not with graphite dies.

2.2.3 Transfer tape method

Worth noting is a third manufacturing process for the fabrication of ceramic BGA packages referred to as the *tape transfer method,* which is currently being pursued by Pacific Hybrid Microelectronics of Portland, Oregon. The process is similar to the multilayer cofiring process as described in Sec. 2.2.1, except for the use of low-temperature-cofired ceramic (LTCC) dielectrics.

The first step of the process is to cut and laser-drill all hard features in the prefired substrate and in the transfer tape (unfired). This would include all via holes, cavities, and alignment holes. The next step would be to print the metallization on the prefired substrate. This layer would be a ground or power plane; then the substrate is fired. The nonfired transfer tape is then laminated and fired to the first layer. Via holes are metalized, filled, and fired. The assembly is screen-printed and fired again for the conductor traces. The next transfer tape layer is aligned to the assembly and the process is repeated until the desired completed design.

The laminating process is performed at 70°C and at a laminating pressure of 56 kg/cm^2 for 4 minutes. Firing of the transfer tape is achieved with a standard thick-film resistor furnace profile of 850°C, except the ramp-up, peak, and ramp-down times are doubled. This process can be used with a variety of different dielectric tape systems, including dielectrics with higher dielectric constant and better Q values for high-frequency wireless communication applications.

2.2.4 Other ceramic materials

As mentioned previously, alumina is the material of choice for BGA packages. However, alternative ceramic materials are needed to meet the ever-increasing requirements of BGA packaging, including VLSI devices with higher integration, higher device speeds, larger chip sizes, and higher power dissipation properties. To keep pace with these VLSI circuit requirements, new ceramic materials are being investigated for BGA packages. Ceramic materials exhibiting lower dielectric constants, lower thermal coefficient of expansion, and higher thermal conductivity are becoming necessary.

Listed in Table 2.1 are the characteristics of these ceramic materials compared to conventional multilayer cofired alumina. The thermal coefficient of expansion is closer to that of single crystal silicon (3.3), with the exception of borosilicate glass compositions (7.9). Aluminum nitride and silicon carbide have thermal conductivities of 150 and 260 w/mik, respectively, at temperatures between 40 and 400°C, which are 10 to 15 times higher than alumina. Meanwhile, mullite and glass ceramics have dielectric constants of 6.8 and 5.0 to 7.9, respectively, representing a significant improvement over alumina.

TABLE 2.1 Characteristics of New Ceramics

	ALUMINA A440	MULLITE ML-750	AIN	SIC	LTCC	LEC
(ELECTRICAL)						
VOLUME RESISTIVITY (Ohm, cm, RT)	$>10^{13}$	$>10^{14}$	$>10^{14}$	$>10^{13}$	$>10^{14}$	$>10^{14}$
DIELECTRIC CONSTANT (RT, IMHz)	10	6.4	8.5	45	7.9	5
DIELECTRIC LOSS TANGENT ($X10^{-4}$, IMHz)	13	17	4	4		
(THERMAL)						
THERMAL COEFFICIENT OF EXPANSION (10^{-6}/°C, 40-400°C)	7	4.2	4.6	3.7	7	4.0
THERMAL CONDUCTIVITY (W/m. K, RT)	18	5	150	270	2.2	2
(MECHANICAL)						
FLEXURAL STRENGTH (Kg/mm^2)	28	20	35	45	28	19
YOUNG'S MODULUS (10^4 Kg/mm^2)	2.6	1.8	3.2	4.1	1.1	0.9

Mullite. Multilayer mullite is composed of a single compound of alumina-silica (Al_2O_3-SiO_2) binary system (Fig. 2.7) which consists of three moles of alumina and two moles of silica. Crystallographically, mullite is an orthorhombic crystal system. It can form solid solutions between 71 and 74 percent alumina compositions, which separates to corundum and liquid phase at temperatures over 1828°C.

Figure 2.8 illustrates the x-ray diffraction intensities of the crystalline phase of mullite ceramics against the CaO content of additives. When CaO exceeds a certain amount, anorthite and alumina phase increase in proportion to the quantity of CaO. Therefore, mullite ($3Al_2O_3 \cdot 2\ SiO_2$) would be decomposed to anorthite ($Al_2O_3 \cdot CaO \cdot 2SiO_2$) and alumina according to the following reaction:

$$3\ Al_2O_3 \cdot 2SiO_2 + CaO \longrightarrow Al_2O_3 \cdot CaO \cdot 2SiO_2 + Al_2O_3$$

Mullite ceramics were delivered for multilayer cofired packages because they exhibit lower dielectric constants and a thermal coefficient of expansion closer to silicon than packages made of multilayer cofired alumina. A key advantage of multilayer mullite packages is that they are fabricated using cofired technology (Fig. 2.9) with tungsten or molybdenum metallization similar to that of alumina systems. Mullite ceramics is finding its way into BGA packages that require material properties with a thermal coefficient of expansion closely

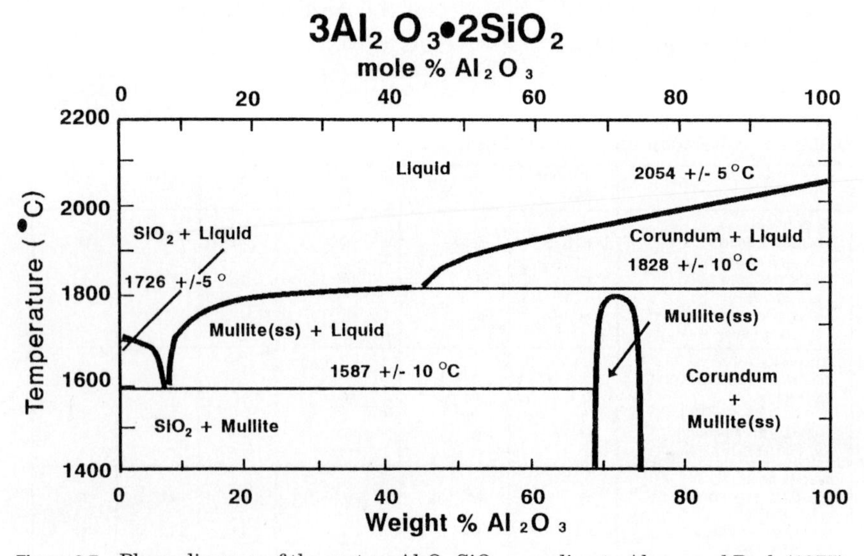

Figure 2.7 Phase diagram of the system Al_2O_3-SiO_2 according to Aksay and Pask (1975).

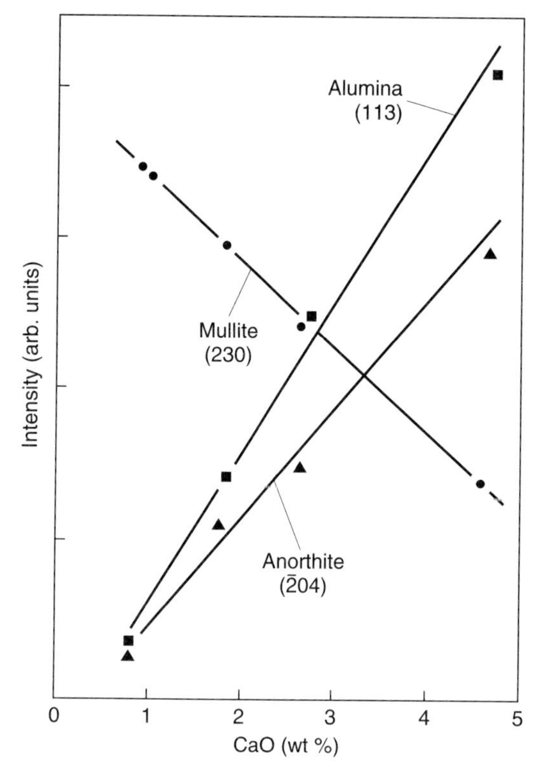

Figure 2.8 X-ray diffraction intensity of mullite, anorthite and alumina against CaO content in 90% mullite.

matched to large-size RISC microprocessor and gate array silicon chips. Due to low thermal conductivity properties, mullite BGAs are typically used with flip chip or flip tab assembly. In using these assembly techniques, thermal management can be accomplished through the back side of the device, as in the case of (Fig. 2.10) the Hitachi M880 mainframe system packaging (the MCC—Micro Carrier for LSI Chip).

Aluminum nitride (AlN). The most promising ceramic material for BGA packages will be aluminum nitride. This is partially due to the fact that, because AlN thermal conductivity and TCE characteristics make it a very desirable material for BGA package configuration, AlN is available as either a pressed ceramic material for single-layer BGA packages or as a multilayer cofired construction. A common problem of various customary packages for thermal management is the thermal coefficient of expansion mismatch between the silicon chip and heat sink material bonded to the package. Also, most heat sink materials

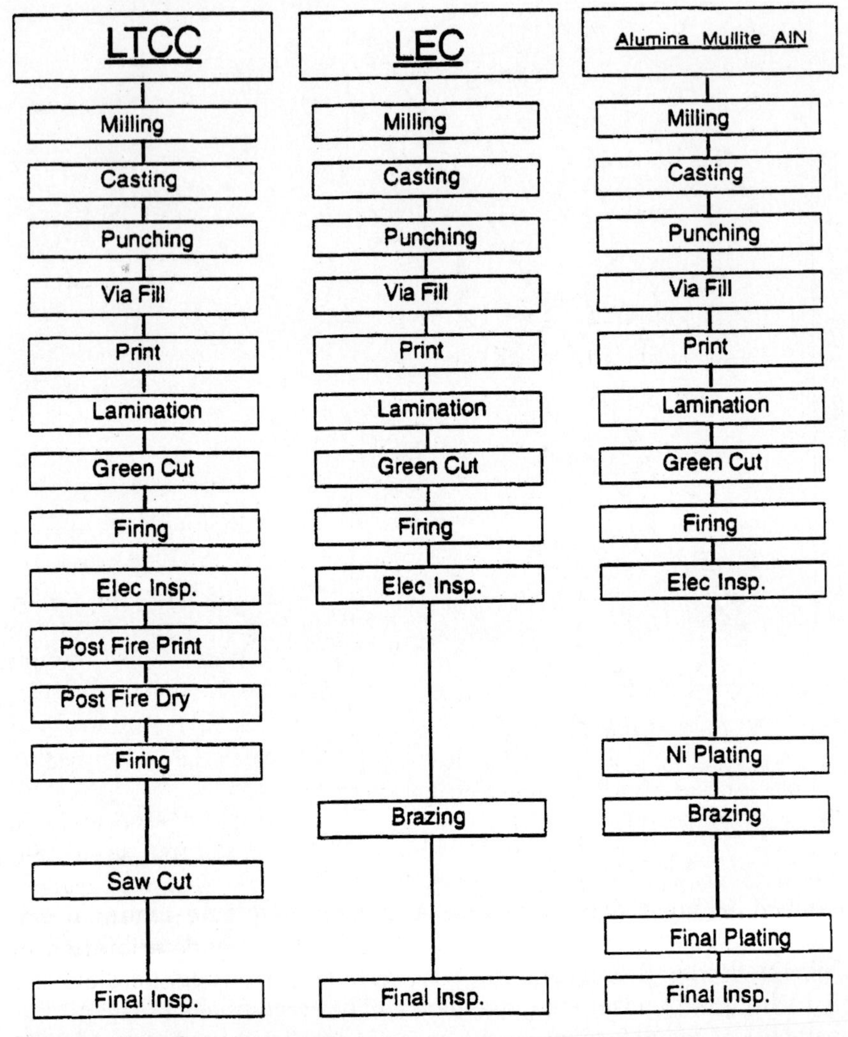

Figure 2.9 Manufacturing process.

are electrically conductive, which is undesirable for many VLSI devices. Measures are being taken to electrically isolate these materials from the device. This increases the thermal performance and deteriorates the use of the heat sink material. This action has necessitated the use and acceptance of AlN.

High-thermal-conductive aluminum nitride ceramics (thermal conductivity range: 70 to 270 w/mk) are obtained either by pressureless sintering or by the hot press using sintering additives, or by the multilayer cofiring process using refractory conductors such as tungsten.

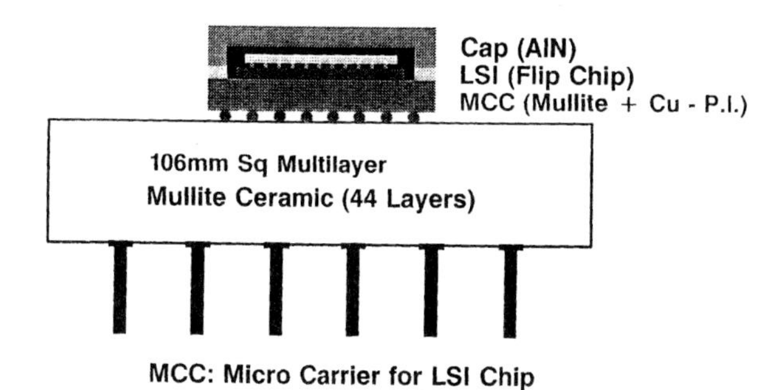

Figure 2.10 Cross section view of system.

High-purified AlN is colorless and translucent, but a dark AlN ceramic body with high thermal conductivity can be obtained through the addition of inorganic metal oxidizers. These additives do not significantly deteriorate the thermal conductivity of the AlN ceramics. Common elements for darkening AlN include W, Nb, Ta, Zr, and Ti, which are distributed intragranually and give AlN the black color. Meanwhile, Fe, Co, and Cr are always located at the grain boundary and give AlN a gray color. It appears that the electrically conductive devices dispersed in AlN are the cause of the shading mechanism. This phenomenon is believed to be caused by the optical absorption and the multiple reflections around these particles. Hence, the absorption spectra over the visible light range are flat. TiO_2 and WO_3 are the most effective shading additives. The gray AlN ceramics have a thermal conductivity of 170 to 270 w/mk. Also, there is not any deterioration in other AlN electrical properties as in dielectric strength and dielectric loss in microwave regions.

Sintering temperatures for hot press high-thermal-conductivity AlN are higher than for alumina ceramics. This, of course, is impractical for the processing of integrate-circuit BGA packages. The sintering temperature can be lower with the introduction of fluoride additives. Fully dense AlN has been achieved by pressureless sintering at 1400°C for two hours, using YF_3, as a sintering aid. During sintering, YF_3 reacts with the oxide impurities of AlN to form YOF and the gas AlF_3. YOF reacts with alumina impurities in the AlN to form yttrium aluminate grain boundary phase with fluoride ions. This mechanism lowers the densification temperature of AlN.

Conductors can be directly attached to the AlN substrate either by direct bond copper (DBC), by screening refractory Mo, or by thin film depositing Ti, TiW, TiMo, followed by Pt or Ni/Au or other compatible elements. Titanium is used as the adhesive metal.

Cofiring AlN manufacturing process is depicted in Fig. 2.11. AlN is mixed with PVB binder, organic solvents and normally yttrium oxide as a sintering aid. The mixture is cast into green sheets of designed thickness in preparation of cofiring. Tungsten refractory metal paste is screen-printed to the desired pattern, which has been previously cut to the appropriate size and shape. The organic binders in the tape and tungsten paste are removed in a dewaxing operation that must take place in a now-oxidizing atmosphere at an elevated temperature. The printed tape is laminated and sintered in a flowing nitrogen gas atmosphere at around 1800°C for several hours; after cofiring, electroless nickel containing boron is plated onto the tungsten metallization.

AlN BGAs can also be fabricated by tape transfer method, provided that AlN tape can be obtained. Most AlN conductive metallizations, except for cofired or thin-film processing, need an oxide or adhesive dielectric layer to enhance conductor adhesion to an AlN substrate. In the case for thick film processing, a glass phase dielectric is used and for DBC or refractory molybdenum an oxide layer is formed before metallization. These additive processes tend to impede the thermal conductivity of AlN with a thermal barrier adhesion layer.

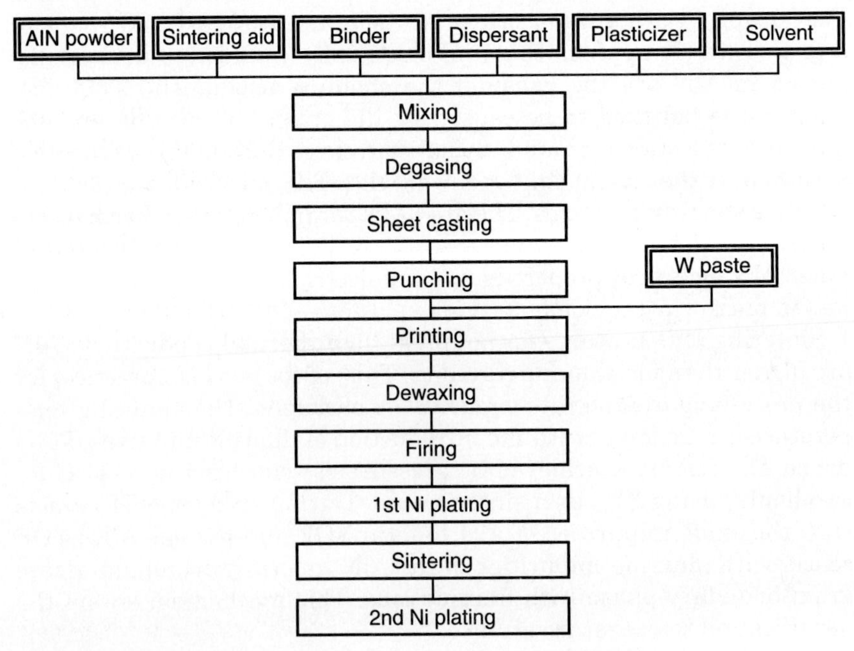

Figure 2.11 Manufacturing process of the W cofired AlN substrate.

Silicon carbide (SiC). Thermal conductive silicon carbide is manufactured by hot pressing silicon carbide with small amounts of beryllium oxide (less than 2 percent). This material exhibits favorable thermal properties such as high thermal conductivity (260 w/mk) and a thermal coefficient of expansion very close to that of silicon. However, silicon carbide has extremely unfavorable dielectric properties, such as high dielectric constants and high dielectric loss factor. With BGA design considerations to cover such dielectric disadvantages, BGAs can be manufactured by combining thin-film/organic dielectrics to compensate for the poor dielectric properties of SiC.

2.3 Glass Ceramics

A series of new low-temperature-cofired glass ceramics are being developed for integrated circuit packaging applications. This class of ceramics is referred to as *low-temperature-cofired ceramics* or LTCC. There are basically two classes of materials in this family. The first of the new LTCC materials is fired at a much lower cofiring temperature than alumina. Its main composition is made up of alumina and quartz glass with a borosilicate (lead, calcium, etc.) as the main constitutent. The conductor metallization system is thick-film gold or silver which has a low electrical conductor resistivity. Due to the nature of this conductor system, this class of LTCC materials is mainly used for high-frequency applications and hybrid circuit replacements where a user can convert their thick-film print-and-fire assembly line to this product. The cofiring sintering temperature of a LTCC BGA multilayer structure is around 850°C in an air atmosphere. This class of LTCC materials usually exhibits fairly high dielectric properties, but they are normally compatible with thick-film passive components which can be buried in the interlayer structure.

The second low-temperature glass ceramic materials are characterized by their favorable dielectric properties, usually with a dielectric constant of around 5. These materials are composed of alumina and quartz, with a basic glass that yields a cordierite crystal phase. The biggest advantages of these materials are their dielectric properties and lower thermal coefficient of expansion (4.4/ppm°C).

The conductor system for these materials is copper, although gold is now also being utilized. A disadvantage of this system is that it is difficult to control the oxidation of copper during sintering and soldering of leads, so lead attachment reliability is a concern. Table 2.2 summarizes the foregoing materials and their most commonly utilized conductor systems, firing temperatures, and firing atmospheres.

TABLE 2.2 Package System Low Temp vs. High Temp

Cooling			Immersion
			Forced Liquid Conv.
			Very Fine Coolant Channels
			Forced Air Conv.
			Natural Conv.
Second Level Packages			Thin Film & Polyimide
			Fine Glass Ceramic
			PCB
			High Density Multilayer
First Level Packages	Multichip Packages		Copper Polyimide
			Multilayer Ceramic
			Multi Function
			Metallized Multilayer
			Fine Pitch

Advantages of the LTCC materials are:

1. Most low-K ceramic materials have dielectric constants lower than that of alumina, about 9 to 10, and approach about 4 (that of silica), which is the inorganic material with the lowest dielectric constant.
2. LTCC materials have a lower TCE and a value closer to that of silicon, about 3.3 ppm/C.
3. As pointed out, the sintering temperature has been maintained at less than 1000°C, thus allowing for high-conductive metals to be used.
4. Except for copper, sintering of LTCC materials can be done in air.
5. If gold, silver, Ag/Pd metallization is used, subsequent plating operations can be omitted.

Disadvantages include:

1. Mechanical and flexural strengths are weaker than high-temperature-cofired ceramics.
2. Thermal conductivities are almost always poorer and lower than alumina.

Figure 2.12 illustrates the relationship between these materials and their thermal conductivities versus dielectric constants. It appears that material with the best dielectric constants has the poorest thermal conductivity and vice versa; this is explained by the crystal appearance of these compounds. Starting with the silica end, the crystal structure changes from a relativity open to a relatively dense composition for alumina, with a corresponding increase in dielectric constant. The lower dielectric constant silica is in line with the fact that a vacuum has a dielectric constant of 1. This same trend is observed with modules, density, and phonon mean free path on one hand to crystal structure on the other.

Thermal coefficient of expansion for these LTCC ceramics is very favorable for die attaching or mounting large area devices. Figure 2.13 illustrates the reliability of different materials for fatigue life. It is obvious that the LTCC ceramic is superior for this parameter for silicon-mounted devices. Conversely, since most LTCC BGAs will be solder-attached to organic printed circuit boards, the LTCC mismatch will be greater for these packages over the alumina BGA packages. Hence, size limitation of alumina BGAs for fatigue life will even be greater for LTCC BGAs. So, not only will the BGA size be smaller, but the thermal conductivity will be worse.

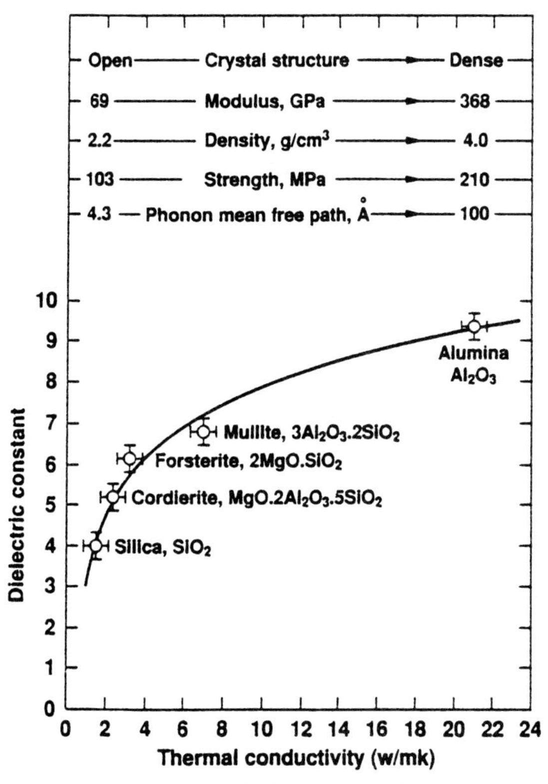

Figure 2.12 Empirical relation between thermal conductivity and dielectric constant of several dielectric materials of interest.

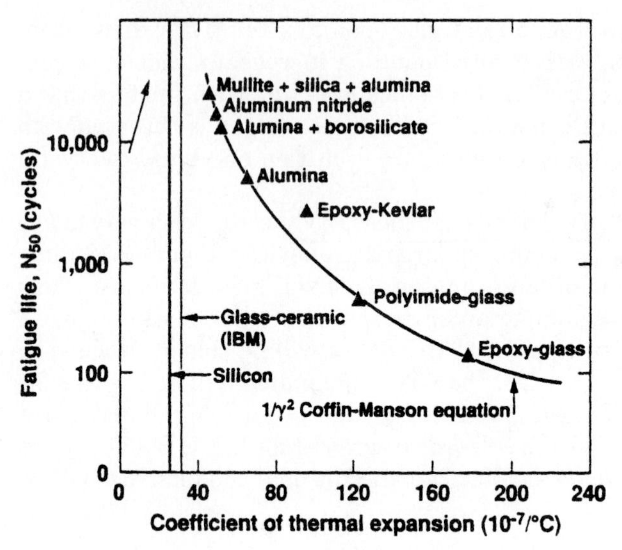

Figure 2.13 Relation between fatigue life failure of solder bumps and TCE of dielectric (Tummala).

2.4 Ceramic BGA Substrate Circuit Density

Multilayer ceramic technology appears to have greater circuit pattern density for BGA packages over PCB (printed circuit board) technology mainly because a higher number of layers and finer design rules are easily achieved. Design rules for cofired multilayer ceramic technology for line pitches and via hole diameters are shown in Figs. 2.14 and 2.15, respectively. Design rules of 200 microns pitches, 100-micron via hole diameters with layer numbers approaching 20 are easily applied to today's cost-driven BGA products. The cofired technology with 200-micron-pitch design rules was established over 10 years ago. In addition, the most remarkable feature of cofired multilayer technology is that filled stacked vias are part of the normal multilayer process used to give a higher circuit density as pictured in Fig.

It is believed that a screen-printing method used for the cofired ceramic technology limits the circuit width to 100-micron geometries. However, the 100-micron-pitch rule with 50 micron widths and 50 micron spaces has been established by using the screen-printing method as shown previously in the figures. Furthermore, 50-micron-diameter via holes have also been established by using standard punching methods. The finer printing and punching technology was developed to give a lower-cost effective technology to replace thin-film technology design rules, which are relatively expensive.

The order of wiring density per layer is the following: thin film> MLC (multilayer ceramic)> PCB. Furthermore, the total number of circuit

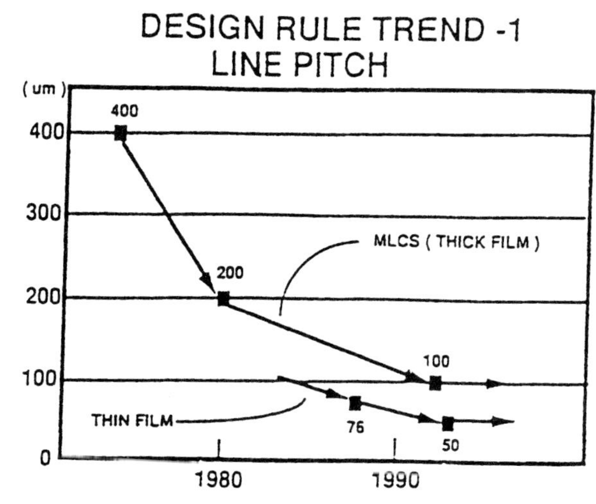

Figure 2.14 Design trend (line pitch).

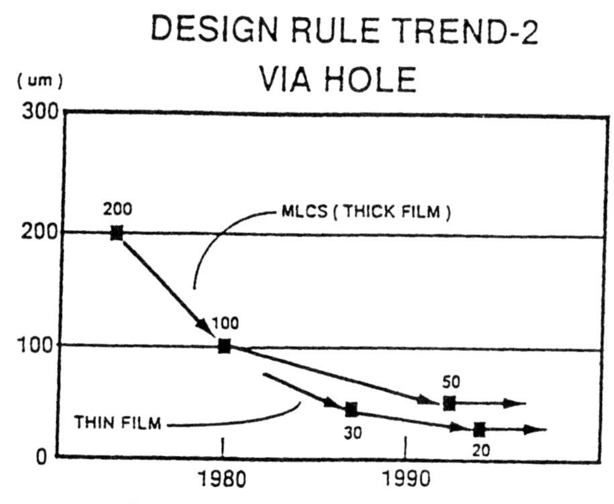

Figure 2.15 Design trend (via hole).

layers being applied to a BGA products is of equal importance considering the total wiring density of the interconnections of the BGA. For the number of layers, order of complexity is: MLC> PCB> thin film. Fifteen to twenty layers of interconnects are easily accomplished for MLC products at a reasonable cost. On the other hand, the number of layers for thin-film technology is normally around five, and for standard PCB technology is between six and eight layers. As always, there are exceptions to these examples. There have been PCB applications with layer counts as high as 40 for supercomputer boards. But, as a norm, most

PCB applications today are around six to eight layers. The wiring density can be illustrated as a function of the following equation:

$$\text{Total wiring density} = \text{wiring density/layer} \times (\text{layer number})$$

Figure 2.16 shows the wiring density calculated by the preceding equation. As depicted, the MLC technology has the highest wiring density. The order of total wiring density when taken into account these conditions is: MLC> thin film> PCB.

Design rules for via hole diameters and via hole pitches are an important consideration when designing device layouts and BGA substrate sizes. Figure 2.17 illustrates PCB standard design rules of 650 microns for via pitches. The PCB standard via hole configuration is processed normally through drilling practices, causing an open hole which limits the routability of the conductor lines in all PCB layers. As a result, integrated circuits that require a smaller form factor for the BGA package may be limited by the fan-out design rules for the PCB BGA. Also noted, via hole pitches for MLC fabrication are based upon 250 microns geometry. Also the via holes are blind and stacked if needed, since each layer is processed separately. Hence, MLC technology can be designed and laid out to approximately the same design device layout as the device if flip-chipped.

With CBGA technology, pads for the solder ball attachment can conceivably reach pitches of 200 to 250 microns. Of course, attachment to the PCB may present a problem if cost-effective finer geometry and routability issues are not resolved.

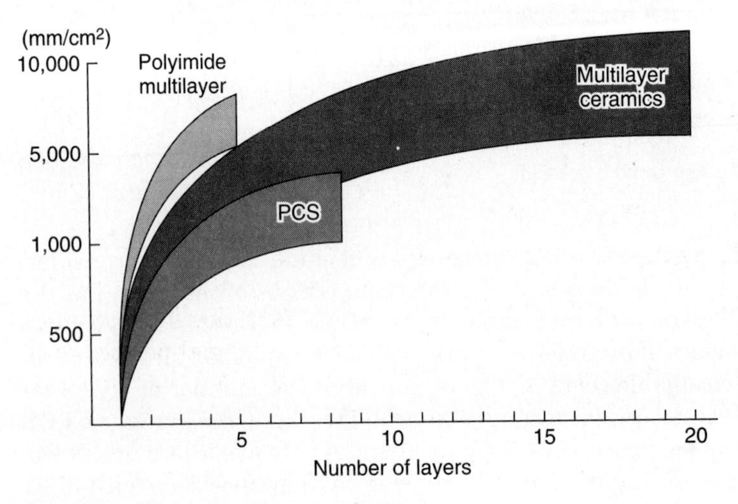

Figure 2.16 Wiring density comparison.

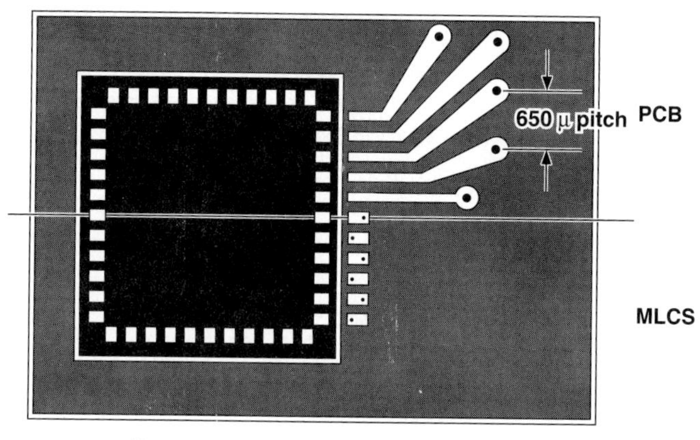

Figure 2.17 Design rule comparison (MLCS versus PCB) (IC pad pitch, 250μm).

CBGA package electrical performance surpasses PBGA's due to the short circuit paths for power and ground pins/pads, plus the ability to use blind and buried vias for I/O connections. CBGA substrates have a maximum inductance of 3.4 nh for signal pins, which will encourage IC designers of high-speed devices to choose CBGAs.

2.4.1 Thermal dissipation

The thermal dissipation of BGA packages varies with the package style. Heat dissipation is typically divided into two directions. The first is through the solder balls to PCB and off the package. The second is from the chip to the package to the environment (in most cases into the airflow). The division of thermal dissipation depends on the thermal conductivity and thermal planes of the BGA and PCB. PBGA designs rely on the thermal vias in the center of the package to carry the heat from the device through the package to the solder balls through the board power and ground planes.

CBGA packages have inherently excellent heat performance because of the thermal conductivity of ceramic materials. CBGA can be designed with cavity-down features or with flip chip designs to enhance thermal dissipation.

Although CBGA substrate materials are considered to be superior thermal materials to PBGA, the PBGA designs take this into consideration, and PBGA designs do provide equivalent thermal performance to the CBGA when connected to a PCB with a ground plane or heat spreader pad. The CBGA also allows the option of attaching an additional external heat fin to the metal lid, which dramatically improves the thermal performance. PBGAs do not have this flexibility since they are normally overmolded with epoxy mold compounds.

2.4.2 Advantages and disadvantages

The greatest benefits of the PBGA package are its low cost and TCE match to organic PCB materials (boards). The most important benefit of CBGA is that it is cost competitive with PBGA and offers higher density and finer pitch connection that reduces board space requirements. CBGA reliability data documents clearly show that TCE mismatch is not an issue for ceramic substrates 25 mm square or less. PBGAs have several technical concerns that CBGA overcomes. Good TCE match between silicon devices and CBGA packages allow for reliable mounting of large-area devices, a real problem for BT resin-based PBGAs. Recent failure analysis studies of PBGA reveal die-to-package failures at the edge of the device. CBGA has no moisture sensitivity eliminating the "popcorn" cracking or delamination problems typical of overmolded packages. Electrically, as stated previously, CBGAs specifications are far superior to PBGAs due to the "die to edge to die" routing required in plastic laminates. Finally, it is important to note that standard CBGA designs incorporate power and ground planes for improved performance, a significant cost adder to PBGAs currently. Table 2.3 summarizes advantages and disadvantages of both CBGAs and PBGAs packages.

TABLE 2.3 CBGA versus PBGA (after Ken Wood).

	Advantages	Disadvantages
Ceramic BGA Package	* Highest I/O per area * High power > 2 watts * Low L > 4nH * Low profile < 4 mm * Hermetic (if needed) * No "popcorn" issues * High speed * Large die sizes, TCE * Pwr/Gnd planes * Proven pkg. tech. * MCM capability	* TCE mismatch, PCB limits package size to 25 mm SQ without epoxy underfill. * Slightly higher cost than PBGA
Plastic BGA Packages	* Good TCE to PCB * Low cost * Strip assembly * Self-aligns to PCB * Low profile < 2 mm * MCM capability * Most reliability data for board attach	* Lower I/O per area * Less power < 2 W * High L > 10 nH * Routes to edge * Mold problems * High cost pwr./gnd. * Nonhermetic * Big TCE mismatch to silicon * Package camber * "Popcorn" problem

2.5 References

1. R. R. Tummala, E. Rymaszewski, *Microelectronics Packaging Handbook,* Van Nostrand Reinhold, N.Y., 1989.
2. R. R. Tummala, "Electronic Packaging in the 90's—A Perspective from America," *Proceedings from the Electronic Components and Technology Conference,* 1990, pp. 9–15.
3. R. E. Sigliano, K. Gaughan, "Ceramic Material Options for MCM's," *Proceedings from the International Conference on Multichip Modules,* 1992, pp. 291–299.
4. H. Wessely, O. Fritz, P. Klimbke, W. Koschnick, K. H. Schmidt, "Electronic Packaging in the 90's—A Perspective from Europe," *Proceedings from the Electronic Components and Technology Conference,* 1990, pp. 16–33.
5. R. E. Sigliano, D. Grooms, "The Future Packaging Market—The Demise of Ceramic Packaging? The Dominance of Plastic?," *Fourth Annual KII Group Technical Conference,* April 1993.
6. R. E. Sigliano, "Single Chip Packaging Past, Present and Future," *Proceedings from Semicon East,* October 1993.
7. R. W. Rice, J. H. Enloe, J. W. Lau, E. Y. Luh, L. E. Dolhert, "Hot Pressing—a New Route to High-Performance Ceramic Multilayer Electronic Packages," *Ceramic Bulletin,* vol. 51, no. 5, 1992, pp. 751–755.
8. P. J. Holmes, R. G. Loasby, *Handbook of Thick Film Technology,* Electrochemical Publications Limited, 1976.
9. J. J. Burke, E. N. Lenoe, R. N. Katz, "Ceramics for High Performance Applications—II," *Proceedings of the Fifth Army Materials Technology Conference,* 1977.
10. M. Sugiura, *The Ceramic Story,* Kyocera Technical Book Series 1, 1989.
11. J. Tanaka, S. Kajita, M. Terasawa, "Mullite Ceramics for the Application to Advanced Packaging Technology," presented at Material Research Society Spring Meeting, K 6.3, 1989, pp. 233.
12. J. Tanaka, S. Tanahashi, M. Terasawa, R. E. Sigliano, "A Multichip Module-Polyimide on a Mullite Ceramic," *Proceedings from NEPCON West,* 1990, pp. 1371–1372.
13. K. Banerji, "Development of the Slightly Larger than IC Carrier (SLICC)," *Proceedings from Nepcon West,* 1994, pp. 1249–1265.
14. D. B. Walshak, Jr., H. Hashemi, "Thermal Modeling of a Multichip BGA Package," *Proceedings from NEPCON West,* 1994, pp. 1266–1276.
15. M. Terasawa, S. Minami, J. Ruben, "A Comparison Thin Film, Thick Film and Cofired High Density Ceramic Multilayer," *Int. Jour. Hybrid Microelectronics,* vol. 6, 1983, pp. 1–11.
16. C. Park, "Ceramics: Cool for Chips," *Electronic Engineering Times,* July 1991, pp. 55.
17. Y. Fujioka, T. Fujisaki, I. Sakaguchi, Y. Yamakuchi, S. Shimo, K. Onitsuka, N. Fujikawa, "Multilayer Ceramic Substrate with Inner Capacitors," *Proceedings from IMC,* 1992, pp. 355–359.
18. P. Bindra, "Moto Gets 'SLICC' with Packaging," *Electronic Engineering Times,* February, 1994.
19. M. Cole, T. Caulfield, "BGA's Are Extending Their Connections," *Electronic Engineering Times,* February 1994.
20. G. Derman, "Interconnects and Packaging," *Electronic Engineering Times,* February 1994.
21. M. Fukui, M. Hori, C. Makihara, M. Terasawa, "Multi-Layer Ceramic Substrate (MLCS) the Alternative Printed Circuit Board Technology, *Proceedings of International Symposium on Microelectronics,* 1993.
22. N. Kuramoto, "Thin-Film and Co-Fired Metallization on Shapal Aluminum Nitride," *Advancing Microelectronics,* January/February 1994.
23. N. Iwase, F. Ueno, T. Yasumoto, H. Asai, K. Anzai, "ALN Substrates and Packages," *Advancing Microelectronics,* January/February 1994.
24. W. Minehan, D. Horn, W. Weidner, J. Volmering, "Current Technology for Hot Pressed, Co-Fired ALN Electronic Packages," *Advancing Microelectronics,* January/February 1994.

25. R. Imura, N. Ito, M. Terasawa, "Development of Aluminum Nitride Multilayer Packages," *Advancing Microelectronics,* January/February 1994.
26. P. Garrou, "Aluminum Nitride for Microelectronic Packaging," *Advancing Microelectronics,* January/February 1994.
27. P. Danner, "High Density Ceramic Modules," *Proceedings from International Conference on Multichip Modules,* 1992, pp. 325–328.
28. C. Brown, "Advances in Screen Printing Technology," *Proceedings from NEPCON West,* 1994, pp. 1856–1861.
29. J. Lideen, A. O. Dahl, "Printing Techniques for Fine Pitch Screen Printing," *Proceedings from NEPCON West,* 1994, pp. 1862–1876.
30. L. Drozdyk, "Capacitors Buried in Green Tape," *Proceedings International Symposium on Microelectronics,* 1993, pp. 209–214.
31. T. Goodman, H. Fujita, Y. Murakami, A. T. Murphy, "A Low Temperature Co-Fired Ceramic Land Grid Array for High Speed Digital Applications," *Proceedings from Electronic Components and Technology Conference,* 1993, pp. 425–430.
32. T. K. Gupta, "In Search of Low Dielectric Constant Ceramic Materials for Electronic Packages," *Microcircuits and Electronic Packaging, First Quarter,* 1994, pp. 80–97.

Chapter

3

Plastic Substrates for Ball Grid Array Package

Osamu Fujikawa and Motoji Kato

3.1 Introduction

It is only recently that printed wiring board is used as IC packaging material. Plastic material inherently has quite a few weaknesses as IC packaging material. However, there are a lot of merits in plastic material; therefore, its usage and application has been expanded over the last 20 years.

In this chapter, we are going to discuss the merits as well as demerits of plastic materials as ball grid array substrate, so that you may understand the virtues of plastic BGA substrates. We will focus on the plastic substrate material and process which is widely available in today's market.

3.2 General Overview of Plastic BGA Materials

3.2.1 Structure of plastic BGA package

Substrate materials, called *laminates,* generally consist of organic polymer resin, reinforced fiber, and conductive material. These components are important to the BGA package characteristics requirements.

Figure 3.1 is a cross section of two of most common structures of the BGA package. As shown on the sketch, the substrate has to fit the total process of assembly, which includes IC assembly, encapsulation, solder ball application, and the package assembly onto a PWB.

In addition, it must meet the general requirements for a high-density IC package, such as good electrical performance and physical reliability. The relationship between the requirements to BGA package and substrate characteristics is shown in Fig. 3.2.

3.2.2 High-temperature resistivity

Generally, laminates used for raw material for PWB are required to meet the soldering temperature stability for electrical components or device assembly. The BGA package will also be exposed to the same type of thermal stress during die attach, interconnect, encapsulation, and solder ball application process. Most laminate available in the

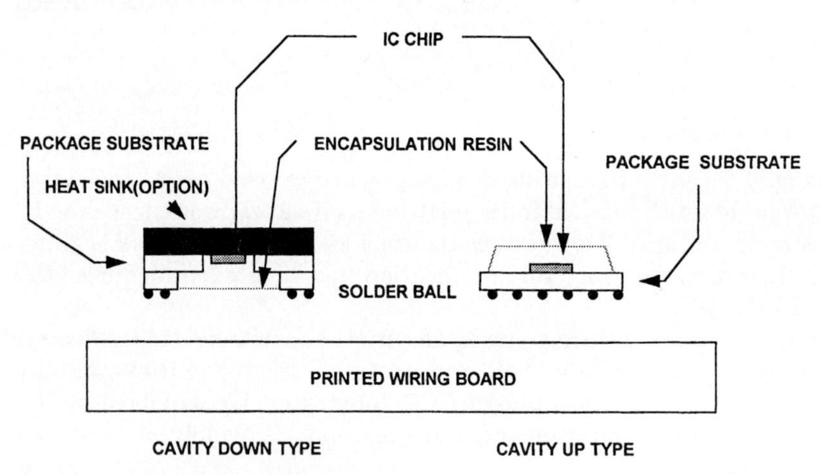

Figure 3.1 Typical plastic BGA package.

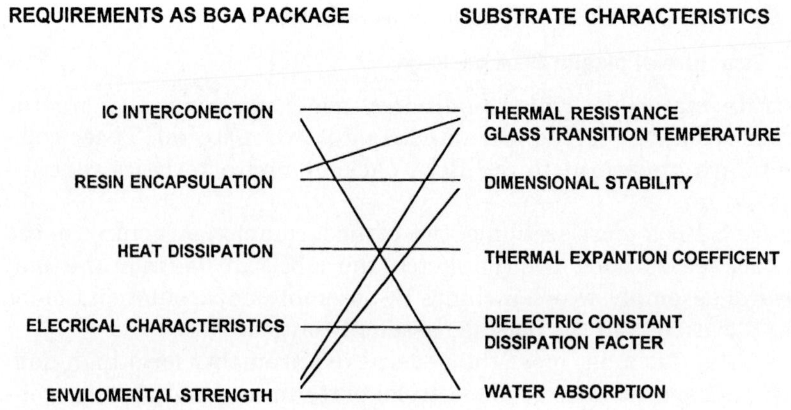

REQUIREMENTS AS BGA PACKAGE

IC INTERCONECTION
RESIN ENCAPSULATION
HEAT DISSIPATION
ELECRICAL CHARACTERISTICS
ENVILOMENTAL STRENGTH

SUBSTRATE CHARACTERISTICS

THERMAL RESISTANCE
GLASS TRANSITION TEMPERATURE
DIMENSIONAL STABILITY
THERMAL EXPANTION COEFFICENT
DIELECTRIC CONSTANT
DISSIPATION FACTER
WATER ABSORPTION

Figure 3.2 PKG/Substrate relationship.

market should survive 260°C solder bath test or IR reflow furnace test for 60 seconds. However, actual process conditions and methods vary in each IC assembly and interconnect assembly line, so high-temperature resistance material is preferable when a high-temperature assembly condition is expected.

But it is more important to pay attention to the substrate storage condition than the substrate high-temperature characteristics. If the substrate temperature is elevated rapidly after absorbing a lot of moisture, it will cause blisters. Therefore, the lower rate of water absorption is another important point (as well as the high-temperature characteristics) if the substrate is used in a high-humidity environment.

Figure 3.3 is a plot of catalogue data of a solder temperature resistivity test (rapid heating) and high-temperature storage test among several manufacturers' laminates. All values are guaranteed minimum temperatures; therefore, it may be possible to use the material at a higher temperature than indicated in the data in actual applications.

If the substrate is used under well-maintained conditions, most of the material will survive high-temperature assembly. But in most cases, 280°C or higher temperature will cause problems.

3.2.3 Dielectric constant

Plastic laminate is a composite material of resin and reinforced fiber. Consequently, the dielectric constant falls in the middle of dielectric constants of those two components.

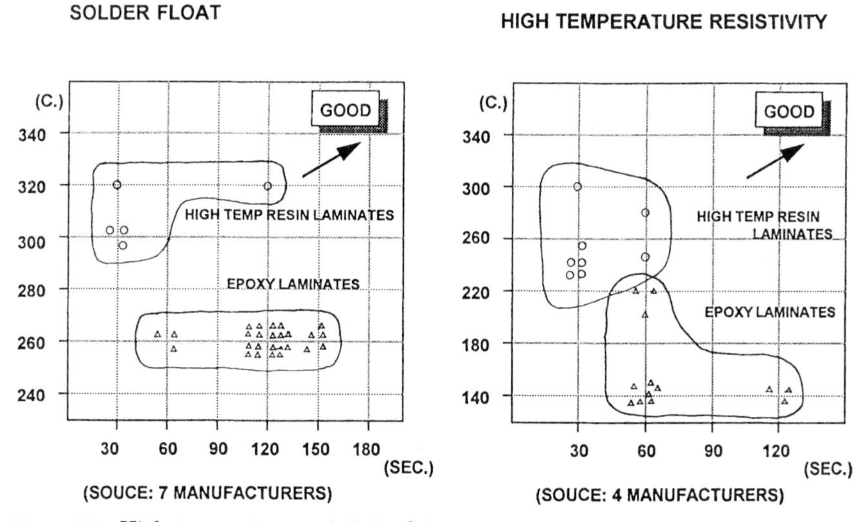

Figure 3.3 High-temperature resistivity data map.

The dielectric constant is desired to be smaller in high-switching-speed IC applications because it is the coefficient of capacitance which causes propagation delay and signal deformation. Plastic material usually has a lower dielectric constant and dissipation factor than ceramic material, so it is a preferable material for high-speed applications.

The dissipation factor indicates the internal power loss when an alternate electric field is applied to dielectrics. The value is indicated as tan δ because it is generally a small number.

Figure 3.4 is the plot of dielectric constant and tan δ based on laminate manufacturers' data.

3.2.4 Dimensional stability

Surface dimension stability in plastic laminate under heat is affected more by reinforced fiber than by resin.

Resin expansion or shrinkage is an issue when *z*-axis dimension stability is important. *Z*-axis dimension stability affects the through-hole reliability which runs through the substrate and interconnects the top and bottom of the board. If the thermal expansion is large along the *z* axis, the plated conductor inside the through holes will receive great stress along the temperature cycling. In the worst case, this type of thermal stress causes cracks on the through-hole shoulder or barrel and creates package function problems.

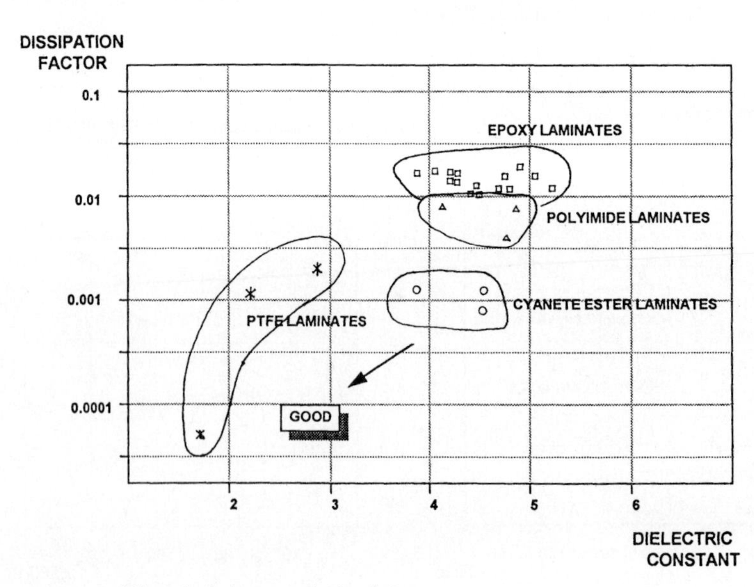

Figure 3.4 Electrical characteristics data map.

The surface dimension change affects the thermal stress between the silicon chip and substrate, and creates cracks at the die attach area or solder joint in case of flip chip bonding. Typical thermal expansion coefficient (TCE) of the silicon chip is 4 ppm/°C; alumina ceramic is 7 ppm/°C; and plastic laminate is 15 to 16 ppm/°C. Although lower TCE laminate is attainable when lower thermal expansion reinforced fiber is chosen, current studies have shown that regular TCE laminate is applicable for flip chip joint in most cases.

3.2.5 Water absorption

Water absorption is a unique characteristic for resin and reinforced fiber. Plastic laminates are "breathing."

The water absorption rate depends on the environment humidity, and in some cases, absorbed water creates blister or delamination when the package is heated, so it is important to select material which has a lower water absorption rate. Water absorption is governed by the composition ratio of inorganic reinforced fiber and organic polymer, and is different whether the fiber is organic or inorganic. Phenol resin, which is inexpensive resin, has bond water in its structure; therefore, it is not quite suitable material for packaging applications. Figure 3.5 is the water absorption data of several different types of materials. It was taken from the manufacturer's catalogue.

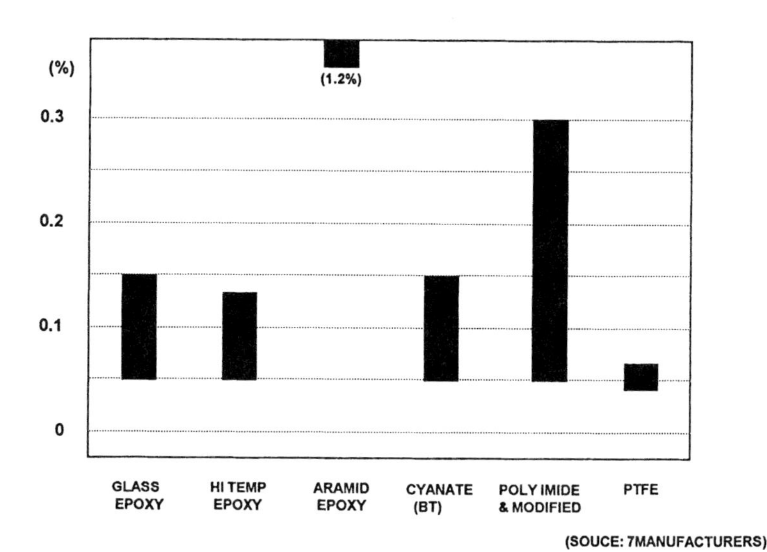

Figure 3.5 Water absorption data map.

3.2.6 Resin material

3.2.6.1 Epoxy resin. Epoxy resin is made by curing a compound which has two or more epoxy radicals being made from the reaction between epichlorohydrin and bisphenol A or cresol novorac.

Bisphenol A type epoxy

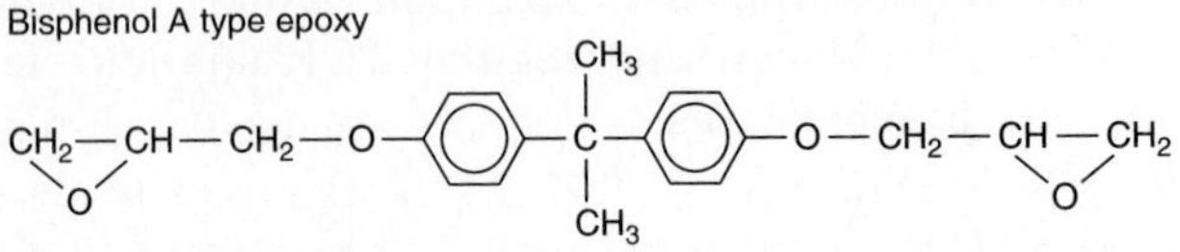

Cresol novorac type epoxy

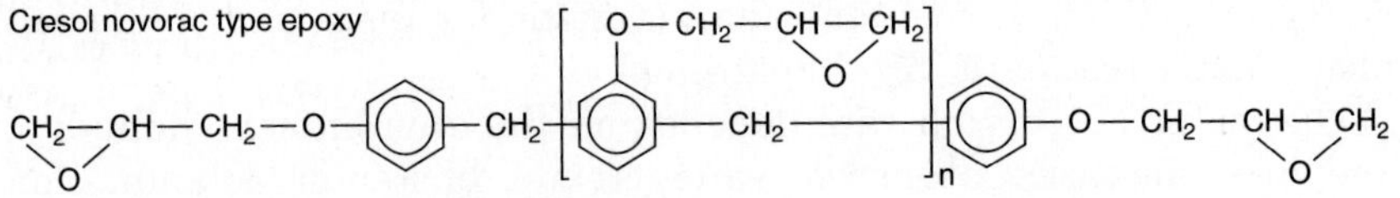

Tg of bisphenol A type epoxy is 110 to 140°C, and cresol novorac type epoxy is 150 to 190°C. If an imide radical is contained in epoxy, the high-temperature characteristics shall be improved. Tg is about 198°C and the solder limit is 300°C for more than 3 minutes. Epoxy is superior material in terms of electrical and thermal characteristics. Also, epoxy has strong bonding strength with many kinds of metal. Overall, epoxy is very good material for IC packaging applications.

3.2.6.2 Imide resin. Resin which has an imide radical in the molecular chain is called *polyimide resin.* There are two types of polyimide: one is film-shaped polyimide such as Kapton (DuPont) and the other is the one that forms 3-D polymer when it is thermally cured. For the packaging applications, the latter type is mainly used.

In the early days, Rhone-Poulene's polyamino bismaleimide resin (Kerimide) was introduced as laminate application resin.

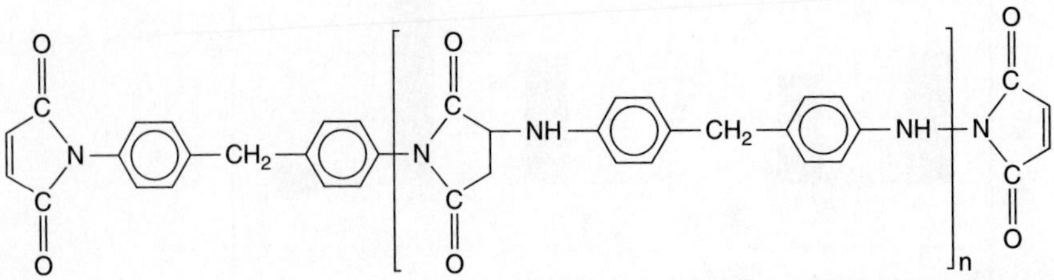

This resin has advantages in high-temperature characteristics and dimensional stability, but there are also weakness in mechanical processability, bonding strength to metal, and water absorption. Most polyimide laminates that are available in the market today have been improved above weak points by means of modifying with epoxy resin.

3.2.7 Reinforced fiber

Glass cloth is most widely used for plastic laminate reinforcing fiber. Quartz fiber and aramid fiber is also used when the thermal dimension stability is important. Paper-base laminate is also used in some areas but it cannot be used for package substrate applications due to the poor water absorption, dimensional stability, and difficulty in multilayer processing.

Glass cloth has several different types of composition, and each has different electrical characteristics and TCE. Table 3.1 shows the electrical characteristics comparison of several types of fibers. Both Q-type glass cloth and aramid mat paper have good electrical characteristics and thermal dimensional stability (actually, aramid has a negative TCE), but they have a weak point in mechanical processability.

3.2.8 Other material

A new material has been introduced into the market, that is, porous ceramics is a base material and epoxy resin is impregnated in the base material. It can be considered as a plastic material because resin is impregnated all across the board including porous ceramics. Properties of this material are in between those of ceramics and plastics; therefore, thermal conductivity and dimensional stability are better than those of regular plastic material.

3.2.9 Material treatment

Sometimes, resin requires additional treatment to pass the UL Flame Retardant Regulation. Simply, it can be done by adding bromine to the molecular structure. So, there is a chance that bromine ion (Br^-) dissolves from the flame retardant material.

Glass cloth surface is treated with organic silane compound coupling. Silane compound couples with glass surface as well as bonding strongly with resin and creates strong coupling action between resin and glass cloth.

TABLE 3.1 Characteristics of Reinforce Fibers

Type	Q	D	S	E	Aramid
SiO_2	99.9	75–76	62–65	53–56	—
Al_2O_3		<1	20–25	14–18	—
CaO		<1	—	20–24	—
MgO		<1	10–15	<1	—
B_2O_3		19–20	<1	5–10	—
ε_0	3.7	4.3	4.5	5.8	3.4
tan δ	0.00015	0.0007	0.002	0.001	0.007
TCE (ppm/°C)	0.5	3.2	—	4.5	–3.5

3.2.10 Delamination

If plastic material is heated rapidly after water is absorbed, the absorbed water will be vaporized, increasing inside pressure, and eventually will cause a delamination. The areas where delaminations are probable inside a BGA package are indicated in Fig. 3.6. Typically, delamination occurs along the boundary of different materials and expands along the boundary.

To avoid this delamination problem, it is effective to store the package material under a dry condition, to prebake before processing the package, and to avoid overheating exceeding 280°C.

3.3 Substrate Characteristics

3.3.1 Epoxy (FR-4)

Epoxy substrate is inferior to other high-temperature resin in high-temperature characteristics. However, many kinds of high-temperature epoxy have been introduced to the market along with the progress of COB or plastic package development. (See Fig. 3.7.)

Improved Tg enables stable dimension at high-temperature operation, and is effective in preventing the capillary's sinking into the substrate during the wire bonding process. Wire bondability is affected by the hardness of the wire bonding pad when the pad is heated; therefore, if the pad surface metallization is hard enough to prevent the capillary from sinking, high-Tg material is not essential. To give a hard metallization surface, increasing the thickness of hard metal (such as nickel) plating is very effective.

Key advantages of epoxy resin are lower water absorption and wetability to epoxy encapsulant. The water absorption rate indicated

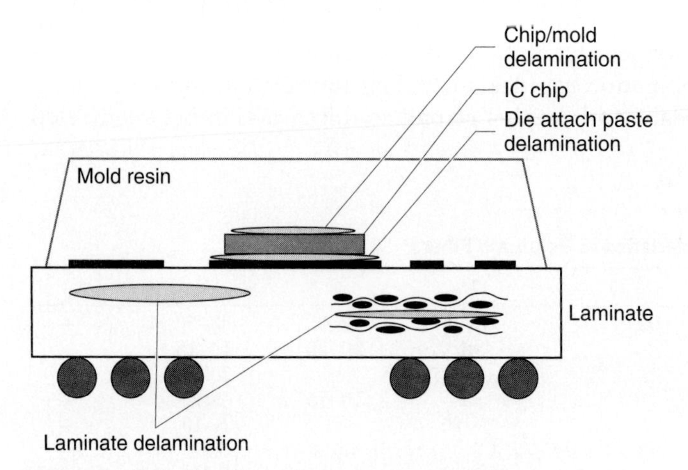

Figure 3.6 Delamination at heterogeneous junction.

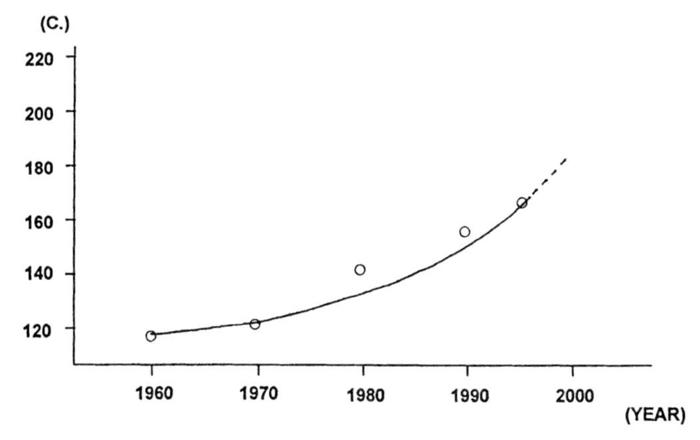

Figure 3.7 Improvement of T_g in epoxy.

in Fig. 3.5 shows the maximum and minimum value of seven different laminate manufacturers. Further lower water absorption rate substrate (0.05 to 0.1 percent) is also available.

Epoxy laminate whose Tg is around 140°C has been used for high-volume leadless chip carrier and plastic PGA in last 15 years, and performance has been confirmed.

3.3.2 Cyanate ester (BT material)

Cyanate ester substrate production was started in late 1970s mainly by Mitsubishi Gas Chemical. It is commonly known as Bismaleimide Triazine (BT).

Since cyanate ester has very good electrical characteristics, for example, ε_0: 4.2 to 4.5 and tan δ: 0.006 to 0.009, it has excellent potential to be used widely for future high-speed signal device applications (Table 3.2).

In particular, cyanate ester has advantages in reliability in humid conditions. For example, resistivity drop after a humidity test is one digit less than that of epoxy substrate. So it is assessed that the humidity reliability as a package substrate is very good. (See Fig. 3.8.)

Initially, cyanate ester had a problem in the adhesion strength of copper foil, but it has been maintained well (Fig. 3.9) and has been used for high-performance plastic PGA (PPGA) material for more than 10 years.

3.3.3 Aramid epoxy

Aramid fiber is superior in high-temperature characteristics and its TCE is a negative number that is about –3.5 ppm/°C. When aramid is

TABLE 3.2 Substrate Characteristics

	BT Materials				Epoxy Materials			Polyimide Material		
	702	800	810	870	G-10	FR-4	FR-5	Pure PI	Modified A	Modified B
ANSI Grade			GPY	GPY	G-10	FR-4	FR-5	GPY	GPY	GPY
Dielectric Const.	4.6	4.4	4.3	3.8	4.7	4.7	4.9	4.7	4.7	4.7
Dissipation Fact.	0.01	0.007	0.006	0.002	0.017	0.017	0.013	0.007	0.009	0.006
Tg (°C)	180	210	200	180	—	130	170	230	190	200
TCE (ppm/°C)	14–15	14–15	14–15	15–18	12	13	13	12–14	13–15	13–17
Insuration Resis. (Ω)	1×10^{14}	1×10^{14}	1×10^{14}	1×10^{14}	1×10^{14}	1×10^{14}	1×10^{14}	4.4×10^{14}	5×10^{14}	1×10^{14}

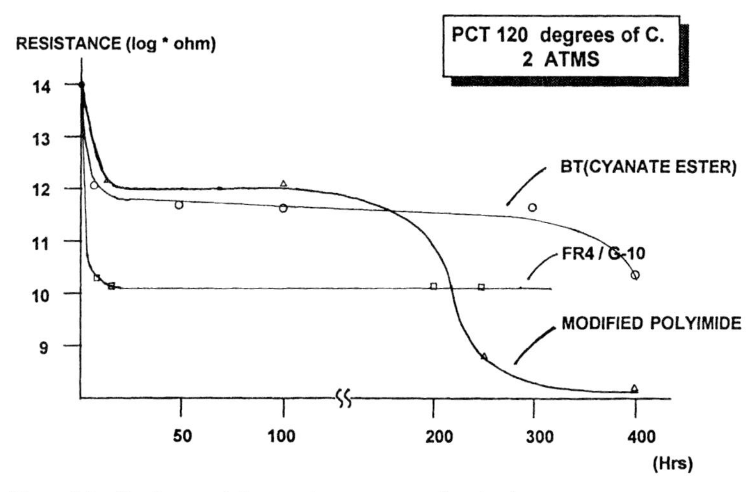

Figure 3.8 Surface resistance at pressure-cooker test.

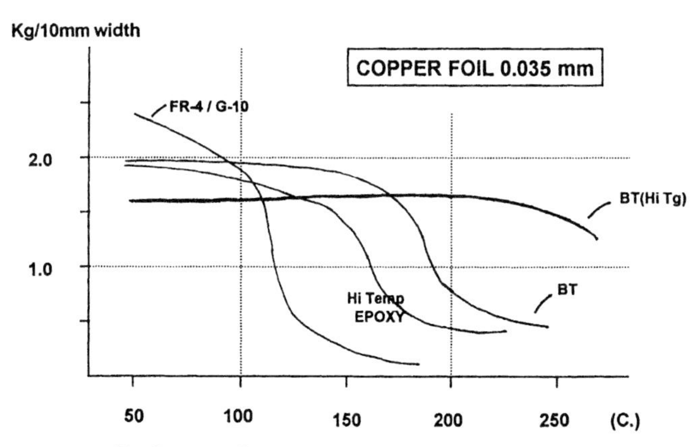

Figure 3.9 Peel strength.

used in epoxy laminate, substrate TCE is 6 to 8 ppm/°C and that is about half of the TCE of conventional FR-4 or BT substrate.

Mechanical processability of aramid epoxy is bad. Therefore the aramid laminates sold today are using unwoven mat aramid (instead of woven cloth) to improve the processability.

Thermal mismatching between IC and the substrate is smaller in the case of aramid than other conventional laminates; however, there is a mismatch to PWB that doesn't exist in conventional laminates. So it is not quite suitable material for BGA applications.

3.3.4 Ceramic core substrate (Ceracom)

Ceracom is the laminate where epoxy resin is impregnated into porous ceramic-base material and copper foil is laminated. When cordierite is used as ceramic core, the TCE of the finished laminate falls around 4 ppm/°C, which is close to silicon.

Mechanical processability is also good because it employs porous ceramics; therefore, normal PCB process (such as dry drilling) is applicable. Electrical performance is mainly determined by the impregnated resin. It is not quite suitable material for BGA applications due to the thermal mismatching with PWB, the same reason as aramid epoxy.

3.4 Reliability as Package Substrate

Reliability under humid conditions is another weak point of plastic material, in addition to high-temperature resistance. The severest test for humidity reliability is the PCT (Pressure Cooker Test). Figure 3.8 shows the reduction of the surface resistance of patterned test coupons along the PCT.

More PCT reliability data after IC is assembled on substrate is indicated in Fig. 3.10. BT material has more failures than epoxy or modified imide.

Comparing the surface traces of a package substrate uncovered with epoxy solder mask with a covered one shows that the uncovered one has a higher failure rate. The reliability improvement with solder mask coating should be based on the better bonding between solder mask surface and encapsulation resin instead of between metal traces and encapsulation resin.

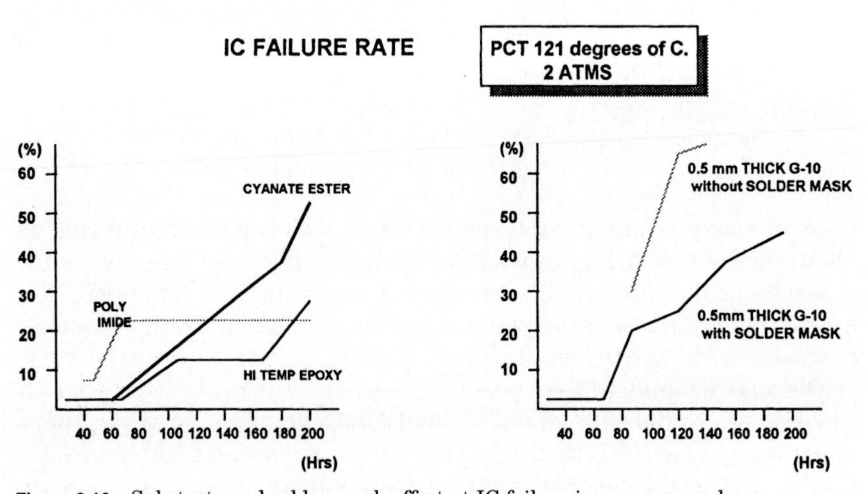

Figure 3.10 Substrate and solder mask effect at IC failure in pressure cooker.

The amount of extracted ions from several different substrates stored in the PCT chamber is shown on Table 3.3. Ion extraction data varies due to measurement conditions, so the value shown in Table 3.3 is not an absolute number—just for reference. BT and modified imide have more free chloride than epoxy, in general. Chloride exists in epichlorohydrine of epoxy resin and there is chloride which does and does not relate to coupling. It is considered that extracted chloride is dissolved chloride that does not relate to coupling.

PCT reliability may not directly represent the environmental reliability in actual usage; however, it is often used to check the package reliability quickly because it is an extremely accelerated condition.

3.5 Fabrication Process

3.5.1 Typical fabrication processes

The process of BGA substrate is almost identical to conventional plastic package substrate or printed wiring board (PWB). Currently, the majority of the BGA package is a simple double-sided single-chip structure; however, in the future, multilayer BGA or BGA-MCM should become very popular to meet high pin count and better electrical performance requirements; thus, the BGA substrate should become a higher-density design. The fabrication process has to be changed to keep up with such specification changing.

Subtractive process is today's most popular plastic package and PWB fabrication process, and it is a well-matured technology. On the other hand, to meet to the future high pin count and MCM requirements several additive processes which enable better pattern geometry limit have been developed and introduced to the industry.

3.5.2 Subtractive process

There are two types of subtractive process, as indicated in Fig. 3.11.

The panel plating method is a simpler process and easy for controlling etching accuracy, so it is a less expensive and effective process for the line and space as small as 0.1 mm. However, thick copper plating creates a wide range of copper thickness variation within a panel, so that it is difficult to control etching for patterns smaller than 0.1 mm.

TABLE 3.3 Ion Extraction Data (Unit: ppm)

Substrate	BT Material		Modified Polyimide			Epoxy		
Ion	Tg = 150°C	Tg = 180°C	Vendor A	Vendor B	Vendor C	FR-4	FR-5	Hi-Temp
Na^+	0.09	0.04	0.01	0.03	0.01	0.05	0.09	0.03
Cl–	0.14	0.15	0.14	0.11	0.20	0.07	0.10	0.23
NO_3–	0.21	0.24	—	0.10	—	0.04	0.04	—

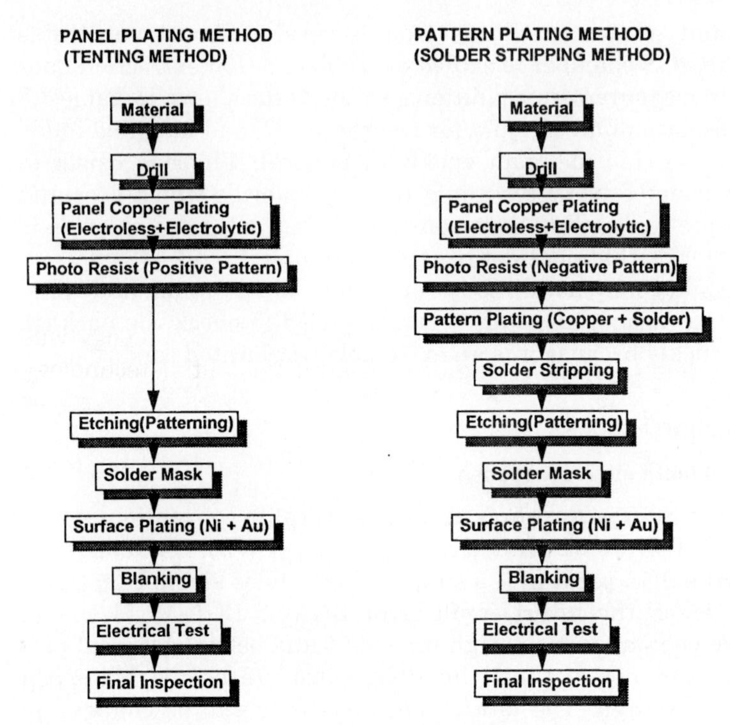

Figure 3.11 Substractive process flowchart (double-sided board).

Although pattern plating is a little more complicated and costs more, it enables thinner panel plating to be applied; therefore, total fine pattern capability is better than the panel plating method, and this method is the most common process in the PWB industry. Most single-chip BGA substrates are designed at 0.1-mm (or wider) traces today, so the subtractive process is a good choice for such applications.

Another factor to limit pattern density is through hole and its pad. In BGA substrate, it is easier to reduce the drill diameter because the board thickness is relatively thinner (0.2 to 0.5 mm) than standard PWB. Minimum drill diameter is 0.25 to 0.3 mm, and pad diameter is 0.5 to 0.6 mm. Smaller-diameter through-hole reliability is worse than bigger, as indicated in Fig. 3.12, but 0.25 mm diameter is considered to be reliable for most applications. Pad diameter is generally larger than drill diameter by 0.2 mm or so to compensate for the drill misregistration, laminate expansion/shrinkage, and photo tool expansion/shrinkage. For a through-hole pad of inner layers in a multilayer structure, it is more difficult to control the positional relationship between drill hole and pad because inner-layer pads are not visible from the outside, so the pad diameter needs to be 0.5 to 0.8 mm larger than the drill diameter. Clearance for the inner layer has an additional limiting factor. That is, there is a risk of microcracking in resin during the drilling operation, and copper may penetrate into such microcracks during the copper plating operation. In this case, electrical resistivity goes down, or sometimes it results in a short circuit. To minimize such electrical problems, standard design rule for antipad is 0.7 to 1.0 mm larger than drill diameter.

3.5.3 Additive process

Additive process is not a very new concept, as CC-4 technology was developed by Photo Circuit a long time ago. Early additive technology had a problem in electrical resistivity due to synthetic rubber content in adhesive layers and also had difficulties in fine line metallization due to the limited masking material availability against strong alkaline plating bath.

But, recently, some reliable additive processes have been introduced and developed for fine pitch applications. Figure 3.13 is an example of an additive process. This process does not employ synthetic rubber particles in the adhesive layer, and that improves insulation resistance under high-humidity conditions.

Minimum line and space width with the additive process is 0.050 mm at the laboratory level, and it is better than the subtractive process

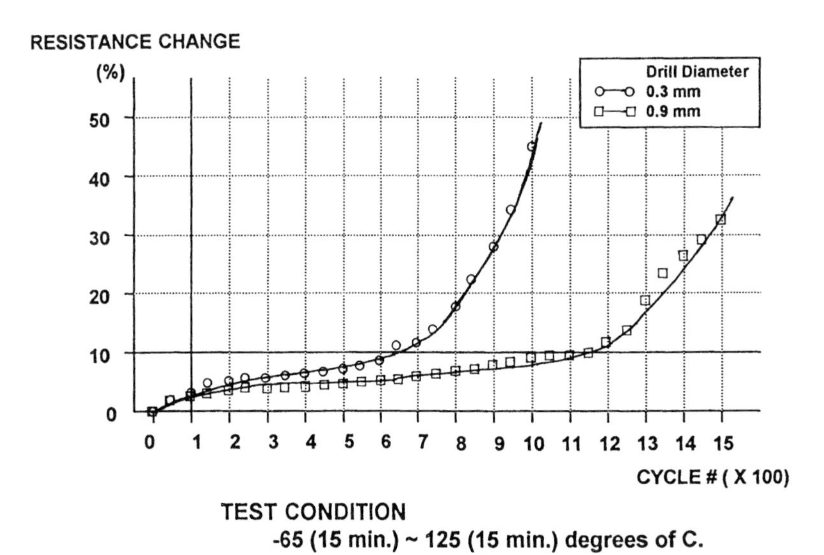

Figure 3.12 Through-hole reliability data.

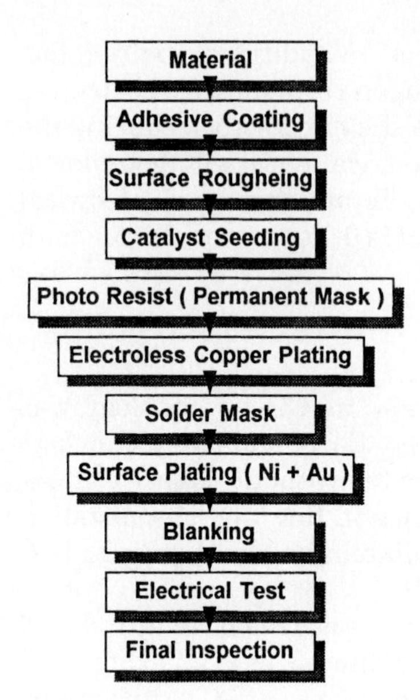

Figure 3.13 Additive process flowchart (double-sided board).

by far. The importance and necessity of the additive process is getting more significant because the technology trend is heading toward the higher-density assembly.

If conventional drilling is applied to additive process, then the same design rule and same limitation for drill, pad, and clearance diameter as subtractive is still applicable. To go around this limit, there is a process to make multilayer packages without using conventional lamination press and drilling methods. The process consists of (1) dielectric layer coating; (2) via hole formation with photolithography or laser cutting; (3) patterning with additive or sometimes subtractive method. (See Fig. 3.14.) One example of this type of process is SLC by IBM. There are other names for this process, such as *build-up process, sequential process,* or *photo via process,* and all are basically identical. Minimum line width and space is expected to be 0.025 mm and minimum via hole diameter and pad diameter is as small as 0.1 and 0.125 mm in the near future.

3.6 Plastic Substrate for Direct Chip Attach

3.6.1 Historical requirements

One of the oldest plastic package substrate examples was the LED module. The substrate was requested to be black to minimize the reflection from the substrate when the LED is on. So the black epoxy

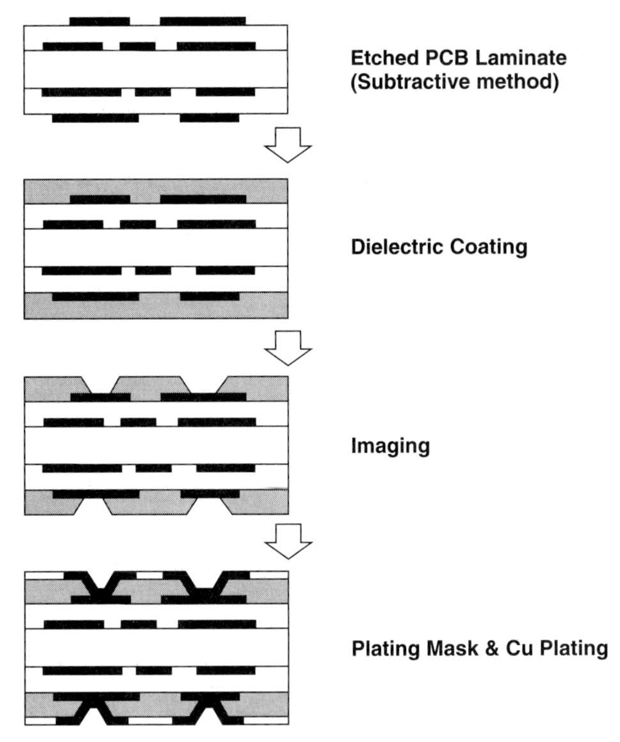

Figure 3.14 Process for build-up.

and black cyanate ester material became very popular for COB applications and is now widely used even for non-LED-type applications. Also, black material, when it is kept at die attach and wire bonding temperature around 150°C for a while, does not get discolored; natural color epoxy (light green) would be browned on the contrary.

Thin COB substrate was also developed for smaller-size packages to contribute to the miniaturization of consumer electronics applications such as a watch, camera, and calculator. Typical thickness of these COB substrates is 0.2 to 0.5 mm versus 1.2 to 1.6 mm conventional PWB.

3.6.2 Requirements for direct chip attach

3.6.2.1 Wire bonding. In the case of thermosonic gold wire bonding, the substrate is usually heated to around 150°C; therefore, the substrate that doesn't get softened at such a high temperature is preferred. BGA substrate is overmolded in most applications now, and mold temperature is also around 150°C. So the high-temperature resistivity is important. Cyanate ester or BT material just fits such high-temperature requirements.

Before the wire bonding process, a chip has to be fixed on the die attach pad of the substrate. It is typically done by dispensing die attach material that is an adhesive such as silver-filled epoxy, putting the chip over it and curing it in an oven. Thermal mismatching between the chip and plastic substrate is fairly large; therefore, stress relief or absorption by the die attach material is the key to obtaining a highly reliable package.

3.6.2.2 Flip chip bonding. It cannot be said that flip chip technology is widely available or established at this time; however, it has great potential to become a major interconnecting method in the future because of the following merits:

1. Easy to achieve high pin count IC because there's almost no limitation in the pad or terminal locations on the IC
2. Reworkable when defect is found
3. Lower inductance because of a shorter path for the interconnection.

Flip chip bonding to plastic substrate has some issues to be solved. First of all, there is thermal mismatching between chip and substrate. This can be compensated for or absorbed by choosing proper underfill material (usually epoxy resin) or by designing solder bump dimensions and material composition properly. Of course, lower TCE substrate is another option to compensate for the TCE mismatching. But if it is chosen, another thermal mismatch problem between package and PWB arises. Currently, many studies have chosen standard TCE substrate and epoxy underfill and confirmed the reliability up to certain-size chip, e.g., 9 mm a side.

The next issue is, since flip chip is a high-density interconnecting method, signal trace routing on the package should also be high density. Additive or build-up process will soon be essential for this application.

The next is, although the reworkability is one of the important merits of flip chip, repeated reworks may cause a problem to the substrate. Generally, the reflow temperature is somewhere between 230 and 260°C, so high-temperature substrate is, of course, desirable. Not only that, but also improving production yield to minimize the number of rework times is the key to minimizing the damage on the substrate.

Metallization specification for the flip chip pad has not been standardized yet. Typical specification is electrolytic tin lead plating on the pad for high-temperature solder bumped IC.

The combination of BGA and flip chip is a hopeful package for future high pin count and high-speed applications.

3.7 Plastic Substrate as BGA Substrate

3.7.1 Typical structure

In Fig. 3.1 typical single-layer and multilayer structures are illustrated. A single-layer package is encapsulated with overmolding in most cases. In the multilayer structure, ground, or power plane may be added for the inner layers so that electrical performance can be improved.

Most of MCMs require multilayer structure anyway, because of the high interconnecting density. Six layers (or higher) layer count is possible if the higher-density interconnection is required; however, increasing the layer count results in the thickness of the substrate increasing, and it may not fit the total miniaturization requirement, such as low profile electronic module.

Both cavity-up and cavity-down options are feasible and, in general, the plastic package has fairly good design flexibility.

3.7.2 Thermal enhanced structure

One of the plastic package's weak points is thermal dissipation. But adding a copper heat slug can help to improve the heat dissipation. Copper-slug-type plastic package has about the same or even better heat dissipation property than that of ceramic package.

Table 3.4 shows a heat dissipation (thermal resistance) comparison with and without heat slug for plastic BGA package.

Thermal via option also enhances the heat dissipation from the package, and the magnitude of the heat dissipation effect is in between no slug design and with-slug type.

3.8 Material Stability

BGA substrate is usually as thin as 0.2 to 0.5 mm, and easily gets warped from thermal stress. Once it is warped, solder balls will not

TABLE 3.4 Heat Dissipation (Thermal Resistance) Comparison (Unit:°C/W)

	Air Flow Speed (m/s)		
	0	1.5	3.0
Without Cu Slug	50	30	26
With Cu Slug	27	25	14

Sample: Package size: 35.56 × 35.56 mm
Slug size: 18.0 × 18.0 mm

stay on the same plane, and this will create problems to the assembly process as well as reliability. The checkpoints to minimize the risk of warp in BGA substrate design are as follows:

1. Select proper thickness of substrate according to the encapsulation method.
2. Design the package symmetrically with reference to the center plane.
3. Balance the remaining copper amount with reference to the center plane.

Also, during the substrate fabrication process, careful material selection (choose less shrink material or shrink about the same for all the layers across) is important to avoid a board-warping problem.

3.9 Solder Mask Material

Solder mask material selection is critical in BGA package because it affects the reliability, especially at the boundary area of solder mask surface and mold resin, if the BGA is overmolded. If the adhesion strength between solder mask and mold resin is not very good at the boundary, the package tends to have delamination or to be water-trapped. Most of solder mask material has good wetability with mold resin, but with some solder mask types, surface roughening with, for example, plasma etching is necessary to improve the adhesion.

In BGA packages, solder mask positional tolerance is tighter than standard plastic packages; therefore, it is almost essential to choose photo imagable solder mask whether dry film or wet type. If the via hole coverage is required, dry film solder mask provides a perfect solution, but mostly the wet type is used in the industry because it has better reliability against humidity and costs less. There are also alternatives to cover or fill via holes with resin materials other than solder mask.

3.10 Summary

3.10.1 Current requirements

BGA development is still a relatively new activity. And, currently, plastic substrate has been chosen in some applications for classical reasons such as lower cost, easy handling, light weight, or good mechanical strength and dimensional stability. High-temperature resistivity, thermal expansion mismatching between IC chip and plastic substrate, and thermal resistance are issues for plastic substrate, but there are relatively easy solutions for them; at this point, these problems are small compared to the merits. Many simple applications will be developed in next few years, and plastic BGA will establish a certain position as an option of the IC package family.

3.10.2 Future requirements

Those classical reasons (merits) are still applicable and true for future applications as well, for some extent. But when IC speeds keep going up and lead count increases, requirements for plastic BGA substrate will be tougher and more demanding. And plastic substrates will play an increasingly important part in the IC packaging field.

The reasons are: (1) plastic substrate has great potential in high-speed IC packaging because of better electrical performance than alternative materials; and (2) plastic substrate has solutions for higher-density interconnecting or routing at reasonable cost.

Actually, some complicated multilayer plastic BGA developments have been started, and those types of more complicated, high-performance packages will be more popular in the near future. To achieve such high-performance BGA packages, most of the sophisticated technologies described in this chapter should be incorporated into BGA production. Of course, there may be some unexpected hazards or obstacles in the way, but there must be solutions to overcome them. The more development activities go on, the more effective solutions for better-performance plastic BGA packages will be available.

3.11 References

1. Rai, A., Doita, Y., Nukii, T., and Ohnishi, T., "Flip-Chip COB Technology on PWB," *IMC 1992 Proceedings,* pp. 144–149.
2. Berg, K., "The Sequential Process Advantage in MCM-L Construction," *Proceedings of 1993 International Conference and Exhibition on Multichip Modules,* pp. 190–199.
3. Freyman, B., "Plastic BGA Overview," *Workshop on Ball Grid Arrays & Advanced Packaging,* August 1993.
4. Tsukada, Y., and Tsuchida, S., "Surface Laminar Circuit, A Low Cost High Density Printed Circuit Board," *Surface Mount International,* September, 1992, pp. 537–542.

Chapter

4

Printed Circuit Board Routing Considerations for Ball Grid Array Packages

Patrick Hession

4.1 Introduction

Ball grid arrays (BGAs) represent a major innovation in component packaging and a significant challenge to printed circuit board (PCB) design. The advantages in assembly and packaging with BGAs can be offset by the difficulties created in PCB routing. The BGA package needs to be carefully chosen with PCB routing in mind. The wrong BGA package selection can result in routing complications that increase the PCB layer count or reduce the manufacturability of the bare PCB. However, BGAs can often be used effectively with minimal impact to the cost and producibility of the bare PCB if the packaging decision considers the impact to PCB layout. In many cases the BGA can even increase the producibility of the bare PCB compared to other peripheral leaded packages like quad flat packs (QFP).

The use of a BGA component will have an impact on two key aspects of the PCB design; PCB "real estate" and PCB layer count. As a grid array package, the BGA offers a high pin count solution in a minimum amount of PCB surface area. The BGA package and required fan-out area are significantly smaller than that required for a peripherally leaded package like TAB or QFP. However, because the BGA has such a high pin count density compared to peripherally leaded devices, the BGA will often increase the PCB layer count required to effectively route the interconnects using the same density and PCB technology.

4.2 PCB Area Requirements

Understanding the impact on PCB board area requirements for a BGA package is simple. As an array package, the pins are densely packaged under the device rather than around the perimeter. The number of pins available on a BGA is a function of the package area and the ball pitch. With peripheral leaded devices, the pin count capability is a function of the circumference of the package and the pin pitch. As a result, peripheral leaded packages must grow in size faster to accommodate increased pin counts compared to an array package.

In addition to the component footprint, every type of package requires extra vias to interconnect the device pins to the rest of the PCB. These vias are typically referred to as fanout vias. With a BGA device, it is possible to fan out the pins to vias entirely underneath the package. Peripheral leaded devices require additional board area to accomplish the same fanout. Due to the fine pitch of the leads on these devices and currently manufacturable PCB via pitches, a significant portion of the fanout must be done outside the package outline.

Figure 4.1 shows a comparison of PCB surface area required for various package types. For example, a 225-pin BGA package requires 1.1 square inches of board area for the package. A 208-pin QFP requires 1.5 square inches. The BGA package itself is about 25 percent smaller than the QFP at those pin counts. Figure 4.2 shows the board real estate required for the component package and a typical fanout area including vias. The 225-pin BGA requires 1.1 square inches of board real estate compared with 2.3 square inches for the 208-pin QFP fanout. The BGA results in a 50 percent board real estate savings. To accomplish the fanout in these areas for both package types, the amount of available area on the opposite side of the PCB for component placement will be reduced due to the high density of via pads required.

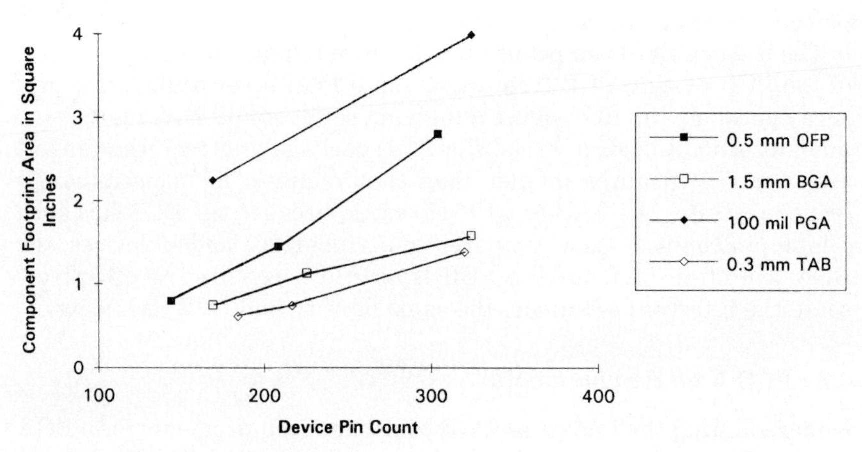

Figure 4.1 Component footprint area.

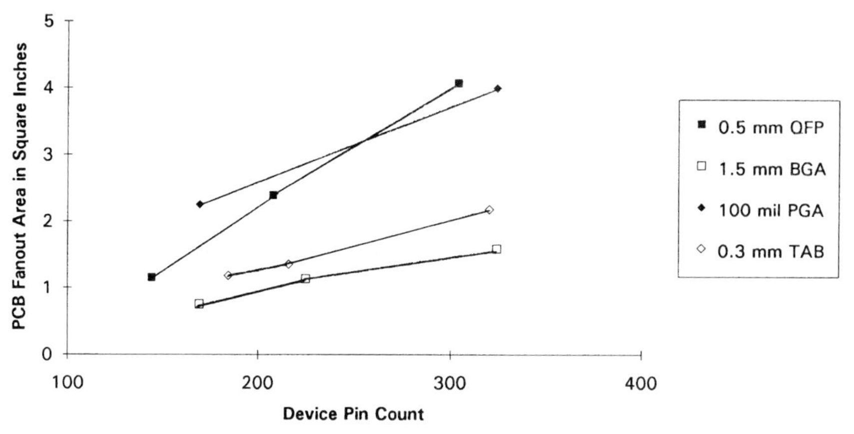

Figure 4.2 Typical component fanout areas.

4.3 Routing Complexity

The PCB surface area requirement is only one of the parameters to consider for PCB routing with a BGA. The device pin density and PCB routing grid chosen are also critical to determining the level of routing complexity. Figure 4.3 shows the pin density for common package types. Intuition tells us that more pins in less area will require either finer PCB geometry or more layers to effectively connect all of the pins. Experience with BGA routing confirms this intuition.

Thad McMillan in his presentation to the IPC entitled "The Effect of High Density Components on the Bare PCB"[1] reported on a technique he developed to quantitatively predict the PCB routing complexity based on pin density and routing technology. McMillan referred to his

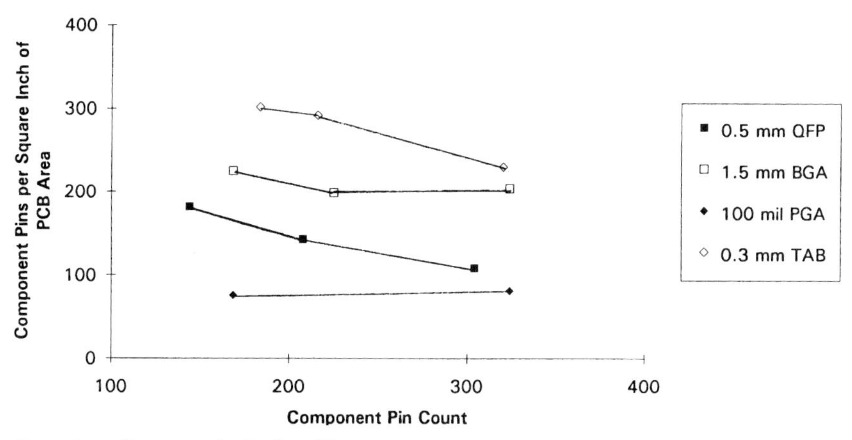

Figure 4.3 Component pin densities.

model as the *layer factor*. The layer factor compares the number of vias required for a package to the number of routing channels available for the package on a single PCB layer. McMillan's layer factor is a tool for comparing the relative complexity of various packages for PCB routing. The formula for generating the layer factor is shown below.

$$\text{The layer factor } (L) = \frac{(N/4) - O}{R}$$

where N = number of pins in the device
O = number of via grid points along one side of the device
R = number of routing channels per layer on one side of the device

Figure 4.4 shows McMillan's data for layer factors of various package types on typical via routing grids. His data clearly shows that a BGA can present the need for more layers than peripheral leaded packages like TAB and QFP. For example, the layer factor for a 208-pin, 0.5-mm pitch QFP is less than 0.5. A comparable pin count BGA layer factor is approximately 1.0. This would suggest that the BGA would generally take twice as many layers to connect using the same routing technology and via grid. This may not mean that in every design the signal layer count will double when a BGA is used to replace a QFP. However, it does mean that the density of routing in the BGA area may double and force an increase in the layer count depending on the amount of available routing channels.

A similar approach to quantifying the density of routing for various packages was used in a paper entitled "A Feasibility Study of Ball Grid Array Packaging"[2] by Randy Johnson et al. presented at Nepcon East in 1993. Johnson's paper used an indices called the route factor. The route factor attempted to draw general conclusions about the number

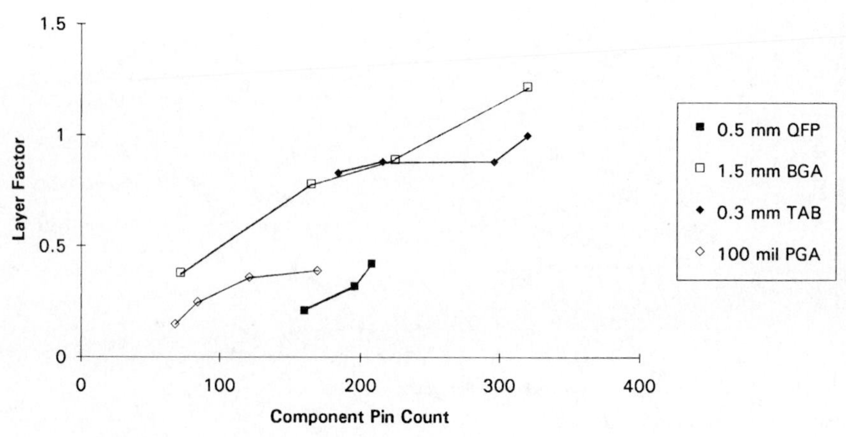

Figure 4.4 Layer factors for various packages.

of layers required for various packages and routing technologies. Johnson's data is shown in Figs. 4.5 and 4.6. Again the density data clearly shows that a fully populated array will be more difficult to route than conventional peripheral packages. However, Johnson's data did show that a depopulated grid on the device could be used to reduce the complexity and required layer count when using a BGA. Johnson also reported that proper management of the location of power and ground pins on a BGA could also ease the difficulty of routing out from the device. Johnson recommended grouping the power and ground pins at the center of the array. This would allow the center pins to be connected directly to a plane layer by a via. This technique clearly decreases the number of signals that must be brought out from the center of the device for connection.

4.3.1 1.5-mm-pitch BGA routing

The most common BGA format to date has been the 1.5-mm-pitch BGA. The 225-pin version of this package has been supplied for production volumes to a limited number of end users and is under development by many packaging houses. The 225-pin BGA offers an alternative package to the common 208-pin, 0.5-mm-pitch QFP. The 208-pin QFP at 0.5-mm pitch offers direct access to all pins for routing on the top surface of the board. This creates the ability to route out, or in, from the device pin to a via for connection to a different layer. The 208-pin QFP can be routed with 8-mil lines and spaces or less.

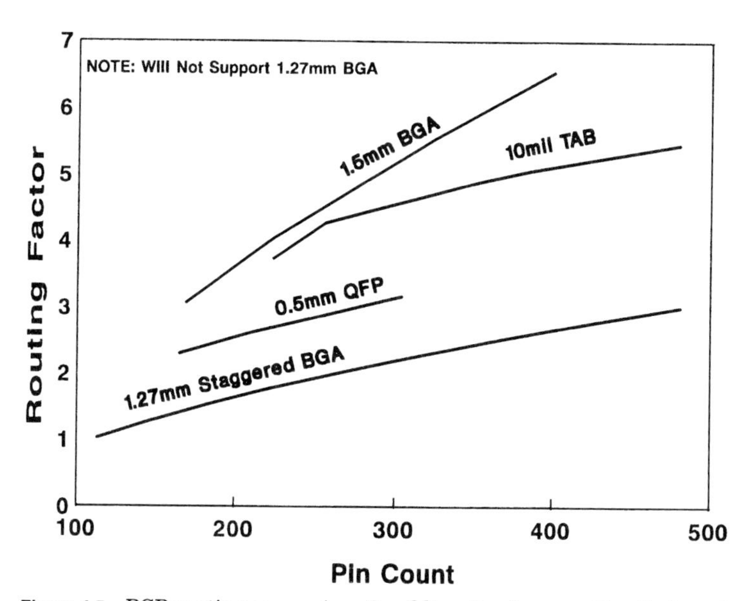

Figure 4.5 PCB routing comparison 8-mil line, 7-mil space, 32-mil via pad.

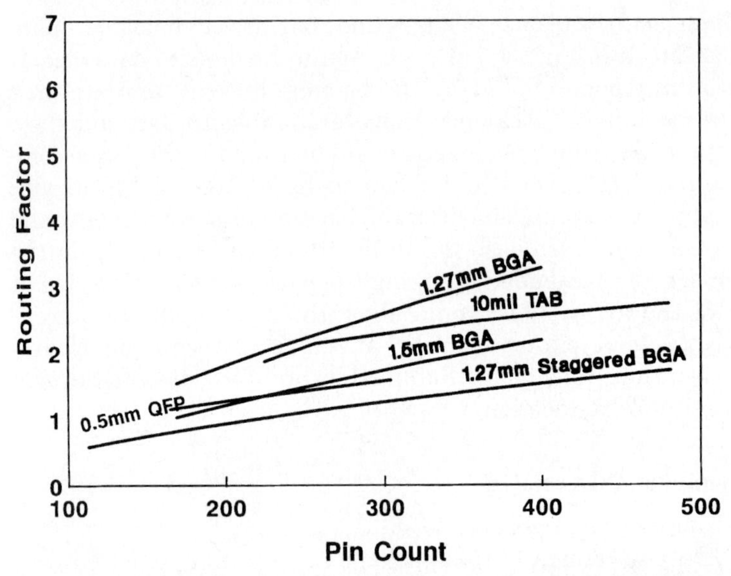

Figure 4.6 PCB routing comparison 5-mil line, 5-mil space, 24-mil via pad.

The 225-pin BGA does not provide direct access to all pins on the top layer with conventional routing technologies. A 1.5-mm ball pitch translates approximately into a 59-mil pitch. Assuming a typical land for solder attachment of the balls is 25 mils in diameter, that leaves a 34-mil channel for routing. Eight-mil lines would allow only single-channel routing, or access to the outer two rows of pins. This would leave 121 pins to be connected through vias to other layers. Six-mil lines and spaces would allow two-channel routing, or access to the outer three rows directly. This would leave 81 pins to be connected through other layers. To achieve greater than two-channel routing through the BGA pads, 4-mil lines would be required. While this technology is starting to become commercially viable in the PCB industry, its use is still limited and its cost can be prohibitive. Figure 4.7 shows a corner of the 225-pin BGA footprint with two-channel routing depicted.

The remaining center pins of a BGA that are unconnected on the surface layer will have to be routed to a via and connected to another layer for routing. As Johnson suggested in his paper, if the power and ground pins can be concentrated in this location of the package, they could be directly connected to the power or ground layer through a via. This would reduce the number of unconnected pins by the number of power and grounds in the package. The remaining pins can be routed out on a different layer. If the vias are placed interstitially to the BGA lands, then they will be on a similar 1.5-mm (59-mil) grid. The size of the via and land must be chosen to allow the remaining balls to be routed through the available channels successfully. If single-channel routing is

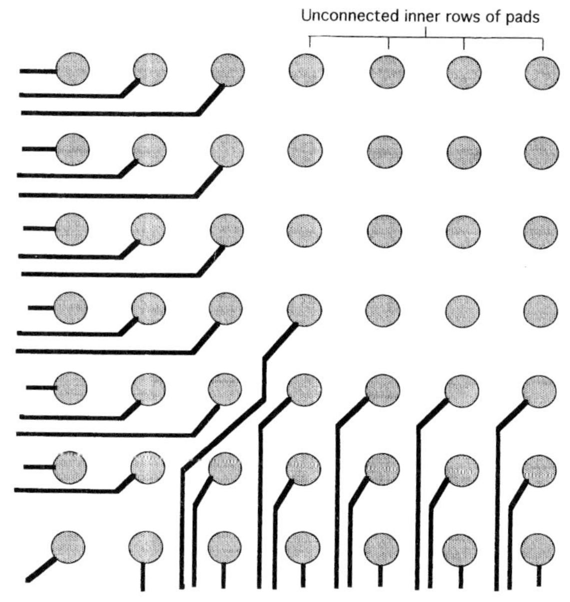

Figure 4.7 Routing from the footprint of one corner of a 225-pin, 1.5-mm BGA. 49 pads from the corner of a 1.5-mm-pitch BGA footprint showing several inner rows of pads that remain unnconnected when using two-channel routing on a single PCB layer.

all that is required to connect the remaining signals, then the pad can be as large as 35 mils in diameter. If two-channel routing is needed, then the pad will have to be less than 29 mils using 6-mil lines and spaces. Typical manufacturing tolerances for plated holes currently require that the via pad diameter be approximately 15 mils larger than the drill size to ensure tangency of the hole to the pad. Therefore, for a 29-mil pad, the drilled via hole size would have to be 14 mils or less. Figure 4.8 shows a via pad placed interstitially to four BGA lands.

Assuming a typical power and ground pin count is about 10–15 percent of the total pins, theoretically it is possible to route out of a 225-pin BGA on two signal layers. In practice, this is not typically the case. Experience has shown that using 6-mil lines, 6-mil spaces, and 29-mil via land sizes does not even guarantee that routing in four signal layers will be easy. As always, consideration of the complexity of the remaining design and where the signals from the BGA need to terminate is needed. Figure 4.9 shows the relative impact to PCB cost with increased layer counts and finer PCB geometry. In general, using a 1.5-mm, 225-pin BGA will typically require 6-mil line spaces maximum, 29-mil pads maximum, and four signal routing layers (a common six-layer board with two plane layers) to effectively connect to a design with a medium level of density.

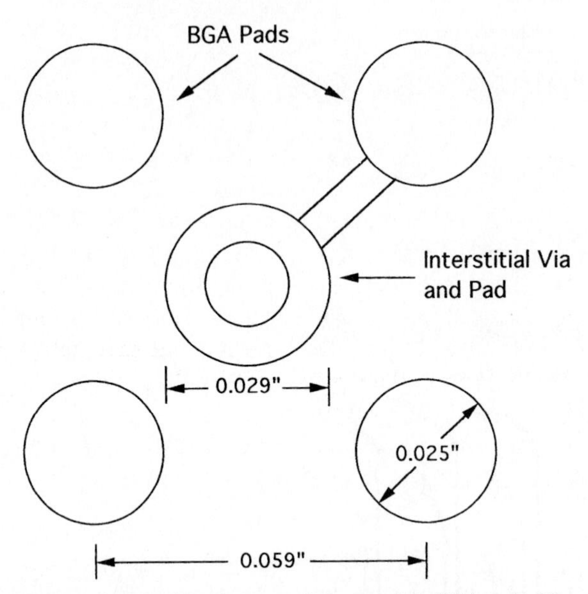

Figure 4.8 29-mil via between 1.5-mm-pitch BGA pads. Via and pad located interstitially between four BGA pads.

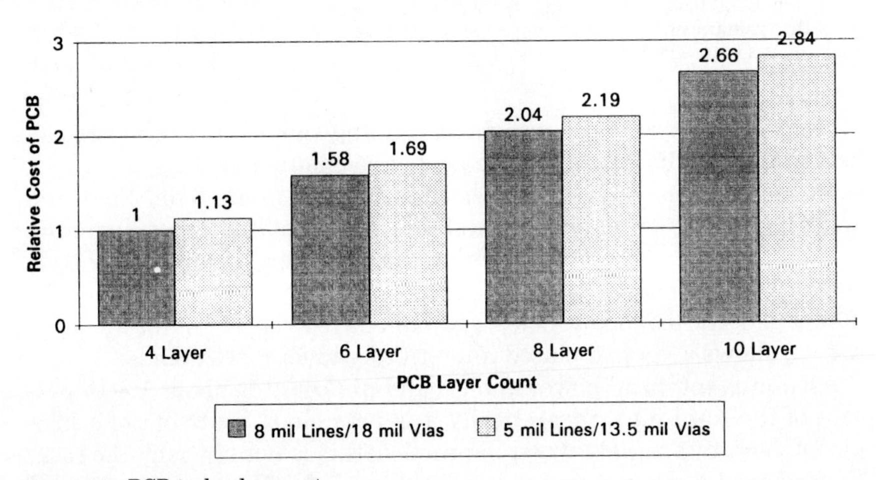

Figure 4.9 PCB technology costs.

4.3.2 1.27-mm-pitch BGA routing

BGA components with 1.27-mm pitch add a greater level of complexity compared to 1.5-mm pitch components. The challenge of routing from an array of pins remains, however the width of the available routing channels has been reduced. The same 25-mil BGA land will result in a routing channel of only 25 mils. This channel will barely support single-channel routing with 8-mil lines and spaces. Two-channel routing can be obtained with 5-mil lines and spaces; however, you will have the minimum space of 5 mils between the BGA land and an adjacent trace. Three-channel routing could only be completed with 3-mil line technology. This level of technology is not readily available today, particularly on the outer layers of the PCB.

A 225-pin count BGA using a 1.27-mm pitch would require 5-mil line and space technology to route on four signal layers. Via land sizes must also be reduced. The via land must be no greater than 25 mils in diameter in order to support two-channel routing on the inner layers using 5-mil lines. It is important to note that, using the normal drill tolerances stated previously, the hole must be drilled at 10 mils in diameter to support tangency with this size pad. With a larger drilled hole size, breakout of the hole from the pad may result. This will decrease the space between the hole and the adjacent trace below the minimum design requirement. Figure 4.10 demonstrates the effect of breakout on the spacing to an adjacent trace. Ten-mil diameter drill bits can be three to four times the cost of larger drill sizes today. Drill throughput is also reduced due to the shorter drill flute length requiring reduced PCB stack heights on the drill machine. Generally, 13.5-mil drill bits are considered the cost-effective limit of today's commercial PCB technology.

It is technically feasible to route a 225-pin, 1.27-mm pitch BGA on two signal layers depending on the number of power and ground pins. Four-signal layers are more likely to be required by most designs due

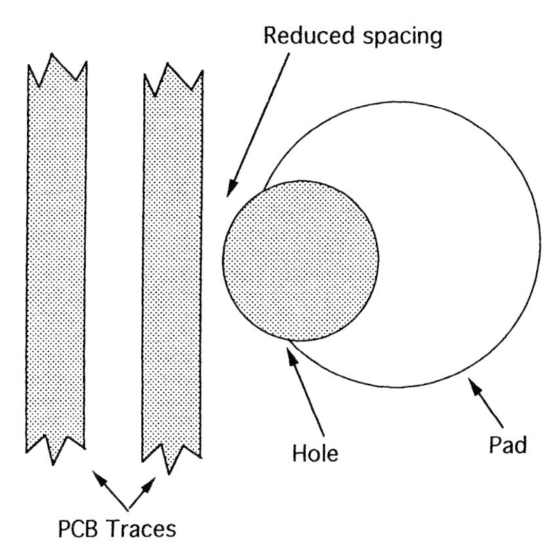

Figure 4.10 Effect of breakout on minimum spacing. Mislocated PCB hole (breakout) causing reduced spacing to the adjacent circuit trace.

to other constraints in the layout. A design tradeoff will have to be considered for high-pin-count, 1.27-mm pitch BGAs. The tradeoff should examine increased layer counts due to single-channel routing on internal layers, smaller drilled holes to accommodate two-channel routing, and reduced spacing between the hole and adjacent traces (breakout). In high-volume, low-cost applications, this is not a tradeoff that a designer should be forced to make. In lower pin counts, these tradeoffs may not be required if all of the signal pins can be routed out without the use of two-channel routing.

4.3.3 1.0-mm-pitch BGA routing

As the reader would expect, a 1.0-mm BGA is very difficult to route out of on a PCB. At 39-mil centers, 25-mil pads would only leave 14-mil-wide routing channels. This would require the use of either smaller BGA lands or 4-mil lines for single channel routing. The result would be very high layer counts, small via lands, very small drilled holes, and a high level of power and ground plane perforation. In most applications today, a fully populated 1.0-mm array is impossible to route cost effectively, except for very low pin counts. Figure 4.11 outlines the

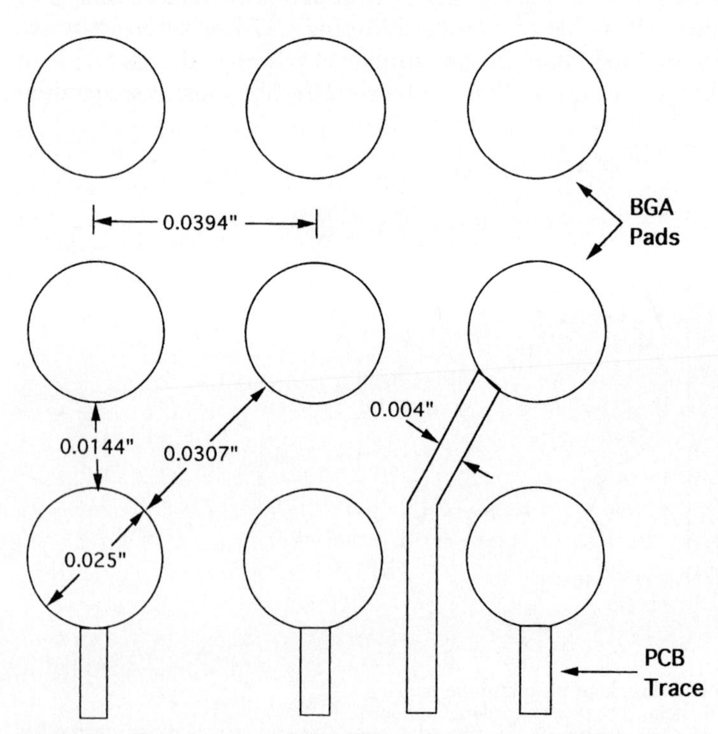

Figure 4.11 Typical 1.0-mm BGA PCB geometry. Typical BGA pad and trace geometries on a PCB for 1.0-mm-pitch BGAs.

spacing available for routing traces and placing via lands for this type of geometry. It is not recommended to attempt to place a 1.0-mm BGA array on a conventional technology PCB. A 1.0-mm-pitch BGA device would be a better routing match with PCB technology consisting of 3-mil lines, buried and blind vias, microvias, or landless vias. These types of PCB technologies can add significant costs to the bare PCB, and not all of these technologies are readily available in volume today.

4.4 Depopulated Arrays

Fully populated BGA packages clearly present some significant routing challenges. It will be very difficult to use 1.27- and 1.0-mm arrays with commercial PCB technologies. It should be easy to adapt 1.5-mm-pitch arrays to six-layer (and greater) PCB designs. However, four-layer PCB designs are predominant in many types of commercial electronic systems. High-pin-count, fully populated BGA arrays do not lend themselves easily to this level of technology. However, several component packaging houses are beginning to offer depopulated BGA arrays as a solution to the routing challenges described above. Two depopulated alternatives to consider are staggered grid and peripheral pad BGAs.

Staggered grid BGAs only use half of the available ball sites. Every other pad site is vacant, creating a surface mount footprint similar to the through hole footprint of staggered grid PGAs. Staggered grid BGAs can be made from any of the available packages; however, the predominant types to date appear to be depopulated 1.27-mm grids. Many packaging houses have selected the staggered grid approach for pin counts in excess of 225. Figure 4.12 shows a typical footprint for a staggered BGA. This example is a 313-pin device using a depopulated 1.27-mm pitch.

A staggered-grid, 1.27-mm-pitch package provides an effective minimum pitch of 70 mils along the diagonal of the package. If you assume a typical solder attach pad diameter of 25 mils, this would leave a routing channel that is 45 mils wide. This is sufficient for three-channel routing using 6-mil lines and spaces, or even four-channel routing with 5-mil lines. The interstitial via locations that are available provide generous room for most standard via land diameters. Figures 4.5 and 4.6 show the impact on routing with 1.27-mm staggered BGAs. The Route Factor is significantly reduced from those of 1.5-mm BGAs, and even lower compared to conventional 0.5 mm QFPs.

Peripheral pad BGAs offer a different approach to depopulating the array than staggered BGAs. A peripheral pad BGA depopulates the center of the device, leaving a fully populated array around the peripheral of the package. Figure 4.13 shows a peripheral pad BGA footprint for a 256-pin, 1.27-mm pitch package. This package is able to put 256 pins into the same body size of a 225-pin, 1.5-mm device, and is potentially routable

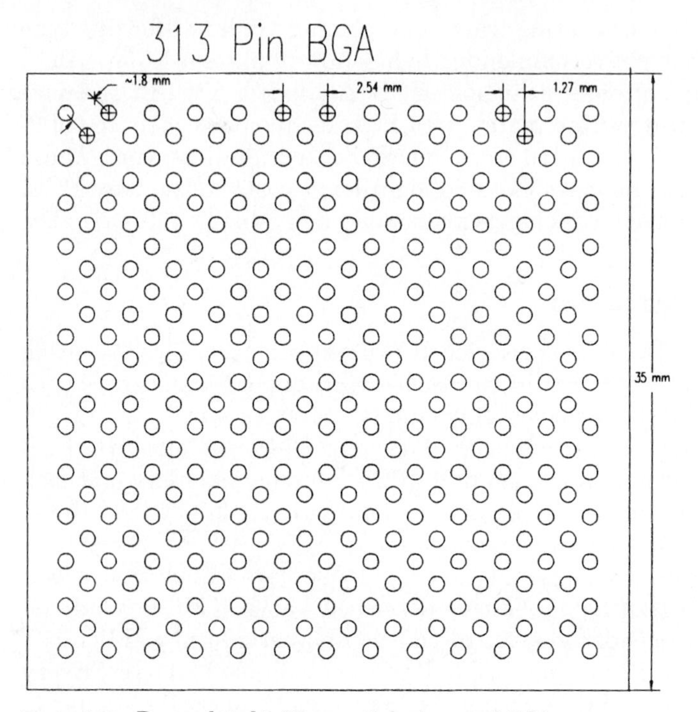

Figure 4.12 Depopulated 1.27-mm pitch staggered BGA.

on four-layer boards using two-signal layers. There does not appear to be any standard peripheral pad formats to date. Most current designs use a depopulated 1.27-mm pitch that is three to five rows deep. Depopulating the center sites eliminates those pins that are most difficult to route. The depopulated center also allows for some of the array pins to be fanned in to reduce the via pad congestion that typically happens inside the footprint of a BGA This will help reduce the power and ground plane perforation under the device caused by clearances around signal vias.

A 1.27-mm peripheral pad BGA allows for single channel routing with 6-mil lines and two-channel routing with 5-mil lines. If the array is only three or four rows deep, six-mil lines should allow the device to be routed effectively on a four-layer design with two-signal layers. If the array is five rows deep, the design may require 5-mil lines to route on two-signal layers. This would depend on the number of power and ground pins, their location, and the surrounding complexity of the PCB design.

Both depopulated array formats offer intelligent choices for higher pin count BGAs. Both depopulated array formats give significant improvements in routability. Depopulated BGAs can decrease the density of vias required for routing. This can increase the potential to place passive devices on the bottom layer directly under the BGA package. The periph-

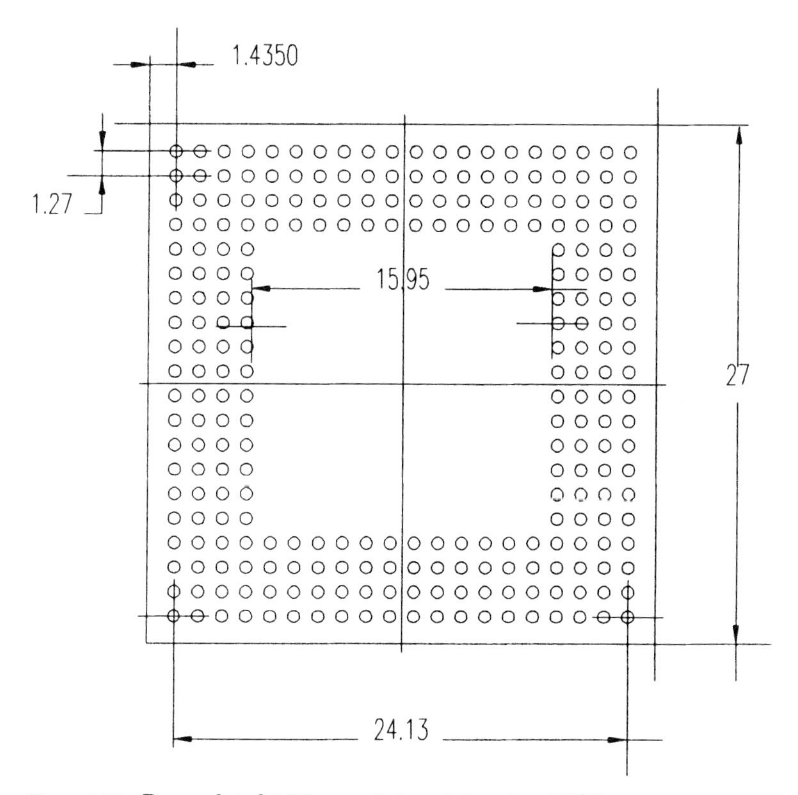

Figure 4.13 Depopulated 1.27-mm pitch peripheral pad BGA.

eral pad BGA approach appears to offer the highest pin densities combined with improved PCB routability. Both of these depopulated BGA formats should be considered in lower-layer-count, less complex designs.

4.5 BGA Impact on PCB Fabrication

The layer count and feature size of the PCB can be influenced by the use of a BGA device in the board design. One additional consideration when designing with a BGA package is the impact on other areas of the PCB fabrication process. The BGA device can potentially offer benefits in fabrication compared to other fine-pitch, peripheral-leaded devices. The increased PCB feature size of the array lands compared to fine-pitch-component PCB pads can increase the producibility of the board in several areas.

The only potential case for a BGA to require process enhancements at the fabrication house is in board test. If a single board design has several high-pin-count BGAs, the fabricator may have to upgrade their testing capability to support the higher pin densities that a BGA provides. However, this may also be true if the design uses several high-pin-count, peripheral-leaded devices. This requirement depends entirely on the age and capability of the equipment installed at the fabrication site.

In most cases, a BGA design will have less stringent requirements than a comparable design using fine-pitch QFPs or TAB. This is often true in regards to the solder surface finish requirements. Many assembly houses have implemented strict solder thickness and planarity requirements to support solder stencil printing of 0.5 mm QFPs. For TAB, where assembly houses want to avoid stencil printing altogether, the solder thickness requirements can force a fabricator to make significant process changes to avoid the use of hot-air solder leveling. A high-pin-count BGA should present no unusual requirements for solder finish. In this regard, the BGA should provide for a more producible bare PCB than fine-pitch packages. The main concern for a BGA should be only that the surface is solderable.

4.6 Summary

The use of BGA packages in a design can increase the level of complexity of that design. BGA packages have a very high pin density creating a challenge for the PCB layout. More pins have to be connected to the design in a smaller area, compared to conventional peripheral-leaded packages. It is important to understand the impact of this design challenge relative to the complexity of the design being considered. Failure to consider the BGA package type can result in an increase in the layer count of the design, or a decrease in the PCB geometry required to route the BGA device. Both of these results will significantly increase the cost of the bare PCB.

In general, higher-pin-count, fully populated 1.5-mm-pitch BGAs will require four-signal layers to connect to medium-density designs. Lower-pin-count, 1.5-mm-pitch BGAs should be routable in most designs. Fully populated 1.27-mm arrays should be used only after careful consideration of the impact on the design and the potential PCB cost implications. Fully populated 1.0-mm pitch BGAs should be avoided on conventional technology PCB designs. In all BGA designs, careful attention to the placement of power and ground pins can help simplify the PCB routing challenges. The center of the array is the best region to locate power and ground pins to minimize the difficulty of routing to these pins.

Depopulated array options are becoming more popular for high-pin-count BGA solutions. A staggered-grid BGA design helps to widen the routing channels that are available to connect to the internal BGA pins. A perimeter pad BGA helps to reduce the number of signals that must be connected through the available routing channels. With the depopulated arrays, it may be possible to connect a high-pin-count BGA to a four-layer design with two-signal layers, depending on the number of power and ground pins, their location, and the complexity of the design. Depopulated BGA options offer an excellent solution that provides most of the benefits of a BGA without the significant penalty to PCB routing.

In many cases, a BGA can be implemented into the PCB design with minimal impact to the cost of the PCB. It is imperative to understand the type of PCB geometry and layer counts that will be required to rout to the device before selecting the package. If the BGA package selection is carefully matched to not exceed the complexity of the PCB required for the remainder of the design elements, the BGA can have a positive impact on the overall PCB technology. The use of the BGA should increase the producibility of the solder surface finish due to less stringent height requirements compared to fine-pitch, peripheral devices. BGA packaging technology should also result in a more robust and cost-effective assembly process.

4.7 References

1. McMillan, Thad, "The Effect of High Density Components on the Bare PCB," Technology Roadblock Forum, *IPC 36th Annual Meeting,* San Francisco, calif., May 1993.
2. Johnson, Randy, et al., "A Feasibility Study of Ball Grid Array Packaging," *NEPCON East Proceedings,* June 1993.
3. Nelson, Brent, "Plastic Pad Array Carrier Routing Considerations," Presentation to the JEDEC BGA Task Group in Dallas, Tex., July 1992.

Chapter

5

An Overview of Ceramic Ball and Column Grid Array Packaging

Thomas Caulfield, Marie S. Cole, Frank Cappo, Jeff Zitz, and Joseph Benenati

5.1 Introduction

Starting in the late 1970s and through the 1980s surface mount technology had established itself as the lowest-cost, highest-function card-assembly format, well positioned for the product requirements of the 1990s. The arduous path of developing a widespread and universally accepted assembly technology infrastructure from tooling to assembly processes, including package formats, has been forged. The results of this effort are formidable, SMT continuously and routinely produces high assembly yields and reliable subassembly products. However, as ASIC and Microprocessor packaging requirements of the 1990s are driving interconnection-intense packages,[1] SMT has reached a critical cross-roads in defining a technology roadmap that will capture packages with a high number of interconnections (>200 I/O's).

Today, the industry routinely assembles 0.65-mm pitch perimeter-leaded packages, and when needed, struggles through 0.5-mm pitch configurations. To achieve higher interconnection counts, perimeter-leaded packages will need to migrate to finer pitches, e.g., 0.4-mm pitch going to 0.3-mm pitch configurations. For lower lead count applications, these fine pitches may be compatible with existing surface mount processes and tooling, but new invention and tooling will be needed for the high-lead-count fine-pitch (0.4- and 0.3-mm) packages. On the other hand, Ball Grid Array (BGA) packages allow for both high lead count and high interconnection density at relatively coarse pitches (ranging from 1.0 to 1.5 mm) in applications requiring greater than 200 interconnections. The interconnection density advantage of area array packages over leaded packages is shown in Fig. 5.1. Accordingly, to fulfill high interconnection requirements, the new standard package for the 1990s will be surface mount area array packages including Plastic Ball Grid Arrays (PBGA), Tape Ball Grid Arrays (TBGA), and Ceramic Ball and Column Grid Array Packages (CBGA and CCGA).

In this chapter, an overview of CBGA and CCGA technology will be presented, including descriptions of the interconnection structure, package assembly, card layout and assembly, electrical and thermal attributes, reliability, and example applications.

5.2 Technology Description and Overview

Ceramic Ball Grid Array (CBGA) technology makes possible the direct attachment to industry standard epoxy glass cards of multilayer ceramic (MLC) SCM and MCM packages with a high number of interconnections. This attachment is accomplished using standard surface mount processes. Ceramic Column Grid Array (CCGA) technology is an extension of CBGA where the solder balls are replaced by solder columns for improved thermal fatigue resistance.

The interconnection structure consists of either a high melting point solder ball or solder column, joined to the ceramic chip carrier and the epoxy glass card with Pb-Sn eutectic solder.[2] An alternative solder column structure joins the high melting point solder directly to the ceramic chip carrier. Figure 5.2 is a schematic cross section of CBGA and the two varieties of CCGA structures assembled to cards. Figure

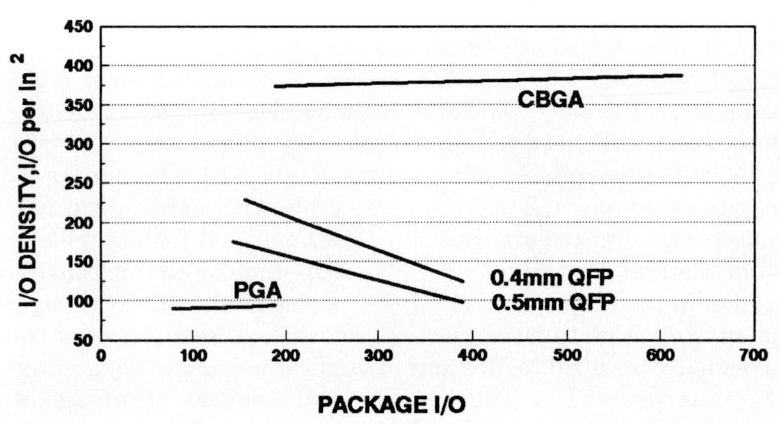

Figure 5.1 Packaging real estate advantage of area array over perimeter-leaded packages.

5.3 is a metallographic cross section of an assembled CBGA interconnection. The solder ball and column consist of a high melting point (>300°C) 90Pb-10Sn alloy, which does not reflow during package or card assembly. This hierarchical solder structure creates a consistent and reproducible gap or standoff between the ceramic package and the card. This standoff accommodates the mechanical strains generated during normal machine power cycling, caused by the thermal mismatch between the ceramic carrier and the card, and thereby provides the thermal fatigue reliability of the CBGA or CCGA assembly. In addition, the standoff also provides clearance for cleaning after card assembly, and accommodates the strains associated with card torquing and flexing resulting from card handling and box assembly.

Current applications of CBGA and CCGA structures are on a 1.27-mm (50-mil) grid array and assemble to cards using exactly the same standard SMT processes and card specifications. The CCGA and CBGA packages can be extended to higher interconnection densities by using a 1.00-mm (40-mil) grid which requires a smaller ball and column diameter. Conversely, CBGA and CCGA packages can also be configured in coarser pitches. The JEDEC standard pitches for CBGA and CCGA are 1.0, 1.27, and 1.5 mm, while body sizes (in both square and rectangular formats) range from 11 to 45 mm. Figure 5.4 contains a subset of the various JEDEC body sizes and I/O counts as a function of interconnection pitch for CBGA and CCGA packages.

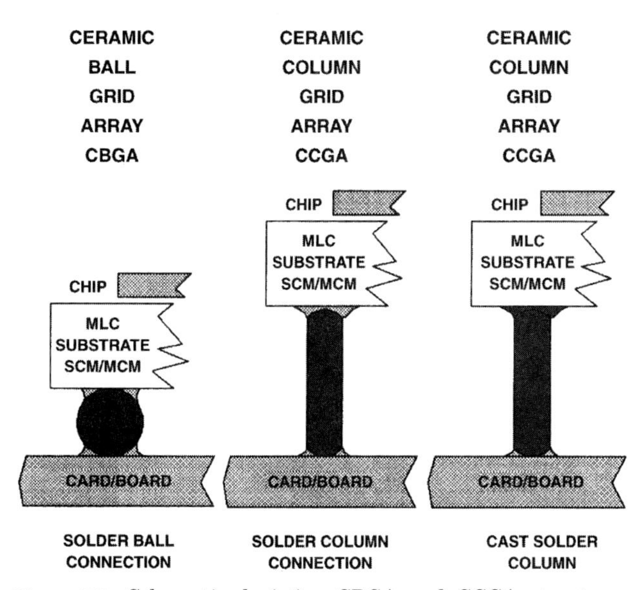

Figure 5.2 Schematic depicting CBGA and CCGA structures assembled to a card.

Figure 5.3 Metallographic cross section of an assembled CBGA interconnection.

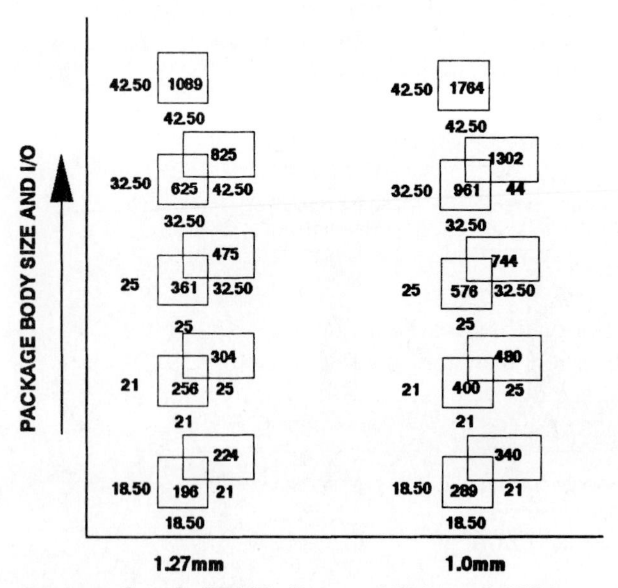

Figure 5.4 Sample JEDEC offerings of CBGA and CCGA.

For the 1.27-mm pitch, the CBGA ball diameter is 0.89 mm (35 mil) and the CCGA column dimensions are 2.2 mm (87 mil) or 1.27 mm (50 mil) in height by 0.5 mm (20 mil) in diameter. The utilization of CBGA versus CCGA is application dependent. The CCGA structure is about 10 times more reliable in thermal fatigue than the CBGA structure for the 2.2-mm-height column and about 5 times more reliable for the 1.27-mm column. In general, applications requiring up to 500 package interconnections (up to 32.5-mm body size) can utilize both CBGA and CCGA formats. Applications with greater than 500 interconnections (body sizes from 32.5 to 45 mm) may require CCGA structures to meet the interconnection reliability. The selection of the ball versus column interconnection is also heavily dependent on the application use conditions dictated by the system requirements. Typically, higher machine on-off cycles coupled with high-temperature machine-on operating conditions will drive the interconnection configuration into CCGA. However, CBGA routinely meets the application requirements of PC and Workstation systems in a 32.5-mm package configuration.

CBGA is an extension of area array flip-chip (C4) die interconnections, which IBM has successfully practiced for over 25 years.[3] The advantages found in high interconnection density, self-alignment of the solder joints, mass reflow of a large number of interconnections, and thermal fatigue-resistant structures are shared by both the flip-chip and CBGA/CCGA technologies. Much of the fundamental modeling, methodology, and understanding of area array solder fatigue established for C4 interconnections has been found to be directly applicable to CBGA and CCGA. The relatively shorter signal and power paths of area array interconnections coupled with strategic placement of signal and power connections within the array result in enhanced electrical performance over perimeter-leaded interconnections.

The combination of using MLC (Multi-Layer Ceramic) carriers with BGA provides cost-effective, high interconnection density package leverage. MLC technology can accommodate both MCM and SCM applications and is compatible with all types of chip interconnections, namely wirebond, TAB, and flip chip. MLC carriers have inherent excellent electrical and thermal characteristics, including built-in reference planes, low inductive power connections with multipower paths. In general, packaging applications that demand greater than 50 MHz speed and simultaneous switching of 80 or more drivers, require the electrical attributes inherent in MLC chip carriers.

Package applications that range below about 200 interconnections and do not require any electrical performance other than power and signal distributions, i.e., no ground plane requirements, are best met with standard plastic QFPs or single-layer BGAs as the lowest cost solution. When applications require ground planes, thermal dissipation and greater than about 200 interconnections, then multilayer BGA packages like CBGA/CCGA become cost-effective solutions. When applications demand greater than 400 interconnections, CBGA/CCGA becomes the most cost-effective solution. Finally, above 600 interconnections, CCGA may be the only packaging choice that can be assembled with standard SMT processes.

5.3 CBGA/CCGA Module Assembly

5.3.1 Package overview

CBGA/CCGA packages use a cofired alumina ceramic substrate which can support either multichip modules or single-chip modules. The ceramic chip carrier is fabricated by screening refractory conductor paste into lines and vias on alumina-based green sheets. The various green sheets are laminated together and sintered at high temperatures. Once sintered, the terminal interconnection pads are plated with nickel and gold such that die and component assembly can be performed. A more detailed description of ceramic chip carrier fabrication can be found in Ref. 4.

Ceramic chip carriers can support all types of chip interconnection (wire bond, flip chip, and/or TAB), plus a variety of lid-sealing or encapsulation techniques. The ceramic material poses no restrictions for package or card assembly processing since it is compatible with high temperature processing and a variety of chemicals. The substrate can be comprised of a varying number of ceramic layers depending on the complexity of the chip being packaged and the package performance requirements. A thin ceramic BGA/CGA substrate might be four or five layers (approximately 0.8 to 1.0 mm), while a thick ceramic BGA/CGA might have as many as 25 layers and be as thick as 4.5 mm. A flat ceramic substrate can be used for joining a flip chip, wire-bond chip, or TAB chip. A wire-bond cavity (for chip-up and chip-down configurations) may also be incorporated for ease of wire bonding or improved cooling (chip-down) of a wire-bond chip. The standard metallurgy of plated gold over nickel is also compatible with all types of chip interconnections. Various gold thicknesses can be accommodated to match either the specifications needed for soldering or wire bonding. With a Thermal Coefficient of Expansion (TCE) of approximately 6.5 ppm, alumina ceramic is closely matched to silicon whose TCE is approximately 3 ppm. This similarity in TCE provides high reliability and compatibility for all types of die interconnection.

5.3.2 Chip interconnection

Various chip interconnection technologies can then be combined with several encapsulation options in CBGA or CCGA packages. Flip chip provides the most advantageous interconnection method for both performance and interconnection density. The multilayer structure of the ceramic substrate allows the wiring of dense interconnection arrays, both for redistributing large numbers of signal connections while simultaneously providing multiple power and ground connections. For flip-chip joining, the MLC substrate is metallized with an array of plated pads to match the array of solder bumps on the flip chip. Distortion and camber control of the ceramic substrate across the chip site are important for high yields and reliable joints. During flip-chip attach, the solder bumps are melted during reflow, typically 350°C for 97Pb/3Sn. Because of the high temperature reflow, rosin-based flux and solvent cleaning are required today. Replacement of the rosin-based flux with a no-clean flux is desirable and possible in the future.

Wire-bond chips can be joined directly to the ceramic surface or a plated die attach area by a variety of techniques/materials including silver-filled polymers, silver-filled glasses, and gold-silicon eutectic. Gold ball bonding, gold wedge bonding, or aluminum wedge bonding can then be used to connect the chip to plated wire-bond pads on the ceramic substrate. Bonding can be achieved in a single orthogonal row, single radial row, or multiple rows, depending on the chip bond pad pitch and tool throughput and capabilities. The die attach and wire-bond area can be the flat, top surface of the ceramic substrate, or a cavity can be built into the ceramic. The most common driver to using a wire-bond cavity in a CBGA/CCGA substrate is to allow a cavity-down configuration, taking advantage of the thermal path through the ceramic from the backside of the wire-bond die by attaching a heat sink to the top of the ceramic package.

TAB chips can be joined to CBGA/CCGA substrates using either a solder reflow or a thermocompression bonding process. Any one of these chip interconnection techniques can be used for an SCM or any combination can be chosen for an MCM. Capacitors or other passives may be joined with solder or a conductive polymer and combined with any of these SCM or MCM options. Process temperature hierarchies must be considered when combining different chip interconnection techniques in one MCM.

5.3.3 Package encapsulation

CBGA flip-chip applications are typically nonhermetic resulting in the lowest cost configuration. An epoxy is used to underfill the gap between the flip chip and the ceramic. The filling epoxy completely surrounds all of the flip-chip interconnections and makes a strong bond to both the device and the ceramic chip carrier. The underfill enhances the fatigue life of the solder interconnection by as much as 10 times and also provides environmental protection. In capless configurations, a heat sink can then be attached directly to the backside of the flip chip with a thermally conductive adhesive.

Another option is to build on the capless package structure by placing an aluminum alloy cap over the chip, sealing it to the substrate with an adhesive, and filling the gap between the backside of the chip and the inside of the cap with a thermal compound to reduce the internal thermal resistance. The adhesive joining of this cap need not be hermetic, but must pass a gross leak test. A heat sink can then be joined to the cap with a conductive adhesive. The heat sink, in either the capless or capped option, typically can only be attached to the package after it is joined to the card. This process flow reduces the thermal mass of the package during card assembly reflow, simplifying the thermal profiling and fixturing of both the initial join process and the rework process. These two configurations are shown in Fig. 5.5. Wire-bond or TAB applications can be hermetic or nonhermetic. Nonhermetic encapsulation processes include a glop-top coating or the attachment of an aluminum alloy cap with an adhesive. A nonhermetic option for a cavity package is a ceramic lid joined with an adhesive to reduce the TCE mismatch between the ceramic substrate and the cap. Hermetic encapsulation of a chip-up, noncavity package uses a stamped gold-plated kovar cap joined with AuSn, either seam-sealed or reflowed. A cavity package can use either the gold-plated kovar lid joined with AuSn or a ceramic lid sealed with a glass frit.

CHIP ENCAPSULATION

INTERCONNECTION (PINS, BGA, LEAD FRAME, LGA)

CAPLESS: DIRECT ATTACH HEATSINK

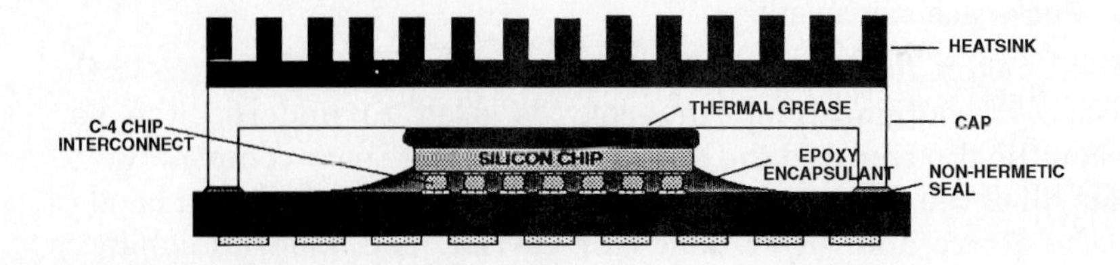

INTERCONNECTION (PINS, BGA, LEAD FRAME, LGA)

Figure 5.5 Flip-chip encapsulation, capless and capped approaches.

5.3.4 Module level burn-in

Once the chip is joined and encapsulated, the solder balls are attached to the ceramic substrate. Burn-in may be done prior to ball attach by contacting the plated substrate pads. Burn-in at this point prevents any damage to the solder balls and reduces the value-added package processes prior to determining the goodness of a chip. Burn-in, along with module test, can also be done after the solder balls have been attached.

5.3.5 Ball and column attach

The 90Pb/10Sn solder balls are joined to 0.86-mm (nominal) diameter nickel- and gold-plated pads with eutectic solder paste. The geometry and composition of this structure was determined by numerous experiments and supported by finite element modeling. The process used to join the solder balls was determined by evaluating a variety of techniques and subsequent optimization of the most promising technique.[5]

The diameter of the solder ball is driven by the need to have the highest standoff possible to accommodate the strain that results from the TCE mismatch between the ceramic substrate and the standard epoxy glass card and the desire to have high yields in both the package assembly and card assembly processes. The need for a high standoff corresponds to a large ball, but the diameter is limited by what can be consistently processed on an array of 1.27-mm pitch. A 0.89-mm diameter was chosen as optimum for the ball. The substrate pad diameter needs to be large enough to capture several vias. The card pad diameter needs to be small enough to allow wireability on the top surface. FEM and reliability testing indicate a nearly balanced geometry of a 0.86-mm pad diameter at the substrate side and 0.72-mm pad diameter at the card side, with sufficient eutectic solder in both fillets, results in the optimized interconnection configuration as described elsewhere in this book (Chap. 8) and in other references.[6]

The solder alloys in the joint structure were also optimized based on FEM and experimental results. The result of this study,[7] clearly indicates that a 90Pb-10Sn ball with a Pb-Sn eutectic fillet give the highest interconnection reliability.

One method for attaching 90Pb-10Sn balls to ceramic chip carriers uses a fixture or boat made of graphite or similar material that is drilled with holes in a pattern matching the desired solder ball array. The 90Pb-10Sn solder balls are loaded into a boat designed to maintain good planarity and centrality of the ball array after attachment. Eutectic solder paste is screened directly onto the array and the ceramic substrate is aligned to the ball array, placed on top of the balls, and reflowed in a furnace at a profile that melts only the eutectic solder.

Either rosin or water-soluble solder paste can be used followed by solvent or deionized water cleaning. If a CBGA package with a good chip is removed from a card, the package can be returned for a new array of balls to be attached.

Finished CBGA packages can be packed in vacuum-formed or injection-molded JEDEC outline stacking trays that are compatible with automated card assembly placement equipment. The tray pockets are designed for a particular size package and foam may be used to line the bottom of the tray, protecting the solder balls.

CCGA interconnections are built onto ceramic chip carriers in a similar fashion to CBGA. Solder wire columns are loaded into graphite boats. Eutectic solder paste is screened onto the array. The substrate is then aligned to the column array in the boat, placed on top of the columns, and reflowed in a furnace at a profile that melts only the eutectic solder. After reflow, the CCGA is extracted from the graphite boat, preventing column bending. Column ends are sheared to the final length, ensuring planarity of the array. A 2.2-mm-high, 0.5-mm-diameter column has a fatigue life approximately 10 times greater than CBGA. A 1.27-mm-high column CCGA is also offered and has a fatigue life approximately 5 times greater than CBGA.

Because both the CBGA solder balls and CCGA solder wire are joined to the ceramic with eutectic solder, the joint to the substrate reflows again during the card assembly process. During card assembly rework, balls or columns can be left behind when a CBGA or wire CCGA package is removed. From a card rework perspective, a cast CCGA is preferred since the columns remain with the ceramic substrate during package removal from the card. The cast CCGA joins the 90Pb/10Sn column directly to the ceramic substrate. Solder wire preforms are loaded into a graphite boat. Flux is applied to the substrate. The boat is aligned to and placed on the substrate. The furnace reflow, now set above the melting range of 90Pb/10Sn, melts the wire preforms and the solder takes the shape of the graphite boat opening. The cast CCGA is extracted and planarized in the same manner as the wire CCGA. Equivalent assembly and reliability performance of wire and cast columns has been demonstrated.

While the cast CCGA is preferred for its ease of card rework, there are applications that are better suited to the wire CCGA. The cast column process is compatible with flip-chip packages by casting the columns after chip joining, but before encapsulation. It can be difficult to have the cast column process mesh with the temperature hierarchies of wirebond packages. The cast columns are not fully reworkable after the CCGA package has been completed because of the temperature hierarchies. In some cases, a cast CCGA package can be reused after removal from a card, but the array of cast columns typically cannot be replaced.

Example process flows for flip chip with CBGA, wire CCGA, and cast CCGA are shown in Fig. 5.6.

5.4 CBGA/CCGA Card Assembly and Wireability

While the card assembly process for BGA packages is thoroughly described elsewhere, this section will highlight card assembly process parameters specifically associated with CBGA and CCGA.

5.4.1 Card top surface descriptions

As with other BGA packages, a dog-bone card attach pad configuration is required. Figure 5.7 is a schematic diagram showing the key attributes and dimensions of card dog-bone pad. Direct joining of BGA package to a card via is not recommended because the solder runs into the via, depletes the solder fillet, and can result in a defective solder joint. There are possible techniques for maintaining a filled via for direct joining, but none are practiced in production today.

A minimum card pad diameter of 27 mils (nominally 28.5 +/– 1.5 mils) is required for CBGA/CCGA solder joint (Fig. 5.7); however, larger card pad diameters are desirable. Contact to inner card layers is made through a via that is connected to the surface pad by a copper trace. Solder mask is used to prevent solder from flowing from the copper pad along the trace. Typically, a solder mask opening larger than the surface pad is preferred. Entek-coated or solder-leveled cards may be used. Any standard epoxy glass PCB material with a TCE of less

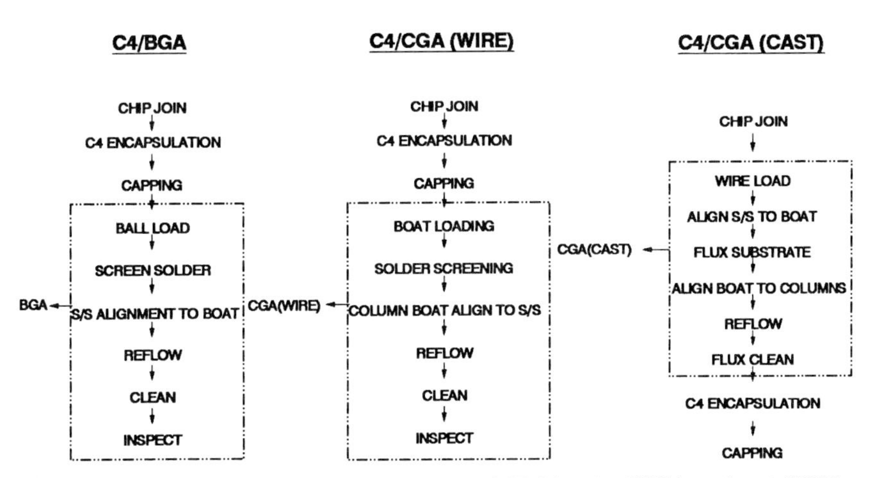

Figure 5.6 Example process flows for flip chip and CBGA, wire CCGA, and cast CCGA.

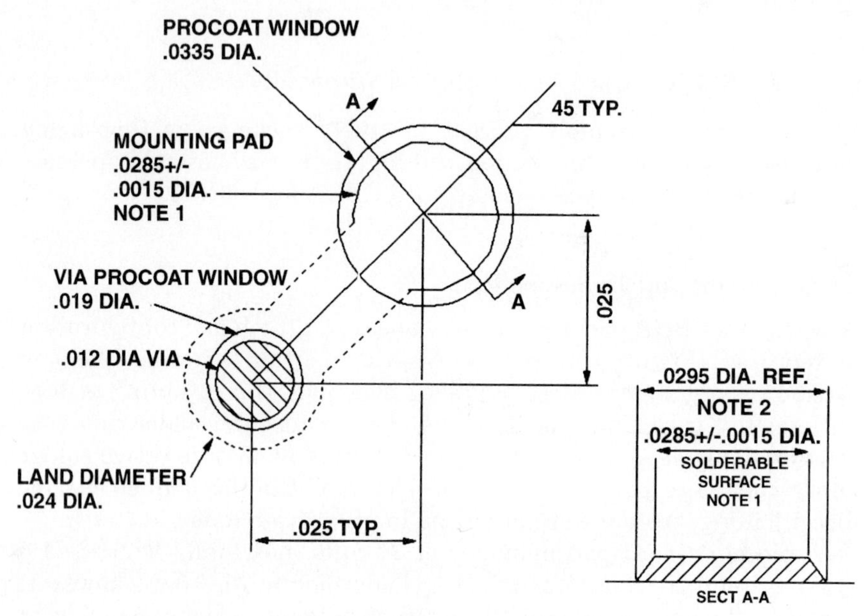

Figure 5.7 Schematic of dog-bone card pad.

than or equal to 21.5 ppm may be used and will provide a reliable solder joint. A clear zone of 200 mils around the CBGA/CCGA card site is required to allow for rework.

5.4.2 Card assembly process flow

The card assembly process and equipment for CBGA/CCGA packages reflects standard SMT requirements. Key steps include solder paste screening, placement, solder reflow, cleaning, and inspection. In Fig. 5.8, the key card assembly process steps and critical parameters are highlighted. Process control monitors or inspections at several steps ensure low defect levels (high yields), and high reliability. In particular, paste screening should be monitored to ensure proper volume and quality of print and the reflow profile should be monitored to ensure the proper temperature profile at the center solder joints. With the proper controls CBGA/CCGA assembly has routinely demonstrated exceptionally high assembly yields in high-volume manufacturing.[8] Figure 5.9 shows the relatively high yield of CBGA packages as compared to 0.65 and 0.5-mm-pitch QFP's.

CRITICAL CARD ASSEMBLY PRODUCT / PROCESS SPECS

PRODUCT REQUIREMENTS	PROCESS CONTROL MONITOR	INSPECTION/TEST
• MODULE SOLDERBALL PLANARITY < 6mils		INCOMING MODULE INSPECTION
• SOLDERABLE CARD PAD SIZE 28.5 ± 1.5mils		INCOMING CARD INSPECTION
• SOLDER DAM 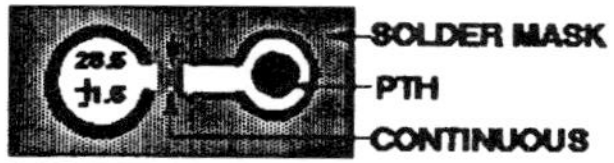 (28.5 ±1.5; SOLDER MASK; PTH; CONTINUOUS)		INCOMING CARD INSPECTION
• SOLDER PASTE SOLIDS 90%		INCOMING PASTE INSPECTION
• SCREENING MASK THICKNESS - 8mils HOLE DIAMETER - 32mils		INCOMING MASK INSPECTION
• SCREENED PRINT HEIGHT - MIN 7mils - NOM >8mils VOLUME- NOM 7000 cubic mils - LOW LIMIT 4800 cubic mils	SAMPLE HEIGHT OPTICAL MEASUREMENT OR LASER TRIANGULATION	
• REFLOW TEMP 220 ± 20°C	CARD PROFILE MONITOR WITH REPRESENTATIVE INTERNAL & EXTERNAL BALLS THERMAL COUPLED	
• EXTERNAL ROWS OF PADS POSITIVE SOLDER FILLETS	SAMPLE VISUAL INSPECTION	100%, IF SAMPLE DEFECTIVE

Figure 5.8 Critical card assembly product/process specifications.

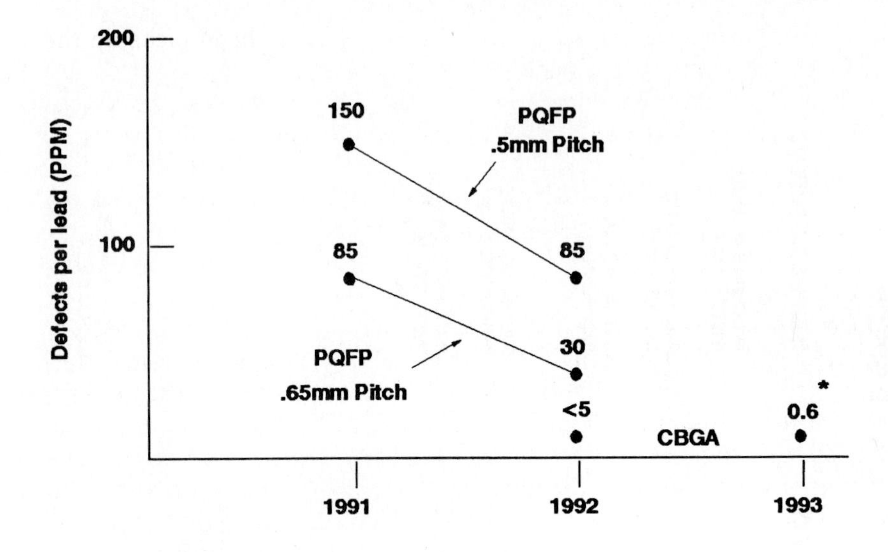

Figure 5.9 Card Assembly Yields, QFP compared to CBGA.

The first requirement of any card assembly process is to have a package that is compatible with a standard process flow. The solder ball array of a CBGA package, or the solder column array of a CCGA package, is specified to have a coplanarity of 6 mils or less and a radial error of 3 mils or less. Solderability, a typical concern with leaded packages, is of little concern with BGA packages. There are no intermetallics present near the soldering surface and the process is soldering to solder. The flux in the solder paste must overcome only surface oxides to form a good solder joint.

CBGA/CCGA packages do require a minimum solder volume for a reliable solder joint. The goal is to screen a nominal of 7000 cubic mils of solder paste and maintain a minimum of 4800 cubic mils of solder paste at each BGA location. These volumes can be achieved with either an 8-mil-thick stencil with a 32-mil-diameter opening or a 10-mil-thick stencil with a 30-mil-diameter opening. Solder volume can be monitored by an automated inspection tool or a manual inspection by an operator. CBGA/CCGAs have been demonstrated to be compatible with water-soluble, rosin, and no-clean solder pastes.

Placement of CBGA/CCGA packages is done by the same tools that place 50-mil-pitch and fine-pitch packages. There is a choice of using "black body" PLCC vision (essentially using the outline of the ceramic body), vision of the array, or mechanical means to align the package during the placement operation. The tool's vacuum head picks up the CBGA/CCGA on the backside of the chip or the cap from the feeder location, performs the alignment operation, and places the package on the card. The CBGA/CCGA package can be placed up to 50 percent off the card pads and still self-align during reflow. A CBGA or wire CCGA package will align to the center of the card pads. A cast CCGA package will align so that the columns are within the card pads, but not necessarily center. Reliability of the solder joint has been shown to be independent of whether the columns are at the center or edge of the pad.

Infrared, IR/convective, or vapor phase reflow can be used for CBGA/CCGA card assembly. The ceramic substrate is inherently flat, ranging from 1.5 to 2.5 mil/inch, and does not warp during reflow, maintaining the good planarity of the array. The balls or columns do not melt during card assembly reflow, giving a consistent component height. Depending on the solder paste chemistry, nitrogen is recommended with IR or IR/convective reflow. A typical SMT reflow profile is used, as recommended by the paste manufacturer. It is important to ensure that the center solder joints reach a minimum temperature of 200°C. The CBGA/CCGA package is not moisture-sensitive, guaranteeing that there will be no "popcorn cracking" or package delamination. CBGA/CCGA packages can endure an absolute maximum temperature of 260°C. The maximum recommended solder joint temperature is 240°C with limited package exposure above 220°C.

Once the components are assembled to the cards, cleaning can easily be accomplished because of the 35-mil (or greater) standoff and the 50-mil pitch. CBGA/CCGA packages are compatible with deionized water or a solvent to match the solder paste chemistry. There is no concern for delamination from moisture absorption, but the drying time/temperature may need to be adjusted to eliminate water from underneath the CBGA/CCGA package in a deionized water process.

The final visual inspection for CBGA/CCGA can be a sample visual of the outer row or a complete automated inspection using transmission x-ray, depending on the inspection philosophy of any given manufacturing facility. A minimum fillet diameter of 24 mils at the card side should be ensured by either the process control at the solder paste screening step or final visual inspection.

5.4.3 CBGA/CCGA card level rework

A rework process to remove and replace CBGA/CCGA packages can be developed using standard SMT rework tools, with some modifications. First, a hot gas tool with an appropriately designed heat head is used to remove the CBGA/CCGA package. A bias preheat of the entire card is recommended prior to the localized reflow of the CBGA/CCGA package. Next, the rework site is dressed to remove excess solder and any solder balls remaining behind after package removal. A cast CCGA package will guarantee that no solder columns remain behind on the card during rework. Solder paste then needs to be applied prior to joining a new package to ensure the minimum solder volume requirements. Solder application can be accomplished in a variety of ways. A single site screen printer can be used to screen paste onto the solder balls or solder column tips of the replacement package, a microscreener can be used to screen paste directly onto the card site, solder paste can be dispensed onto the card site, or preforms can be used. Finally, the hot gas rework tool is again used to place and reflow the replacement package.

5.4.4 Card wireability

The MLC substrate base of the CBGA/CCGA package allows flexibility in wiring. The location of signal, power, and ground interconnections on the chip carrier can be optimized for package performance and card wireability. The card wireability for CBGA/CCGA packages is a function of electrical requirements, package design, and card ground rules. The electrical requirements of the CBGA/CCGA package as interconnected to the card, the location of the signal and power interconnections within the CBGA/CCGA package array, and card ground rules such as the line and space dimensions, pin-thru-hole (PTH) capture pad size, and the dog-bone size and placement all contribute to the card-wiring capability of any CBGA/CCGA package site. Given appropriate design choices, a simple, inexpensive card cross section can be used to fan out and interconnect CBGA and CCGA packages.

An 18.5-mm CBGA with 196 total interconnections on a 1.27-mm grid will be used to demonstrate these points. This MLC chip carrier design has two voltage planes and two signal fanout layers. With two voltage planes in the chip carrier, the voltage interconnections could be assigned to the center of the ball array and the signal interconnections assigned to the outer rows of the array, as shown in Fig. 5.10. There is no performance sacrifice with this layout since the signal fanout lines are referenced to the power planes for noise immunity and multiple interconnections are used in parallel to feed the power planes, reducing the ground bounce (dI/dT).

A card to fan out this package can be designed using two lines per channel on the 50-mil grid, except for the top surface where only one line per channel can be accommodated because of the dog-bone card pad design, shown in Fig. 5.7. The ground rules in this case are 5-mil lines with 5-mil spaces and a 14-mil drilled hole with a 24-mil capture pad,

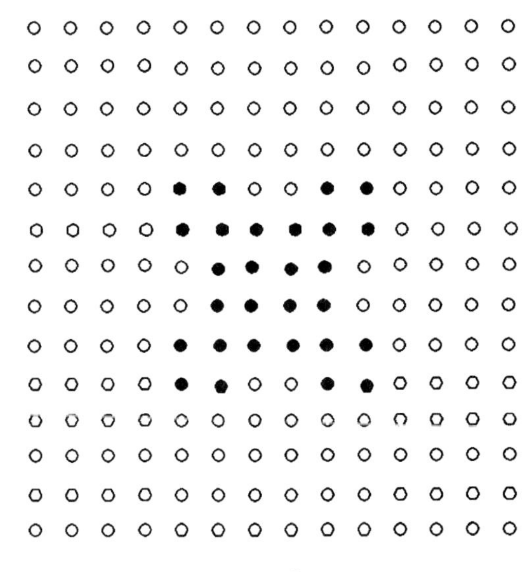

Figure 5.10 Power and signal interconnection locations in an 18.5-mm CBGA with 196 interconnections on a 1.27-mm grid.

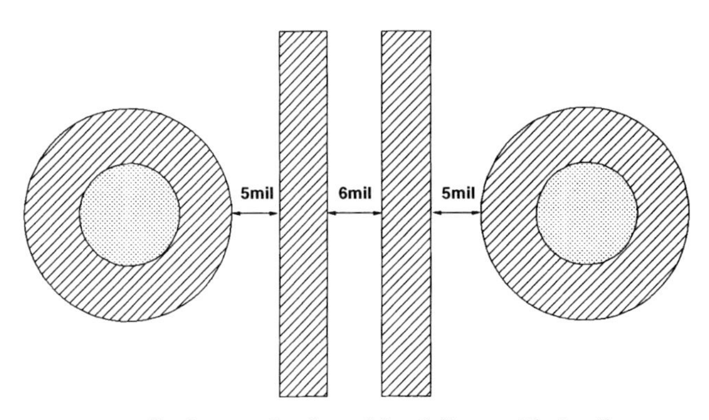

Figure 5.11 Card ground rules of 5-mil lines with 5-mil spaces and a 14-mil drilled hole with a 24-mil capture pad.

as shown in Fig. 5.11. This design can be wired using a 2S2P cross section, Fig. 5.12, where 168 interconnections are devoted to signals and 28 interconnections are devoted to power (14 ground and 14 voltage).

Table 5.1 illustrates the number of signals that can be escaped for various CBGA/CCGA packages on a 1.27-mm grid, using similar ground rules. With larger packages (greater than 25 mm), eight signal interconnections can be traded for power interconnections to improve the signal

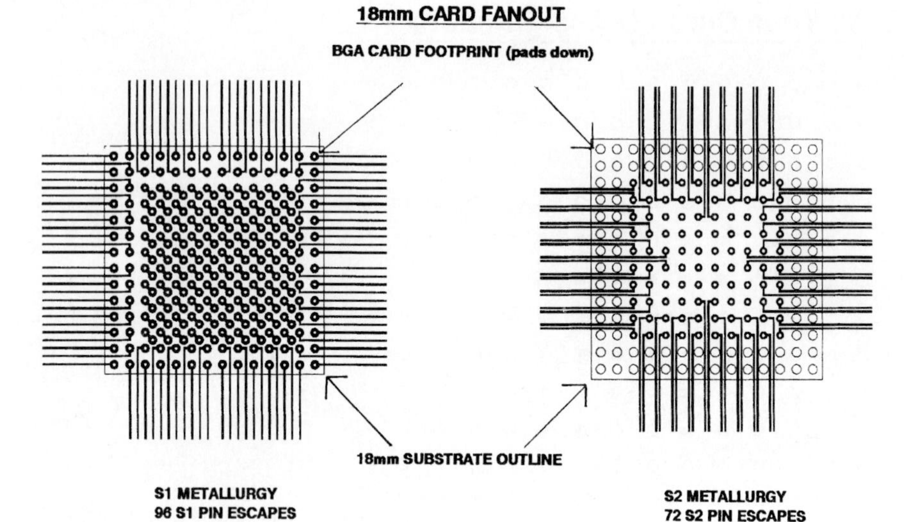

Figure 5.12 Card wiring of an 18.5-mm CBGA using a 2S2P card cross section.

return currents on the package. One of each voltage would replace a signal connection in the outer corners of the array, while maintaining the same card cross section.

In most CBGA/CCGA applications, no card cross section penalty will be incurred for the use of this technology given the card cross sections typically used in the industry today, as shown in Table 5.2. The CBGA/CCGA package will occupy the smallest card area compared to QFPs or PGAs.

TABLE 5.1 Signal Escape for Various CBGA/CCGA Package Sizes

Package size	Card cross section	Signal I/O	Power I/O	Total I/O
18.5 × 18.5 mm	2S 2P	168	28	196
21 × 21 mm	2S 2P	180	45	225
	3S 2P	209	16	225
21 × 25 mm	2S 2P	224	80	304
	3S 2P	288	16	304
25 × 25 mm	2S 2P	248	113	361
	3S 2P	328	33	361
32.5 × 32.5 mm	3S 2P	472	153	625
	4S 2P	560	65	625
42 × 42 mm	4S 2P	824	265	1089
	5S 3P	952	137	1089

TABLE 5.2 Typical Card Cross Sections for Various Applications

System type	Card cross section	Function
Commodity	2S0P, 2S2P	Planar boards and feature cards
Personal systems	2S2P, 4S2P, 6S2P	Planar boards and feature cards, server boards
Work stations	4S2P, 6S4P	Planar boards and feature cards
Midrange and high-end computers	4S4P, 8S8P	Planar boards and feature cards

5.5 Electrical Attributes of CBGA and CCGA Packages

Historically, the only electrical requirement of chip carriers was to provide signal and power space transformation from the relatively tight interconnection pitch of the die to the coarser pitch of the next level of assembly, typically the card. In this context, all chip connections were treated equally independent of their nature—namely, signal, power, or ground. The ever-increasing silicon integration of ASICs and Microprocessors has resulted in an increase in the number of chip circuits and clock speed of these devices. This silicon functionality migration drives an increased number of chip drivers and decreased switching voltages. From a package perspective, this increased chip functionality drives the need for higher lead counts, and in certain applications, electrical requirements in the form of signal line integrity, simultaneous switching noise control, and multiple power and ground feeds. In applications, where only an increase in interconnections is required, with no package electrical and thermal enhancements, the best cost/performance packaging option is a migration from PQFP (Quad Flat Pack) to single-layer BGA (Ball Grid Array). However, if the system and device application requires package electrical and thermal performance in the form of signal line integrity, simultaneous switching noise control and multiple power and ground planes, then a further package migration from single-layer BGA to multilayer BGA is the appropriate package selection. Figure 5.13 gives a visual representation of this package cost/performance migration. In general terms, applications requiring clock speeds in excess of 50 MHz coupled with greater than 80 simultaneous switching drives is the transition point where package electrical performance becomes necessary. In this section the electrical attributes of area array balls and columns will be compared to gull wing and pin interconnections. In addition, the electrical attributes of multilayer ceramic chip carriers to meet applications requiring package electrical performance will be discussed.

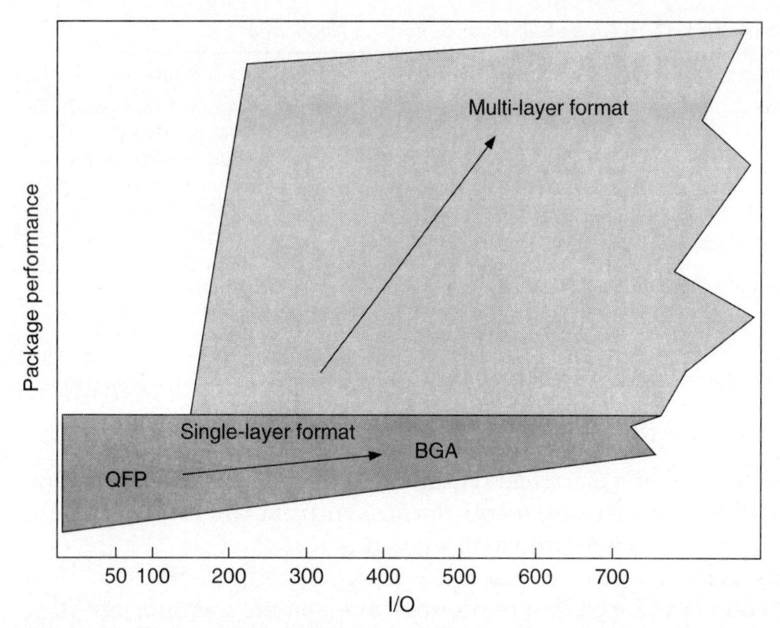

Figure 5.13 Segmentation of packaging applications by cost and performance requirements.

5.5.1 Comparisons of interconnections

The primary method to achieve increased electrical performance is to reduce inductance. The critical package path is from the die to the board, for both power/ground and signals.

Several formats for connecting the package to the card can be considered. For pin-in-hole (PIH), a 0.018-in-diameter pin that is 0.090-in long is chosen. This length represents the typical height between the package and the card. Two pin pitches are evaluated, a 0.100-in array and a 0.100-in staggered interstitial array. For the gull wing, a lead 0.26 mm wide by 0.15 mm thick and 3.25 mm long is chosen. Two lead pitches are evaluated, one 0.4 and the other 0.5 mm.

In the BGA case, a 0.035-in-diameter ball is used. For the CGA cases, a single-column diameter of 0.020 in is used, with two different lengths: 0.050 and 0.090 in. These choices cover a broad range of application conditions.

A comparison of inductance between interconnection schemes is shown in Table 5.3 to assess when it is appropriate to migrate from one technology to the next.

Table 5.1 clearly shows that the BGA array has less than half the inductance of the next best option, the short CGA array column. All of

TABLE 5.3 Inductance of Various I/O Configurations

	Pin	Pin	CGA	CGA	BGA	Gull wing	Gull wing
Configuration	Array	Interstitial	Array	Array	Array	Perimeter	Perimeter
Length	0.090 in × 3.25 mm	0.090 in × 3.25 mm	0.050 in	0.090 in	0.035 in	0.26 mm	0.26 mm
Pitch	0.100 in	0.100 in	0.050 in	0.050 in	0.050 in	0.4 mm	0.5 mm
Self (nh)	1.071	1.071	0.443	1.028	0.191	1.64	1.64
Leff (nh)	1.753	1.615	0.646	1.358	0.259	1.48	1.708

the SMT arrays (BGA and CGA) clearly have lower inductance than either of the PIH or gull wing choices.

Decreasing the pitch on the gull wing to 0.4 mm shows improvement over the 0.5-mm pitch. This improvement is created by the increased mutual inductance between leads, but this benefit is lost to increased trace length caused by the increased carrier size needed to support devices with higher lead counts.

The clear winner for a low inductance interconnection is the BGA.

5.5.2 MLC carrier performance

The electrical benefits of an MLC carrier with internal voltage power and signal planes are astounding. These benefits are even further enhanced with the use of flip-chip technology (C4).[3] Because MLC is a multidimensional packaging strategy, internal power and ground layers can be used for two purposes simultaneously. One use for the internal power and ground layers is to provide a low-inductance path from the fine chip pitch to the coarse package pitch. The second is to reduce coupling coefficients between both vertical and horizontal signal traces. And the third is to create a triplate structure providing an impedance-controlled environment for signal traces. MLC's flexibility permits optimization of the I/O assignments. Power and ground I/Os can be placed directly under the device, as in the case of flip chip, or under the wire bond fingers in wire-bonded devices. This flexibility provides a low vertical inductance path through the carrier. Depending on the application, package size and cost, the ratio of signal-to-power/ground I/O assignments becomes an important factor. In general it is best to achieve a ratio of between 2:1 and 4:1 of signal to power. Figure 5.14 shows the model results for a specific case where various signal-to-power ratios are used to show the effect on package-effective inductance. The effect of added discrete decoupling capacitors on module-effective inductance is also depicted. In this model, L_{eff} is the calculated value of inductance derived from the noise generated on a quiet driver with 25 active drivers surrounding it.[9]

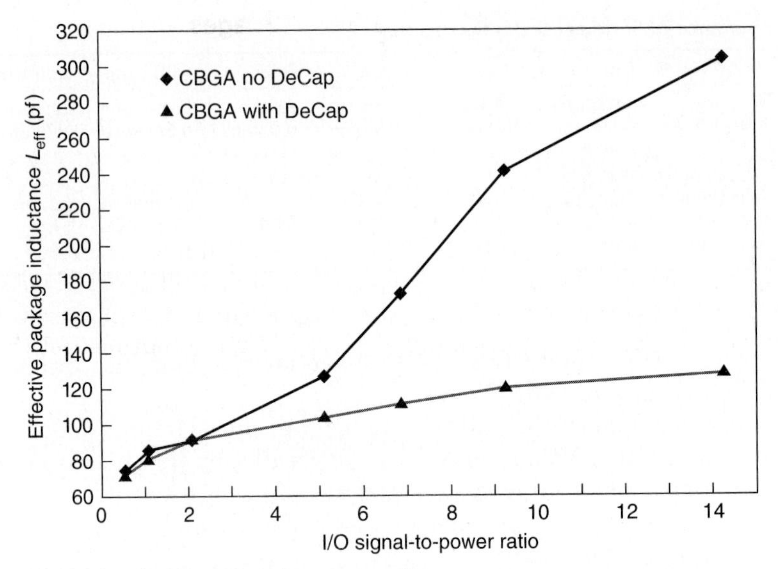

Figure 5.14 Package signal-to-power ratio.

$$L_{\text{eff}} = V \Big/ \left(\frac{\text{di}}{\text{dt}}\right) \tag{5.1}$$

where V = voltage on the quiet driver divided by the number of simultaneous switching drivers

$\frac{\text{di}}{\text{dt}}$ = change in voltage over time of the active drivers

As the signal-to-power ratio increases, package-effective inductance increases. This relationship holds true for the chip interconnection as well. In general, as the number of power and ground contacts at the device level increase, inductance decreases. This trend highlights the benefits of flip-chip interconnections with their area array, as opposed to wire-bonded devices with their bond pads at the periphery of the device. Because of this low-inductance path, flip-chip (C4) interconnection inductance is orders of magnitude lower than wire-bond interconnections which typically induce 1nh of inductance. Another advantage of flip-chip interconnections is that I/O cells are allowed to be strategically placed in the device to minimize on-chip delays. At the same time, power and ground C4's can be positioned within the device area to reduce power bussing in the device. MLC's design and ground rule flexibility supports this C4 strategy with multiple vias, vertical contacts, connecting the C4 contacts to the internal power and ground planes, providing very low inductance power paths.

5.6 Thermal Attributes of CBGA and CCGA Packages

5.6.1 Overview

CBGA and CCGA are undoubtedly the most desirable package technologies in the industry today when cooling considerations are important. Thermal performance with wire-bonded chips will be at least equivalent to Pin Grid Array (PGA) counterparts using expensive Cu/W heat slug or spreader technology. But a significant thermal advantage is realized when combining CBGA/CCGA with C4 chip attach technology, where typical power dissipation capability increases of 50 percent over alternate packaging, such as PGA, are routine. Both CBGA and CCGA interconnection will provide equal thermal performance on any one package. A variety of internal and external thermal enhancement techniques as well as capped and capless package configurations allow the designer to tailor a module's thermal performance as necessary. Structural integrity of the package allows for more robust heat sinking than peripheral leaded packages. The reduced package size as well as the ability to maintain a low package profile offer increased latitude for module placement, heat sink configuration, and second-level packaging density.

The attributes of these packages which combine to provide outstanding total thermal performance can be classified as both internal and external to the module. Internally, ceramic packaging and substrate geometry provide a low-resistance thermal path out of the chip through the die-bond adhesive for wire-bonded chips or through the solder bump array for C4 mounted chips. Additionally, heat can be further extracted through a C4 chip backside to the external surfaces of the module using a thermally conductive paste, or can be conducted directly to a heat sink with direct chip-to-heat sink attach. A parallel heat flow path out of the chip minimizes the temperature drop internal to the package while uniquely positioning the large high-temperature surfaces of the module where the heat can be dissipated most effectively, the top surface and the interconnection array.

By design, heat is dissipated from the CBGA/CCGA package primarily through the top surface to the air and through the I/O interconnection to the card power planes. When encapsulated, the aluminum cap spreads the heat across the package to provide a large uniform temperature surface. This provides a most effective surface for direct convection to the cooling air or, if a heat sink is used, minimizes the heat-sink-to-cap interface resistance. A capless configuration minimizes the internal interfaces by attaching the heat sink directly to the chip backside for maximum performance at the lowest cost. Heat transfer to the card is also maximized with CBGA/CCGA packaging due to the large array of highly thermally conductive connections from the uniform temperature substrate to the card and card power planes. Again, a parallel external path for heat flow results is a robust thermal path, highly conductive and relatively insensitive to changes in the individual components of each path.

5.6.2 Product performance of CBGA and CCGA packages

Ceramic Ball Grid and Column Grid Array packages are offered in a variety of sizes, and thermal performance will vary with package size, heat sink selection, and card heat load. Figures 5.15 through 5.19 show the projected thermal performance of these packages with C4 chip interconnect throughout the 18 to 44-mm size range as a function of airflow, both with and without heat sinks under typical use conditions. The approximate power dissipation capability of each package can be estimated from these figures.

Thermal performances of CBGA/CCGA packages using wire-bond chip technology are equivalent to or slightly better than other packaging technologies such as PGAs. Both cavity-down and chip-up/aluminum cap configurations are used, and the relative thermal performance of each is dependent on package geometry and materials selection as well as external cooling constraints. For instance, if a significant heat load can be dissipated into the card, such as in a portable computer application, a chip-up/aluminum cap package design will outperform a cavity-down design. The opposite would be true in a high-air-flow, large-heat-sink environment, where most of the heat dissipation potential is through the top of the package. In all cases, the heat dissipation potential with a wire-bond chip will be less than that of an equivalent C4 mounted chip, which is illustrated in the following example.

5.6.3 Microprocessor packaging example, C4/CBGA versus wire bond/SPGA with heat spreader

Overall module thermal performance can be impacted significantly by the choice of the first-level package. Power density is the greatest in the vicinity of the chip and it is important to move this heat away from the chip in three dimensions through several parallel, low-resistance thermal paths.

A common microprocessor package used in the industry today is the cavity-down (staggered) Pin Grid Array (PGA, SPGA). It is widely used to package wire-bonded chips because of the short thermal path it provides to a localized portion of the top surface of the package. Due to the

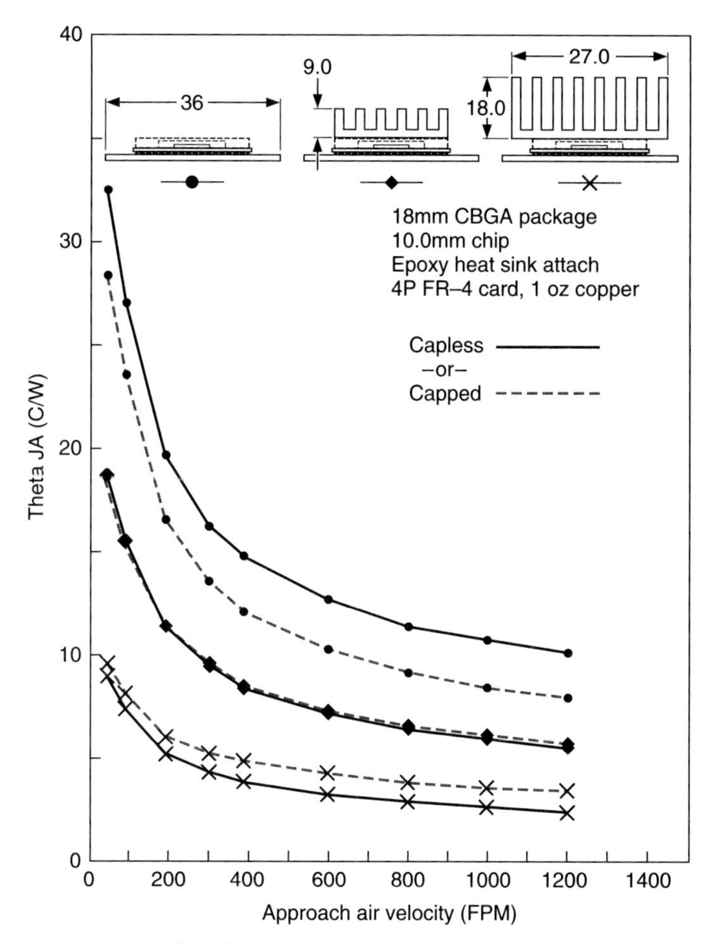

Figure 5.15 CBGA/CCGA thermal summary—18-mm CBGA.

cavity-down configuration, the package tends to be quite large in order to achieve the required off-module I/O. A large chip, which necessitates a large cavity, further aggravates the package size problem. Another disadvantage is that a large temperature gradient on the top surface of the package exists because the heat cannot spread well through the package due to the low ceramic thermal conductivity. Often heat spreaders and/or heat slugs are used to minimize this problem, but the periphery of the package in the vicinity of the pin array is usually much cooler than the center of the package. This impacts the heat flow in both directions, to the heat sink because of the localized "hot spot" on the package surface and to the card because of the low temperature difference between the pins and the card.

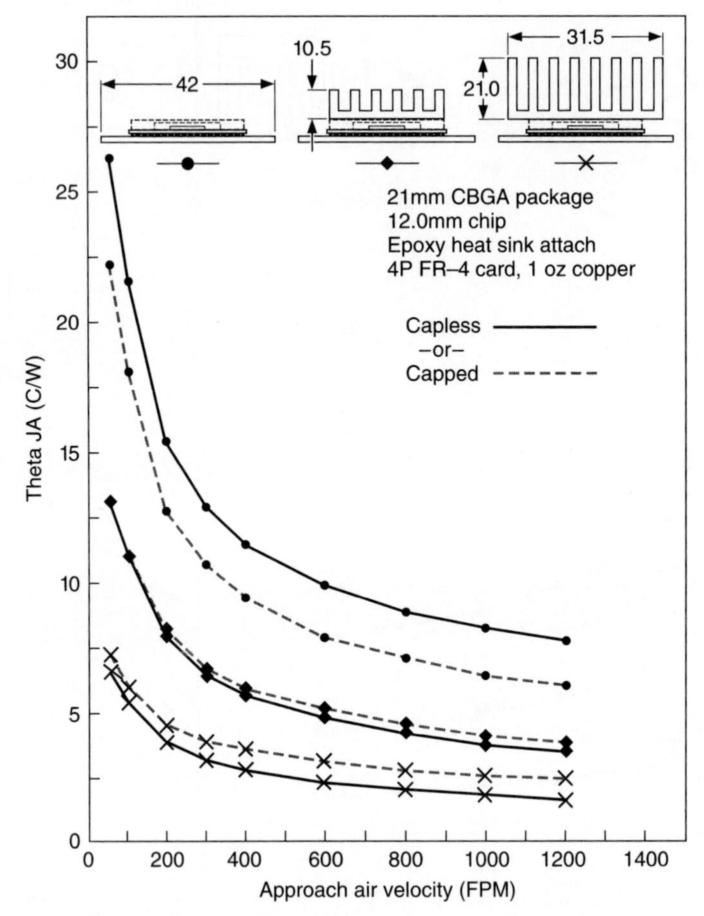

Figure 5.16 CBGA/CCGA thermal summary—21-mm CBGA.

An alternative packaging solution is to "bump" the same chip for C4 interconnection and attach it to a capless CBGA module. Typical package size is about one-fifth the size of the required PGA. Direct attach of the heat sink to the chip is possible and results in an extremely low-resistance thermal path to the heat sink. Additionally, the C4/CBGA combination provides an excellent thermal path into the card power planes. In personal computers and workstations, where the high heat dissipating microprocessor is surrounded by primarily low power components, the card path is usually significant. The direct path to the heat sink and the card combine to offer a significantly better package, often which can dissipate 50 to 100 percent more heat than the PGA counterpart.

An illustration of this impact can be seen in Fig. 5.20. Results for a current microprocessor application packaged in both a 50-mm, cavity-

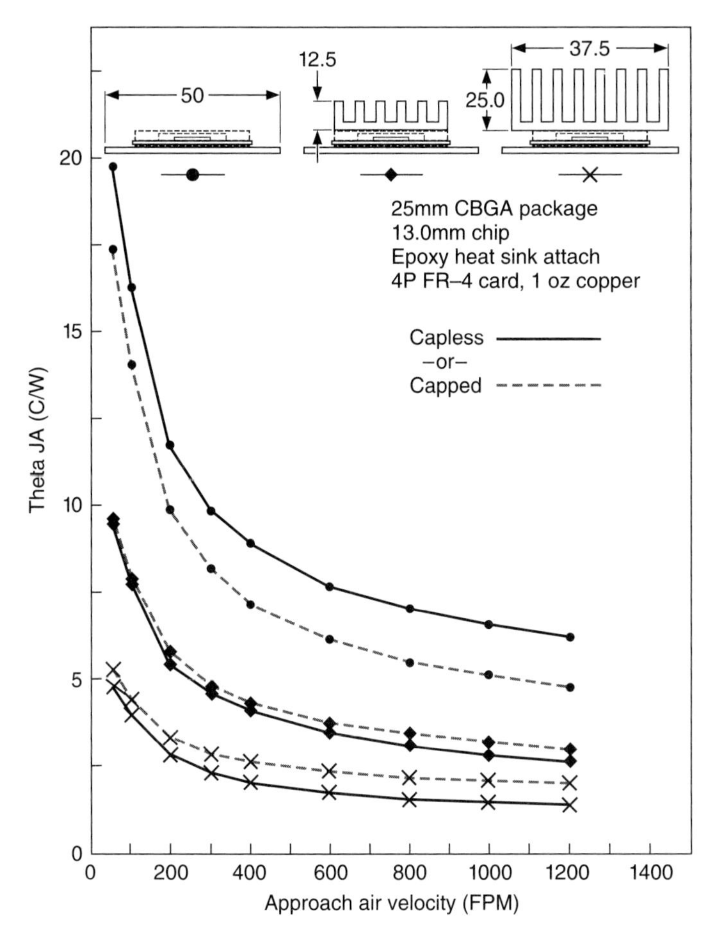

Figure 5.17 CBGA/CCGA thermal summary—25-mm CBGA.

down SPGA and the equivalent 21 × 25 mm CBGA package are shown, along with the package information necessary to show equivalence. Even with an SPGA package 4.7 times the size of the CBGA equivalent, the 50-mm SPGA cannot dissipate the heat as effectively as the 21- × 25-mm CBGA. The CBGA solution translates directly to additional power handling capability, 60 percent more at the typical PC/workstation operating point of 200 FPM airflow.

5.7 CBGA/CCGA Interconnection Reliability

Ceramic ball and column grid arrays are direct extensions of area array flip-chip (C4) interconnections, which IBM has successfully practiced for 25 years. Much of the fundamental modeling, methodology, and understanding of solder fatigue established for area array flip-chip

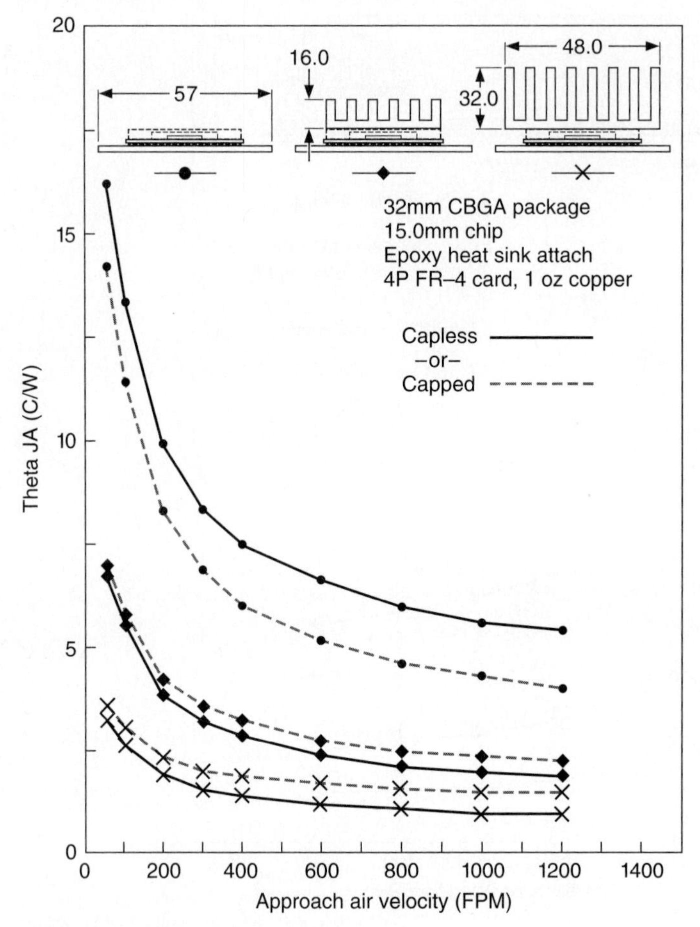

Figure 5.18 CBGA/CCGA thermal summary—32-mm CCGA (1.27-mm columns).

interconnections has been found to be directly applicable to CBGA and CCGA interconnections. In this section, reliability data for both CBGA and CCGA interconnections is presented. Emphasis is placed on how fatigue data collected under accelerated conditions in the laboratory is bridged to various system applications conditions.

Thermal fatigue of most metals including Pb-Sn-based solders is best described by the Coffin-Manson relationship, Eq. (5.2).

$$N_f \, \alpha \, \varepsilon_p^{\,c} \tag{5.2}$$

Through empirical and physical analysis this relationship has been extended and modified to address the unique aspects of thermal cycle fatigue in Pb-Sn-based solder interconnections. These modifications

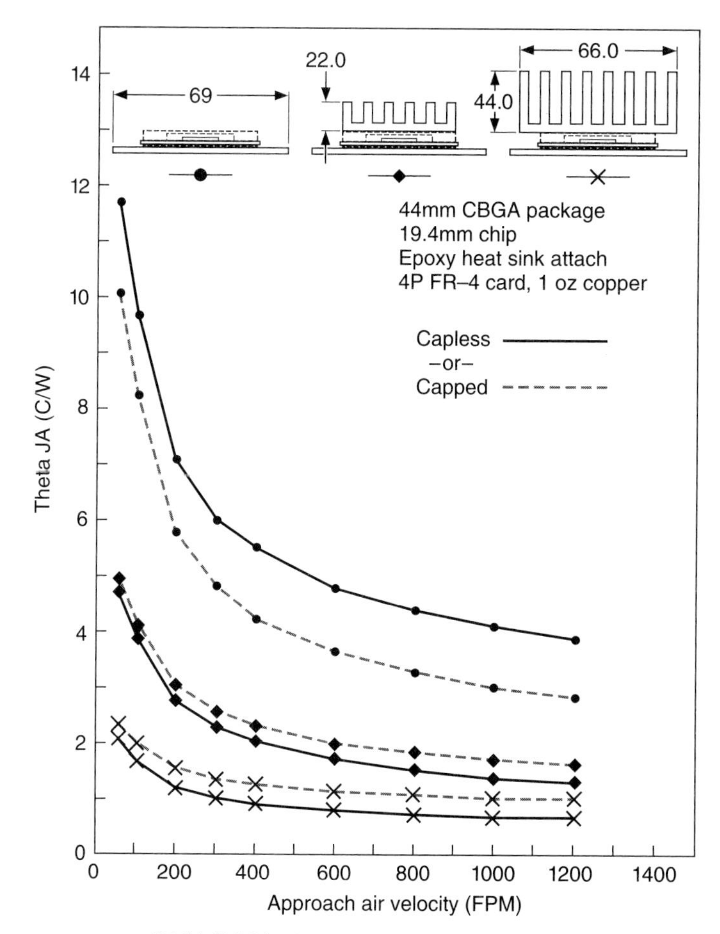

Figure 5.19 CBGA/CCGA thermal summary—44-mm CCGA (2.21-mm columns).

include frequency and temperature effects. A detailed account of solder fatigue behavior can be found in Ref. 4. The modified Coffin-Manson relationship describing solder fatigue behavior is given in Eq. (5.3). Where N_f is the number of cycles to failure, A is a constant, E_p is the strain per cycle, f is the frequency, K is Boltzmann's constant in eV, and T_{max} is the maximum temperature of the cycle in kelvins.

$$N_f = (A/\varepsilon_p)^{1.9} \cdot f^{1/3} \cdot \mathrm{EXP}[0.123/KT_{max}] \qquad (5.3)$$

For solder interconnections of fixed geometry and materials, the strain term can be reduced to the delta temperature, which comprises each cycle. The modified Coffin-Manson equation is not intended to predict fatigue life a priori of any reference data for a given interconnection

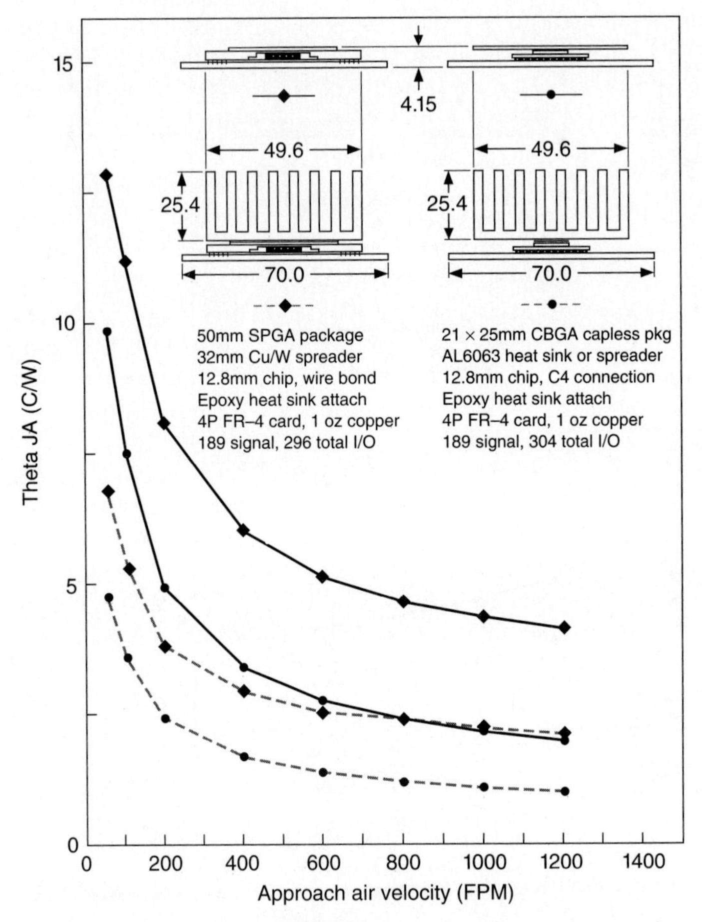

Figure 5.20 CBGA/C4 vs. SPGA/WB thermal performance comparison.

structure. Rather it is best used to determine the acceleration factor (AF) between two sets of fatigue conditions. In this manner, the acceleration factor is a convenient method for bridging accelerated lab stressing of both CBGA and CCGA of actual system field application conditions. The acceleration factor, Eq. 5.4, is simply the ratio of the cycles to failure, N_f, for two sets of fatigue conditions as given in Eq. (5.3).

$$AF = N_{f(F)}/N_{f(L)} \tag{5.4}$$

$$AF = [\Delta T_L/\Delta T_F]^{1.9} \cdot [f_F/f_L]^{1/3} \cdot EXP[1414(1/T_F - 1/T_L)]$$

The terms T_L and T_F are the delta temperature excursions for the lab and field thermal cycles, T_L and T_F are the peak temperatures for the lab and field cycles, and the f_L and f_F are the frequencies for the lab and field cycles, respectively.

Two example acceleration factors are calculated in Table 5.4. In the first example, lab stressing from 0 to 100°C is compared to a representative system field application where the CBGA or CCGA component cycles between 25 and 51°C during system on-off cycling. The acceleration between these two fatigue conditions is ten times. This means that every lab cycle is equivalent to 10 machine on-off cycles in the field! For example, 25-mm CBGA modules typically experience interconnection fatigue fails at about 500 cycles of 0 to 100°C stressing. For CBGA components cycling at 25 to 51°C in a system, the interconnections would withstand over 5000 on-off cycles. In the second example calculated in Table 5.4, 0 to 100°C thermal cycling is compared to 20 to 80°C thermal cycling. For this case, 0 to 100°C thermal cycling accelerates fatigue about three times greater than 20 to 80°C thermal cycling.

In all cases, the fatigue failure mode for CBGA interconnections is found to bc crack propagation in the eutectic joints of the interconnection. However, in CCGA interconnection the failure mode is crack propagation in the 90PB-10Sn column next to the eutectic solder fillets. Figure 5.21 contains metallographic cross sections of CBGA interconnections before and after accelerated fatigue stressing at 0 to 100°C. These cross sections illustrate the aforementioned failure mode.

In an effort to verify that the modified Coffin-Manson expression accurately predicts CBGA/CCGA thermal fatigue behavior, a series of controlled experiments was performed. In the experiments, CBGA and CCGA interconnections were thermally cycled between 0 to 100°C and 20 to 80°C. Figures 5.22 and 5.23 are the CBGA and CCGA thermal fatigue data, respectively. This empirical data shows that both CBGA and CCGA interconnections fatigue about three times faster at 0 to 100°C than at 20 to 80°C as predicted by the Coffin-Manson–derived acceleration factor. Accordingly, data collected on various laboratory accelerated thermal cycling conditions can be bridged reliably to any system field conditions using the modified Coffin-Manson expression.

TABLE 5.4 ATC Acceleration Factors

	Typical Field Conditions vs. 0–100°C		
	Field	Lab	Acceleration
Delta T	26°C	100°C	
Peak T	51°C	100°C	10×
Frequency	6 Cycles/day	72 Cycles/day	
	20–80°C vs. 0–100°C		
	Field	Lab	Acceleration
Delta T	60°C	100°C	
Peak T	80°C	100°C	3.2×
Frequency	72 Cycles/day	72 Cycles/day	

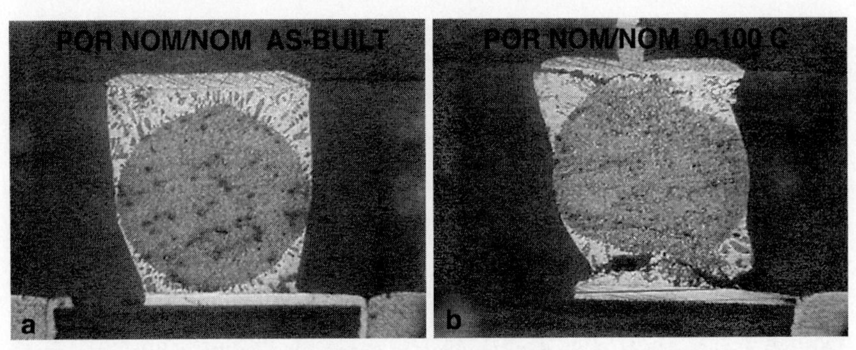

Figure 5.21 Metallographic cross sections of CBGA interconnections before (*a*) and after (*b*) accelerated fatigue stressing at 0 to 100°C.

An example application space for CCGA interconnections is illustrated in Fig. 5.24. The application space is defined by the number of system on-off cycles a CCGA or CBGA component can accommodate for a given temperature delta associated with each system on-off cycle. For statistical reasons, 100 ppm module fails at end of product life was chosen as the boundary for supporting a system application use condition. As expected, the larger the delta temperature associated with a field on/off cycle, the smaller the number of on/off cycles supported. Typical field use conditions for various types of Personal Computers is also included in Fig. 5.24. Microprocessor junction temperatures, T_j, for desktop PCs are typically specified to 100°C; this results in a CCGA package delta temperature of 60°C per system on/off cycle. As seen in

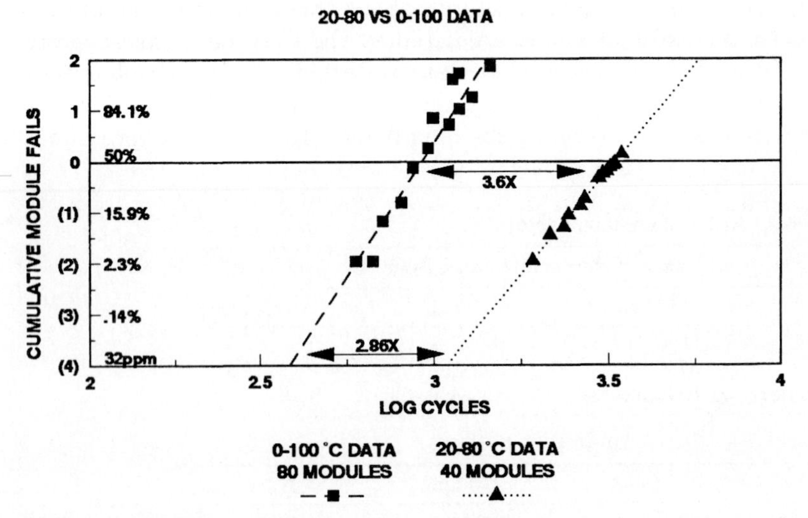

Figure 5.22 CBGA module failure rate as a function of cycles for 20 to 80°C and 0 to 100°C thermal cycling (taken from Ref. 10).

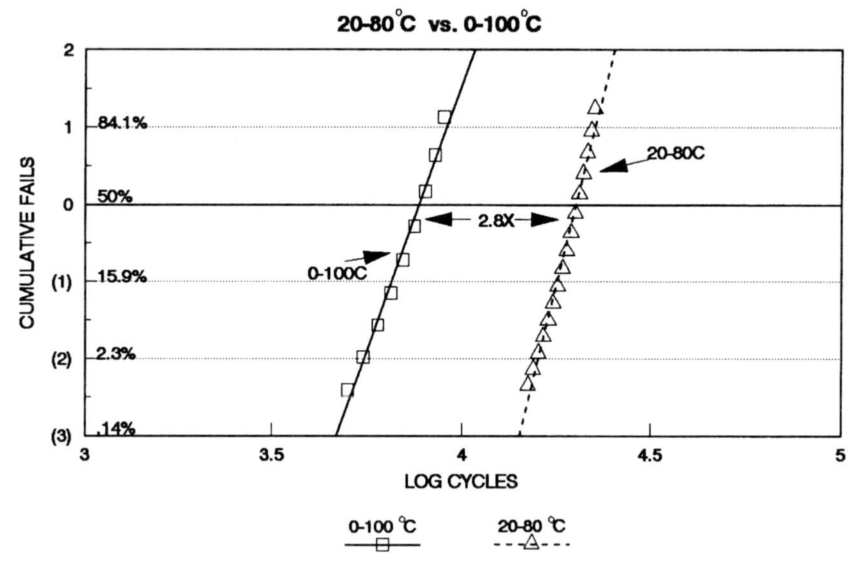

Figure 5.23 CCGA module failure rate as a function of cycles for 20 to 80°C and 0 to 100°C thermal cycling (taken from Ref. 10).

this example, a 533-I/O CCGA package can support up to 6000 of these 60°C delta temperature on/off cycles, while typical desktop PCs require only a total of 3500 on/off cycles to meet the product reliability objectives. Similarly, PC laptops and portables have microprocessor case temperatures specified not to exceed 85°C. This specification would result in a CCGA component system level on/off cycle, delta temperature of 45°C. A 533-I/O CCGA package can support up to 15,000 such field cycles.

Last, Fig. 5.25 shows the reliability improvement CCGA offers over CBGA structures. In Fig. 5.25 CBGA and CCGA module failure rates are plotted as function of field on/off cycles for similar field use conditions and package size. The data shows that CCGA is 10 times more reliable than CBGA. Accordingly, for aggressive field use conditions and high product I/O requirements (i.e., large delta temperature per on/off cycle and many cycles, and greater than 500 package I/Os), only CCGA structures can achieve high thermal fatigue reliability.

5.8 Example Application Using CBGA

In this section a review of the cost/performance tradeoffs of converting a CQFP to CBGA is quantitatively assessed.

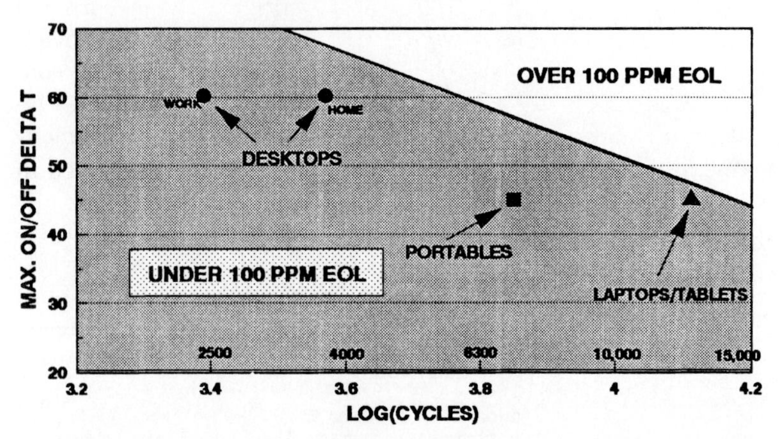

Figure 5.24 Use conditions (on/off delta temperature vs. on/off cycles) supported for a 32mm-533 signal CCGA (taken from Ref. 10).

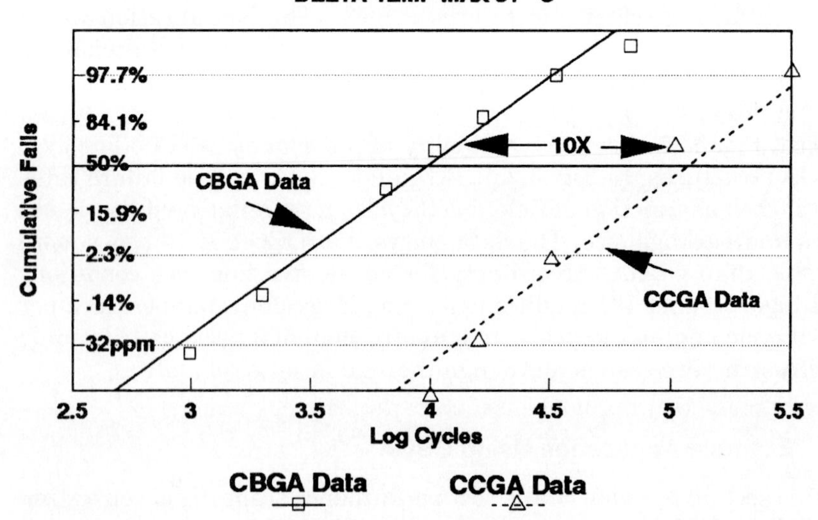

Figure 5.25 Thermal fatigue comparison of CBGA vs. CCGA surface mount connections (taken from Ref. 10).

Expanded Random Access Memory in IBM's RISC 6000 workstation is provided in the form of a pluggable card subassembly. This subassembly consists of banks of SIMMS and three memory controller modules, which control data streams between the microprocessor and the SIMMS modules. In the original system board level design, the memory controller modules were packaged in 40-mm CQFPs with 304 I/Os (0.5-mm lead pitch). However, assembly yield problems drove a package conversion. The present configuration of the memory controller modules is a 21 × 25 mm Ceramic Ball Grid Array package. Figure 5.26 is a comparison photograph of the two forementioned memory card subassemblies with the memory controller modules assembled in CQFP and the CBGA package formats.

The advantages gained by migrating the memory controller modules from a CQFP package to the CBGA are many and range from the package size, electrical performance, cost, package robustness, and card assembly. Table 5.5 gives a quantitative summary of all the advantages the CBGA package solution provided over the CQFP option for this memory controller application.

5.8.1 CBGA package size advantage

The CBGA size advantage is striking. The CQFP occupies 16 cm^2 of card real estate per module, while the CBGA package occupies only 5.25 cm^2. This represents a 3× package shrink for the CBGA over the CFQP. Smaller packages not only save card real estate but also drive the cost of the chip carrier down as well. In this particular comparison, the CBGA offered a 20 percent reduction in package cost.

5.8.2 CBGA package electrical performance advantage

CBGA packages offer many electrical performance advantages over the QFP packages. From a form factor perspective, CBGA provides a much shorter chip-to-card signal path as illustrated in Fig. 5.27. This chip-to-card signal path includes package wiring and interconnection length. Shorter signal paths result in less electrical parasitics which impact signal time of flight. Quantitatively, the CBGA package provides a minimum of two times less time of flight delay in this application. A major contributor to this improvement results from the fact that the CBGA interconnection provides a 3× improvement in capacitance, and a 30× improvement in inductance over the lead frame interconnection scheme of QFPs. Overall, the electrical advantages in inductance, capacitance, and shorter signal trace length of the CBGA package provided 2.25× decrease in signal line noise over the CQFP package, as determined by measuring the noise spike on the worst case signal lines for each package.

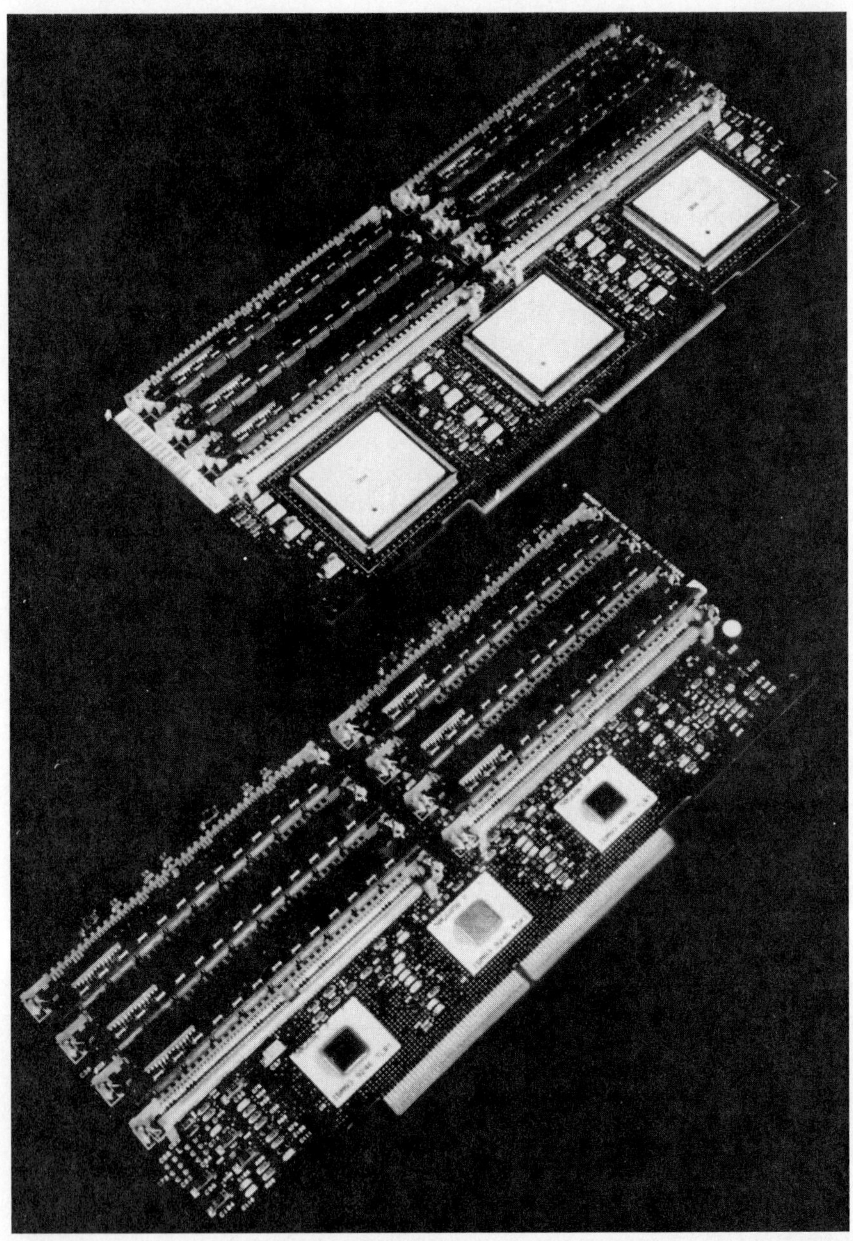

Figure 5.26 Photograph of IBM's RISC 6000 memory card in both CQFP and CBGA versions.

TABLE 5.5 Workstation Memory Card Conversion

	Memory controller modules →		CBGA advantage
Chip carrier type	(3) 40mm CQFP (0.50mm pitch)	(3) 21mm × 25mm CBGA (1.27mm (50 mil) pitch)	
Chip carrier size	16cm² 40mm × 40mm	5.25cm² 21mm × 25mm	3X package shrink
Chip carrier cost	1x	0.8x	20% package cost
Chip carrier signal noise	2.25x	1x	130% quieter package
Planar	4s4p (4 lines/channel)	4s4p (4 lines/channel)	
Card assembly yield	100 ppm/lead	0.6 ppm/lead	170X higher assembly yield
Component reject rate	7% lead damage	0% reject	More robust package

5.8.3 CBGA assembly yield and quality advantages

Perhaps the most compelling aspect of the CBGA advantage over CQFP for this application is found at the card assembly comparison. The card assembly yields for CQFP have been found to be 100 ppm/lead defects which translates into 3 percent module defect rate. By comparison, the CBGA is running at 0.6 ppm/lead defects which translates into 0.02 percent module defect rate. This represents a 170× improvement in assembly yields. Last, the fragility of the leads on CQFP packages drives an exceptionally high component reject rate, (7 percent), at component placement. CBGA packages are very robust and have demonstrated zero component rejection rate.

When viewed collectively, the migration of the memory controller module from a CQFP to a CBGA solution had many advantages. Although this is an example of a single application, it is representative of the benefits that can be realized for many similar product applications.

5.9 Summary

CBGA and CCGA technology offers package cost and performance leverage for applications requiring greater than 200 interconnections,

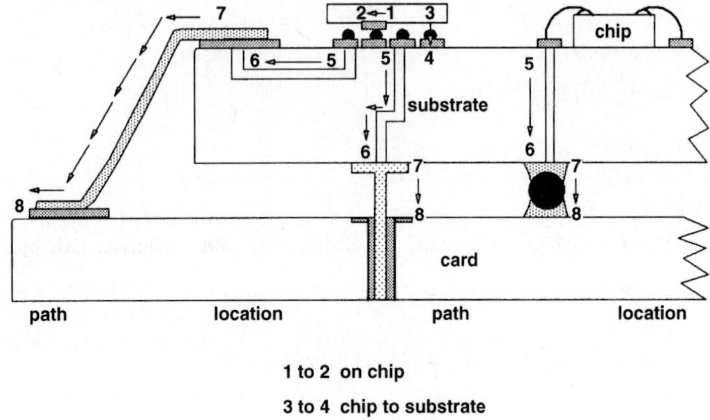

Figure 5.27 Schematic depicting electrical paths for various chip and package interconnection schemes (taken from Ref. 11).

50 MHz speed, and greater than 80 simultaneous switching drivers (i.e., when a package reference plane is required). This packaging technology also offers excellent thermal performance, high card assembly yields using standard SMT assembly processes, and high reliability. CBGA and CCGA technologies offer the highest number of interconnections at the greatest packaging density of any SMT compatible technology. This technology is compatible with all types of chip interconnection methods and can accommodate a variety of SCM and MCM applications, ranging from thin and light chip carriers for PC applications to complex packages for high-end workstations, midrange computers, and mainframes. CBGA and CCGA packages set the standard to meet the interconnection-intense requirements in packaging the ASICs and Microprocessors of the 1990s.

5.10 References

1. *BPA System 2000 Report,* "Advanced Integrated Circuit Packaging," 1992.
2. Behun, J. R., et al., "Interconnection Structure and Test Method," U.S. Patent 5,060,844, Oct. 29, 1991.
3. Miller, L. F., "Controlled Collapse Reflow Chip Joining," *IBM Journal of Research and Development,* 13, May 1969, pp. 239–250.
4. Tummala, R. R., and E. J. Rymaszewski, *Microelectronic Packaging Handbook,* Van Nostrand Reinhold, N.Y., 1989.

5. Behun, J. R., et al., "Improvement in Packaging Manufacturing Through DMI," *IBM SMT Conference Proceedings,* Austin, Tex., Mar. 1991, pp. 595–600.
6. Corbin, J. S., "Finite Element Analysis for Solder Ball Connect (SBC) Structural Design Optimizations," *IBM Journal of Research and Development,* Sept. 1993, pp. 585–596.
7. Caulfield, T., et al., "Thermal Fatigue Reliability of Surface Mountable Solder Ball Connections," *IBM SMT Conference Proceedings,* Mar. 1991, Austin, Tex., pp. 416–423.
8. IBM and Compaq, *Workshop on Ball Grid Arrays & Advanced Packaging,* Aug. 1993, Sunnyvale, Calif.
9. Cappo, F. F., J. C. Milliken, and J. M. Mosley, "Highly Manufacturable Multi-Layered Ceramic Surface Mounted Package," *Proceedings of IEEE/CHMT Symposium,* Sept. 1991, pp. 424–428.
10. Caulfield, T., et al., "Surface Mount Array Interconnections for High I/O MCM-C to Card Assemblies," *Proceedings of the 1993 ICEMM,* April 1993, Denver, CO, pp. 320–325.
11. Puttlitz, K. J., and W. F. Shutler, "Extendability of Key Interconnection Elements to Meet Future Packaging Requirements," *Proceedings of the 4th International Symposium on the Physical and Failure Analysis of Integrated Circuits,* Singapore, Nov. 1993, pp. 146–151.

Chapter

6

Ceramic Ball Grid Array Assembly

Donald R. Banks, Karl G. Hoebener, and Puligandla Viswanadham

6.1 Introduction

The ever increasing demand for high-density and high-performance packages in the electronics industry in recent years has resulted in several innovations at the module level as well as in the interconnection scheme. Conservation of card real estate through increasing wiring and interconnection densities has become a significant as well as attractive trend. Some recent interconnection developments evolved from prevalent chip-level technologies such as Controlled Collapse Chip Connection (C4). One such second-generation leadless package that is an extension of flip chip technology utilizes solder balls to provide interconnection from the package to the pads on the carrier.[1] One feasibility study suggests that for designs requiring more than 200 I/O, ball grid array packages will be the preferred surface mount design.[2] Ceramic Ball Grid Array (CBGA) packages, when compared to plastic packages, have advantages such as low inductance, improved signal-to-noise ratio, and better cooling performance. From a cost perspective, they may be considered more economical due to their nonhermetic, pinless structure, with less ceramic and no gold brazing. Current applications of CBGA packages include, among others, IBM PS/2* Model 95, RISC System 6000* workstation, and AS/400* System.[3] The PowerPC† family of microprocessors is offered in CBGA package versions.[4] Area array packages in general enable high-function component assembly

* Trademark of the IBM Corporation.

† PowerPC is a trademark of the IBM Corporation and has been licensed by Motorola, Inc.

with low-tech assembly processes and offer better real estate utilization; hence they are likely to be more prevalent in future second-level packaging.[5] Coplanarity and skew problems that are generally associated with the leaded packages are largely absent in CBGA packaging, so it is an attractive surface mount package option.[6, 7]

6.2 Environment

With increasing sophistication in electronics and attendant increase in packaging density and functionality, powerful mobile computing machines are becoming pervasive. Due to the portability of these units they are required to operate in much harsher environments than the machines of yesteryears. Typical operating environments for electronic hardware can be categorized as military and avionics and automotive under the hood environment, space and telecommunications, and consumer and commercial electronics. Typical temperature and humidity ranges for these environments are shown in Table 6.1.

Increased portability also increases the chances for mechanical damage. The above classification is by no means rigid, and considerable overlap of the use environments is possible. Products are also likely to experience various levels of corrosive gases, particulates, fungus, and harmful radiation that could impact their reliability.[8]

6.3 Carrier Geometries

Because of their multichip module and high I/O capabilities, ceramic ball grid array packages are generally assembled on high-density multilayer (4-12) boards of 1.32 to 1.83 mm containing 0.30 mm or larger vias. A 1.27-mm pitch ceramic ball grid array package can provide 62 I/O per cm^2, compared to 31 I/O per cm^2 in a 0.3-mm pitch 400-lead quad flat pack. The escape or fanout wireability on the card depends on the grid array pitch. In general the higher density of the CBGA requires wireability of 197 cm/cm^2.

TABLE 6.1 Typical Operating Environments for Electronic Assemblies

Environment	Temperature (°C)	Humidity (percent)	Product life (years)	Annual cycles
Automotive under-the-hood, avionics, and military	–55 to 125	0 to 100	up to 10	3000
Space and telecommunication	–40 to 85	0 to 95	up to 20	9000
Consumer and commercial	–20 to 60	0 to 90	up to 5	1500

CBGA carrier pads utilize different designs including the dog bone, where the vias are separated from the pads. Alternatively, the via-in-pad design could be used. Vias could be filled with an appropriate material and overplated with copper to form the pad. This design would increase wireability on the top layer and reduce the potential for solder bridging. However, this may or may not be a cost-effective option. Figure 6.1 shows the two types of pads. While the via-in-pad design facilitates the use of larger pads, provides open-wiring channels, and marginally better electrical performance than the dog-bone design, it has significant process implications. Molten solder migrating from the joint to the vias can cause reliability exposures. Unfilled vias can thieve solder from the interconnection and produce considerable variation in solder joint structures and hence weaker solder joints. Current card technologies that can accommodate CBGA packages assume 0.127-mm lines on 0.152-mm spacings, 0.356-mm plated-through holes with 0.508-mm caps, 0.762-mm pad diameters, one line per channel on top side, and two lines per channel internally and on the bottom side. A component clearance spacing of 5.08 mm is recommended to facilitate for board-level rework.

A carrier technology based on thin film polyimide with dry-etched and plasma-etched microvias (PEV) has recently been described.[9] The technology enables 3–8 ppm carrier CTE capability. The carrier is amenable to multilayer lamination and processability in a conventional manner, and is versatile for rigid flex applications. Vias as small as 75 to 100 microns with a density of 100 vias/cm^2 and line widths of 75 microns were demonstrated. Application to multichip module assembly was indicated. These advanced substrates are likely to be potential candidates for finer pitch area array tape-automated bonding (TAB), ceramic ball grid array packages, and flip chip attach assemblies in high-density packages.

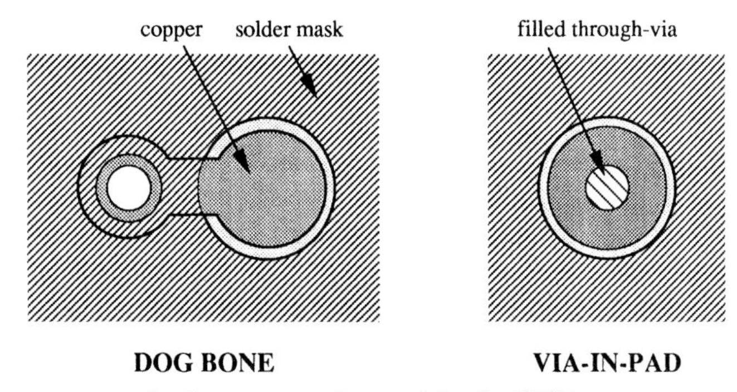

Figure 6.1 Surface mount pad geometries for CBGA.

6.4 The CBGA Package

Ceramic ball grid array packages currently consist of both rectangular and square packages. Body sizes range from a nominal 18 to 44 mm and the lead counts encompass a broad range of 196 to over 1000 leads. Table 6.2 shows the prevalent body sizes, the lead counts, and the lead pitches as proposed by the Joint Electron Device Engineering Council (JEDEC) Committee.

Ceramic ball grid array package substrates are generally made of cofired alumina with molybdenum metallization. Though wire-bonded chips and cavity-down configurations are possible, die are typically C4-mounted on the top side, permitting a full array of carrier interconnections on the bottom side of the substrate. Interconnection pad grids are provided on a regular array with spherical standoffs composed of 90/10 Pb/Sn spheres. These spheres are attached to the backside of the substrate with eutectic (63/37 Sn/Pb) solder. The eutectic solder joint at the ceramic interface should be controlled in such a manner as to reduce lead (Pb) dissolution into the eutectic solder. This allows the top joint to melt at typical assembly temperatures, allowing the joint to equilibrate to optimum planarity and centrality. Figure 6.2 shows a typical ceramic ball grid array package. According to JEDEC guidelines mentioned previously, the ceramic substrate should have a flatness to within 0.10 mm, with a ball planarity to within 0.15 mm across the package for satisfactory yields.

The high I/O available in the CBGA package allows for a greater number of power and ground connections, and for shielding and thermal dissipation into the inner planes. Currently, cooling capacity of a typical 25-mm CBGA package is from 2.6 to 2.9 watts without forced air. It is possible to cool a 3.2- to 11-watt chip with forced-air cooling, based on a junction temperature of <85°C and inlet temperature of 30°C.

TABLE 6.2 CBGA Packages

	Square Packages		Rectangular Packages	
Pitch (mm)	size (mm)	leads	size (mm)	leads
1.27	18.5 × 18.5	196	18.5 × 21	224
	21 × 21	256	21 × 25	304
	25 × 25	361	25 × 32.5	475
	32.5 × 32.5	625	32.5 × 42.5	825
	42.5 × 42.5	1089		
1.00	18.5 × 18.5	296	18.5 × 21	340
	21 × 21	400	21 × 25	480
	25 × 25	576	25 × 32.5	744
	32.5 × 32.5	961	32.5 × 44	1302
	42.5 × 42.5	1764		

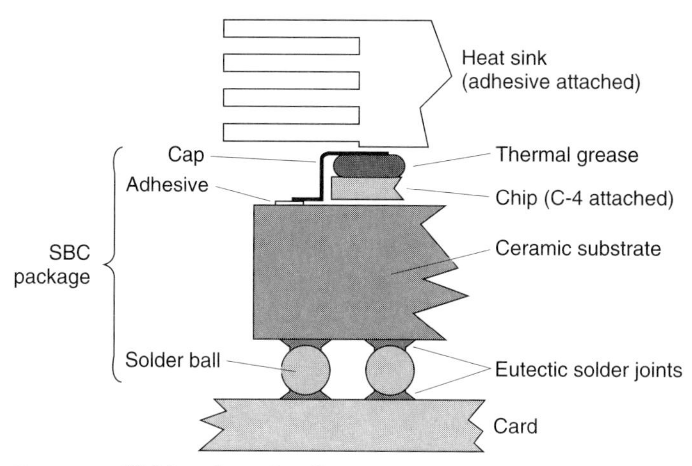

Figure 6.2 CBGA package detail.

6.5 Assembly Process

The assembly process, as is customary in many developmental activities, was determined by using test vehicles that represent worst case scenarios in regard to component density and complexities to exercise design extremities.[10] Typical 25.4 × 38.1 cm test cards with four to eight layers, having four to 51 CBGA packages per card were used. Ceramic ball grid array package wiring pattern layouts within the ceramic substrate and on the card were designed to provide concentric rings or daisy chains of test networks. In some cases, each network was associated with a specific distance from the neutral point (DNP). Biasing alternate nets with direct-current voltages and monitoring current leakages allowed evaluation of cleaning efficiency under the modules through determining the propensity for electromigration due to inadequate cleaning. This limited the number of testable joints in a given module, but electrical test access to more solder joints was accomplished by wiring joints into a larger network on a separate test vehicle to assess assembly yield and solder joint integrity under accelerated thermal cycling (ATC).

It is possible to populate both sides of a printed circuit card with CBGA modules, but most current applications use CBGA on only the topside. Assembly process qualifications generally include a 2× rework to encompass occasional card rework in the actual manufacturing environment.

Solder ball and solder paste interactions before and after placement during the process development may be studied using modules with optical glass substrates instead of alumina substrate.

Table 6.3 shows a typical assembly process that incorporates ceramic ball grid array packages. The processes are essentially the same as for conventional SMT and SMT/PIH processes.

Placement errors are commonplace in surface mount assembly processes. However, ceramic ball grid array package placement is generally very forgiving compared to peripheral leaded packages. The propensity for CBGA self-alignment is due to the large surface area of the 90/10 Pb/Sn ball and the surface area minimization of the molten 63/37 Sn/Pb solder. A misregistration recovery of as much as 0.35 mm is possible. Normally, a 50-percent misregistration is permitted in coarse-pitch SMT assembly operations; this is tolerated by the CBGA packages quite well. A typical placement accuracy profile for a manual split optics placement tool and a high-speed automatic placement tool is shown in Fig. 6.3. It can be seen that package placement accuracy is well within a –0.10- to +0.18-mm range. Satisfactory ball-to-pad spacings can be obtained with a placement force in the range of 1.5 to 5 pounds and the placement accuracy is independent of the force in that range. Solder paste holds the CBGA package in place under normal handling and processing.

The interconnection can be made using either a vapor phase, infrared, or convection reflow process. Vapor phase reflow for array packages has recently been discussed by Wasielewski.[11] Ceramic ball grid array packages require heat under the packages, and hence a preheat step prior to vapor phase reflow is desirable. Since vapor phase reflow will carry the flux with the condensed fluid off the board, in the high-heat capacity CBGAs, fluid drag-out could be a problem. Infrared heating was successfully applied in the reflow of ceramic ball grid arrays. A typical infrared reflow profile for the packages is shown in Fig. 6.4. It depicts, respectively, the card surface temperature and the

TABLE 6.3 Hybrid SMT/PIH and Double-Side, Double-Pass Infrared Reflow Processes

SMT/PIH hybrid process	Double-side/double-pass SMT process
Components and cards	Components and cards
Screen solder paste (top)	Screen solder paste (bottom)
Place topside components	Place bottomside components
Infrared reflow	Infrared reflow
Clean	Clean
Dispense adhesive	Screen solder paste (top)
Place bottomside surface mount components	Place topside surface mount components
Adhesive cure	Place topside PIH components
Place topside PIH components	Reflow
Wave solder	Clean
Clean	Test and repair
Test and repair	

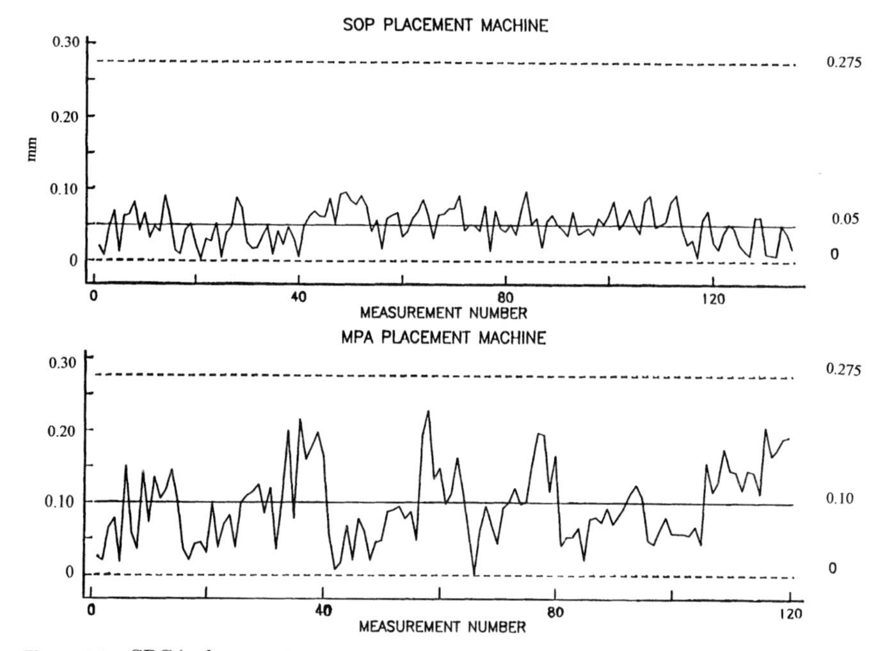

Figure 6.3 CBGA placement accuracy.

temperature under two different module types (ceramic and plastic) at three locations on the card. In CBGA applications where the packages are poorly balanced across a printed circuit card, special care in reflow profile characterization must be exercised. Such applications may be more easily manufactured in a full forced-gas convection furnace. As with other SMT joints, oxygen levels in the reflow oven maintained in the 100- to 1000-ppm range will yield better solder joints with little or no surface graininess.

6.6 Cleanability

Compared to conventional surface mount components, CBGA packages have a relatively high standoff of 0.89 to 1.0 mm. Cleaning under CBGA packages is generally not a problem, and CBGA packages present no surface insulation resistance failures or electromigration problems. Current ball pitches of 1.0 mm and 1.27 mm ensure cleanability between CBGA leads. Also, solder mask is present between joints, providing little chance for adherence of contaminants. The most important factor in cleaning CBGAs is the large number of interconnections under the module. A significant volume of water or solvent under the component is necessary to dissolve residues and remove them from

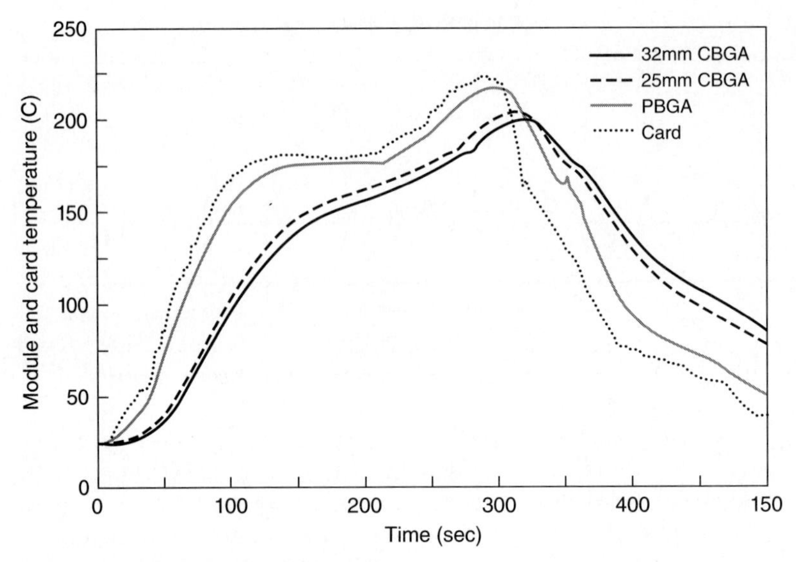

Figure 6.4 BGA infrared reflow profile.

under the component. Drying the module can be problematic. Adequate heat and air flow is required to remove all water or solvent.[12]

6.6.1 No-clean flux assembly

In any cost-competitive environment, reducing process steps, material costs, energy costs, as well as eliminating capital-intensive equipment and tooling are manufacturing advantages. No-clean soldering materials to eliminate post-soldering cleaning and subsequent use of ozone-depleting chlorofluorocarbons are increasingly coming into use.[13] CBGA technology works well in a no-clean manufacturing scheme. Assembly processes with no-clean fluxes are not expected to present insurmountable technical challenges to ceramic ball grid array packages. However, owing to the low activity of some fluxes, the solderability of the high-melt solder ball and pad wetting should be carefully examined. Process optimization will be required to migrate from high-activity solder pastes to relatively low activity no-clean systems. Satisfactory results have been reported with no-clean solder pastes on benzotriazole-coated and solder-leveled cards for both ceramic ball grid arrays and ceramic column grid arrays by adhering to the paste screening and reflow guidelines provided by the manufacturer.[14]

6.7 Inspection

The majority of ceramic ball grid array solder interconnections are located under the module and are not amenable for easy optical inspection. Using a capital-intensive x-ray inspection system is one way to perform total inspection. However, if a quality assembly process with requisite process parameters and controls is developed, then a periodic inspection audit to verify that the process is in control may be all that is needed to produce a quality product. An x-ray tool could be used for both process development as well as for the audit. A philosophy of inspecting the process rather than inspecting the product would serve the assembler well in a cost-competitive environment.[15] It is obvious that once reflow has been performed, interior solder joints can neither be inspected nor touched up.[16] Verification of the different stages of the assembly process such as paste screening, placement, and reflow would essentially alleviate the difficulties of inspection. Accurate paste volume measurement should take into account *x*-, *y*-, and *z*-axis precision and accuracy of the tool, the number of data points on the surface of the paste, and estimation of the zero height (i.e., the base). Consideration must also be given to the variation of the card surface itself in the area of measurement. This may be influenced by the solder mask thickness. The sampling frequency, if not 100 percent, should be chosen to minimize events such as paste stencil clogging and degradation. An in-line 100 percent solder paste inspection, either visually or with a three-dimensional laser inspection tool, is one way of assuring good yields.

Common solder joint inspection tools for process developmental work are x-ray laminography, scanned beam laminography, tomosynthesis, and the like. Typical solder defects are solder bridging, solder voiding, opens, poor wetting, solder balls, contamination, and misregistration. Opens are caused by severe nonwetting of the solder pads or screening deficiencies. Shorts are caused by misregistration of the package, poor handling, and/or poor wave solder process control in the SMT/PIH hybrid process.[14] Figure 6.5 shows a schematic representation of the different solder joint defects applicable to CBGA assemblies. Cross-sectional slicing at different planes (pad level, ball center level, and carrier level) provides valuable information on the solder joint integrity and quality.[17]

Successful assembly processing depends on monitoring and controlling key carrier, module, and process parameters. Parameters and requirements are shown in Table 6.4.

6.8 Rework and Repair

The board-level rework process for CBGA modules is straightforward and is shown sequentially as follows:

1. Remove module
2. Dress site
3. Flatten site (if needed)

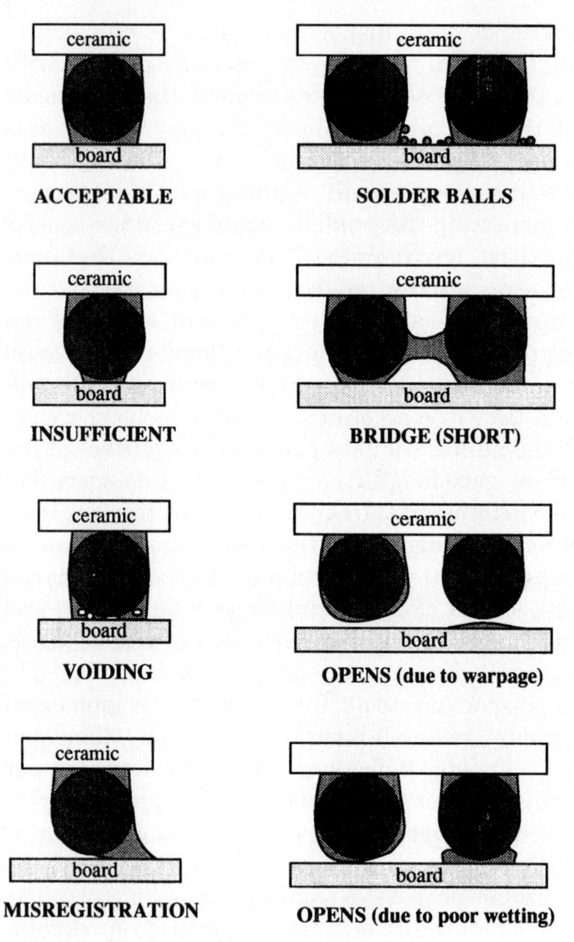

Figure 6.5 Potential CBGA solder joint defects.

TABLE 6.4 Key Assembly Parameters for Successful CBGA Assembly

Category	Parameter	Requirement
Module	Solder ball planarity	0.15 mm maximum
	Ceramic substrate planarity	0.10 mm maximum
Carrier	Pad diameter	0.75 ± 0.04 mm
Assembly	Solder paste	90 weight percent solids
	Stencil thickness	0.20 mm
	Stencil aperture diameter	0.81 mm
	Paste print height	0.20 mm nominal
	Paste volume (per joint)	0.115 mm^3 nominal
		0.079 mm^3 minimum
	Reflow temperature	220 ± 20°C
	Dwell (above 183°C)	45 to 145 seconds

4. Apply solder
5. Place module
6. Reflow
7. Clean (if needed)

The large thermal mass and the large heat capacity of the module pose a special challenge in the rework process. Also, if the spacing between the CBGA package and neighboring components is not adequate, removal of neighboring components becomes a prerequisite for the CBGA rework process. In one rework process flowing hot gas is employed. The entire card is preheated and the site to be reworked is subjected to hot air or nitrogen with a hot gas rework tool. Generally, off-the-shelf rework tools are suitable, and custom equipment is not needed to rework CBGA packages. Care must be taken in selecting rework tools, however, to ensure that adequate heating and air flow is available to handle high thermal mass packages on thick carriers.

Since eutectic solder is used on either side of each ball, high-melt solder balls are usually left randomly attached to the card. The site may be exposed to a solder fountain to remove the balls as well as the excess solder. Local deformations of the site (oil-canning), caused by the solder fountain, require a site-flattening step to prepare the site for module replacement. Using a vacuum site dressing tool will achieve site dressing without local deformations and can eliminate the site-flattening step.

As in initial assembly, it is important to control reworked CBGA joint solder volume to ensure yield and reliability. Addition of solder to the site to provide adequate solder volume in the joints on the replacement package may be accomplished in one of several ways. Solder paste may be screened onto the surface mount pads of the newly-dressed site with a miniscreener. Alternatively, a liquid deposition tool may be used to put solder paste individually on each surface mount pad. Using custom tooling, solder paste may be screened directly onto the balls of the replacement module with a stencil having appropriate size openings. The amount of paste applied to the module can be controlled by weighing the module before and after paste screening. Another technique is to apply a solder preform having the proper dimensions and volume, and flux the site before package placement.

The new package is typically placed onto the board with the aid of split optics vision present in many rework tools. Side camera vision also works; also, a mechanical placement nest may be used. The replacement package is reflowed with hot gas, cooled, and cleaned, if necessary. Nitrogen reflow is again recommended for improved solder joint appearance. Peak reflow temperature is critical; it should be hot enough to ensure reflow but not excessively hot to cause local board distortion or reflow of nearest neighbor components.[11, 18] A detailed description of the rework of the ceramic ball grid array packages is described elsewhere in this volume.

Also, a reball/recolumn operation was successfully demonstrated for rework and repair.[19] High value CBGA packages that have been removed from a printed circuit card may be reconditioned with fresh solder balls and reused, if the application permits. The reconditioning operation consists of removing residual balls and excess solder through an elevated temperature operation. Next, the redressed ceramic substrate is put through the original ball attach procedure at the package manufacturer. Careful process characterization is needed to recondition CBGA packages. Some CBGA users may find that reconditioned packages present an unacceptable risk in their applications.

6.9 Finite Element Modeling

Assembly process development is significantly enhanced through the use of three-dimensional modeling. This is important given the noninspectability of the interconnection array of solder joints under the components. Areas that benefit from modeling include joint strain estimates for different card-pad and module-pad diameters under different loading conditions and solder joint configurations that facilitate the choice and optimization of appropriate pad dimensions.[20] Distribution of plastic strain maxima in a 25-mm CBGA/FR-4 system was estimated by Corbin for a corner joint at 100°C with an 0.86-mm diameter module pad and 0.61-mm card pad. The risk of fatigue fail was calculated to be greater in the card-side joint. The maximum percent strains were reported to be 1.91, 0.92, and 4.31 for eutectic solder on the module side, solder ball, and on the eutectic solder on the card side, respectively. This was expected since, in this case, the card-side pad was smaller than the module-side pad. This estimation was borne out in that actual fails were indeed observed in the eutectic solder in the joint nearest the card. Structural optimization can thus be achieved by designing the two pads in the joint with diameters as close as possible within design and manufacturing constraints.

Finite element modeling also enables identification of preferential failure locations on the module side and card side within the system. It was estimated that axial deformation from the bending of the assemblies results in corner joint fails.

Another area where modeling came to the rescue of process development was in solder joint reflow. For example, in the infrared reflow process the radiative and convective conditions experienced by the card and module continually change as the assembly travels through the reflow oven. The time-dependent radiative exchanges were modeled to derive module and card temperatures at various locations, including center and corner solder balls, and the card area near and away from the module. Such modeling studies enabled the optimization of reflow profiles. A typical predicted and observed reflow profile is shown in Fig. 6.6. The temperature differentials within the CBGA module are rather small, and the card area away from the module heats and cools faster than the module owing to lower thermal mass and higher emissivity.[21]

6.10 Reliability Testing/Results

Solder joint reliability depends upon component quality, carrier quality, and assembly process controls. Due to high (0.89- to 0.99-mm) standoff and coarse pitch, corrosion and electromigration are almost nonexistent. Solder joint reliability issues are predominantly due to the nature of the package construction and several assembly process conditions. These include package-side solder volume, solder ball planarity, flatness of the substrate and carrier, lead (Pb) dissolution into the solder volume, and attachment reflow time. In any process development it is important to be cognizant of these factors and evaluate them in each of the circumstances. In process development, when card and package parameters were controlled, the reliability effects of card-side solder volume were tested by varying the solder paste volumes within the confines of the process limits. Once the optimum card-pad diameter was established, minimum fillet diameter of the card-side joint became the controlling factor for reliability. Thus, the printed paste volume is the process variable that becomes the critical factor.

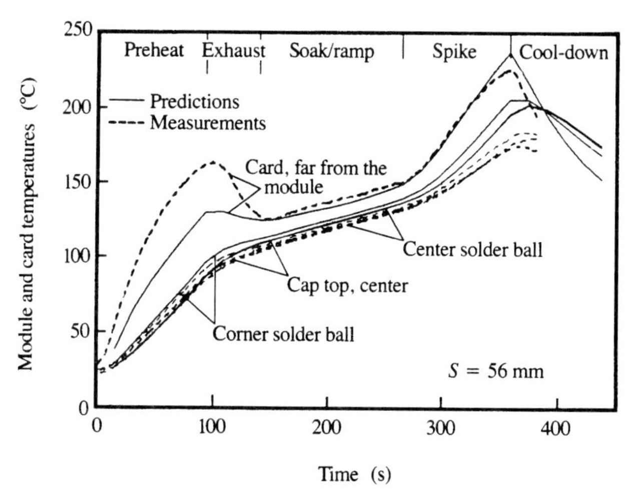

Figure 6.6 Modeled and actual reflow profiles for CBGA assemblies. *(Photo courtesy of Dr. V. Mahaney of IBM Austin, Texas.)*

Secondary process conditions such as wave solder and adhesive curing operations are less likely to have a reliability impact.

Accelerated testing has been performed over many temperature ranges: –40 to +125°C, 0 to 100°C, or 20 to 80°C, for example, with a frequency of one to three cycles per hour depending on test-chamber capability. To determine if a given package can meet application requirements, multidimensional plots of field cycles versus the coefficient of thermal expansion (CTE) of the assembly and the temperature differential at various defect levels are useful. Figure 6.7 shows a series of such plots at 3.4, 100, 500, and 1000 ppm for a 25-mm CBGA package with a 13.74-mm DNP. The number of cycles to which the modules are subjected is dependent on the expected product life. Caulfield et al. provided a comparison of reliability testing results between 0 to 100°C and 20 to 80°C and showed that fatigue life in the 20 to 80°C range is longer than at 0 to 100°C by at least a factor of 3.[22]

6.11 Failure Modes and Analysis

Solder joint integrity and reliability are aspects of primary concern in CBGA assemblies. The wearout mechanism is generally characterized during process development by examining the solder joints through

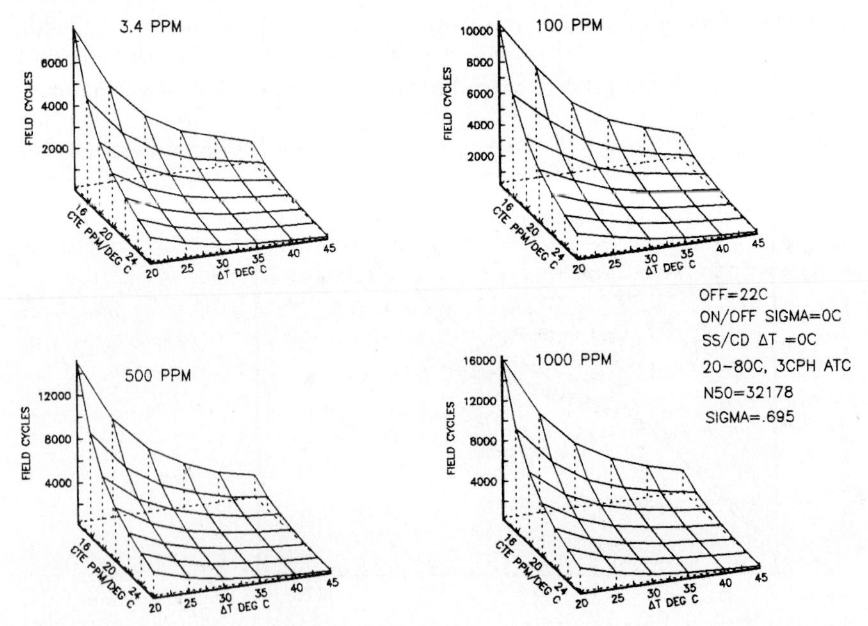

Figure 6.7 Field cycles as a function of coefficient of thermal expansion (CTE) and ΔT for a 25 mm CBGA assembly (DNP = 13.74 mm).

accelerated thermal cycling. Cross sections of the packages after card assembly enables examination of a particular row of joints for fatigue failures. Also, separation of fatigue-failed joints by optical and scanning electron microscopy permits the examination of the entire crack surface. A new image processing technique, the computational Fourier transform Moiré method was developed to provide whole field strain maps from digitized images from the interference fringes using laser Moiré interferometer.[23, 24]

Fatigue fails generally begin in the eutectic solder regions on the package and card side after a certain number of cycles and tend to propagate in that region. Complete failure (separation) was found to occur most often on the card-side joint as shown in Fig. 6.8*a*. For comparison, a "good" joint is shown in Fig. 8*b*. The path of propagation is not along the joint interface, but in the coarsened Pb-rich phase regions in the solder and toward the ball. No fails through the solder ball have been observed. Considerable movement of the ball from its initial position has been seen. The ball tends to remain with the side of the joint having the largest solder fillet, as solder relaxation occurs at elevated temperature during thermal cycling. During creep, the ball is carried along with the solder to a new position. The ball thus moves incrementally in each cycle as the thermal cycling advances. The magnitude of this shift is dependent on the distance of the ball from the neutral point. This phenomenon of the ball movement during the stress relaxation is called the "racheting" effect and is shown in Fig. 6.9. Cracks appear to grow more during the cooling cycle than during the heating cycle. Actual strain distributions are superpositions of global and local strains. Both shear (as a thin interlayer) and (column) bending strains are encountered. Height and diameter dictate the dominant strain mode. For shorter heights the shear mode is predominant, while in longer columns the bending mode is predominant.

CBGA solder joints assembled to a printed circuit card and then epoxy-underfilled were found to have a failure mechanism when tested at −40 to 125°C different from those tested at 0 to 100°C. The final crack propagating through the eutectic solder was at the ceramic side of the high-melt solder ball, probably due to adhesive failure of the underfill to the ceramic.

6.12 The Ceramic Column Grid Array Package

Ceramic Column Grid Array (CCGA) packages have been evaluated as a logical extension of CBGA technology.[25] CBGA packages, especially those with body sizes ≥32 mm have inherent reliability limitations in some applications. One way to enhance reliability is to increase stand-

Figure 6.8*a* Typical CBGA solder fatigue failure.

off to more than the 0.89 to 0.99 mm of CBGA. Reliability was tested on CCGA packages having 0.50-mm diameter, 2.21-mm-tall columns. Two different column types were tested: a 90/10 Pb/Sn wire column, attached to the ceramic substrate with eutectic solder, and a cast-in-place 90/10 Pb/Sn column, with no eutectic solder joint between the column and substrate. It was demonstrated that both column packages can be assembled and reworked exactly the same as CBGA, although more care must be taken in handling. The cast column package was more aesthetically appealing and resistant to handling damage. The wire column package exhibited better solderability and the ability to have bent columns straightened.

A direct comparison of accelerated thermal cycling tests at 0 to 100°C showed the CCGA packages had N50 values five to seven times higher than CBGA packages. Projections to field use conditions indi-

Figure 6.8*b* Good joint.

cated that modules with columns would fail at a rate up to two orders of magnitude lower than identical CBGA packages. No statistical difference in reliability was observed between the two column types. A failure mechanism different from that of CBGA was present. As shown in Fig. 6.10, CCGAs failed through the column, not the eutectic solder. Failures were independent of screened solder paste volume, unlike CBGA. The different failure mode meant more predictability in joint lifetime because the diameter of the column is easily controlled. CCGA has a higher compliance, allowing the columns to undergo bending to compensate for thermal mismatch. Column-length, column-diameter, and pad-size optimization work is under way. Potential problems include high package manufacturing costs, shipping and handling damage, and excessive joint creep in applications requiring heavy heat sinks.

Figure 6.9 CBGA joint "racheting" caused by accelerated thermal cycling.

6.13 Future Trends

Application of CBGA technology to surface mount technology is already proving to be a significant step in accommodating high-density packages on existing carriers without using fine-pitch peripheral-leaded packages that can be difficult to assemble. Obvious ceramic ball grid array advances include denser arrays, and work is underway on both interstitial and finer (0.50- to 0.65-mm) pitch packages. Further enhancements to CBGA technology include a combination of array and peripheral-leaded technology in multichip module (MCM) packaging. This concept was used to enhance routing resources, decouple the signal and power/ground distribution and also provide good thermal coupling between package and board. Peripheral leads are used for functional I/O signals while the short-path CBGA interconnections are used for power/ground and clock lines with low resistance and inductance

Figure 6.10a Ceramic column grid array: as-soldered structure.

paths.[26] This mixed technology provides some advantages, but may present assembly challenges in solder reflow and cleanability. Another development is the Slightly Larger than IC Carrier (SLICC). This idea takes advantage of the high volume efficiency of the flip chip package, combined with the simplicity of the array package for assembling to a printed circuit board. It is applicable for low I/O (≤100) package applications, and can enhance testability and repair of the package.[27] While these developments are aimed at plastic ball grid arrays, they can be extended to include ceramic ball grid arrays as well.

6.14 Summary

Ceramic ball grid array packages offer an efficient alternative to finer-pitch peripheral-leaded packages that are increasingly difficult to assemble in terms of component fragility, placement, paste screening,

Figure 6.10*b* After ATC-induced failure.

reflow, cleaning, and rework operations. Comparing pin grid array (PGA), TAB, and quad flat pack (QFP) with surface mount array packaging indicates that significant savings in card real estate can be gained by using CBGA. A 1.27-mm pitch CBGA can accommodate a 620-I/O package in 10.3-cm^2 area while only a 115 I/O PGA or QFP can be accommodated in a similar area.[22] Assembly solder defects are significantly lower for CBGA assemblies than fine-pitch quad flat packs; an order of magnitude defect-level improvement is achievable. Ceramic ball grid array packages can easily be assembled along with other SMT components without major changes to assembly equipment and processes. Reliability of CBGA assemblies is equivalent to or better than conventional surface mount packages if appropriate process controls are incorporated. Noninspectability of the packages is not seen as a serious impediment to producing reliable assemblies with high yields once the assembly process is developed with key parameters well defined and controlled. Use of ceramic ball grid assemblies is expected to be pervasive in second-level electronic packaging for some years to come.

6.15 Acknowledgments

The work reported in this chapter was based on the efforts of many individuals within the International Business Machines Corporation, and Motorola, Incorporated. The helpful discussions and suggestions we received are gratefully acknowledged.

6.16 References

1. Buschomb, M. L., W. H. Schroen, and E. R. Wolf, "Fine Pitch Packaging for Surface Mount," *Handbook of Fine Pitch Surface Mount Technology,* J. H. Lau, ed., Van Nostrand Reinhold, New York, New York, 1994, pp. 55–80.
2. Randall, J., et al., "A Feasibility Study of Ball Grid Array Packaging," *Proceedings NEPCON East,* February 1993, pp. 413–422.
3. Tuck, J., "The BGA: The Next Chapter," *Circuits Assembly,* **4**(8), 1993, pp. 24–25.
4. Banks, D. R., T. E. Burnette, R. D. Gerke, E. Mammo, and S. Mattay, "Reliability Comparison of Two Metallurgies for Ceramic Ball Grid Array," *Proceedings of the International Conference and Exhibition on Multichip Modules,* April, 1994, pp. 529–534.
5. Cadigan, M., "Trends in Electronic Packaging," *Circuits Assembly Market Supplement,* **4**(9), 1993, p. 19.
6. Gilleo, K., "The SMT Chip Carrier: Enabling Technology for the MCMs," *Electronic Packaging and Production,* **33**(9), 1993, pp. 88–89.
7. O'Brien, K. O., "Will BGA Live Up to Its Billing?," *Surface Mount Technology,* **7**(8), 1993, p. 40.
8. Viswanadham, P., "Reliability Aspects of Fine Pitch Assembly," *Handbook of Fine Pitch Surface Mount Technology,* J. H. Lau, ed., Van Nostrand Reinhold, New York, New York, 1994, pp. 598–636.
9. Schmidt, W., "A Revolutionary Answer to Today's and Future Interconnection Challenges," *Proceedings Sixth Printed Circuit World Convention,* 1993, pp. T12-1–T12-7.
10. Ries, M. D., D. R. Banks, D. P. Watson, K. G. Hoebener, "Attachment of Solder Ball Connect Packages to Circuit Cards," *IBM Journal of Research and Development,* **37**(5), 1993, pp. 597–608.
11. Wasielewski, J., "Vapor Phase Reflow for Fine Pitch Assembly," *Handbook of Fine Pitch Surface Mount Technology,* J. H. Lau, Ed., Van Nostrand Reinhold, New York, New York, 1994, pp. 308–332.
12. Bartley, J., D. Best, and P. Isaacs, "CBGA: A Packaging Advantage," *Surface Mount Technology,* Nov. 1993, pp. 35–36, 40.
13. Melton, C., "Non CFC Cleaning (No Clean) of Fine Pitch Assemblies," *Handbook of Fine Pitch Surface Mount Technology,* J. H. Lau, ed., Van Nostrand Reinhold, New York, New York, 1994, pp. 479–487.
14. Gerke, R. D., "Ceramic Solder Ball Grid Array Interconnection Reliability over a Wide Temperature Range," *Proceedings NEPCON West,* February 1994, pp. 1087–1094.
15. Zweig, G., "BGA: Inspect the Process, Not the Product," *Circuits Assembly,* **50**(2), 1994, p. 92.
16. Beck, M. R., and R. J. Weldon, "Solder Paste Inspection for Ball Grid Array (BGA)," *Proceedings NEPCON West,* February 1994, pp. 686–692.

17. Adams, J. A., "Using Cross Sectional X-ray Techniques for Testing Ball Grid Array Connections and Improving Process Quality," *Proceedings NEPCON West,* February 1994, pp. 1257–1265.
18. Weldon, R. J., "3-D Inspection of Solder Paste for High Quality BGA Assembly," *Proceedings First International Symposium on Flip Chip Technology,* San Jose, February 1994, pp. 52–56.
19. Banks, D. R., G. Phelan, and M. Cole, "Letter to the Editor," *Circuits Assembly,* **4**(8), 1993, p. 8.
20. Corbin, J. S., "Finite Element Analysis of Solder Ball Connect Structural Design Optimization," *IBM Journal of Research and Development,* **37**(5), 1993, pp. 585–596.
21. Mahaney, V., "Thermal Modeling of Infrared Processes of Solder Ball Connect," ibid., pp. 609–619.
22. Caulfield, T., J. Benenati, and J. Acocella, "Cost Effective Interconnections for High I/O MCM-C to Card Assemblies," *Surface Mount Technology,* July 1993, pp. 18–20.
23. Guo, Y., C. K. Lim, W. T. Chen, and C. G. Woychik, "Solder Ball Connect (SBC) Assemblies under Thermal Loading: I. Deformation Measurement via Moiré Interferometry, and Its Interpretation," *IBM Journal of Research and Development,* **37**(5), 1993, pp. 635–647.
24. Choi, H.-C., Y. Guo, W. La Fontaine, and C. K. Lim, "Solder Ball Connect (SBC) Assemblies under Thermal Loading: II. Strain Analysis via Image Processing, and Reliability Considerations," ibid., pp. 649–659.
25. Banks, D. R., C. G. Heim, R. H. Lewis, A. Caron, and M. S. Cole, "Second-Level Assembly of Column Grid Array Packages," *Proceedings Surface Mount International,* San Jose, August 1993, pp. 92–98.
26. Hashemi, H., M. Olla, D. Cobb, P. Sandborn, M. McShane, G. Hawkins, and P. Lin, "A Mixed Solder Grid Array and Peripheral Leaded MCM Package," *Proceedings 43rd IEEE Electronics Components and Technology Conference,* June 1993, Orlando, pp. 951–956.
27. Banerjee, K., "Development of Slightly Larger than IC Carrier," *Proceedings NEPCON West,* February 1994, pp. 1249–1256.

Chapter

7

Thermal and Electrical Management of Ceramic Ball Grid Array Assembly

Y. C. Lee, Jay J. Liu, Chi-Taou Tsai, Jeffrey A. Zitz

7.1 Introduction

Thermal and electrical management is critical to the performance of ceramic ball-grid-array (CBGA) assemblies. Thermal management controls junction temperatures and temperature gradients among structural elements. Electrical management controls clock speed and various inductive and capacitive noises. In general, the management schemes for CBGA assemblies are not significantly different from those developed for the assemblies with pin-grid-array (PGA) packages and ceramic chip carriers. As a result, a general introduction of different management schemes will not be repeated in this chapter; it can be found in many review articles and books.[1–4] Instead, the chapter will address some specific issues important to thermal and electrical management of CBGA assembly.

For thermal management, experimental results of several CBGA test assemblies will be reported and discussed. At the present time, there are very few publications reporting thermal performance curves with respect to different CBGA configurations. Most of results had a limited application range. For example, Coors and Buck studied a CBGA-on-ceramics SEM Format E module with experimental, analytical, and numerical analysis.[5] Unfortunately, the results were for a conduction-only case due to the SEM-E requirements, so they could not be used for most commercial CBGA assemblies. The experimental results to be reported will provide critical reference data for a CBGA-on-PWB assembly with a C4-connected chip. Five different internal and external heat transfer enhancement schemes will be discussed.

Although five schemes are considered, the reference data still cannot cover every possible CBGA assembly configuration. A generalization method is needed to apply these results for other CBGA designs. A three-dimensional, finite-difference modeling technique will be introduced for the generalization. The model can be calibrated by the experimental data, and after calibration, it can be used to study new configurations. Such a generalization technique has been used to study the effects of package materials and number of thermal vias in a printed wiring board (PWB), and the results are to be presented and discussed. The effect of thermal vias in the CBGA itself was not studied because these vias are not commonly used for most ceramic packages. However, the effect of the package material is indirectly related to the package's thermal vias. If the package thermal conductivity has a significant effect on the thermal performance, then the package thermal vias are expected to be critical also. For these assemblies, we can use thermal vias instead of high-conductivity material to enhance the thermal performance.

For electrical management, two issues are to be addressed. The first study will focus on the comparison of electrical characteristics of CBGA, plastic BGA, plastic quad flat pack (PQFP), and PGA. Due to the cost factor, most CBGA assemblies are to be used for high-speed systems that demand accurate electrical designs. The comparative study will provide a valuable understanding of the performance improvement when the packages are changed from PQFPs or PGAs to BGAs.

The second study will discuss the effect of chip connection schemes such as wire bonding and C4 as well as the scaling effect for high I/O packages. One of the most appealing features of BGA is its high I/O connection capability. The electrical characteristics to be reported with respect to different connection schemes or densities will help designers choose right device and package assembly technologies.

7.2 Thermal Management

In general, thermal management includes internal and external cooling schemes. Packaging designers usually pay attention to internal thermal management by synthesizing different thermal elements for efficient conduction. On the other hand, board designers need to identify the correct external cooling scheme for the packages to be used. Although there is a strong interaction between the internal and the external cooling considerations, the partition between these two schemes is technically reasonable and closely related to design activities.

7.2.1 Internal thermal design

Internal thermal design involves those components within a package that can affect its thermal performance. Some examples are die bond materials, thermal compounds, heat slugs, and encapsulation. Conduction is the dominant mode of heat transfer; thus the designer aims to use high thermal conductivity materials to transfer heat through large cross-sectional areas and over very short distances. Experimentally, the internal thermal resistance parameter is measured as the difference in a representative package temperature (T_{pkg}) and junction temperature (T_j) normalized by the module power dissipation (P_{chip}) and can be written as:

$$R_{int} = \Theta_{JC} = T_j - T_{pkg}/P_{chip}$$

An optimized internal thermal design will achieve three objectives at the lowest possible cost. The primary objective is to get the heat to the surface of the package with a minimum temperature drop. For C4-attached chips, thermal compounds between the chip and the cap can achieve this objective effectively. In a cavity-down wire-bonded (WB) package, the judicious choice of the die bond material is key, in addition to the (costly) addition of a heat slug/spreader for small chips or high-power density situations. Cavity-up wire-bonded chips have no direct path to the package surface and usually cannot achieve this objective. As a result, the R_{int} of these packages can be the dominating part of the total package thermal resistance.

The secondary objective is to spread the heat, striving for large, flat isothermal surface to which a heat sink can be attached. A capped C4-attached CBGA module achieves this most effectively through the use of a well-designed aluminum cap. It is difficult to achieve this in a cavity-down WB package in a cost-effective manner due to difficulty spreading heat through the low-conductivity alumina substrate. The addition of an integrated heat slug or heat spreader will help, but is a costly choice. A cavity-up WB package may achieve this objective by judicious design of the indirect internal heat-flow path, but it is usually of little help without the prior achievement of the primary objective explained previously.

The last objective is to set up a parallel heat flow path to the external environment. While a heat sink or the top of the package may be the path of least resistance, other important heat-flow paths can exist, most notably the conduction path into the card. Indeed, for some low-profile packages conduction into the card is the primary path and convection off the top surface secondary. So it is important to investigate other parallel paths, and explore modifications to the thermal design that will exploit them as well. Some techniques for improving the heat-flow path to the card would be the use of thermal vias in PWB, the selection of C4 or cavity-up WB chip attach, and CBGA module I/O to minimize the downward conduction thermal resistance. An optimum substrate thickness also exists and can be determined easily through modeling.

7.2.2 External thermal design

External thermal design involves those components outside a package that can affect its thermal performance. Some examples are heat sinks and adhesives at the package level; conduction and convection along a PWB; and air-moving devices, ducting, venting and active cooling devices at the system level. Both conduction and convection are the dominant modes of heat transfer, but radiation can also be important in natural convection situations. Again, the thermal designer aims to minimize conduction resistance, but the primary effort shifts to improving convection heat transfer. Large surface-area-per-unit volume and high-heat transfer coefficients are the key focus elements. Experimentally, the external thermal resistance parameter is measured as the difference in a representative package temperature and the cooling air temperature (T_{air}) normalized by the module power dissipation and can be written as:

$$R_{ext} = \Theta_{Ca} = (T_{pkg} - T_{air})/P_{chip}$$

The most common path utilized is through the top surface of the package. Typically the air stream in this vicinity is at its lowest temperature and highest velocity; thus exists the greatest potential for convection heat transfer. The choice of fan or blower in a forced-convection system has the greatest convective effect, secondary to venting, ducting and other measures to balance the air flow within the system. If the external thermal resistance of the module is unacceptable after reasonable system-level modifications, usually the most cost-effective method of further external thermal enhancement is the addition of an off-the-shelf heat sink attached to the module with a thermally conductive adhesive.

Heat transfer from the package itself to the surrounding environment can be maximized by designing the package to be as isothermal as possible. The sides and edges of the package often have better than average heat transfer coefficients due to geometry, impinging air, and local turbulence. Care should be taken during the card layout to ensure the package is not shrouded by passives or other components which will divert cooling air.

Heat flow into the card can have a significant effect on overall thermal performance, but only if it can be transported through the card and removed somewhere else. Copper power planes within the card are the primary vector of the heat, and generally more copper in the card translates to a decrease in the thermal resistance of this path. If the number of copper levels in the card is fixed, the thermal resistance can still be improved by increasing the copper thickness. Common thicknesses are ½ oz (0.0175 mm), 1 oz (0.035 mm) and 2 oz (0.07 mm). The power planes must have a good thermal connection to the module, and maximum via or plated-through-hole contact area, as well as high-module power/ground I/O counts improve this gate in the card path.

A strong linkage between internal and external thermal design is apparent. The achievement of each of the three internal thermal design objectives improves the internal thermal paths as well as enabling optimum performance of the external thermal paths. Balance must be achieved for an optimum overall thermal design. Each thermal path can be thought of as a chain, the links of which are the individual thermal resistance in the path. A weak link or high thermal resistance in any one path may render it almost useless while multiple chains or paths will result in a strong overall thermal design, insensitive to changes in any of the individual paths. The following experimental study will discuss these considerations further.

7.2.3 Experimental study on a 32-mm CBGA with a single C4 mounted chip

7.2.3.1 Test vehicles and experiment. A 32-mm CBGA single chip thermal test vehicle was chosen as the package to represent the CBGA family. Along with the 25-mm package, it is the most common CBGA package in use today, and thus is typical of current products. Both the traditional capped module and recently released capless module were tested in high-profile and low-profile configurations. Five configurations were studied:

1. Low-profile capless CBGA package (see Fig. 7.1)
2. Low-profile capped CBGA package (see Fig. 7.2)
3. High-profile capped CBGA package (see Fig. 7.3)
4. High-profile capless CBGA package-A (see Fig. 7.4)
5. High-profile capless CBGA package-B (see Fig. 7.4)

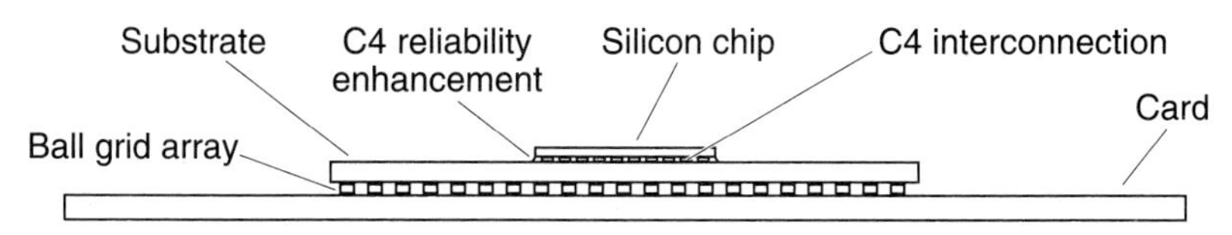

Figure 7.1 Low-profile capless CBGA package.

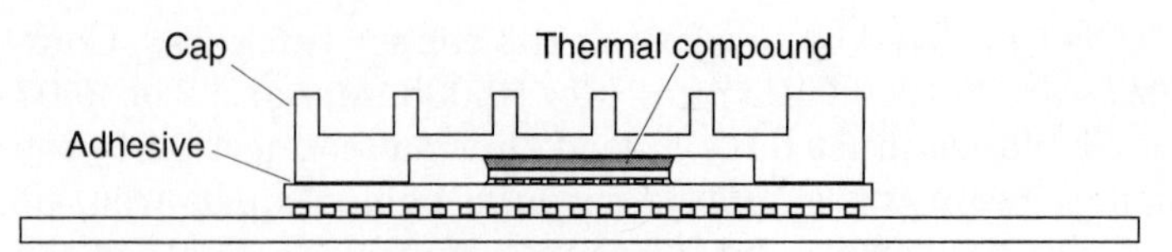

Figure 7.2 Low-profile capped CBGA package.

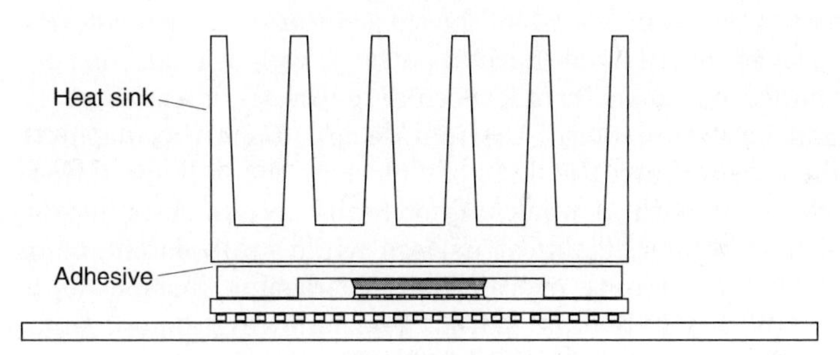

Figure 7.3 High-profile capped CBGA package.

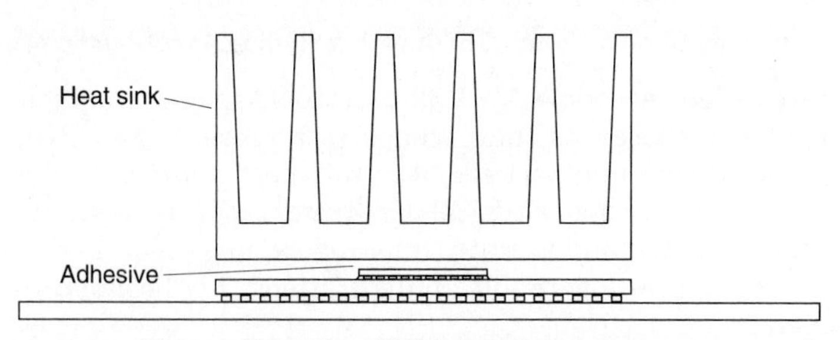

Figure 7.4 High-profile capless CBGA-A package using pressure-sensitive adhesive and CBGA-B package using conductive epoxy.

As shown in Fig. 7.1, the low-profile capless package consists of a $32.75 \times 32.75 \times 3.0$ mm-thick alumina/molybdenum multilayer ceramic substrate with 625 solder ball connection I/O. A single $12.9 \times 12.9 \times 0.625$ mm C4-attached thermal chip is located on the substrate top surface and passivated with a C4 reliability-enhancing epoxy. The number of C4 connections is 1936. The chip power is dissipated within a uniform 10×10 mm area centered on the chip.

Figure 7.2 shows the low-profile capped package. An aluminum alloy cap is placed over the chip and sealed to the substrate with an adhesive. A thermal compound is placed between the chip and the cap to enhance the internal thermal resistance. A cap may be required in a particular design for additional protection of the chip or decoupling capacitors on the module, heat-sink attach, or marking requirements.

For high-profile packages, a 32 × 32 × 23 mm extruded aluminum heat sink with six fins is attached. For a capped package, the heat sink is attached to the cap, as shown in Fig. 7.3. For a capless package, the heat sink is attached to the chip backside directly by pressure-sensitive adhesive (PSA) or by epoxy as shown in Fig. 7.4. The epoxy has a thermal conductivity of 1.06 W/m°C that is better than the thermal conductivity of the PSA (0.46 W/m°C). A detailed summary of the critical thermal elements and the associated dimensions and properties will be listed in Table 7.1.

Each CBGA package was attached to a 107 × 156 × 1.8 mm, 2-signal/4-power (2S4P) card prior to testing. The four 1-oz copper power planes in the card were perforated similar to a product card and 183 of the 625 I/O vias were connected directly to the power planes. Cooling air was forced over the module and module-side of the card, while the backside of the card experienced natural convection conditions throughout testing. Refer to Fig. 7.5; the front surface having heat transfer coefficients of h_1, h_2, and h_4 is cooled by forced convection. The h_4 is for a capless package. The backside of the PWB having a heat transfer coefficient of h_3 is cooled by natural convection.

Testing was performed with an automated air-cooled test facility which stepped through a 2-D matrix of chip powers ranging from 3 to 20 W and air velocities ranging from natural convection to 1200 feet per minute (fpm). The power was adjusted to keep the chip temperature in the typical product range of 40 to 80°C during testing. The air flow was forced over the package and front side of the card inside a 38-mm-high × 150-mm-wide duct. The card and module were oriented vertically in the duct and all tests were conducted at room temperature and +200 ft above sea-level conditions.

As many as three temperatures were measured on each package. The chip temperature was measured using a four-point measurement of an on-chip resistive element (TCR) centered on the chip. Cap and heat-sink temperatures were recorded with a 36-gauge type T thermocouple buried in a drilled hole in the surface and neatly dressed off the package. Approach air temperature is also made with a 36-gauge type T thermocouple fixed in the air upstream of the module.

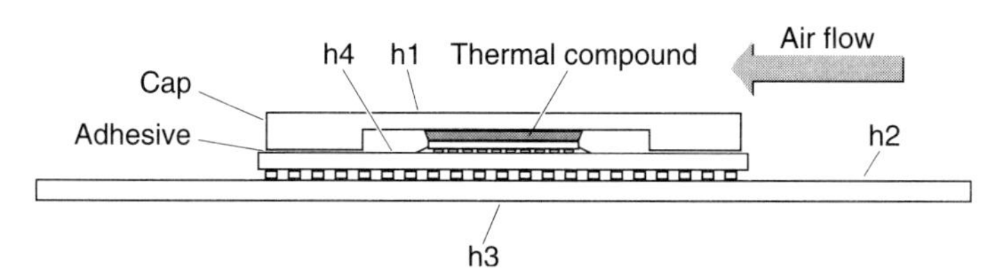

Figure 7.5 Experimental and simulated thermal test assembly.

The thermal resistances were calculated as the differences in these temperatures divided by the total power dissipated in the package. The overall junction-to-air thermal resistance is the temperature difference between the chip center and air per unit power dissipated. The internal junction-to-package thermal resistance is the temperature difference between the chip center and the cap (or the heat sink) per unit power dissipated.

7.2.3.2 Results. The average overall thermal resistance for a cell of six low-profile capless modules are shown as data points in Fig. 7.6. Other curves shown in the figure are simulated results, which are to be discussed later. Due to the low-module surface area for convection, it is likely that the conduction path into the card is very important. At natural convections, the junction-to-air resistance is 10°C/W. If a 60°C temperature rise is allowed, this low-profile capless CBGA assembly can manage a 6-W chip. To improve this performance, a forced convection flow could reduce the resistance to about 6°C/W. Due to insufficient heat spreading from the chip, the forced convection improvement is not substantial.

On the other hand, the improvement, the difference between the maximum and the minimum thermal resistances, reaches a 5°C/W level for a capped package (see Fig. 7.7). Such a significant improve-

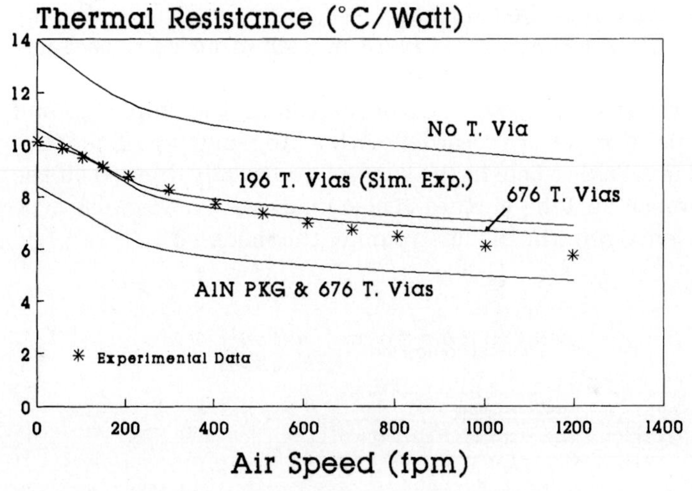

Figure 7.6 Measured and simulated thermal performance curves of the low-profile capless CBGA with different configurations.

ment indicates that R_{ext} is the dominating resistance. Any efforts made at improving the thermal performance of this module should therefore concentrate on external thermal design. Improved thermal pastes, larger chips, and other steps aimed primarily at improving the internal thermal design will have little measurable effect on thermal performance. The measured internal thermal resistance from the chip to the cap confirmed such expectation. The R_{int} changed from 0.47 to 0.58°C/W when the air speed increased from natural convection to 1200 fpm. The internal thermal resistance is very small compared with the overall thermal resistance.

Data for the high-profile versions of the capped and capless modules are shown in Figs. 7.8, 7.9, and 7.10. The reduction in external thermal resistance is significant in all cases, and balance between the internal and external thermal paths is achieved.

The data for the capped high-profile module is shown in Fig. 7.8. The internal thermal resistance from the chip center to the heat sink is 0.7°C/W, which remains constant at different air speeds. Both R_{int} and R_{ext} comprise a significant portion of R_{total}, thus further performance improvements could be internal (better thermal compounds for instance) or external such as an even larger, overhung heat sink. In a typical personal computer (PC) environment (200 fpm, 40°C ambient), this module could dissipate nearly 20 watts before reaching a 100°C junction temperature limit.

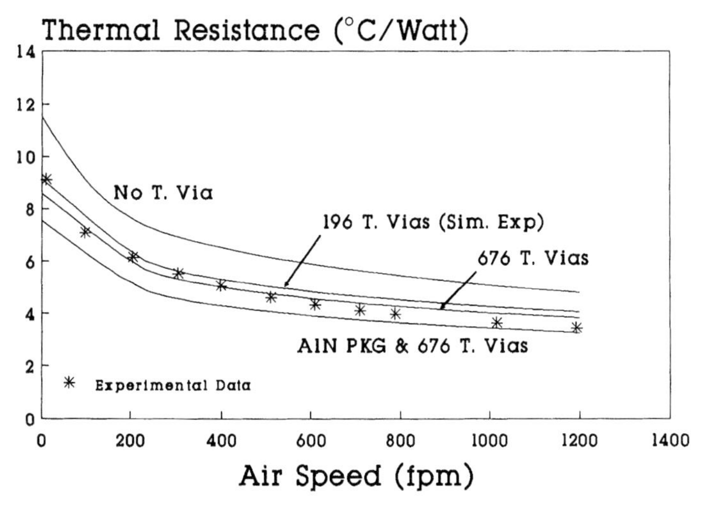

Figure 7.7 Measured and simulated thermal performance curves of the low-profile capped CBGA with different configurations.

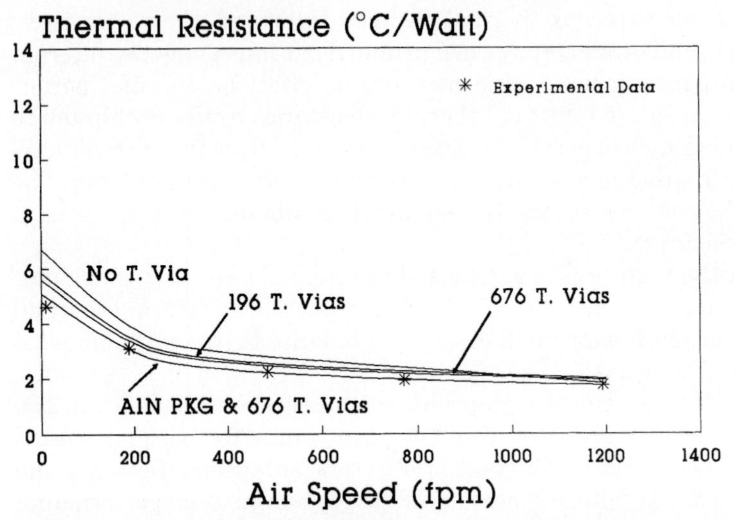

Figure 7.8 Measured and simulated thermal performance curves of the high-profile capped CBGA with different configurations.

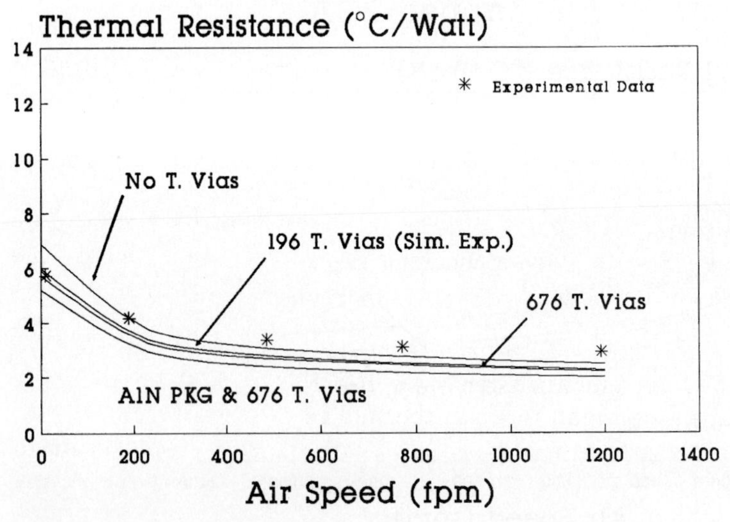

Figure 7.9 Measured and simulated thermal performance curves of the high-profile capless CBGA-A using pressure-sensitive adhesive with different configurations.

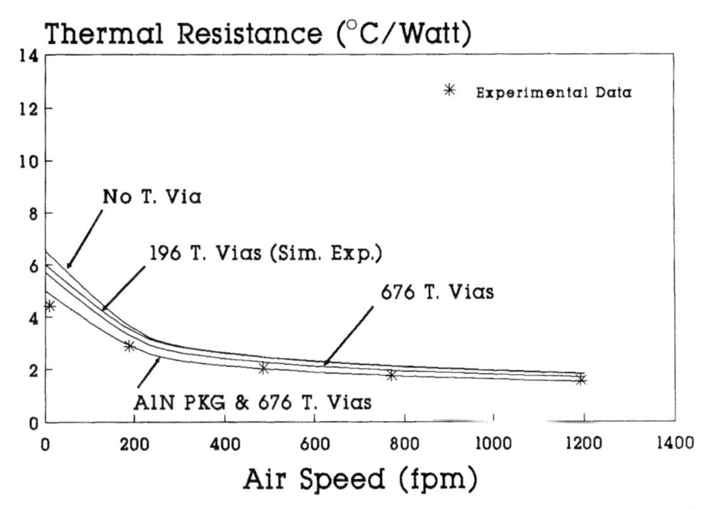

Figure 7.10 Measured and simulated thermal performance curves of the high-profile capless CBGA-B using epoxy with different configurations.

The attachment of a heat sink to a capless module presents a unique problem due to the small chip surface area. Data for a low-cost pressure-sensitive adhesive (PSA) attach is shown in Fig. 7.9. Restricting the heat to flow through an area equal to the chip size requires particular attention to be paid to the heat sink adhesive selection. While the PSA attach is the least costly, its high thermal impedance causes the total thermal resistance to suffer. The R_{int} is 1.8°C/W for this module. At approximately 300 fpm, the internal thermal resistance exceeds 50 percent of the total resistance, which is unusual for an air-cooled module and should only be considered if cost savings and performance objectives justify it.

Figure 7.10 shows data for the same capless module with a thermally improved heat-sink epoxy. The thermal performance has improved significantly over the results shown in Fig. 7.9 and has even surpassed the equivalent capped module (Fig. 7.8) with a smaller overall package. As mentioned earlier, this epoxy's thermal conductivity (1.06 W/m°C) is much higher than the PSA's (0.46 W/m°C), and resulted in a very small internal thermal resistance (R_{int} = 0.5°C/W). Due to the hardware and assembly elimination associated with encapsulation, the capless package also has the most attractive cost. Capable of removing 20 watt from a chip in a PC environment, this capless package is a very good choice for many high-performance air-cooled applications.

7.2.4 Modeling study on the CBGAs and their variations

The experimental results covered five representative configurations for CBGA assemblies. However, there are still many other design factors need to be addressed. A method to generalize these measured results is needed to consider those design factors. A three-dimensional, finite-difference, control-volume based, heat-conduction analysis has been developed for the expansion. The method itself is not new; its details could be found in Ref. 6 or other numerical heat-transfer books or articles. However, the implementation techniques and the results obtained may be helpful to packaging designers, and they will be presented as follows.

Figure 7.5 is the cross section of the configuration modeled. Basically, the configuration simulated the five CBGA assemblies measured. The major elements and the associated thermal conductivities and geometric information are listed in Table 7.1. The dimensions listed are for the lengths in the *X, Y,* and *Z* directions. All the circular surfaces are simulated by squared ones.

7.2.4.1 Numerical simulation. The modeling procedure was to model the experimental configurations first. Then, the calibrated models were used to simulate other cases. One of the major challenges to the modeling was to "guess" reasonable convective boundary conditions. As shown in Fig. 7.5, there are four convection boundary conditions affecting the heat transfer.

The heat transfer coefficient h_1 represents the forced air convection (1) on the chip for the case of a low-profile BGA assembly without the cap, (2) on the cap for the case of a low-profile BGA assembly with the

TABLE 7.1 Major Thermal Elements Modeled

Element	Number of layers or units	Thermal conduc. (W/m°C)	Area × thickness X × Y × Z (mm³)
Printed wiring board	1	0.25	156 × 107 × 1.656
Thermal vias	0, 196, or 676	35.00	0.788 × 0.788 × 1.8
Copper planes	2	387.00	156 × 107 × 0.072
BGA solder joints	676	35.00	0.788 × 0.788 × 0.96
Alumina package	1	21.00	32.25 × 32.25 × 2.9
AIN package (optional)	1	216.0	32.25 × 32.25 × 2.9
C4 solder and encapsulant	1	35.00	13 × 13 × 0.08
Silicon chip	1	160.00	13 × 13 × 0.625
Thermal grease	1	1.10	13 × 13 × 0.153
Al alloy cap	1	100.00	32.25 × 32.25 × 2.0 (with a cavity for the chip and the grease)
Al alloy heat sink (Al 6063)	1	218.00	32.25 × 32.25 × 5.08
Epoxy for heat sink attachment	1	0.66	32.25 × 32.25 × 0.127

cap, or (3) on the heat sink for the case of a high-profile BGA assembly with the heat sink.

The heat transfer coefficient h_2 represents the forced-air convection on the top surface of the PWB. The h_3 represents the natural air convection on the bottom surface of the PWB. And the h_4 represents the forced-air convection on the ceramic package surface for all the cases without the cap.

The h_2 and h_4 could be derived by the averaged heat transfer coefficient of a laminar flow along a flat plate.[7] For a length L, the averaged Nusselt number is

$$\mathrm{Nu_m} = 0.664\ \mathrm{Pr}^{1/3}\ \mathrm{Re_L}^{1/2} = h_m L/k \tag{7.1}$$

For air's Pr (= 0.692), ν (= 0.223 × 10^{-4}) and k (= 0.0313 W/m°C), and the length L (= 107 mm), the averaged heat transfer coefficient

$$h_m = 9.3\ u_{air}^{1/2}\ \mathrm{W/m^{2}{}^\circ C}$$

For different air speeds, this coefficient is 8 W/m$^{2\circ}$C for u_{air} = 100 fpm, 16 W/m$^{2\circ}$C for u_{air} = 200 fpm, 23 W/m$^{2\circ}$C for u_{air} = 400 fpm, 32.5 W/m$^{2\circ}$C for u_{air} = 800 fpm, and 40 W/m$^{2\circ}$C for u_{air} = 1200 fpm. These coefficients as a function of air speeds were used to calculate the heat convection of h_2 and h_4.

The h_3, a natural convection heat transfer coefficient, is usually around 5 W/m$^{2\circ}$C. The h_1, however, was difficult to determined due to different surface conditions. For the case without the cap, it was assumed to be the same as h_2 and h_4. For the low-profile capped module, it was assumed to be 3 times the h_2. The cap had many small pin fins that increased the total area to be about 1.5 of the flat surface. Calibrated by the experimental data, another multiplying factor of 2 was needed to simulate the heat transfer enhancement by the pin fins. The $h_1 = 3\ h_2$ modeled the experimental results well.

For the case with the heat sink, the h_1 was assumed to be 15 times the h_2. About half of the multiplying factor accommodated the area increase, and another half accommodated the heat transfer enhancement.

To simulate the heat dissipation, a thin uniform heat source along the bottom of the flipped chip was assumed. Due to symmetry, only a quarter of the assembly was calculated. The numbers of grid points in X-Y-Z directions were 39-39-22 for the simplest configuration with the low-profile capless BGA assembly, and 39-39-32 for the high-profile assembly with the cap and the heat sink. Refer to Fig. 7.6; the X-direction is parallel to the air flow, and the Z-direction, from the PWB to the cap, is perpendicular to the flow direction.

7.2.4.2 Results. Figures 7.6 to 7.10 show the simulated and measured thermal resistances from the center of the chip to the air. Due to symmetry, the maximum temperature is always at the center. First, we need to compare the experimental and simulation results. To simplify the simulation, 196 thermal vias were placed to simulate the real case with 183 thermal vias. The number difference is very small, and should not contribute any new effects. Figure 7.6 for the low-profile capless assembly demonstrates an amazing accuracy of the model. Since all the heat transfer coefficients were derived by Eq. (7.1) for this case, there was actually no calibration procedure. The numerical results match the experimental data very well. For the capped module, however, the $h_1 = 3\ h_2$ assumption was needed for the convection heat transfer on the cap surface. As mentioned above, the factor of 3 considered the area increase due to the small pin fins and the heat transfer enhancement. With this h_1, the simulation matches the experiment well (see Fig. 7.7).

Considering these two cases again, we can conclude that the averaged heat transfer coefficients derived for a flow along a flat plate can represent the heat transfer boundary conditions well for a low-profile BGA assembly without the cap. On the other hand, if there is any surface enhancement accomplished by the use of pin fins or other extended surfaces, the area increase needs to be considered. In addition, another multiplying factor to represent the heat transfer enhancement needs to be added. The impingement of the air to a pin fin and the flow direction change after the impingement may result in this additional multiplying factor. The calibration of the model by adjusting the heat transfer coefficient is not difficult if some experimental reference data exist.

For a high-profile package, the h_1 was similarly adjusted according to the experimental data. After adjustment, the simulated results match well to the experimental data for the three cases as shown in Figs. 7.8 to 7.10. With these calibrated models, some other effects can be simulated with a great confidence level. As mentioned at the beginning, there are many parameters that affect the performance of the CBGA assembly. Only thermal vias in the PWB and package materials will be discussed as examples.

Thermal vias of the PWB are filled with solder for efficient heat conduction from the BGA package to the copper planes. They can be used to reduce the overall thermal resistance in some cases. Figure 7.6 shows the effect of the number of thermal vias on the overall thermal resistance of the low-profile capless CBGA. Changing from 0 to 196 thermal vias, the resistance is reduced from 11 to 8°C/W for the air speed of 400 fpm. The reduction is substantial. However, adding the number of thermal vias from 196 to 676 does not improve the performance further. When the number reaches a certain level such as 196, the conduction path through these vias is no longer the major barrier to the heat spreading. As a result, further reduction of the thermal resistance across the thermal vias cannot enhance the thermal performance. Other critical elements need to be identified for further improvement.

One of such elements is the package material. As shown in Fig. 7.6, when the package material is changed from Al_2O_3 to AlN, the thermal resistance is significantly decreased. The 8°C/W is reduced to only 4.7°C/W. For practical designs, we may not want to use a new ceramic material for the enhancement. Thermal vias and metal planes can be added to the package to achieve the similar improvement. The developed model can be used to simulate the effects of these additions.

From the above results, it is clear that the use of thermal vias and high-conductivity package material can result in about 5°C/W improvement. For a 1-W chip, such an improvement may not be critical. However, for a 20-W chip, the improvement represents 100°C reduction!

The improvement resulting from these two effects, however, is not important for the high-profile assemblies. As shown in Figs. 7.8 to 7.10, the improvements from the no-via case to the case with A1N+676 thermal vias are in the range of only 1°C/W. Compared with the improvement for the low-profile capless package, this reduction is very small. The reduction for the low-profile capped CBGA assembly is around 2°C/W, which is somewhat in between.

In general, there are two thermal paths to remove the heat. One path is from the chip to the top convection flow (h_1) through the thermal elements such as grease, cap, epoxy layer, or heat sink. Another path is from the chip to the PWB convection flow (h_2 and h_3) through C4 solder joints, package, BGA solder joints, and copper planes. When the top convection is not effective, the second path is the major heat transfer route, and its major elements such as thermal vias and package material become very important.

On the other hand, when the top convection is very effective, the second path is marginally critical to the overall thermal performance. As a result, any changes in the module-to-board path may not result in significant improvements. The configuration of a CBGA assembly affects not only the overall thermal performance, but the roles of the thermal elements.

The thermal design considerations for the CBGA assemblies with a wire-bonded chip were discussed briefly before. But, no detailed experiments and simulations were reported. For the low-profile, capless CBGA assembly, the difference between the thermal resistances of a wire-bonded (WB) and a C4 CBGA should be very small. The conduction through the die-attachment for WB/CBGA or the flip-chip region for C4/CBGA is not the major thermal barrier. As a result, changing the chip-to-package connection scheme would not affect the overall thermal performance. For other assemblies, the conclusions will be difficult to make without simulations. It is difficult to establish two thermal paths for a WB/CBGA as those provided by the C4/CBGA. However, the WB/CBGA can choose either cavity-down or -up configuration for thermal management. Using the aforementioned model, the thermal performance curves corresponding to the possible designs can be predicted with a reasonable accuracy.

In addition, the model can be used to study other effects such as the difference between a plastic BGA and a ceramic BGA assemblies, the difference between single-package and a multipackage assemblies, effect of varying the card size, etc.

7.3 Electrical Management

The objectives of the electrical study are (1) to benchmark the electrical characteristics of CBGA against other package families, (2) to compare wire-bonded CBGA with C4 CBGA, and (3) to assess the effects of high I/O count and small solder pitch on electrical characteristics.

7.3.1 Package comparison

7.3.1.1 Package electrical parasitic calculation. The electrical characteristics of plastic quad flat pack (PQFP), wire bonded (WB) PBGA, CBGA (WB and C4), and PGA are compared in this study. Figures 7.11 and 7.12 show these package configurations and the pin assignment. The physical attributes of these packages are summarized in Table 7.2. The array package with 324 pins is chosen because this pin count is close to 304 that is the highest pin count for PQFP. The solder ball (pin) identification number for a 324 package with CBGA-WB, CBGA-C4, PGA, and PBGA is shown in Fig. 7.11. Due to symmetry, only one-eighth of the balls are shown. The center six balls (pins), corresponding to a total of 48 balls for the whole package, are assumed to be thermal balls which can be connected to power/ground planes. The first pin is the one right next to these thermal balls (see Fig. 7.11).

The connections from the wire bonding pads to these pins were chosen by the router according to the minimum-length criteria. Models have been established to estimate the trace length for these packages. Identical die size of 12 mm is used for each package. Common practiced design rules are used to place wire bond fingers or C4 pads on the package. The connecting trace from a wire bonding finger or a C4 pad to a terminal solder pad is assumed to be straight. This assumption is more

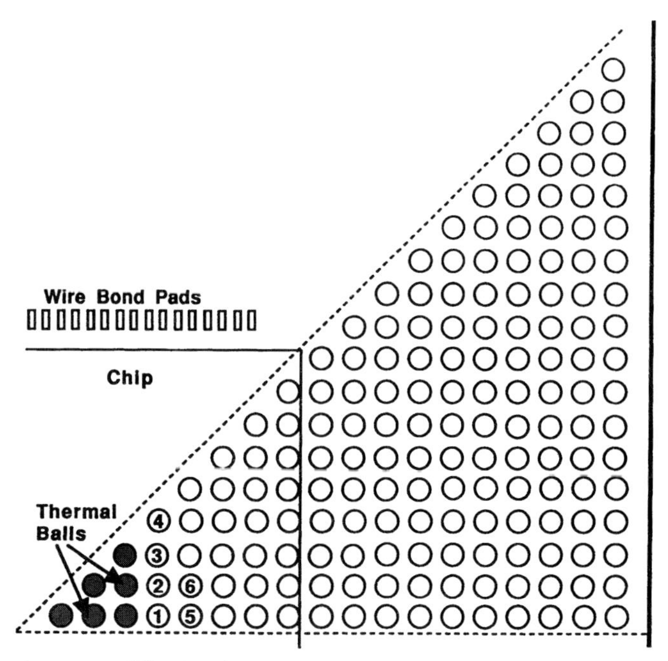

Figure 7.11 Wire bonding and BGA solder pads and the number assignment for the pads.

accurate for CBGA than PBGA, since the ceramic technology allows smaller via size and top-to-bottom connection without through vias. For PBGA, it is likely more than 50 percent of the traces are routed from the wire bonding fingers to the edge of the substrate and then folded backward to connect to the terminal solder pads, generating

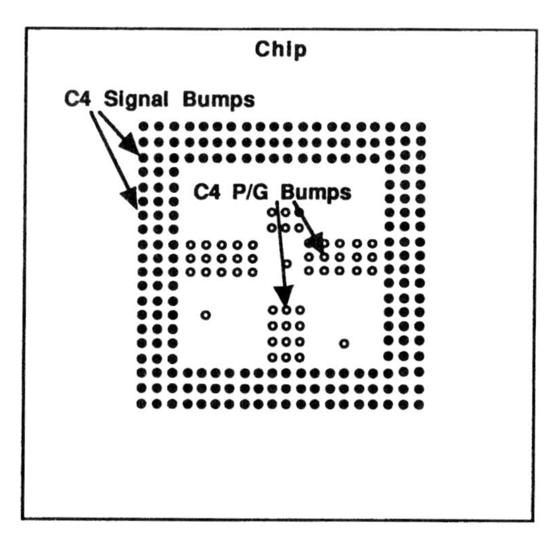

Figure 7.12 C4 solder bumps.

TABLE 7.2

Package family	PQFP	PBGA	CBGA(WB)	CBGA(C4)	PGA
I/O count	304	324	324	324	324
Package size (mm)	40	25	25	25	50
Wire bond/C4 pitch (μm)	216	190	244	254	244
Lead/solder ball pitch (mm)	0.5	1.27	1.27	1.27	2.54
Lead/trace width (μm)		75	100	100	100
Lead/trace spacing (μm)		75	100	100	100
Dielectric thickness (μm)		250	200	200	200
Dielectric constant		4.5	9.5	9.5	9.5

longer trace length. Only one-eighth of the total traces in one-eighth sector of the package are considered in the model, assuming that the fanout pattern of the signal traces is symmetric and is identical for all the eight sectors.

The model generates straight traces to connect all the wire bond fingers or C4 pad to all the terminal solder pads and the trace lengths are calculated accordingly. Only signal traces are used to generate the distribution of the trace length. Among all the connecting traces, 20 percent of them are assigned for power and ground (P/G) connection and are not included in the distribution. For array packages, most of P/G pads are assigned to central rows of solder pads; a few are assigned to the corner or near-corner pads (near to corner of the package). This P/G pad arrangement can ease routing on the mother board and can improve power supply and current return.[8]

For C4 connections, additional P/G solder bumps can be used to enhance the power distribution performance. A typical chip I/O for the C4 connection is shown in Fig. 7.12. The signal bumps are assumed to be located at ¼ from the die edge, and the Power/Ground bumps are assumed to be distributed close to the center.

Different from the above array-type package, the PQFP has pins populated along the periphery. Figure 7.13 shows the pin assignment for a 304 I/O PQFP. Again, only ⅛ of the I/O pattern needs to be presented due to symmetry.

The net electrical parasitic, which is defined as the sum of the parasitics of all the interconnects from the chip pad to the terminal surface mount pad, is used for comparing packages at an equal base. These interconnects include chip-to-package connection (i.e., wire bonding or C4), package trace, connecting vias, and package-to-board connection (i.e., lead, solder ball, or pin). The typical values of the parasitics of chip-to-package connection and package-to-board connection are listed in Table 7.3. The trace parasitics can be calculated if the trace length and the parasitics per unit length are known. Table 7.4 shows the package line capacitance and line inductance which are calculated

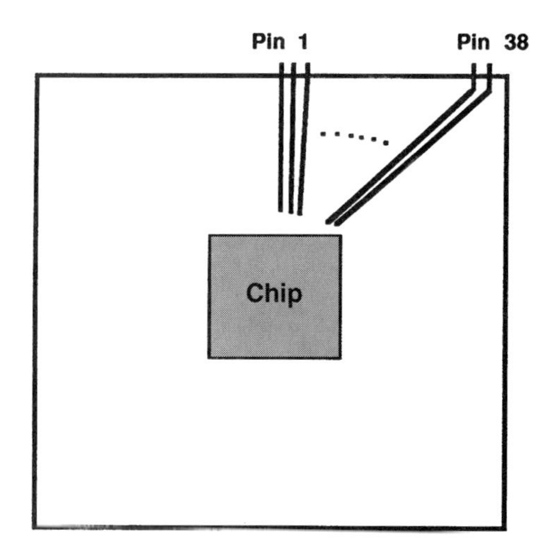

Figure 7.13 Pin assignment for a 304 I/O PQFP.

according to the package attributes as listed in Table 7.2. For the CBGA packages, the signal trace is assumed to be the strip line structure and its neighboring traces are grounded. A substrate with two metal layers is used for PBGA and the signal trace has the microstrip structure.

7.3.1.2 Distribution of trace length. The distribution of trace/lead length of each package is shown in Fig. 7.14 and the minimum, maximum, and average length are listed in Table 7.5. The more or less periodic abrupt change in the trace length with the pin number occurs when the trace is assigned to the next row of the solder pads. The PGA and PQFP have longer lead length, since both have larger body size. PQFP has much narrower distribution of lead lengths, which is typical for a peripheral I/O package, than the array packages.

TABLE 7.3

	Wire bond	C4 joint	Solder joint	PGA pin	QFP lead
Capacitance		0.2 pf			
Inductance	1.1 nH/mm	0.1 nH	0.1 nH	1.0 nH	1.2 nH

TABLE 7.4

	CBGA	CPGA	PBGA	PQFP
Line C. (pF/cm)	0.27	0.27	0.77/0.66	0.19
Line L. (nH/cm)	0.41	0.41	0.12/0.05	0.67
Via C. (pF)	0.35	0.35	0.17	

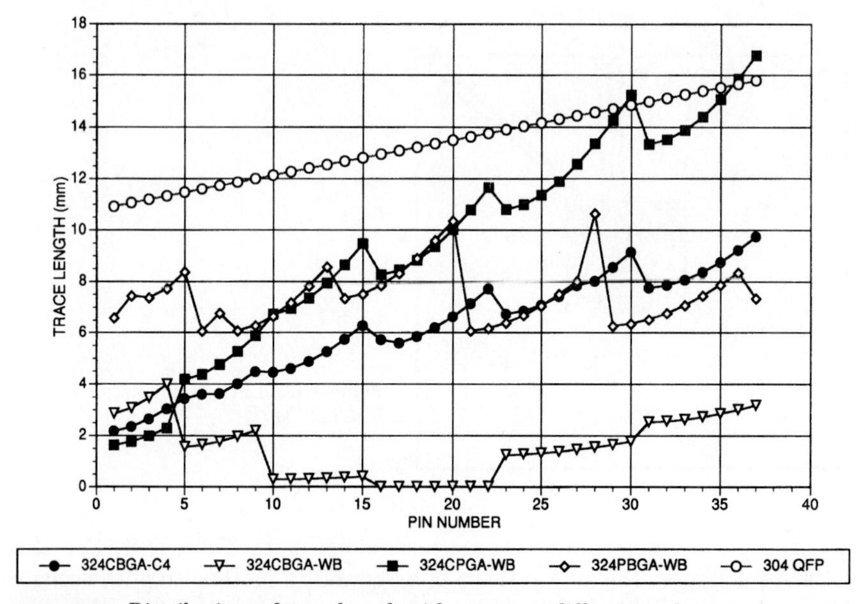

Figure 7.14 Distributions of trace length with respect to different packages.

In general, the wire-bonded (WB) CBGA has shorter trace lengths than the C4 package, since the C4 pads are placed closer to the center of the package than the wire bond fingers. C4 pads are placed around one-fourth of the die size (about 3 mm) from the die edge toward the center, while wire bond fingers are about 1.5 mm (2.5 mm for the outer bond shelf) from the die edge away from the center. The wire bond fingers are placed in such convenient locations, neither far away from inner rows of solder pads nor from outer rows, that connecting traces can run inwards or outwards from the wire bond pads. This makes all the traces of the WB package within 4 mm long. Short trace lengths occur for pin number 10-23 where the wire bond fingers are directly above the terminal solder pads. In contrast, C4 pads are placed close to the center but far away from the outer rows and most of the traces can only run outwards. This makes some traces on the C4 package almost 10 mm long.

The traces on the wire-bonded PBGA are much longer than those on the CBGA because most of them are fanned out to the edge of the package and then folded back connecting to the terminal solder pads. Such an inefficient routing scheme is caused by processing limitations of current PCB technologies with large via land.

7.3.1.3 Signal inductance. Figure 7.15 shows the net inductance of all the nets in one-eighth sector of each package, and the minimum, maximum, and average values are listed in Table 7.5. Among all the pack-

TABLE 7.5

	Lead length (mm)	Total L (nH)	Total C (pF)	Delay (psec)
324CBGA-C4				
Minimum	2.18	1.09	1.15	35.45
Maximum	9.75	4.20	3.23	116.40
Average	6.14	2.72	2.24	77.90
324CBGA-WB				
Minimum	0.02	1.61	0.35	26.30
Maximum	4.00	4.52	1.45	74.37
Average	1.51	3.07	0.76	47.60
324CPGA-WB				
Minimum	1.64	3.07	5.30	130.74
Maximum	16.77	11.00	6.52	265.10
Average	9.46	7.22	5.74	200.92
324PBGA-WB				
Minimum	6.07	5.85	0.64	62.37
Maximum	10.37	9.59	1.22	109.81
Average	7.41	7.17	0.88	79.36
304 PQFP-WB				
Minimum	10.93	11.73	2.11	157.27
Maximum	15.80	15.00	3.05	213.85
Average	13.36	13.36	2.58	185.62

ages, the PQFP has the highest net inductance averaging at 10 nH. The distribution of the net inductance for PGA is rather broad ranging from 3 to 11 nH. Because the rather long traces, PBGA also exhibits high inductance ranging from 6 to 10 nH. The other two BGA packages, namely, C4/CBGA, and WB/CBGA have similarly low net inductance ranging roughly from 1 to 4 nH.

Surprisingly, the wire bond package has about the same net inductance as the C4 one. This observation is quite contrary to traditional thinking that the C4 package gives lower net inductance. It is true that the inductance of a wire bond is much higher than that of a C4 joint. In this case, the wire bond inductance ranges from 1.4 to 3.5 nH and the C4 solder joint typically is around 0.2 nH. However, the long signal traces of the C4 package as described in Sec. 7.3.1.2 really offset the advantage of low inductance of C4 solder joint.

7.3.1.4 Signal loading capacitance. Figure 7.16 shows the distribution of net capacitances for each package, and the minimum, maximum, and the average values are listed in Table 7.5. PGA has the highest net capacitance around 6 pf, followed by PQFP around 2.5 pf. Again, because of rather long traces, the net capacitance of the C4/CBGA is close to that of PQFP, particularly for higher pin numbers. Most of the

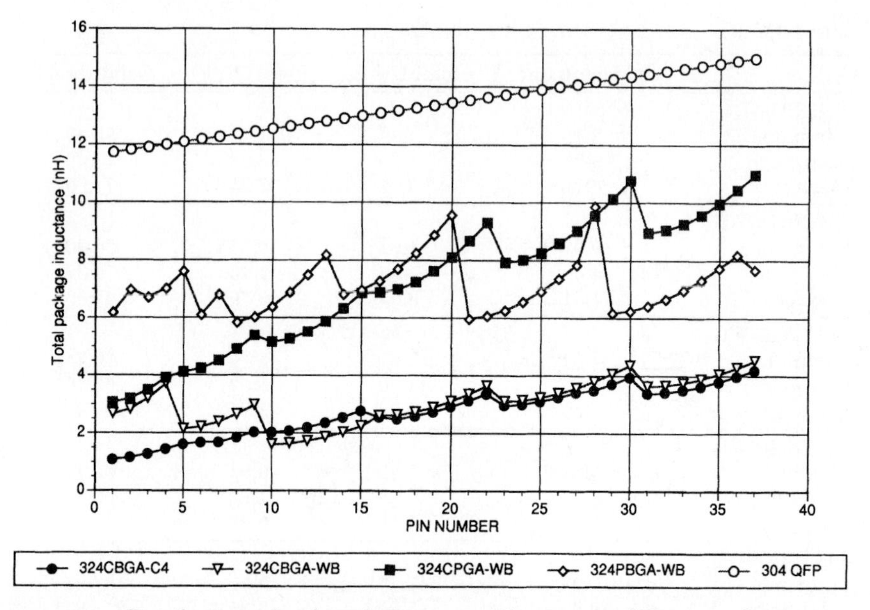

Figure 7.15 Distributions of package inductance with respect to different packages.

nets on both WB/CBGA and PBGA have capacitances less than 1 pf. Although PBGA has long traces, its net capacitance is low because of lower dielectric constant of PCB resin of 4.5 than ceramic material of 9.5. One should note that the net capacitance can be increased when plating stubs are connected to the traces.

7.3.1.5 Signal delay. Figure 7.17 shows the total package delay for each package, and the minimum, maximum, and average values are listed in Table 7.5. To incorporate the effects of various discrete elements such as wire bonds and vias, the delay is set to be equal to the square root of multiplication of the net capacitance and the net inductance. Two distinct groups can be seen in Fig. 7.17: PQFP and PGA in one having high package delay up to 250 psec, CBGAs and PBGA in another having lower package delay of less than 100 psec. C4 bonded CBGA has longer delay than the wire-bonded CBGA. The wire-bonded CBGA offers the shortest delay among all the packages under evaluation, because of short traces.

7.3.1.6 Power and ground inductance. Comparing the trace length distribution of the array package with that of PQFP, one can easily realize that it is easier for the former to have low power and ground (P/G) inductance than the latter. The array package inherently has a larger spread in trace length and the package designer can always select the shorter traces for P/G connection. The P/G lead inductance can be as

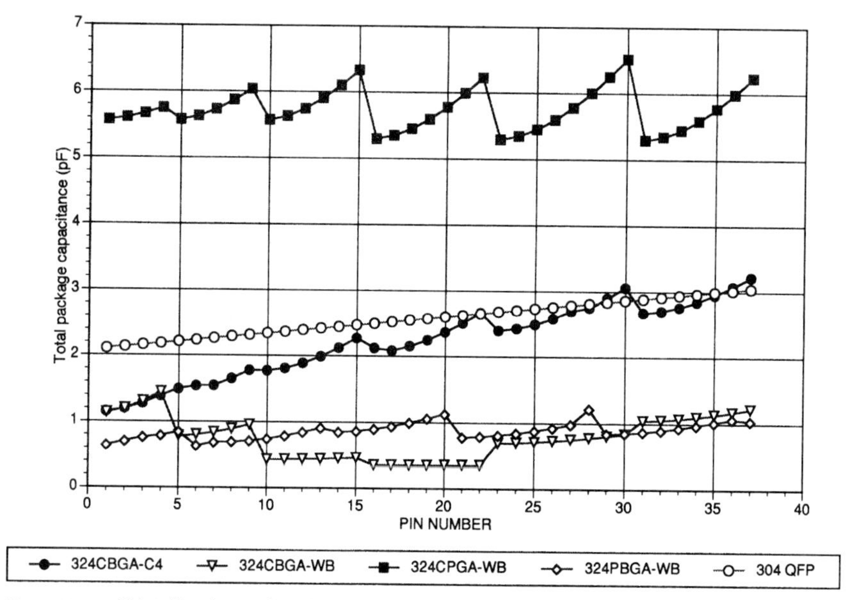

Figure 7.16 Distributions of package capacitance with respect to different packages.

low as about 1 nH. The PQFP has a narrower distribution in trace length and tends to have longer trace than the array package. The P/G inductance of a 40-mm body size PQFP can be larger than 12 nH, which is not acceptable for high-speed applications.

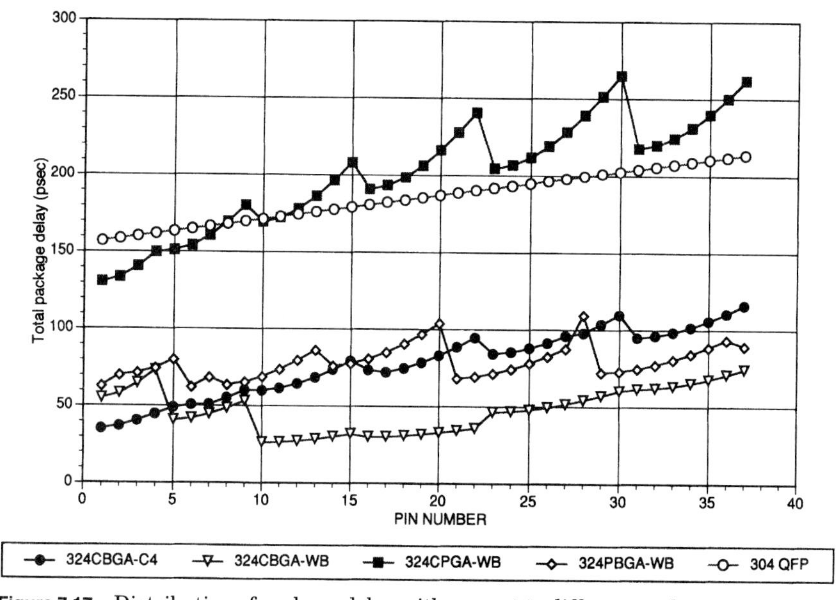

Figure 7.17 Distribution of package delay with respect to different packages.

The multilayer capability further differentiates the array packages from the PQFP. The multilayer structure allows internal power and ground planes which ease power and ground distribution and reduce ground inductance to be less than 1 nH. Multilayer construction is commonly seen in printed circuit boards and ceramic substrates, but the multilayer lead frame technology is yet to become mature for PQFP. The reliability and the cost of the multilayer lead frame PQFP are of concern.

For the multilayer array packages containing internal ground planes, the C4 package provides lower ground inductance than wire bond package for the following reasons. First, the inductance of a C4 joint of about 0.2 nH is much lower than for a wire bond which ranges from 1 to 3 nH. Secondly, the C4 pad can be placed directly above the terminating solder pad and both are connected by a through via, yielding the lowest interconnect inductance possible. Depending on the distance between the current source point and the sink point, the plane inductance can be significantly high. For wire bond packages, the flag area which the die is bonded to is grounded and connected to the underneath terminal solder pads. Unlike the C4 package, on the ground plane, the vias (source points) connected to the wire bond finger have some distance from the vias (sink points) connected to the solder pads (sink point) which are located underneath the die. The plane inductance can be as high as 1 nH.

7.3.2 High I/O CBGA

To understand the effects of high I/O count, solder pad pitch, and chip interconnect methods on the electrical characteristics of CBGA, four cases are studied: combination of C4/wire bond (WB) and 1.00/1.27-mm solder pad pitch with identical I/O count of 1024. In all cases, die size is chosen to be 16 mm, which is reasonable for such a high I/O count package.

The distributions of trace length, net inductance, net capacitance, and net delay for each case are shown in Figs. 7.18 to 7.21, respectively. The maximum, minimum, and average values of each parameter are listed in Table 7.6. Again, the periodic abrupt change in the values of these parameters with the pin number reflects the change in the row number of the solder pads. Generally speaking, the large body size of these packages, 41 mm for 1.27-mm pitch and 33 mm for 1.00-mm pitch, gives long traces, large net parasitics, and long package delay, particularly the 1.27-mm-pitch version. Long package delay (100–200 psec) may exclude these packages (except the WB package with 1.00-mm pitch) from being used for high-performance system applications. Reducing the solder pad pitch from 1.27 to 1.00 mm, as one would expect, the trace length as well as the delay is shorten by about 30 percent.

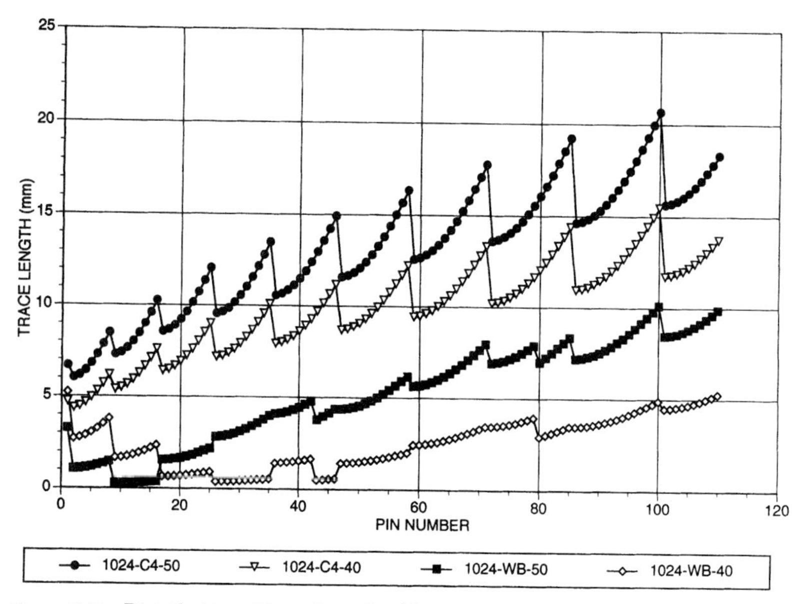

Figure 7.18 Distribution of trace length with respect to different high-I/O packages.

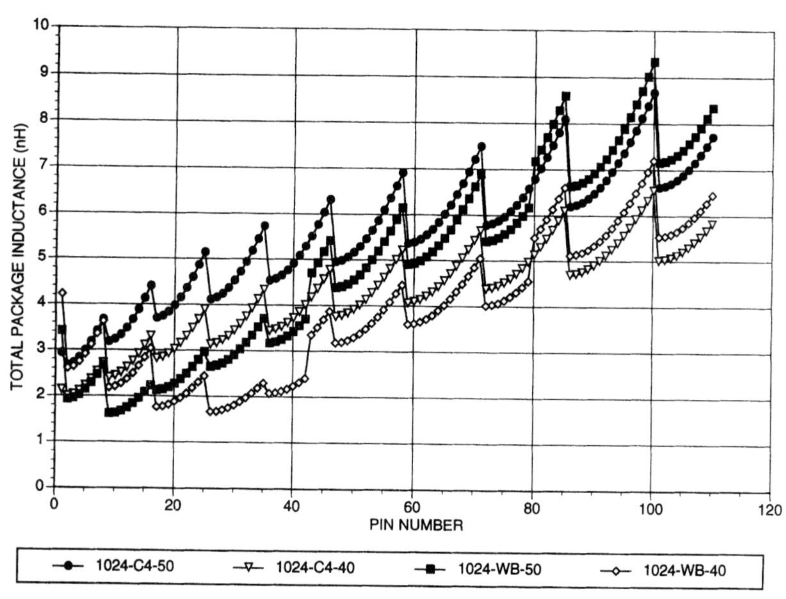

Figure 7.19 Distribution of package inductance with respect to different high-I/O packages.

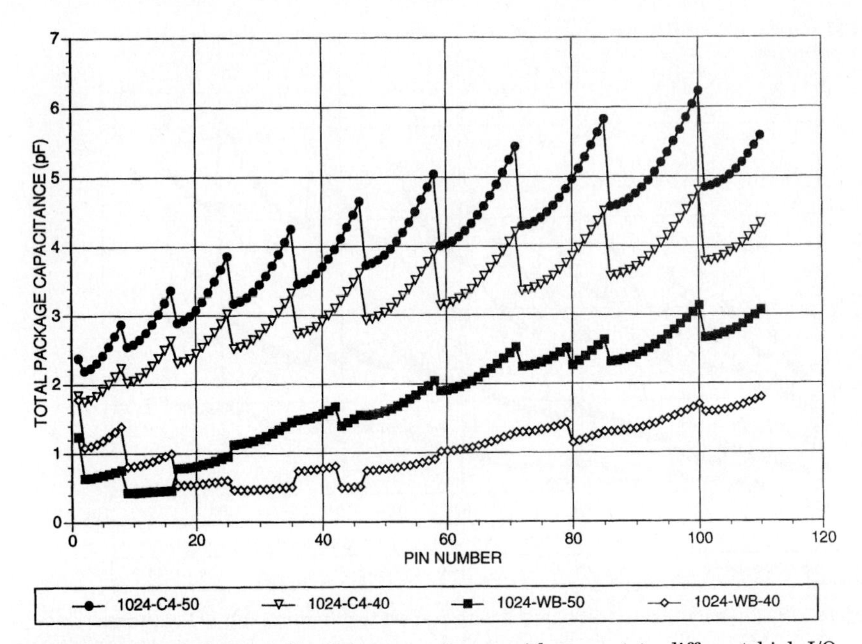

Figure 7.20 Distribution of package capacitance with respect to different high-I/O packages.

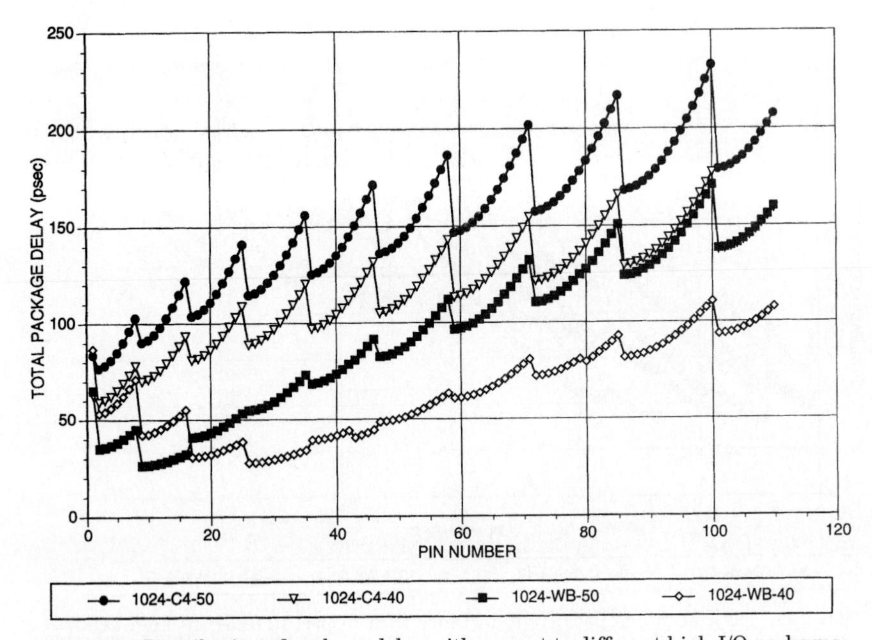

Figure 7.21 Distribution of package delay with respect to different high-I/O packages.

TABLE 7.6

1024 CBGA	Lead length (mm)	Total L (nH)	Total C (pF)	Delay (psec)
C4-1.27 mm solder pad pitch				
Minimum	5.99	2.65	2.19	76.30
Maximum	20.65	8.67	6.22	232.20
Average	13.16	5.60	4.16	152.59
C4-1.00 mm solder pad pitch				
Minimum	4.33	1.98	1.74	58.61
Maximum	15.45	6.53	4.79	176.90
Average	9.82	4.22	3.24	117.05
WB-1.27 mm solder pad pitch				
Minimum	0.26	1.60	0.42	25.94
Maximum	10.10	9.33	3.12	170.71
Average	5.19	4.97	1.77	93.73
WB-1.00 mm solder pad pitch				
Minimum	0.38	1.65	0.46	27.43
Maximum	5.25	7.19	1.79	110.39
Average	2.53	3.88	1.04	63.26

In general, the WB package has lower parasitic levels than those of the C4 package. Using wire bonding can reduce the trace length by more than 50 percent of that of C4 package. The net inductance of the WB package is lower than the C4 package up to about pin number 80. Above pin number 80, the C4 package has slightly lower inductance. On average, the net inductance of the WB package is about 10 percent less than that of the C4 package. Although the wire bond inductance ranging from 1.4 to 5.1 nH is added to the WB package, shorter signal traces really make its net inductance lower than the C4 one. With regard to the net capacitance, the WB package has average values less than half of that of the C4 one. Because of lower inductance and capacitance of the WB package, it has shorter package delay than the C4 package by about 50 percent.

7.4 Summary

Some specific thermal and electrical characteristics of CBGA assemblies have been presented and discussed. Critical experimental data for five different CBGA-on-PWB assemblies were reported. At an air velocity above 800 fpm, the thermal performance of the assemblies could achieve as low as only 2°C/W for the junction-to-air overall thermal resistance. For a typical PC environment (200 fpm, T_{air} = 40°C), the maximum chip power dissipated could be approximately 20 watts without exceeding 100°C. If the high-profile heat sink is not desirable, the low-profile capped CBGA can still manage about 10-watt power.

In addition to the experimental study, a numerical analysis tool has been developed to consider other packaging considerations. Calibrated by the experimental data, the model can accurately simulate other effects in details. For a low-profile capless CBGA, the thermal vias in PWB and package's thermal conductivity are critical factors. A total 5°C/W resistance change can result from different design choices. For a 20-watt chip, such a change may result in a 100°C difference! On the other hand, high-profile assemblies' thermal performance curves are not affected significantly by the thermal vias and the package material. The heat transfer to the heat sink is the predominant thermal path; thus overall thermal resistance is not affected by these two factors. The developed generalization numerical analysis could be used to study many other effects on CBGA's thermal performance.

For electrical management, BGA packages have much better electrical performance than PQFP or PGA. The inherent small package size of BGA not only gives shorter package delay, but also gives short package-to-package delay at the system level. Plastic BGA has the potential to be better than ceramic BGA, because of the low dielectric constant. However, better processing technologies for printed circuit boards are required. The multiple-layer array package can offer internal power and ground planes to decrease power and ground inductance.

Contrary to common thinking, the wire-bonded CBGA has shorter signal delay than the C4-bonded CBGA because of smaller capacitive loading. Although the inductance of the bonding wire is substantial, the saving in trace inductance due to short traces makes the net inductance of wire-bonded packages about the same as that of the C4 package. However, the C4 package has better power and ground distribution capability, resulting in lower simultaneously switching noise. Also, it should be noted here that the C4 array connections reduce on-chip interconnection lengths, which need to be considered for overall electrical performance.

The increase in I/O counts can raise the package signal delay significantly. The solder pad pitch has to be further reduced to be smaller than 1.00-mm pitch to meet future high-speed applications.

7.5 Acknowledgments

Y. C. Lee would like to thank the National Science Foundation (MIP9058409 and ECD9015128) for the support to his thermal and soldering studies that are related to this chapter.

7.6 References

1. R. R. Tummala and E. J. Rymaszewski, *Microelectronic Packaging Handbook,* Van Nostrand Reinhold, New York, 1989.
2. *Electronics Materials Handbook, vol. 1: Packaging,* ASM International, Metals Park, December, 1993.
3. D. P. Seraphim, R. Lasky, and C. Y. Li, *Principles of Electronic Packaging,* McGraw-Hill Book Co., New York, 1989.
4. A. D. Kraus and A. Bar-Cohen, *Thermal Analysis and Control of Electronic Equipment,* Hemisphere Pub. Co., New York, 1983.
5. G. Coors and B. Buck, "Pad Array Carriers—The Surface Mount Alternative," *International Electronics Packaging Conference Proceedings,* 1991, pp. 373–389.
6. Y. C. Lee, H. T. Ghaffari, and J. M. Segelken, "Internal Thermal Resistance of a Multi-chip Packaging Design for VLSI Based Systems," *IGGE Trans. on Components, Hybrids, and Manufacturing Technologies,* 1989.
7. M. N. Ozisik, *Basic Heat Transfer,* McGraw-Hill Book Company, New York, 1977, p. 220.
8. Wayne Huang and Jim Castro, "CBGA Package Design for C4 PowerPC™ Microprocessor Chips: Trade-off between Substrate Routability and Performance," *Electronic Components and Technologies Conference,* Washington, D.C., 1994, May 1–4.
9. Gary Kromann, David Gerke, and Wayne Huang, "A Hi-Density-C4/CBGA Interconnect Technology for a CMOS Microprocessor," *Electronic Components and Technologies Conference,* Washington, D.C., 1994, May 1–4.
10. Vernon L. Brown, "A New Land Grid Array Package Family," *Proceedings of Surface Mount International,* September 1–3, 1992, p. 105.

Chapter

8

Reliability of Ceramic Ball Grid Array Assembly

Yifan Guo and John S. Corbin

8.1 Introduction

Ceramic ball grid array (CBGA) assembly is an area array surface mount technology in which ceramic modules are connected to printed circuit cards by means of solder balls. Because of the coefficient of thermal expansion (CTE) mismatch in the material system, during the product development the thermal fatigue of the BGA solder joints is a major concern in the package reliability. The fatigue life of the solder is usually a function of the magnitudes of the plastic and elastic strains.[1,2] If these strains can be computed mathematically or actually measured experimentally, the fatigue life of the solder and the reliability of the package can be estimated. Many researchers have studied the low-cycle fatigue of solder and its relation to solder strains.[3–5] Although, it was known that, from the previous researches, the strain distributions in solder joints were nonuniform, because of the difficulties of determining the strain distribution in an already small joint, effective strains, such as average shear strain, have been used as parameters in models of the fatigue life of solder joints. If the effective strain used in such a model is close to the strain that actually causes the fatigue damage, its use provides a good approximation. In many cases, especially in solder joints specifically designed for solder fatigue tests, the effective strain does dominate the failure mechanism, such that the average strain can be used to correlate the fatigue life approximately. In most real solder joints, such as the CBGA joints, the strain variations are dramatic and the strain concentrations are significant; the fatigue damage is generally dependent upon the maximum strains instead of the average strains.[6] Therefore, when fatigue life is estimated in a reliability model, it is important to determine and utilize the strain distributions and the maximum strains and their locations.

As electronic packaging technology evolves, solder joints used in electronic devices are becoming smaller and smaller. The determination of strain distributions within solder joints has become more and more difficult. The packaging structures usually consist of several materials and interfaces, all with different mechanical properties, and strain concentrations with high magnitudes are frequently localized in very tiny zones of the solder joints. Such situations make the accurate determinations of strain distributions extremely difficult. In recent years, many new analytical and experimental techniques have emerged in the electronic packaging area. Numerical methods, such as the finite element method (FEM), have become so powerful due to super computers and advanced software that very complicated geometries and material properties can be modeled and simulated. Experimentally, moiré interferometry has evolved into a very powerful tool. Using these new techniques, thermal-mechanical strains can be accurately calculated and measured so that the detailed mechanical behavior of the solder joints can be better understood, and the reliability of the package can be estimated more efficiently and accurately.

In this chapter, the mechanical behavior and the reliability of the CBGA assembly is discussed through the results provided by FEM models and experimental measurements. The FEM models are constructed to view the interactions that occur between the module and card, as well as the resulting deformations that occur in any single CBGA solder joint within the array. The experimental technique, moiré interferometry, is a highly sensitive optical method which provides in situ measurements and graphical display of deformations and strains in the CBGA assembly and individual solder joints. The correlation and verification between the experimental and analytical methods provide a solution with accuracy and confidence. With these powerful techniques, strain concentrations in very localized areas in a solder joint have been determined and used to assess the dependence of reliability on various structural configurations. Finally, a reliability model has been created by correlating the resulting strains with the fatigue life of CBGA assemblies.

8.2 Strain Distribution and Reliability in CBGA Assembly

The reliability of CBGA interconnections is measured by the fatigue life of solder joints, which is closely related to solder strain. The solder strain is usually determined by experiments or calculations under the thermal and mechanical loading which simulates the application conditions. Through accelerated thermal cycling (ATC) tests, the relation between strain and fatigue life can be established, and a reliability model is created by correlating the solder strains with the observed ATC life of CBGA solder joints.

8.2.1 CBGA assembly

A typical CBGA package is a surface mount second-level assembly in which multilayer ceramic (MLC) modules containing one or more chips are directly attached to printed circuit cards by solder joints (here the solder "joint" is considered as the entire structure consisting of solder ball and eutectic solder at both card and module interfaces). The common assembly process is described in detail by Ries et al.[7]

The CBGA structure shown in Fig. 8.1 consists of high-temperature-melting noneutectic solder balls (90%Pb/10%Sn, T_{melt} = 268–290°C) approximately 0.89 mm in diameter. These balls are attached to the modules (on molybdenum pads) and to the printed circuit card (on copper pads) with eutectic solder (63%Pb/37%Sn, T_{melt} = 183°C). The solder balls are arranged in a square array with a typical ball spacing of 1.27 mm (from center to center).

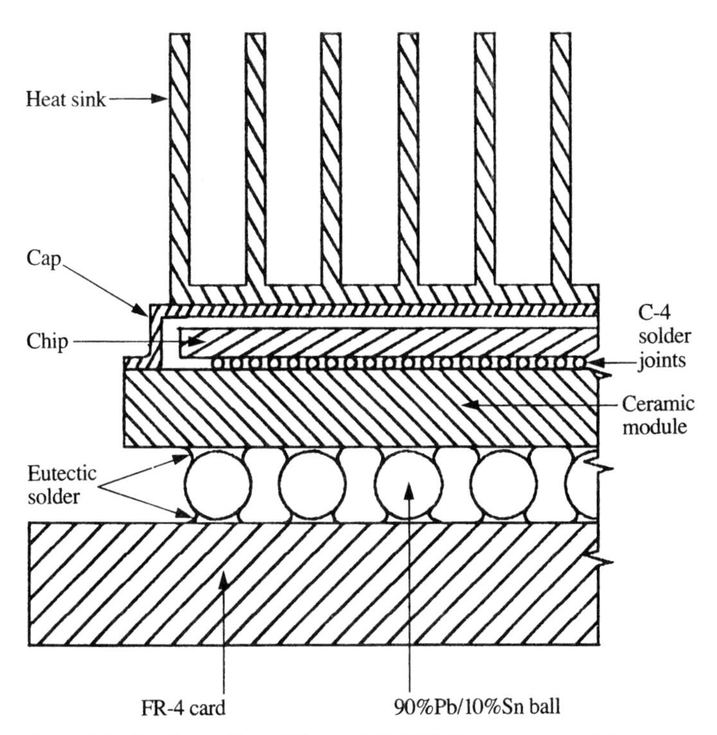

Figure 8.1 Configuration of Ceramic Ball Grid Array assembly.

Module sizes usually range from approximately 18 to 32 mm square, with a total input/output capability of 196 to 625. The first-level package, consisting of the multilayer ceramic module and attached solder balls, is surface-mountable and compatible with existing surface mount technology assembly processes. Eutectic solder paste is deposited onto the pad array on the card by means of a screening process; the module and solder ball package is placed onto the pad array of the card; and the card/module assembly is reflowed by means of a vapor-phase or infrared process. A heat sink is optionally attached to the module cap after assembly of the module and the card.

The experimental results presented in this chapter are from the analysis of a 25-mm (module size) CBGA assembly which has a 19×19 solder ball array with a ball spacing of 1.27 mm. The MLC module thickness is 2.8 mm with a coefficient of thermal expansion (CTE) of 7 ppm/°C. The printed circuit card (FR-4) is 1.5 mm thick and has an average CTE of 20 ppm/°C. On top of the MLC module is an aluminum heat sink attached by epoxy adhesive. The solder balls are 0.89 mm in diameter.

8.2.2 Strain distribution in solder joints

Historically, it is difficult to accurately describe the fatigue life of solder joints from solder strains when an average strain is used in a reliability model, because the average strain usually does not represent the actual strain state in real solder joints. To improve the accuracy and consistency, the maximum strain in the solder joint should be determined and used in calculations of fatigue life.

The strain distribution in a CBGA solder joint is highly nonuniform. If the material properties are constant from joint to joint, the geometry is a major factor in determining the strain distribution. With different strain distributions, the failure mode in the solder joints can be very different.

Figure 8.2 shows a simple analysis using 2-dimensional FEM linear elastic models to characterize the geometry effect in solder joints. The diagrams show the cross-sections of cylindrical geometries. The three simple joint geometries have different aspect ratios of height (h) versus diameter (d). The input displacement applied to the joints is a simple translational displacement, a pure shearing with the top surface of the joint shifted relative to the bottom surface in the horizontal direction by Δx. The same average shear strain is applied to the three joint geometries by keeping the $\Delta x/h$ (h is the height of the joint) constant in all three cases. The arrows in the cross-section areas show the direction and magnitude of the maximum principle strain. It is clear that the strain states are different in these three cases.

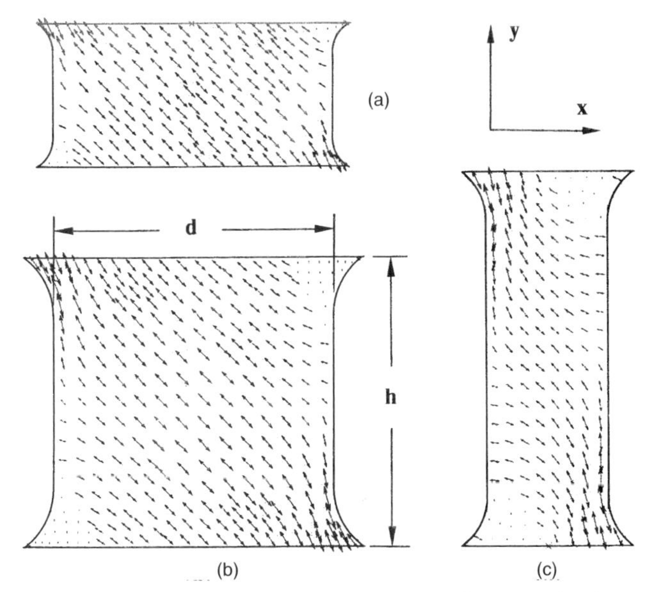

Figure 8.2 Distributions of maximum principle strain in three joint configurations with different height/diameter (h/d) aspect ratios: (*a*) h/d = 0.5; (*b*) h/d = 1; (*c*) h/d = 3.

Figure 8.2*a* is more like a C4 (Controlled Collapse Chip Connect) joint with an aspect ratio of $h/d = 0.5$ (d is the diameter of the joint). When the aspect ratio is small, the shear strain is dominant in the joint, and the principle strains are mostly in the 45° direction. The direction and magnitude are fairly consistent and uniform. In this case, the likely failure mode is due to large shear deformation, or mode II cracking.

Figure 8.2*c* is a joint with an aspect ratio of $h/d = 3$, more like a solder column joint. The strain pattern, which is totally different from a small aspect ratio joint, resembles that due to a bending of a short beam with fixed ends. The maximum principle strain is in the vertical direction at the edges of the joint, which means that tensile and compressive strains are dominant. The failure mode in this case will likely be a bending distortion, or mode I cracking.

The CBGA solder joint has a geometry similar to that shown in Fig. 8.2*b*, with an aspect ratio around unity. This is a mixed case, and the deformation mode is neither dominated by pure shear nor pure normal strain, and the failure mode will be a mixed mode.

In all three cases, when an average shear strain $\varepsilon_{xy} = 0.25\%$ is applied to the joint, the maximum principle strain in the joint is much greater than the average shear strain. The maximum principle strain values are: $\varepsilon_{max.} = 0.44\%$ in joint (*a*), $\varepsilon_{max.} = 0.51\%$ in joint (*b*), and $\varepsilon_{max.} =$

0.39% in joint (*c*). It is surprising that the greatest principle strain value occurs in joint (*b*). Note that if the average shear strain is used in a Coffin-Manson–type reliability model,[1,9] the fatigue life would be overestimated. It is also apparent that even if the average shear strain is the same, different aspect ratios cause significant differences in the maximum strain. To obtain a good correlation between strain and fatigue life, strain distributions and maximum strains need to be determined for the actual geometries of solder joints.

8.2.3 Reliability concerns

The reliability of devices operating in customer environments is a key issue in electronics packaging design. One aspect of reliability that is of particular importance to CBGA technology is the potential premature fatigue failures that can occur in the solder joints. These fatigue failures are induced by the thermal expansion mismatch between the ceramic module and the printed circuit card, which creates cyclic loads on the solder joints as they are thermally cycled during normal operation.

As discussed before, the solder joints may have different failure modes depending on the loads and structural geometries of the joints. For a short joint (small h/d ratio), shear deformation is dominant and fatigue failure is driven by a shearing mode (mode II), while for a long joint (large h/d ratio) a bending mode and resultant normal strain is dominant and the failure is driven by an opening mode (mode I). Because a CBGA solder joint has a h/d ratio close to one, both modes exist and contribute to the fatigue life of the solder joints. The dependency of the fatigue life on the joint geometry is very significant and is an interesting and important topic in the reliability study of CBGA assemblies.[8]

A CBGA structure that has been subjected to ATC cycles is shown in Fig. 8.3. In this figure, the module is at the top and the card at the bottom; the card pad and part of the via and inner-plane structure are evident. The fatigue failure shown in this cross-section photomicrograph indicates cracking of the solder joint along the interface between the card pad and the eutectic fillet. The crack initiates on the side of the solder joint nearer the module center, and propagates along the interface of the solder fillet and the card pad until fracture occurs. In this case, the eventual fracture invariably occurred along the card-side interface with a mixed mode crack.

To optimize reliability, packaging designers must be able to accurately estimate the effect of different sets of materials and geometry on fatigue life. This can be accomplished by using available solder reliability models (e.g., Coffin-Manson or its derivatives[1,9]), all of which require as input some estimate of the cyclic distortions, or plastic strain, that the solder joint will experience.

Figure 8.3 Typical solder joint fatigue failure in CBGA assembly due to accelerated thermal cycling. *(Copyright 1993 by International Business Machines, reprinted with permission.)*

8.3 Methods Used in Analysis

Many techniques and methods have been developed to aid in determining the thermal-mechanical strains in solder interconnections. Because of the small scale, complicated geometry, and highly nonlinear properties of the solder joints, it is difficult to obtain accurate strain distributions in real solder joints. Some newly developed methods used in the analysis of strains in CBGA solder joints are discussed in this section.

8.3.1 Macro-micro approach

In the mechanical analysis of CBGA solder joints, a global-local (also called macro-micro) approach is employed. In this method, a complete structure is analyzed first by a global model which consists of all elements in the structure and all external loads. The global model usually provides an overview of the problem and the locations of the critical elements in the structure. Displacement results from the global model can be applied to a local model to provide more detailed information in the critical elements of the structure. This approach is very applicable to simplifying the analytical process and enhancing the solution accuracy in critical areas.[10] In this chapter, the terminology of *macro* and *micro* is used instead of *global* and *local,* but the meanings are the same. The macro-micro approach is very common in FEM analysis and experimental studies. In some case, it is more effective to combine FEM and experimental methods in a macro-micro approach, such as using experimental results to provide macrodeformations which are applied as the boundary conditions to a FEM micromodel.[11]

The complexity of the thermal deformations in a CBGA package presents a problem class for which the macro-micro approach can be effectively applied. In this approach, the macrosolution describes the global deformations at the level of the entire CBGA package, and each solder joint is treated as an element with an equivalent mechanical property as a real solder joint. The macrodeformation is due to the CTE mismatch between the module and the card (this CTE mismatch is defined as a macro-CTE mismatch in this chapter). In the microsolution, the local deformations in the individual solder joints are studied. The CTE mismatches at the solder-module interface and solder-card interface are defined as micro-CTE mismatches. Because the solder joints interface with components (module and card) which have very different mechanical properties from the solder, the strain variations near the interfaces are significant. The microdeformation describes the strain distributions and concentrations in the individual solder joints and near the interfaces.

In the macroanalysis, the locations of the solder joints which experience the most severe deformations in the package are determined. These are joints most likely to first fail and they are measures of the reliability of the package. In the microanalysis, the displacement solution obtained from the macroanalysis is used as boundary conditions to further investigate the detailed thermal strains in these critical locations.

8.3.2 FEM macro and micro models

In C4 technology, the finite element method has been applied to estimate the thermal strain sensitivity to geometric variables.[12,13] These efforts view only a single solder joint with an applied translational (shearing) displacement. In the work described in this chapter, models are constructed which view the macrolevel interactions that occur between the module and card, as well as the resulting microlevel deformations that occur in any single CBGA solder joint within the array.

The modeling protocol described in this chapter employs both macro- and micromodels. The model geometries are constructed in CAEDS®;[14] the model solutions are generated using ANSYS®.[15] Ideally, one would prefer to construct a single model including an appropriate portion of both the module and the card, with sufficient detail in the solder ball regions of interest to accurately determine the plastic strain distribution. Such a model would, unfortunately, have to be quite large and would impose substantial computational requirements. To make the problem more feasible computationally, a macro-micro approach is used.

In the macro-micro approach, a relatively coarse macromodel is constructed to represent the structural coupling between the module and the card, and to determine the major deformation modes that occur during thermal cycling. These thermal deformations (a shearing deformation and an axial deformation at a particular solder ball of interest) are used as input boundary conditions for a much more detailed micromodel of a single solder joint to determine the strain distribution within the joint.

The material properties used in the macro- and micromodels, excluding the solder materials, are shown in Table 8.1. Temperature dependence is included, and the materials are assumed to be linearly elastic. The properties of the solder materials used in the models include the effects of elastoplastic behavior, temperature, and strain rate.[16] The finite element models are static, however, since no time-dependent effects are included.

The macromodels consist of thin plate elements representing the ceramic module and the printed circuit card, and beam elements that couple the module and the card. A group of beam elements that connects a single module pad to a card pad is termed an equivalent beam, and is designed to replicate the shear, bending, and axial-stiffness characteristics of a single joint structure at its interfaces with both the module and the card (see Fig. 8.4). The initial step is to determine the dimensions and material properties of the equivalent beams.

TABLE 8.1 Material Properties*

Material	Elastic modulus E at 20°C (10^{12} Pa)	Elastic modulus E at 100°C (10^{12} Pa)	Poisson's ratio v at 20°C and at 100°C
Ceramic	.304	.304	0.21
Molybdenum	.321	.315	0.30
Copper	.126	.102	0.29
FR-4	.011	.008	0.28

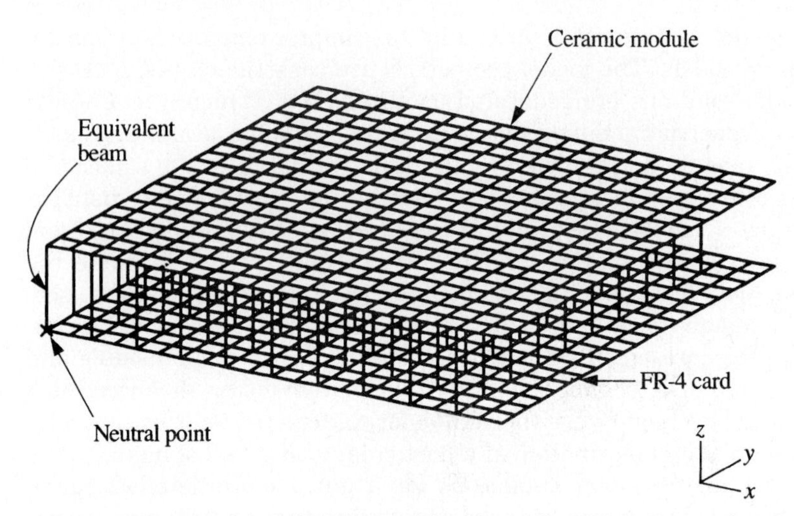

Figure 8.4 Macro model of CBGA assembly. *(Copyright 1993 by International Business Machines, reprinted with permission.)*

Before the equivalent beams can be defined, the nonlinear stiffness characteristics of the solder structure must be estimated. This is accomplished using a micro-level finite element model. A typical micromodel is shown in Fig. 8.5 for a module pad diameter of 0.86 mm and a card pad diameter of 0.61 mm. Only half of the joint structure is modeled, because of symmetry. In this case, the symmetry plane passes through the center of the solder joint and the module center, or neutral point, and is perpendicular to the card. This three-dimensional model represents a ceramic module thickness of 1.40 mm, a card thickness of 1.60 mm, a ball diameter of 0.89 mm, and a card-module standoff (separation) of 0.94 mm. Module and card pads are also represented. Note that in the macromodel, the equivalent beams are structurally connected to the plate elements at points that represent the midplanes of the module and card. Thus, in the micromodels, the module and card elements extend only to midplane, i.e., half thickness.

To estimate the nonlinear stiffness characteristics of the solder structure, the midplane boundaries of the micromodel are displaced in steps sufficiently small to ensure plastic solution convergence. Two independent cases of displacement are considered: a simple shear case, in which the module midplane elements are translated laterally with respect to the card midplane elements, and an axial case, in which the module midplane elements are translated axially (in compression) with respect to the card midplane elements. These cases are considered at 20 and 100°C. The reaction forces and moments (shear force and moment for simple shear displacement, axial force for axial displacement) are determined as functions of the imposed displacements at the two tem-

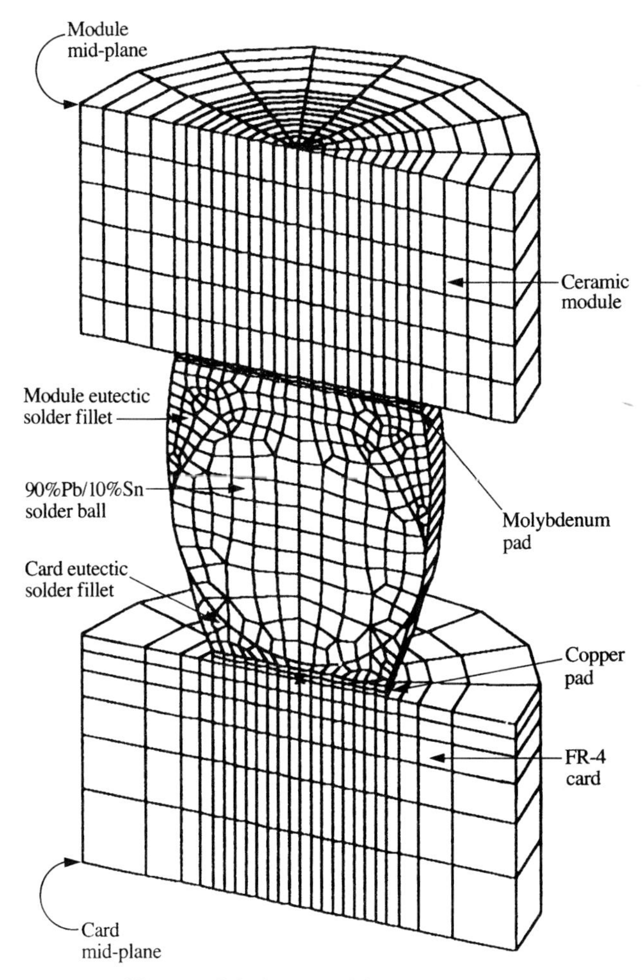

Figure 8.5 Micro model of single solder joint in CBGA assembly. *(Copyright 1993 by International Business Machines, reprinted with permission.)*

peratures. The equivalent beams must reproduce the reaction forces and moments of the micro model when the models are subjected to the same displacement cases (shear and axial) at the two temperatures.

The equivalent beam geometry is derived from linear elastic beam theory applied to a constant–cross-section cantilever displaced independently in simple shear (no end rotation) and in axial compression. Shear effects are included. The equivalent beam geometry (length, area, and moment of inertia) and elastic modulus are determined explicitly in the elastic range by requiring the equivalent beam reaction forces and moments to match those obtained from the micromodel when subjected to the same shear and axial displacements. Beyond the elastic region, the material properties of the equivalent beam are determined iteratively. That is, the tensile stress-strain relationship of the equivalent beam material is adjusted such that the equivalence between the equivalent beam and the micromodels is maintained for the reaction force and moment throughout the region of plastic deformation.

A macromodel for a 25-mm-square CBGA module is shown in Fig. 8.4. This three-dimensional model represents one quarter of a 19×19 array of solder balls spaced 1.27 mm apart (center to center) and consists of two layers of thin-plate elements, one each for the module and card. Only one quarter of the structure is modeled because of symmetry. The module and card are coupled by a group of equivalent beam structures modeled with beam elements, the physical geometry and material properties of which are tailored, as previously described, to exhibit elasto-plastic responses to axial and simple shear deformations equivalent to those determined by the micromodel.

Macromodels simulate the major deformation modes that occur during thermal cycling. The outputs of the macromodel are the relative displacements (shear, axial, and rotational) of each solder connection in the array. These relative displacements are used as midplane boundary values for the micro model in order to determine the stress-strain distribution within the CBGA structure for any chosen solder joint within the array, at any temperature. Plastic strains in the CBGA structure can, therefore, be estimated as a function of structural configuration and material properties.

8.3.3 Experimental method

Moiré interferometry. Determining the strain distribution in a small solder joint is a very difficult task. High-sensitive whole-field experimental techniques are required. Moiré interferometry is an optical technique for determining in-plane displacements and strains,[17] and features very high displacement sensitivity and spatial resolution. The principle of this technique is depicted schematically in Fig. 8.6 together with an equation describing an optical interference of two plane wavefronts. For this method, a high-frequency diffraction grating (1200 lines per millimeter in this work) is needed on the specimen surface at the testing area. This specimen grating is usually a crossed-line grating which has grating lines in the two orthogonal directions. It can be made by a lithography process or a simple replication from a premade grating mold. In experiments, the specimen grating deforms together with the specimen under an applied mechanical or thermal load, such that the deformation in the grating represents the deformations of the specimen. The optical set-up in a moiré interferometry system produces a virtual reference grating of frequency f, which is an interference pattern of the two coherent beams B_1 and B_2 from a laser light source. As shown in Fig. 8.6, the frequency f is a function of the wavelength λ of the laser light and the incident angle α of the two coherent beams B_1 and B_2. The virtual reference grating is superimposed on the specimen grating, and the interaction of the two gratings forms a fringe pattern which is a contour map of fringe order N_x. The fringes are equivalent to geometric moiré fringes. In Fig. 8.6, the two beams B_1 and B_2 intersect in the horizontal plane, and the contour map of N_x is produced. Additional input beams B_3 and B_4 in the vertical plane (not shown) produce the contour map of N_y.

In a fringe pattern, the displacements are proportional to the fringe orders N. Usually, two fringe patterns of two perpendicular fields, N_x and N_y, are recorded at the same time, in order to define the in-plane displacement fields U and V, corresponding to the x and y directions. The equations for determining U and V displacements from the fringe orders can be written as

$$U = \frac{N_x}{f} \quad \text{and} \quad V = \frac{N_y}{f} \tag{8.1}$$

where N_x and N_y are the fringe orders in the corresponding fringe patterns and f is the frequency of the reference grating. Linear strains can be calculated by simply taking the derivatives of the displacements U and V, leading to

$$\varepsilon_x = \frac{1}{f} \cdot \frac{\partial N_x}{\partial x} \qquad \varepsilon_y = \frac{1}{f} \cdot \frac{\partial N_y}{\partial y} \qquad \varepsilon_{xy} = \frac{1}{2f} \cdot \left(\frac{\partial N_x}{\partial y} + \frac{\partial N_y}{\partial x}\right) \tag{8.2}$$

The strain values can also be obtained directly from the fringe patterns by measuring the fringe gradients. The preceding equations indicate that the linear strains are proportional to the fringe gradients.

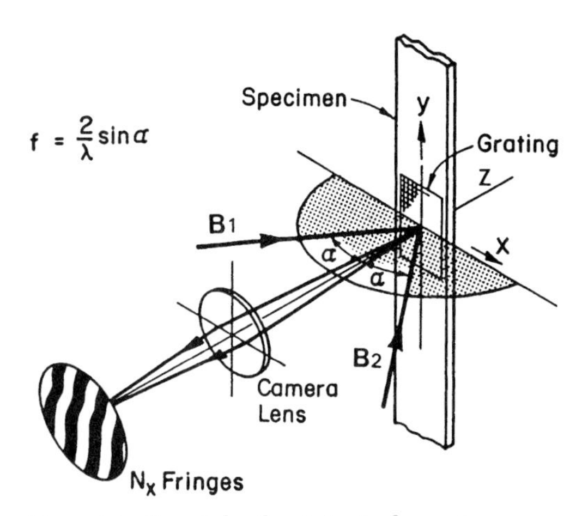

Figure 8.6 Principle of moiré interferometry.

The typical displacement sensitivity of moiré interferometry is 417 nm per fringe order with a reference grating $f = 2400$ lines/mm. The spacial resolution is about 10 μm. With these capabilities, detailed strain distributions can be obtained in a very small and localized area.

Specimen and experimental procedure in moiré interferometry. A cross-section specimen is usually needed in moiré experiments. A schematic diagram of the specimen, a 25-mm CBGA assembly, is shown in Fig. 8.7. In this particular specimen, the pads on the module side have diameters of 0.86 mm and on the card side of 0.71 mm.

The cross section of the specimen is a flat surface that contains cross sections of all the 19 solder balls in that particular row. An index number n is used to identify each solder ball shown in the cross section. A strip specimen is usually used in the experiments to ensure a two-dimensional deformation and to eliminate bending in the y-z plane (Fig. 8.7). Thermal loading can be applied by replicating the specimen grating as the specimen is held at an elevated temperature T_1, and the measurements are conducted at room temperature T_2. The deformations of the specimen grating reflect the deformations experienced by the specimen under the thermal loading ΔT $(T_2 - T_1)$.

In thermal strain measurements in electronic packaging using moiré interferometry, thermal loading can be applied by placing a uniform and known frequency grating (undeformed grating) on the specimen at an elevated temperature, and measuring the deformed grating at room temperature.[18] In order to have an undeformed grating at the elevated temperature, a ULE (Ultra Low Expansion) glass grating mold is used. In the process, a thin layer of epoxy is applied to the ULE grating mold. Because the epoxy has very low viscosity at the elevated temperature, a thin and uniform layer of epoxy can be formed. The specimen and the ULE grating are then pressed together and maintained at the elevated temperature until the epoxy is cured. By separating the specimen from

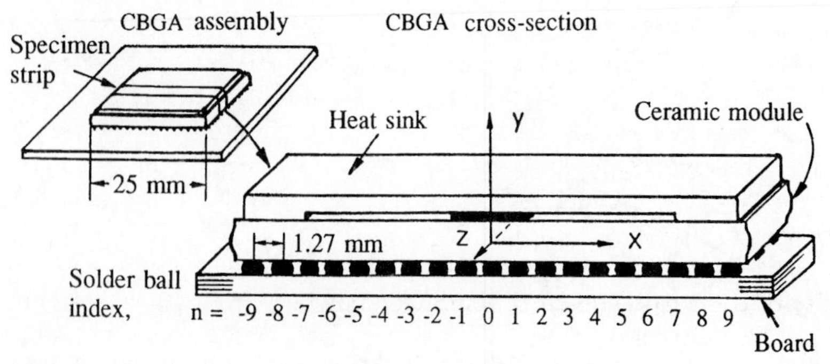

Figure 8.7 CBGA specimen used for the experimental analysis. *(Copyright 1993 by International Business Machines, reprinted with permission.)*

the ULE grating mold, a phase grating profile is transferred to the specimen surface. The specimen is then cooled to room temperature, and measurements are conducted.

In the procedure of producing the specimen grating, the specimen should be held at the elevated temperature for about one hour before the specimen grating is replicated. After being cooled to room temperature, the specimen is also held for an hour to let the solder fully creep before the measurements are conducted. Consideration of dwell time is important in achieving repeatability of measurements.

8.4 Macro (Global) Deformations in CBGA Assemblies under Thermal Loading

In the macroanalysis, the global thermal deformations of the CBGA structure is of interest. In the macroexperimental analysis, the strain distributions in the individual solder joints are not considered, but rather the nominal strains in each solder joint. Each solder joint is treated as a unit with a nominal strain value (the average strain along the centerline of the joint) which is constant over the joint area. In the FEM macroanalysis, each solder joint is treated as an equivalent beam designed to replicate the coupling forces and moments which exist at the module and card interfaces. The macro deformation is a thermal deformation created by the global CTE mismatch between the ceramic module and the printed circuit card.

8.4.1 Experimental analysis

The specimen used in this analysis is a cross section of a 25-mm-square module with a module pad size of 0.86 mm and a card pad size of 0.71 mm. Figure 8.8 shows the thermal deformations obtained by moiré interferometry measurements. The fringes are contour lines of displacements with a contour interval of 417 nm per fringe order. The thermal deformation is induced by cooling the sample from 82°C to room temperature ($\Delta T = 60°C$). When the temperature decreases, all of the components contract in both the x and y directions. The deformations in the ceramic module are very small (indicated by the low fringe density), because the material has a low CTE and high Young's modulus. In the printed circuit card, the y-direction contraction is much greater than the x-direction contraction (shown by the higher fringe gradient in the V displacement fields) because of the anisotropic CTE property of the card. The card bends as shown in the U displacement field. In the x direction, the CTE mismatch results in different displacements at the module and the card. The relative displacement at a point is a function of the distance from that point to the geometric center, or neutral point, of the assembly. The relative displacement, and therefore the stress and strain, reaches a maximum value where the DNP is the greatest. (DNP is the distance to the neutral point which is generally located in the plane of symmetry. In the cross section of the 25-mm CBGA assembly, the DNP of the rightmost, or the leftmost, solder joint is 11.4 mm.) The increase in fringe density in the solder joints with DNP is readily observed in Fig. 8.8*a*. Strains in all 19 solder joints can be determined from the displacement fields using Eq. (8.2).

Macro normal deformation. Under the applied thermal loading, the module and the card bend into different curvatures in the x-y plane (Fig. 8.8). Because of the nonequal bending curvatures, the distance between the bottom surface of the module and the top surface of the card varies. The distance change between the two surfaces is measured by the y-direction relative displacement shown in Fig. 8.9, where the index of solder joint n is used to label the horizontal axis. In this case (temperature decrease), the relative displacements are negative, implying that the two surfaces moved closer together everywhere. The relative dis-

(a) U displacement field

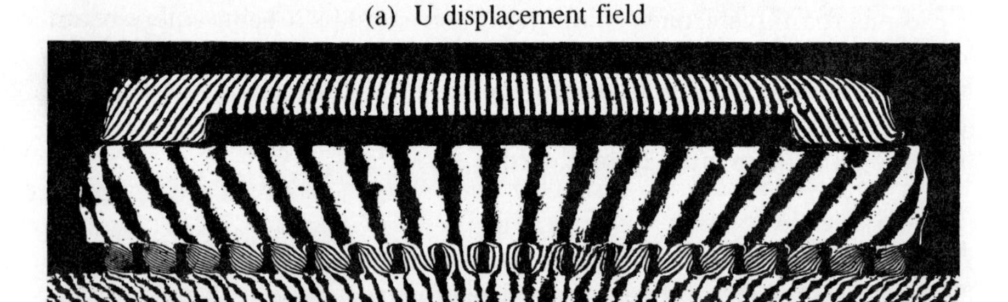

(b) V displacement field

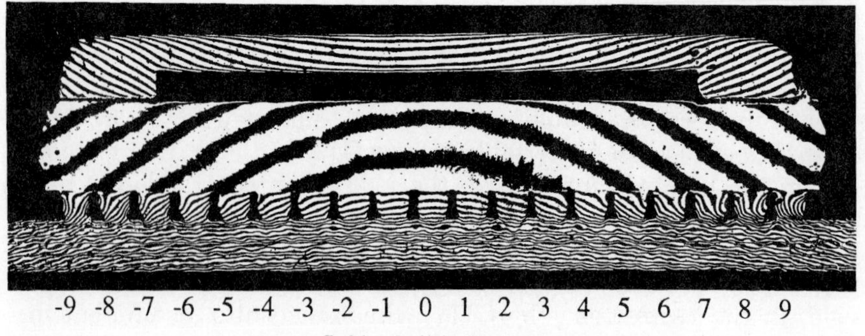

Figure 8.8 Fringe patterns of U and V displacement fields of the CBGA specimen under thermal loading.

placement also describes the y-direction deformation of each solder joint because the solder height equals the distance between the two surfaces. The nominal normal strain (y-direction) in each solder joint is calculated by dividing the relative displacement at that position (the centerline of the solder joint) by the height of the solder joint. The strain values are also given in Fig. 8.9 with the same curve but using the scale at the right which is created by dividing the relative displacement value (shown by the scale at the left) by the nominal solder height.

The solder strain shown in Fig. 8.9 is the total strain, which includes: the thermal strain due to a free thermal expansion of the solder, and the mechanical strain from mechanical constraints. The relation of these strains can be expressed as

$$\varepsilon_{\text{total}} = \varepsilon_{\text{mech}} + \alpha \Delta T \tag{8.3}$$

where $\varepsilon_{\text{mech}}$ is the mechanical strain, α is the CTE value, and $\alpha \Delta T$ is the strain caused by the free thermal expansion.

The strain caused by the free thermal expansion can be obtained from the material properties and the geometry of the assembly. The free thermal expansion of the solder is 28 ppm/°C, which corresponds to −0.17 percent strain for a −60°C temperature change, shown as the horizontal dashed line in Fig. 8.9. The uniform part of the strain equal to the free thermal expansion of the solder can be removed from the total strain. The mechanical strain is actually equal to the difference between the total strain (solid curve) and the free thermal expansion of the solder

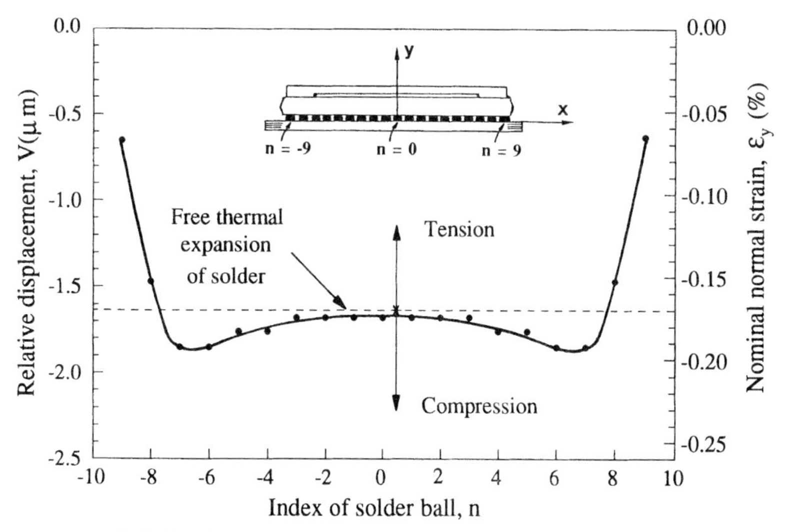

Figure 8.9 Relative displacement between the module and the FR-4 card in the y-direction, and the resulting nominal normal strain in each solder joint.

(dashed line). This is equivalent to shifting the curve so that the dashed line becomes the horizontal axis. As the assembly cools to room temperature, the middle solder joints are subjected to slight compressions and the two solder joints at each end (index $n = \pm 8$ and ± 9) are subjected to tensions with greater magnitudes. Strain signs would be reversed if the temperature were increased. The deformation characteristic shown in Fig. 8.9 is similar to the thermal deformation in the adhesive joint of a two-layer laminate described by Chen et al. in Ref. 19. This similarity is not surprising if the CBGA solder joint array is considered as a continuous layer of joint material between the module and the card. The module-solder-card structure should act like a two-layer laminate joined by a layer of adhesive, which is the solder array in this case.

Macro shear deformation. The relative displacement in the x-direction between the two surfaces (the bottom surface of the module and the top surface of the card) can be obtained from the displacement field shown in Fig. 8.8*a*. The resulting nominal shear strains in the solder joints can also be calculated from the shear displacement (Fig. 8.10). It is obvious that, when the vertical centerline of the assembly is used as the reference point, the x-direction displacements as well as the shear strains should have opposite signs at the two sides of the assembly. In Fig. 8.10, however, absolute values of the displacements and strains are used for simplicity. During the cooling process, the card shrinks more than the module due to the CTE difference. Near the end of the module, the card moves inward relative to the module by about 5 μm. Assuming that the stress in the solder joints was near zero, and the module and the card could deform freely during the temperature change, the relative shear displacement would be entirely dependent on the relative CTE and take the values given by the dashed lines. The difference in the two displacement curves (solid and dashed) is the result of the mechanical constraints that exist between the module and the card through the solder joints, and the residual stresses in the solder joints. A similar situation exists when the package cools from an elevated temperature used in an assembly process; residual stresses remain in the solder joints and the structure.

The nominal shear strain is obtained at the vertical centerline of each solder joint from the U and V displacement fields using the shear strain formula shown in Eq. (8.2). The maximum nominal shear strain occurs in the end solder joints where the macro-DNP is the greatest. Here, the macro-DNP is defined as the DNP of the assembly, and the neutral point is located at the vertical centerline of the center solder joint. The shear strain values for the solder joints with lower macro-DNPs are also given in Fig. 8.10. Because free thermal expansion does not induce shear strain, the total shear strain in the solder joints is equal to the mechanical strain.

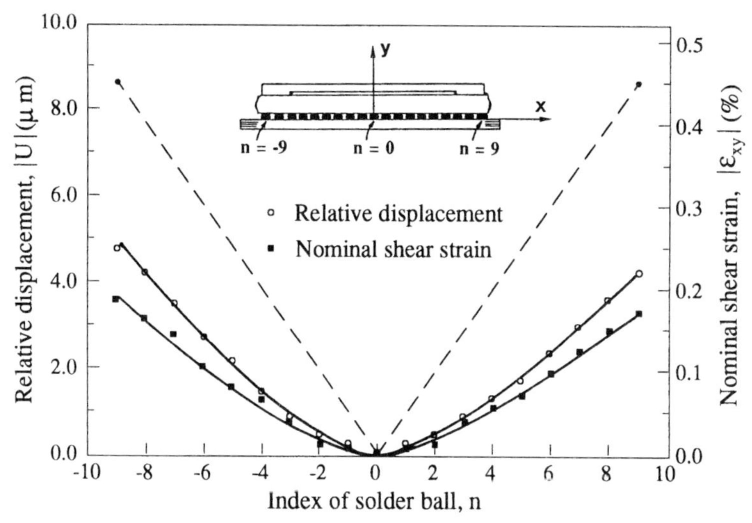

Figure 8.10 Relative displacement between the module and the card in the x-direction, and the nominal shear strain in each solder joint. The dashed line gives the relative x-direction displacement that would occur between the module and the card if the assembly were free of mechanical constraint from the solder joints. Absolute values are used in the plots for simplicity.

Similar shear strain distributions are found in the adhesive layer of a two-layer laminate,[19] and in the encapsulated C4 interconnections in a semiconductor packaging under thermal loading.[20]

8.4.2 FEM macromodel

The macrolevel FEM models provide the global deformations of module and card, from which the net solder joint deformations (shear, normal, and rotational) can be computed. The net deformations are simply the relative motions of the solder interface at the module side of each solder connection with respect to the interface at the card side, and are ultimately applied as boundary conditions to individual solder joints in the micro-FEM models.

In the macromodels, a series of nine temperature steps from 20 to 100°C is chosen to simulate the heating portion of an ATC cycle. These temperatures are applied uniformly to all of the element nodes within the models. The solution of interest is at 100°C, but interim solutions must be generated at each of the applied temperatures to ensure numerical convergence, since the equivalent beams are defined elasto-plastically. The reference temperature (i.e., the temperature at which the thermal stresses are assumed to be zero) is 20°C, resulting in a simulated temperature change of $\Delta T = 80°C$.

There are three net deformation modes, namely shear, normal (normal to the module and the card surfaces) and rotation, for a 25-mm-square module as shown in Fig. 8.11 for the simulated temperature change of $\Delta T = 80°C$. (Note that for ease of visualization, the shear deformation mode is viewed from the module center, while the normal and rotational modes are viewed from a module corner.) The shear mode is effectively linear with respect to the distance from the module neutral point, with the maximum net shear displacement occurring at the corners (maximum distance to module neutral point). The maximum shear displacement in the 25-mm-square module is only 48 percent of the unconstrained shear that would occur were there no coupling between the module and card, indicating a substantial elastic contribution from the module and card. This ratio in shear displacements predicted by FEM macromodel agrees very well with the experimental results shown in Fig. 8.10.

The normal deformation, shown in Fig. 8.11*b*, is large along the module perimeter, is extreme at the module corners, and is minimal in the module interior. This result is analogous to that shown by Chen and Nelson[19] in their analysis of bonded elastic layers subjected to temperature variations.

The net rotation (rotation of the ceramic midplane with respect to the card midplane) is small as shown in Fig. 8.11*c*. Although the gradients appear substantial, the displacements due to rotation are quite

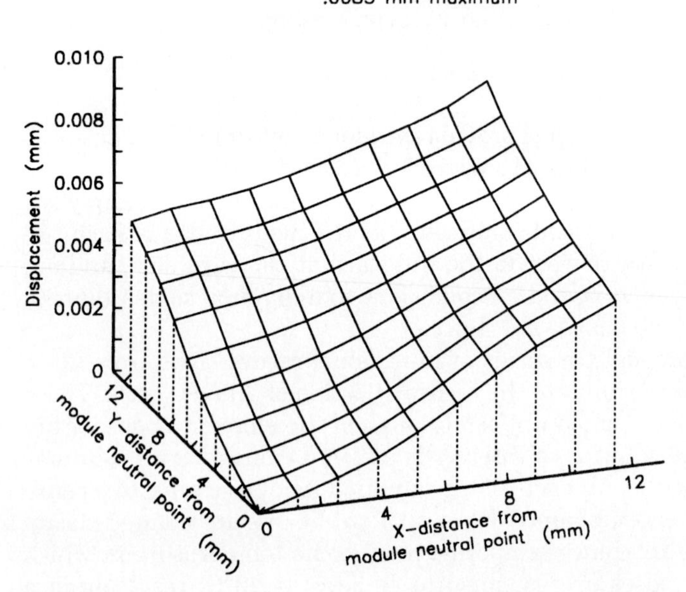

Figure 8.11*a* Thermal deformations of a 25-mm-square module at 100°C: net shear translation. *(Copyright 1993 by International Business Machines, reprinted with permission.)*

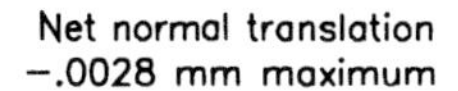

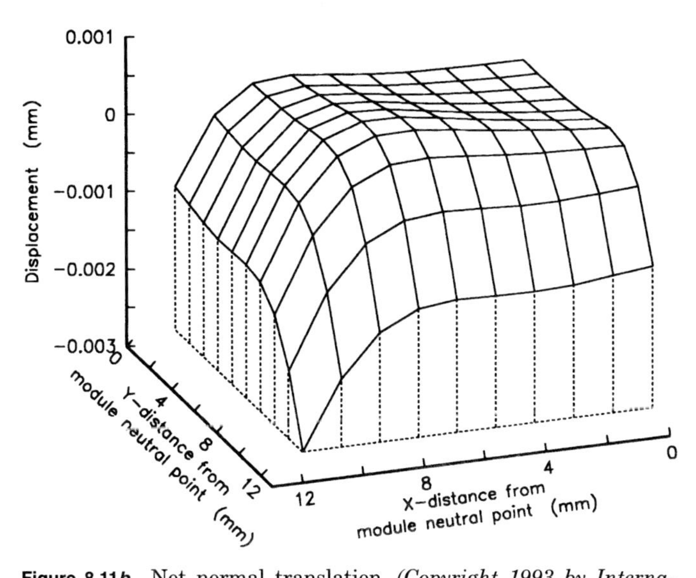

Figure 8.11*b* Net normal translation. *(Copyright 1993 by International Business Machines, reprinted with permission.)*

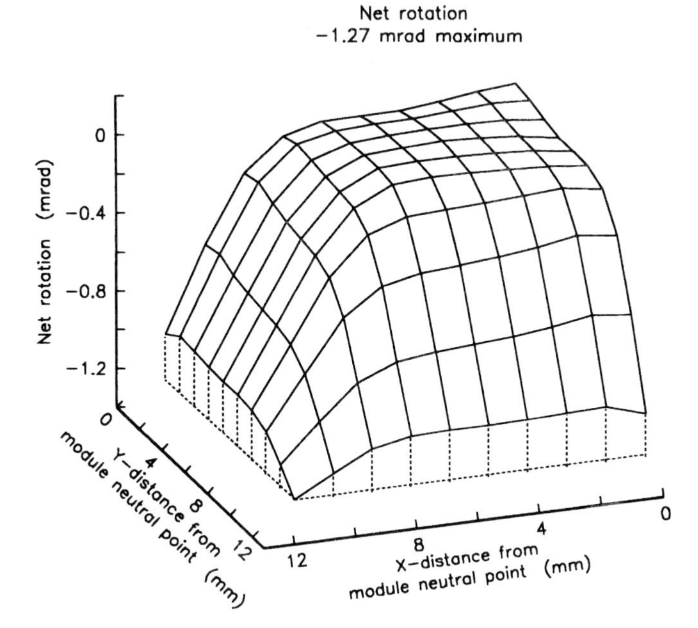

Figure 8.11*c* Net rotation. *(Copyright 1993 by International Business Machines, reprinted with permission.)*

small compared to the shear and normal displacements. These small relative rotations are consistent with those observed by Hall.[3]

The solder joints thus experience two major deformations during thermal cycling, a shear mode and a normal mode. A comparison of the net shear and normal deformations for three different module sizes is shown in Figs. 8.12 and 8.13, respectively. This comparison is viewed along the diagonal of the module, from the neutral point to the corner joint. Note that the maximum (corner) shear displacement is not a linear function of the maximum distance to the module neutral point (DNP), but approximates a square-root relationship. This suggests that reliability projections from a smaller module size to a larger one which assume a linear dependency on the maximum DNP, such as the Norris-Landzberg–modified Coffin-Manson relationship,[1] would significantly underestimate the fatigue reliability.

The net shear and normal deformations for any single solder joint in the array are used as boundary conditions to the FEM micromodel. These are input as a series of displacement steps as required to ensure numerical convergence of the solution.

8.5 Micro (Local) Deformations in CBGA Assemblies under Thermal Loading

The objective of the microanalysis is to determine strain distributions and maximum strains in individual solder joints where strain concen-

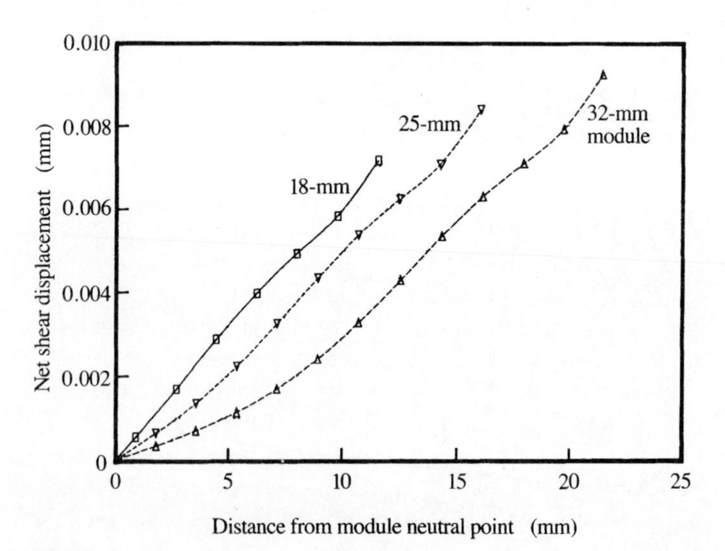

Figure 8.12 Net shear displacement along module diagonal at 100°C for three module sizes. *(Copyright 1993 by International Business Machines, reprinted with permission.)*

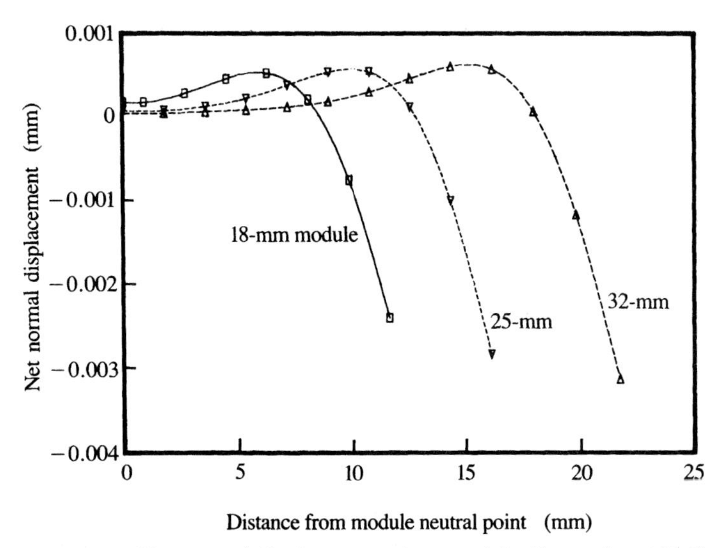

Figure 8.13 Net normal displacement along module diagonal at 100°C for three module sizes. *(Copyright 1993 by International Business Machines, reprinted with permission.)*

trations are usually localized in very small areas. Experimental methods with high spatial resolution and three-dimensional micro-FEM models are the tools described in this section. The results from the microanalysis are used for correlation and prediction of fatigue life of the solder joints.

8.5.1 Experimental analysis

The same specimen as used in the macroanalysis is used in the experimental microanalysis. Figure 8.14 shows the displacements of the last six solder joints ($n = 4 \sim 9$) at the right-hand side of the cross section. Because of the CTE mismatch, high strains are developed in the eutectic solder near the interfaces with the module and the card. Strain concentrations are found at the "corners" of the solder joints.

The force components acting on the two interfaces, between the solder joint and the module and between the solder joint and the card, are schematically depicted in Fig. 8.15. A simplification is used in depicting the cross section of the solder joint as a rectangular shape; the true shape is shown in Fig. 8.3. These forces can be expressed as four components at each interface, as seen in Fig. 8.15: (*a*) a uniform normal force, (*b*) a uniform shear force, (*c*) a moment which is equivalent to a distributed normal force, and (*d*) a distributed shear force. The uniform normal and shear forces are caused by the CTE mismatch between the module and the card, as is the case in most laminated

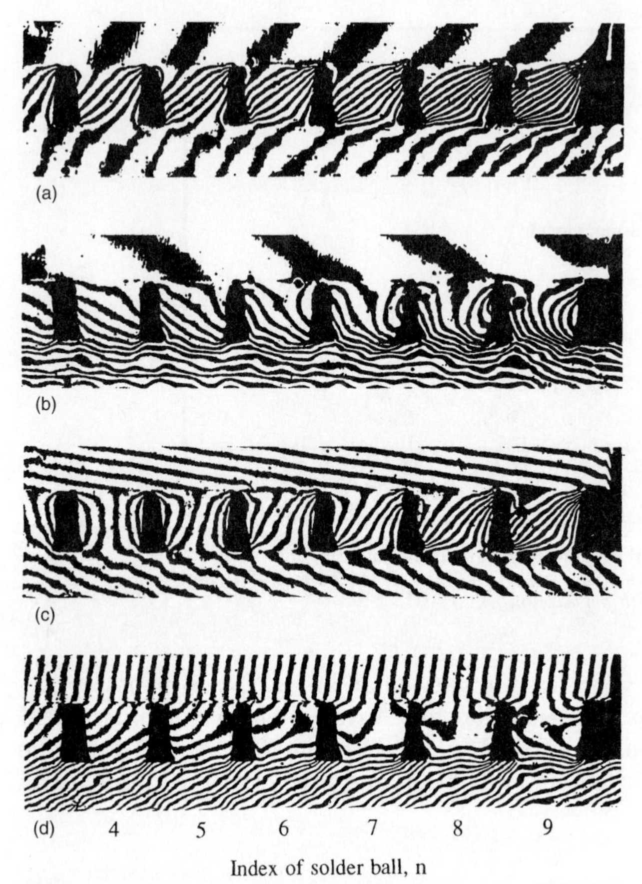

Figure 8.14 Fringe patterns showing the micro effects of the last six solder joints at the right-hand side of the cross section: (*a*) original *U* displacement field; (*b*) original *V* displacement field; (*c*) *U* displacement field with carrier fringes to cancel the rigid-body rotation at the n = 4 solder joint; (*d*) *V* displacement field with carrier fringes to cancel the rigid-body rotation at the n = 9 solder joint. *(Copyright 1993 by International Business Machines, reprinted with permission.)*

structures. The distributed shear forces are caused by the interface CTE mismatches at the two connecting interfaces of the solder joint. The moments in (*c*) are the reacting forces to balance the shear forces in (*b*). The strains in the solder joints are the results of these forces, which vary from solder joint to solder joint.

Normal strains. Figure 8.14*d* is a *V* displacement field showing the rightmost six solder joints of the cross section. It was obtained by adding carrier fringes to the fringe pattern shown in Fig. 8.14*b*. The carrier fringes alter the fringe appearance but retain the original dis-

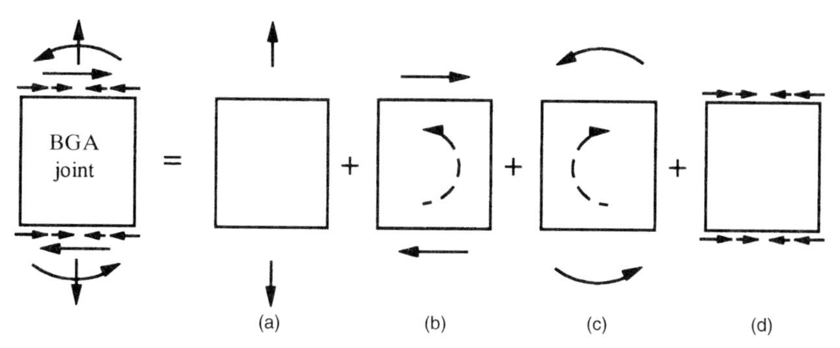

Figure 8.15 Forces acting on a CBGA joint with a nonzero macro-DNP. *(Copyright 1993 by International Business Machines, reprinted with permission.)*

placement information.[21] In Fig. 8.14*d*, the carrier fringes cancel the rigid-body rotation in the rightmost solder joint ($n = 9$, DNP = 11.4 mm), so that the normal strains in the y direction can be observed more easily. The high fringe gradients at the four corners of the solder joint clearly indicate normal strain concentrations. If the four corners of the solder joint are numbered 1, 2, 3, and 4 (Fig. 8.16), normal strains are found to be tensile at corner 2 and 3 and compressive at 1 and 4. This strain distribution is dictated by the interfacial forces shown in Fig. 8.15. The forces acting on the interfaces of the rightmost solder joint are schematically described in Fig. 8.16, where the forces in dashed arrows are the force components (the uniform normal force and the moment) described in Fig. 8.15, and the forces represented in solid arrows are the resulting forces acting on the small elements located at the four corners of the solder joint. The moment components create distributed normal forces which have larger magnitudes near the edges of the interfaces; thus, the maximum effect is located at the corner elements. The distributed normal forces are considered as microeffects because they are related to the solder joint geometries. If the solder were a continuous layer or a one-dimensional thin rod, this local effect would have vanished.

During thermal cycling, the interfacial forces change signs and magnitudes and cause different strain distributions in the solder joints. However, as demonstrated in Fig. 8.16, the normal forces from the macro- and microeffects always have the same sign at corners 2 and 3 (shaded areas), but opposite signs at corners 1 and 4, independent of whether the assembly is being heated or cooled. Thus, no matter how the temperature changes, the maximum normal strain always occurs at corners 2 and 3. For an example, during a cooling process, the macroeffect on the rightmost solder joint is a uniform tension according to the curve shown in Fig. 8.9. The microeffect on the rightmost sol-

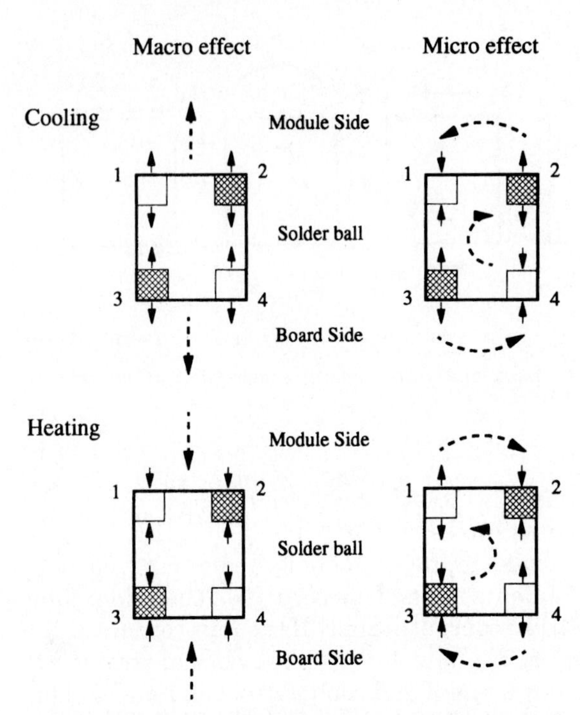

Figure 8.16 Macro- and micro-effects on the normal strain at the four corners of the rightmost solder joint. *(Copyright 1993 by International Business Machines, reprinted with permission.)*

der joint, which is a distributed force at the interfaces, causes strain concentrations at the four corners, corners 2 and 3 in tension and corners 1 and 4 in compression. At corners 2 and 3, since the strains from the macro- and microeffects have the same sign, the superposition results in a larger total strain value. At corners 1 and 4, the strains have opposite signs, such that they cancel each other and result in a smaller total strain value. Since the result is temperature-independent, it is obvious that during thermal cycles, corners 2 and 3 experience larger strain cycles and they will be the locations most susceptible to fatigue failures.

In an individual solder joint, the microeffect has a strong influence on the normal strain concentrations. For the particular solder joint geometry of the sample used in this investigation, the nominal normal strain in the $n = 9$ solder joint is about 0.1 percent strain (the mechanical strain shown in Fig. 8.9). But the maximum normal strain at corner 2 is about 0.5 percent, which signifies a strong normal strain concentration. It also means that the normal strain variation is very dramatic in the solder joints. In this case, it would be a poor approximation if the nominal strain (or average strain) is used to represent the solder strain. Since the strain concentration is closely related to the solder joint geometry (such as the solder fillet shape and size), changing the joint shape can significantly reduce the maximum strain and improve the solder fatigue life. An FEM micro model can be effectively applied to this optimization process.

The detailed strain distributions in the rightmost solder joint are shown in Fig. 8.17. These strain fields are produced from the moiré fringe patterns by an image-processing program introduced by Choi et al.[22] Full field strains and their numerical values are provided by the color contours.

Shear strains. The shear strain in each individual solder joint can be analyzed in much the same way as the normal strains. From the displacement fields shown in Fig. 8.14*a* and *c*, shear strain concentrations are found at the four corners of all the solder joint, even the one having a zero macro-DNP. The *U* displacement field shown in Fig. 8.14*c* was obtained by adding carrier fringes to the fringe pattern shown in Fig.

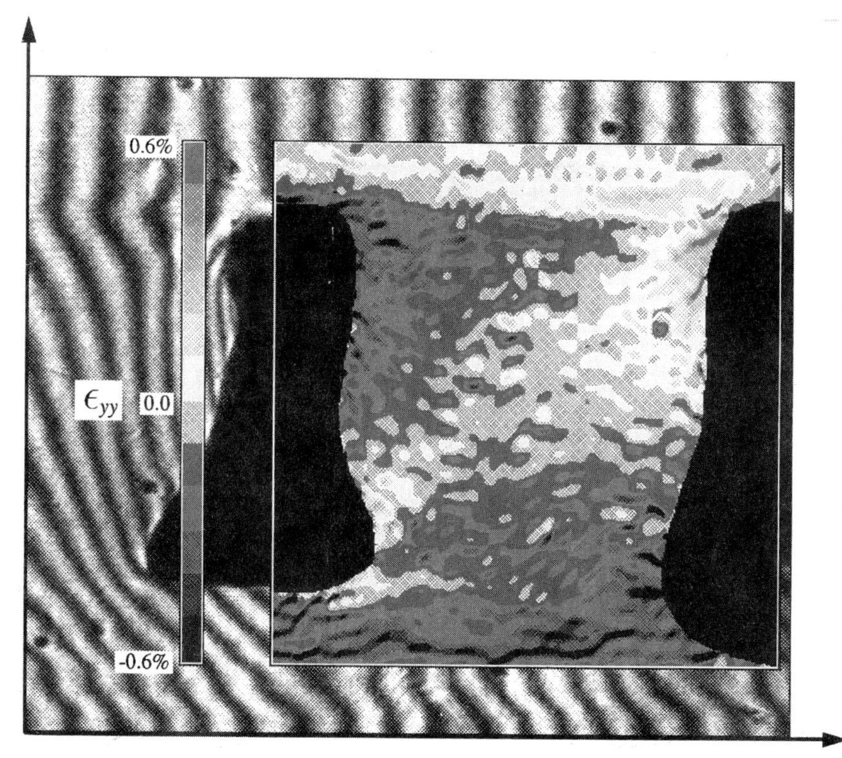

Figure 8.17*a* Normal strain distributions produced from the moire fringe patterns by a computer image processing program. *(Courtesy of H.-C. Choi.)*

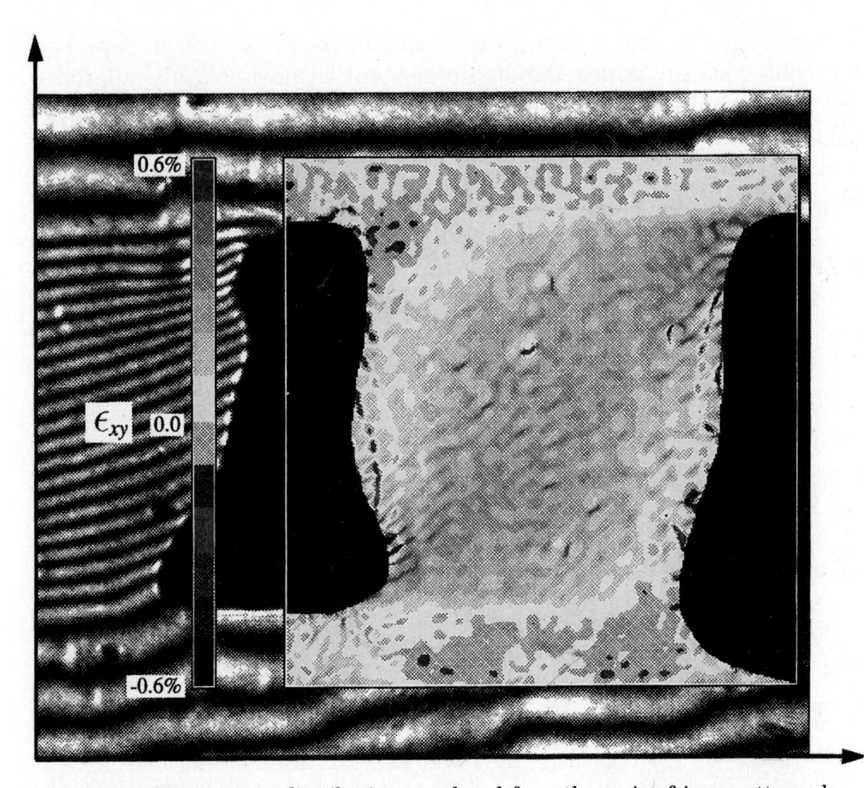

Figure 8.17*b* Shear strain distributions produced from the moire fringe patterns by a computer image processing program. *(Courtesy of H.-C. Choi.)*

8.14*a*. The carrier fringes cancel the nominal shear strain (the uniform part of the shear strain) in the $n = 4$ solder joint (an arbitrary choice), so that the fringes there show only the microeffect—the shear strain caused by the micro-CTE mismatches at the interfaces. Similar to what was discussed in the section on macroshear deformation, the microshear strain is a function of the local DNP (the DNP measured at the interface of an individual solder joint, with the neutral point at the axis of symmetry of the solder joint). At the interface region in a solder joint the micro shear strain is zero at the center of the symmetry (the neutral point) and increases dramatically near the edges of the interface. When the displacement field of the $n = 4$ solder joint in Fig. 8.14*c* is compared with the displacement field of the $n = 0$ solder joint in Fig. 8.8*a*, the one with no macroshear strain, it is clear that the microeffect has the same impact on both solder joints under a thermal load. In fact, since all solder joints have the same nominal size and same interface areas with the module and the card, the resulting shear strains from the microeffect should be identical in all the solder joints, independent of their locations in the assembly. The shear strain caused by the microeffect (micro CTE mismatch) has the distribution characteristic typical of the shear strain at an interface between two materials with different mechanical properties, where the shear strain concentrations are localized only near the edges of the interfaces. If the solder were a continuous layer or a one-dimensional thin rod, as discussed in the previous section on normal strains, the micro effect of the shear strain would vanish as well.

The shear forces and deformations in the rightmost solder joint during a thermal cycle are schematically depicted in Fig. 8.18, where the forces depicted as dashed arrows are the two shear force components described in Fig. 8.15, and the forces depicted as solid arrows are the resulting forces acting on the small elements located at the four corners of the solder joint. With respect to the x direction, under cooling, the solder joint contracts almost freely at the center, but is constrained at the top and bottom interfaces where the micro-CTE mismatches occur. The solder joints which have a nonzero macro-DNP also experience shear strains developed by the macro-CTE mismatch, as shown in Fig. 8.10. The resulting nominal shear strain in the solder joints is a function of macro-DNP.

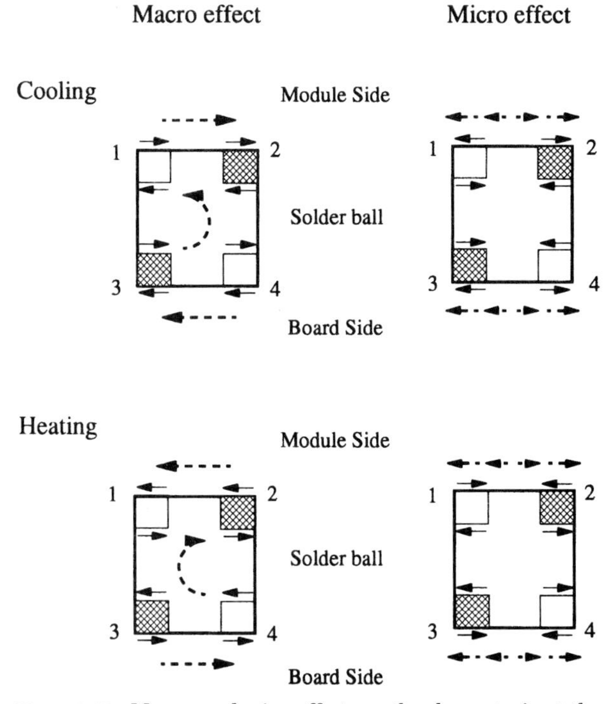

Figure 8.18 Macro- and microeffects on the shear strain at the four corners of the rightmost solder joint. *(Copyright 1993 by International Business Machines, reprinted with permission.)*

The total shear strain in each of those small elements is the sum of the shear strains from the macro- and microeffect (Fig. 8.18). Examining the signs of the forces from the micro- and macroeffects in the rightmost solder joint, the same effect discussed in the section of normal forces is also found in the shear forces. The shear forces from the macro- and microeffects always have the same sign and are additive at corners 2 and 3 (shaded areas). They always have opposite signs and are subtractive at corners 1 and 4. Consequently, large, shear strains are always located at corners 2 and 3 independent of heating or cooling in the thermal cycles. On the other hand, corners 1 and 4 of the same solder joint always have smaller shear strains, no matter how the temperature changes.

There is a great difference in the shear strain values between corners 1 and 2 of the rightmost solder joint as shown in Fig. 8.18. For the condition described above (a 25-mm CBGA package with 0.86-mm/0.71-mm pad sizes under a temperature change of –60°C), in the rightmost solder joint and at the interface with the module, the shear strain caused by the micro effect has almost the same magnitudes as the nominal shear strain caused by the macroeffect. Consequently, at corner 1, by subtraction, the shear strain is close to zero, and at corner 2, by addition, the shear strains reach to the maximum value, more than twice the nominal shear strain value. The shear strain and its distribution in the rightmost solder joint is also shown in Fig. 8.17*b*, where the locations of high shear strains are easily recognized.

8.5.2 FEM micromodel

The highest-risk sites in the solder joints in a CBGA array is the corner joint (DNP = 16.2 mm in the 25-mm-square CBGA assembly) because of the maximum shear and normal deformations. The corner joint is examined by using a FEM micromodel with boundary displacements determined from the FEM macromodel. The equivalent plastic strain distribution within the eutectic solder of the corner joint at 100°C is shown in Fig. 8.19 and Table 8.2. In this model, the module pad diameter is 0.86 mm and the card pad diameter is 0.61 mm.

When the assembly is heated, the plastic zone first appears in the region along the interface between the module pad and the eutectic fillet, on the side of the joint oriented away from the module center (outboard, or corner 2). Plastic strain development quickly follows in the region along the interface between the card pad and the eutectic fillet, on the side of the joint oriented toward the module center (inboard, or corner 3). At 100°C, the plastic zone at the inboard card region is larger than that at the outboard module region, and fatigue failures are more severe at the inboard card region. Reis et al.[7] also observed and reported

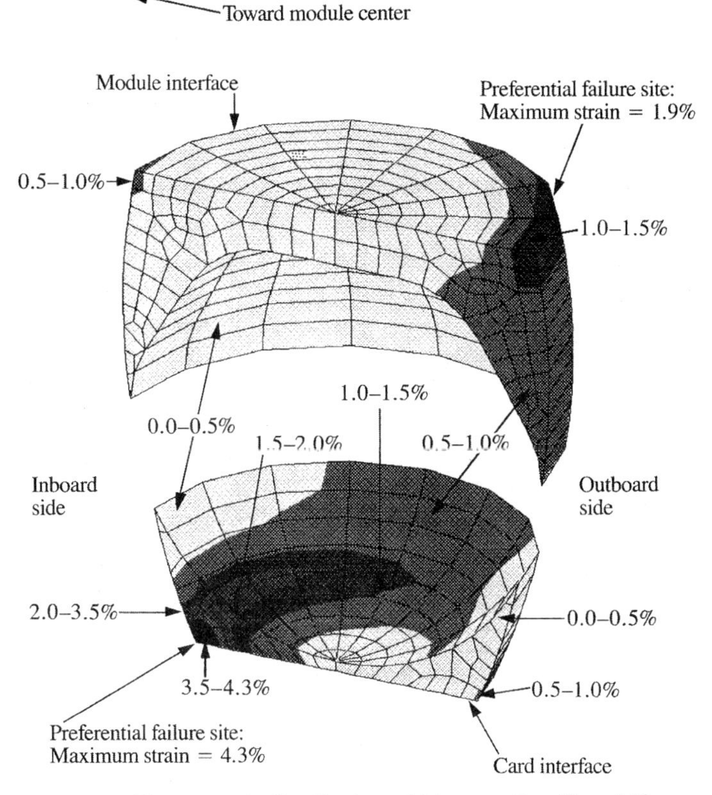

Figure 8.19 Plastic strain distribution within eutectic solder of 25-mm-square module corner joint at 100°C. *(Copyright 1993 by International Business Machines, reprinted with permission.)*

that, in ATC tests, the CBGA assemblies exhibited some module-side cracking but invariably fracture along the interface between the card pad and eutectic fillet.

As seen in the experimental analysis, the preferential failure sites of a solder joint, as shown in Fig. 8.19, are at the two diagonal corners, and are due to the superposition of the stress states caused by the two

TABLE 8.2 Strain Distribution Within SBC Joint*

Solder region	Maximum strain (%)
Eutectic solder at module	1.91
90%Pb/10%Sn solder ball	.92
Eutectic solder at card	4.31

major deformation modes. The shearing deformation mode creates a predominantly normal stress, as described by Robert and Keer[23] in their analysis of an elastic cylinder subjected to a shearing displacement. This normal stress varies across the module and card pads, and is compressive (at elevated temperature) at the module outboard and card inboard sites, and tensile at the module inboard and card outboard sites. The normal deformation mode due to card-module bending creates a uniform compressive stress (at elevated temperature), which is additive at the module outboard and card inboard sites, but is subtractive at the module inboard and card outboard sites. The resulting stress asymmetry creates preferential failure sites at the module outboard and card inboard sites; thus, the normal stress component resulting from bending is significant. The model predicts the same strain distribution in the solder joints as that which was measured in the moiré experiments.

There are some discrepancies noted between the strain magnitudes determined experimentally and those estimated with FEM modeling. These discrepancies are caused by several differences in structures and thermal loads used in the experimental analysis and the FEM modeling: (1) maximum DNP, 11.4 mm in the CBGA sample used in the experimental analysis, versus 16.2 mm in the CBGA assembly modelled, (2) module/card pad size, 0.86 mm/0.71 mm in the experimental analysis, versus 0.86 mm/0.61 mm in the modeled assembly, (3) temperature excursion, 60°C in the experimental analysis, versus 100°C in FEM model (the temperature dependency and nonlinear property of the solder material can create higher strains in the higher temperature regions), (4) model structure, the experimental model is a strip which has a two-dimensional behavior, whereas the FEM model reflects the full three-dimensional CBGA assembly, and (5) the experimental resolution of moiré interferometry is not as high as that allowed with the FEM mesh density used.

A FEM macromodel of the strip specimen used in the moiré interferometry measurements has, for the same temperature excursions, predicted shear and normal deformations that compare extremely well with those measured experimentally, as shown in Ref. 16. In the micro-FEM model, the predicted preferential failure locations also agree very well with the experimental measurements.

The distribution of maximum plastic strain within the CBGA material system is summarized in Table 8.2 for the corner solder joint in a 25-mm-square module at 100°C, with an 0.86-mm module pad diameter and a 0.61-mm card pad diameter. Two apparent observations can be made: (1) The risk of fatigue failure is greater in the card-side eutectic solder than in the module-side solder (this is expected, since the card-side joint area available to support the shear load is smaller than the module-side joint area) and (2) the strain in the ball is small compared to that in the eutectic solder. This suggests that some reduction in peak system strain could be accomplished by a redistribution of compliance within the CBGA structure. That is, the high-melting-temperature solder ball itself could be made softer, to accommodate more of the strain. This might or might not improve the fatigue reliability of the structure, depending on the fatigue characteristics of the material used to accomplish the softening.

8.6 Reliability Optimization

The reliability of the solder joints in CBGA assemblies can be optimized by determining the geometric configuration that both equalizes and minimizes the plastic strains occurring in the module-side and the card-side eutectic solder.

8.6.1 Structural optimization

The structural optimization is accomplished using a designed experiment approach (see, for example, Box et al.[24]) in which the variable combinations that are chosen allow a regression model of specified order to be formed with a minimal number of trials. In the case of CBGA assembly, the objective is to understand the effect of several geometric variables on the maximum plastic strain within the corner solder joints, calculated using the finite element modeling procedure previously described. The geometric variables of interest and their ranges are:

Module pad diameter D_m ($0.51\ \text{mm} \le D_m \le 0.86\ \text{mm}$)

Card pad diameter D_c ($0.51\ \text{mm} \le D_c \le 0.86\ \text{mm}$)

Normalized module-side solder volume V_m ($0.5 \le V_m \le 1$)

Normalized card-side solder volume V_c ($0.5 \le V_c \le 1$)

The module-side and card-side solder volumes can be normalized with respect to a reference volume defined to be that volume contained by a linear fillet extending from the edge of the module or card pad and tangent to the solder ball.

Table 8.3 shows a trial matrix used in an optimization procedure. In this case, a four-variable quadratic designed experiment consisting of twenty trials was chosen. For each trial in the matrix, a three-dimensional finite element micro model was constructed, with inputs consisting of shear and axial displacements based on solutions from macro models. Geometric assumptions include linear solder fillets, a fixed 0.94-mm standoff (card-to-module separation), and fixed 0.025-mm gaps (pad-to-ball separations).

TABLE 8.3 Structural Optimization Trial Matrix*

Trial	D_m (mm)	D_c (mm)	$\bar{V}_m$	$\bar{V}_c$
1	.51	.51	0.50	0.50
2	.86	.86	0.50	0.50
3	.86	.51	1.00	0.50
4	.51	.86	1.00	0.50
5	.51	.51	0.50	1.00
6	.86	.86	0.50	1.00
7	.86	.51	1.00	1.00
8	.51	.86	1.00	1.00
9	.86	.86	1.00	0.75
10	.86	.51	0.50	0.75
11	.51	.86	0.50	0.75
12	.51	.51	1.00	0.75
13	.69	.69	1.00	1.00
14	.86	.69	0.75	1.00
15	.69	.86	0.75	1.00
16	.69	.51	0.50	1.00
17	.69	.86	0.50	0.50
18	.86	.51	0.75	0.50
19	.51	.69	1.00	0.50
20	.51	.69	0.50	1.00

The responses of interest are the maximum plastic strains within the module-side and card-side eutectic solder, and the maximum plastic strain within the solder ball.

The response surfaces from a regression analysis of the responses are shown in Fig. 8.20, in which the sensitivity of the maximum strain to module pad and card pad diameters for constant solder volume is shown for both the eutectic- and ball-solder regions. The response surface in Fig. 8.20*a*, for peak eutectic solder strain, indicates a trough of minimum strain for equal module pad and card pad diameters. (Depending upon the ratio of card pad diameter to module pad diameter, the maximum strain may occur on the card side or the module side.) Along this trough, maximum CBGA-system eutectic solder strain is smaller for larger pad diameters. Hence, the optimum solution for minimizing peak CBGA system eutectic solder strain is to use equal module pad and card pad diameters, with both as large as physically possible. This conclusion is the same as that described by Ohshima et al.,[25] Kamei and Nakamura,[26] and Lin et al.[27] in their analytic investigations of flip-chip thermal fatigue.

It should be noted, however, that a CBGA structure optimized for minimum peak plastic strain will produce an optimally reliable structure only if the metallurgies of the card and module solder interfaces are identical, and therefore have identical fatigue characteristics. In C4 technol-

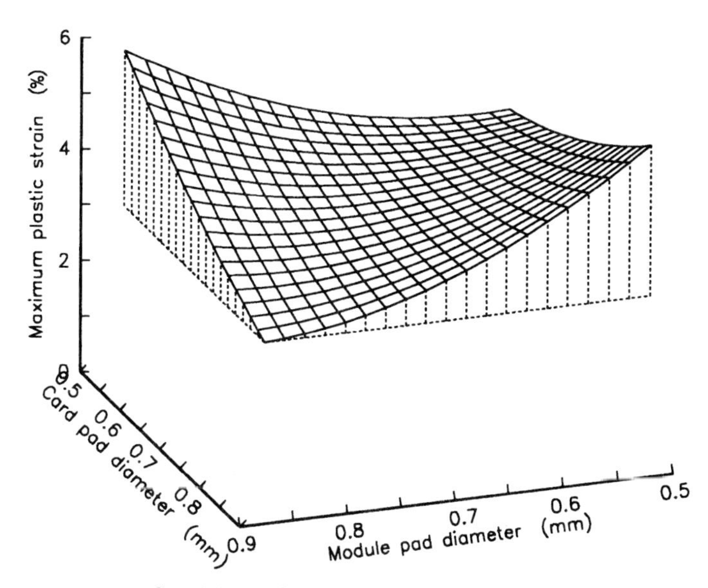

Figure 8.20*a* Sensitivity of maximum plastic strain to diameters of module pads and card pads with normalized module-side and card-side solder volume = 0.75: in eutectic solder. *(Copyright 1993 by International Business Machines, reprinted with permission.)*

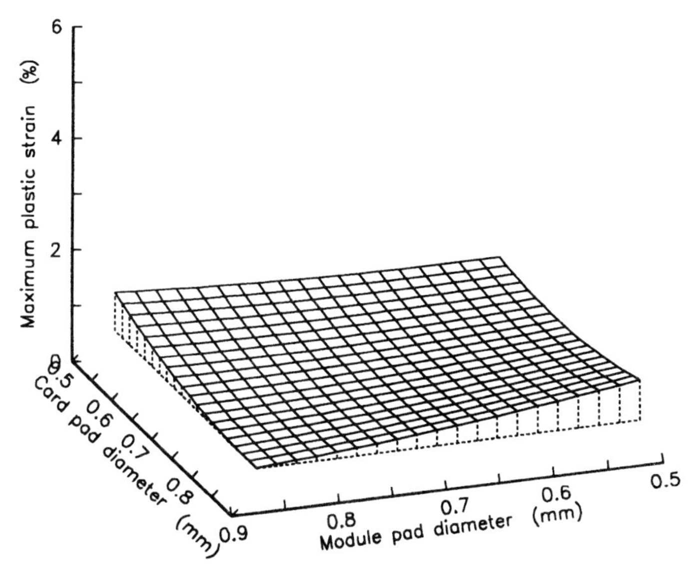

Figure 8.20*b* In solder ball. *(Copyright 1993 by International Business Machines, reprinted with permission.)*

ogy, for example, the differing metallurgies result in an optimal pad-diameter ratio different from unity (Goldmann,[28] Goldmann and Totta[29]).

In CBGA technology, pad-diameter ratios close to unity are sometimes precluded because of manufacturing constraints. Physical limits for the current CBGA technology include an upper limit of approximately 0.71 mm on the card pad diameter (constrained by plating tolerances, the CBGA pitch, and the required number of lines per channel) and an 0.81–0.86-mm lower limit on the module pad size due to drilling considerations.

The strain sensitivity of the solder ball to module and card pad diameter is shown in Fig. 8.20*b*. The shape is similar to that of Fig. 8.20*a*, but the maximum strain amplitude is substantially below that experienced by the eutectic solder. Clearly, the eutectic solder strains dominate, and minimizing peak eutectic solder strain effectively minimizes CBGA-system strain.

The sensitivity of the maximum eutectic solder strain to module and card solder volumes for two different pad-diameter sets is shown in Fig. 8.21. Strain within the eutectic solder is essentially insensitive to the normalized solder volume on both the module side and the card side, if the module-side and card-side pad diameters are equal or close to equal, as shown in Fig. 8.21*a* for the case of $D_m = 0.76$ mm, $D_c = 0.76$ mm. This is an extremely important observation, since a CBGA structure optimized for minimum system strain also offers the important advantage of minimizing the sensitivity to assembly process variables, specifically solder volume. This results in a much more robust and easily controlled assembly process.

As the pad diameters become unequal, however, solder volume does influence the maximum eutectic solder strains, as shown in Fig. 8.21*b* for the case of $D_m = 0.86$ mm, $D_c = 0.61$ mm. For unequal pad diameters, maximum eutectic strain can be minimized by reducing the solder volume on the larger pad and increasing the volume on the smaller pad. The overall influence of solder volume is, however, less pronounced than that of module pad and card pad diameters.

8.6.2 Reliability model

The accelerated thermal cycling (ATC) test is a commonly used technique for evaluating the reliability of electronic packages. Figure 8.22*b* shows the typical shape of a failed solder joint in a CBGA assembly with module pad size of 0.86 mm and card pad size of 0.71 mm after an ATC test. The initial shape of the joint is shown in Fig. 8.22*a*. The temperature range was 0–100°C with a stress rate of three cycles per hour. As shown in Fig. 8.22*b*, cracks appear at two interfaces in the eutectic solder, and propagate almost parallel to the interfaces.

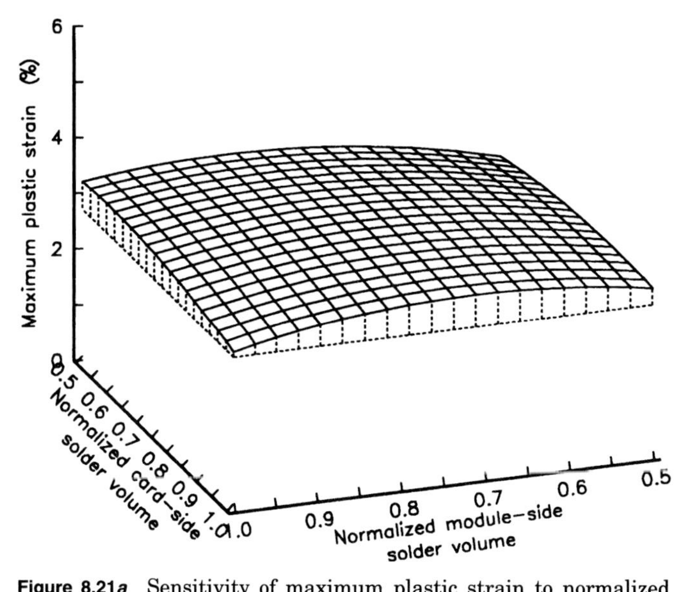

Figure 8.21*a* Sensitivity of maximum plastic strain to normalized module-side and card-side eutectic solder volume: module pad diameter = 0.76 mm and card pad diameter = 0.76 mm. *(Copyright 1993 by International Business Machines, reprinted with permission.)*

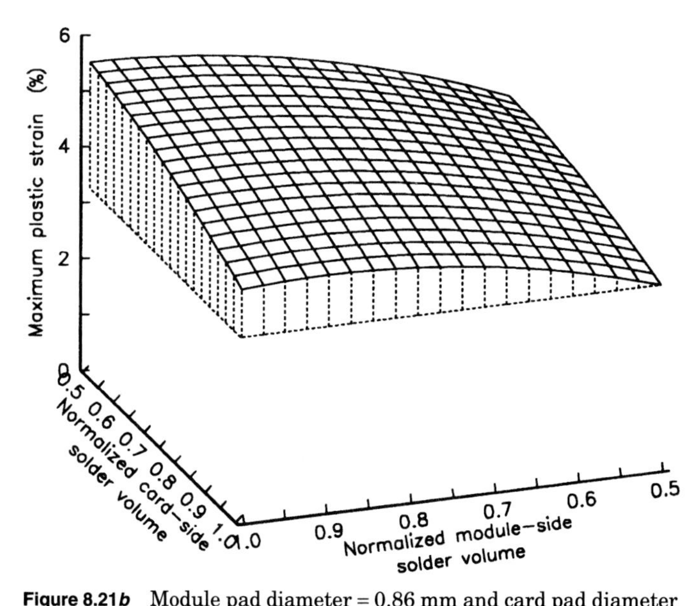

Figure 8.21*b* Module pad diameter = 0.86 mm and card pad diameter = 0.61 mm. *(Copyright 1993 by International Business Machines, reprinted with permission.)*

Figure 8.22*a* Solder joints using module pad diameter = 0.86 mm and card pad diameter = 0.71 mm: before ATC test.

Figure 8.22*b* After ATC test.

A number of structurally different CBGA assemblies have been tested to failure with the same ATC test (0–100°C, three cycles per hour). To confirm the sensitivity to module pad and card pad diameters suggested by the FEM models, three groups of CBGA assemblies, comprising nearly 170 25-mm-square modules, were tested. These groups were representative of three different combinations of module pad and card pad diameters: 0.86 mm/0.61 mm, 0.86 mm/0.71 mm, and 0.79 mm/0.71 mm. Each group had a nominal normalized solder volume of approximately one.

The relative mean number of cycles to failure (failure is defined as an electrical open) for each group was determined and is shown in Fig. 8.23, plotted as a function of the peak eutectic solder plastic strain as estimated by the corresponding finite element model. As can be observed, the data follow a power-law relationship with an exponent of −1.91, very close to the value of −2.00 suggested by Coffin and Manson,[9] and closer still to the value of −1.9 suggested by Norris and Landzberg.[1]

This power-law relationship is the basis of a reliability model used to estimate the influence of various structural modifications and assembly-process variables on fatigue reliability. The ATC cycles to failures are used as a basis of comparison; actual performance is a function of operating environment, cycle frequency, and other factors, and can be estimated using a modified Coffin-Manson approach.[30,31]

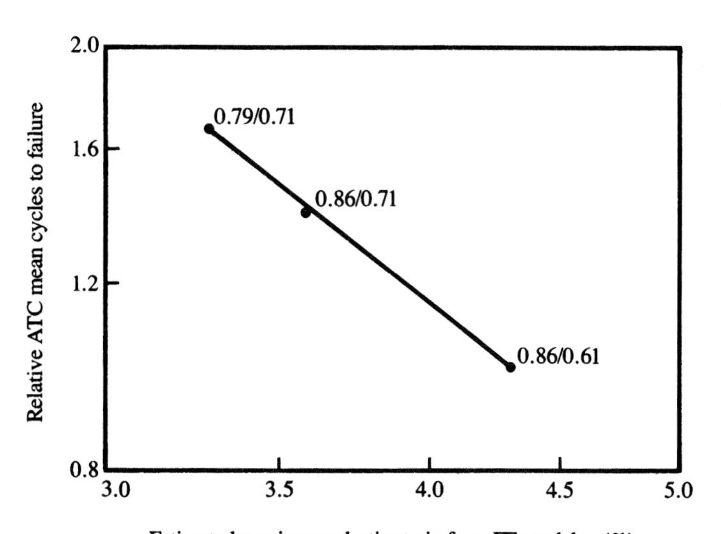

Figure 8.23 CBGA reliability model for 25-mm modules. The combination of pad diameters (in mm) is shown for each point as module/card. The straight line corresponds to the following relation: relative mean ATC cycles to failure = 16.28 × (maximum plastic strain). *(Copyright 1993 by International Business Machines, reprinted with permission.)*

8.7 Summary

The strain distribution in CBGA solder joints is very nonuniform, with strain concentration factors of two to four (Table 8.2). It is very important to used the maximum strain in the solder to correlate or predict the fatigue life and reliability. If the average strain were simply used as the controlling parameter, the results could be very inaccurate. For example, for the three module/card pad size combinations (namely, 0.86 mm/0.61 mm, 0.86 mm/0.71 mm, and 0.79 mm/0.71 mm), the average strains in the corner solder joints of a 25-mm CBGA assembly are almost the same, and the fatigue lives should be equal according to the strain-fatigue life relation. However, in the actual ATC tests, the assemblies with these three pad size combinations showed very different fatigue life (Fig. 8.23). It is when the maximum strain is used that the strain-fatigue life relationship shows good agreement with the Coffin-Manson low-cycle fatigue model.

The FEM models and the experimental measurements clearly indicate the locations of preferential failures in the solder joint array. Globally, the thermal deformations (a shear mode due to macro CTE difference between the module and card, and a normal mode due to module/card bending) are most severe at the module corners, making corner joints the most likely to fail. Locally, because of the high strain concentrations, the fatigue damage will occur first at corners 2 (outboard module-side eutectic solder) and 3 (inboard card-side eutectic solder) in the solder fillets during thermal cycling, both locations consistent with observed failures on ATC-cycled assemblies. Fatigue cracks are initiated at those critical locations and propagate across the entire interface. The strain distributions and the failure modes described in CBGA solder joints are actually common in surface mount packages using solder bumps to connect two components with different CTEs. The preferential failure locations in the solder joints always occur at the outboard corners (corner 2) at the interfaces with the component of a lower CTE, and at the inboard corners (corner 3) at the interface with the component of a higher CTE.

The strain concentrations in an individual solder joint can be analyzed, both experimentally and using FEM techniques, using a macro-micro approach. For a given CBGA package, the macro shear deformation is a function of the macro-CTE mismatch and the DNP in the package. The macronormal deformation is driven by module/card bending, and is extremely pronounced at the corners of the CBGA array. The strain concentrations caused by the microdeformation is a local effect, closely related to the micro-CTE mismatches and the solder geometry. Those strain concentrations are, however, consistent with the superposition of the major macro deformation modes.

Because the strain concentrations are closely related to the solder geometry, it is very important to optimize the shapes of the solder joints under the given structural, electrical, and thermal constraints. The eutectic solder strain can be minimized by designing the module pad and card pad diameters to be as close to each other and as large as possible, consistent with the constraints imposed by manufacturing. Such an optimal CBGA structure provides a robust assembly process; i.e., one substantially insensitive to moderate variations in solder volume. The FEM models can be used extensively to predict geometry and material sensitivities, and provide optimized parameters for general designs of CBGA packages.

Peak strains in the solder ball are substantially lower than those in the eutectic solder fillets; hence, there is a potential reliability enhancement based on redistributing the CBGA joint compliance and allowing the ball to accommodate more strain. This will actually lead to the solder column design in CGA (Column Grid Array) packages.[31]

Finite element model estimates of the eutectic solder maximum plastic strain correlate well with measured accelerated thermal cycle testing fatigue data. These models, therefore, provide an excellent vehicle for predicting the influence of structural modifications on fatigue reliability.

8.8 Acknowledgments

The authors express their appreciation to T. Caulfield and M. Cole of IBM East Fishkill for their generous help, their material properties data, and the solder joint photographs used in this chapter. Also thanks to W. Chen and C. Lim of IBM Endicott for their technical input.

8.9 References

1. Norris, K. C. and A. H. Landzberg, "Reliability of Controlled Collapse Interconnections," *IBM Journal of Research and Development,* vol. 13, no. 3, May 1964, pp. 266–271.
2. Lau, J. H. and D. W. Rice, "Solder Joint Fatigue in Surface Mount Technology, State of the Art," *Solid State Technology,* Oct. 1985, pp. 91–104.
3. Hall, P. M., "Force, Moment and Displacement during Thermal Chamber Cycle of Leadless Ceramic Chip Carrier Soldered to Printed Circuit Board." *IEEE CHMT-7,* no. 4, Dec. 1984, pp. 314–327.
4. Dudderar, T. D., N. Nir, A. R. Storm, and C. C. Wong, "Isothermal Low-cycle Fatigue Testing of Microscale Solder Interconnections," *Experimental Mechanics,* vol. 32 (1), March 1992, pp. 11–20.

5. Wilcox, J. R., R. Subrahmanyan, and Che-Yu Li, "Thermal Stress Cycles and Inelastic Deformation in Solder Joints," *Proceedings of the 2nd ASM International Electronic Materials and Processing Congress,* Philadelphia, Pa., April 1989, pp. 203–211.
6. Solomon, H. D., "Predicting Thermal and Mechanical Fatigue Lives from Isothermal Low Cycle Data," *Solder Joint Reliability—Theory and Applications,* J. H. Lau, ed., Van Nostrand Reinhold, New York, 1991, pp. 406–449.
7. Ries, M. D., D. R. Banks, D. P. Watson, and K. G. Hoebener, "Attachment of Solder Ball Connect (SBC) Packages to Circuit Cards," *IBM Journal of Research and Development,* vol. 37, 1993, pp. 597–608.
8. Phelan, G. and S. Wang, "Solder Ball Connection Reliability Model and Critical Parameter Optimization," *Proceedings of the 43rd Electronic Components and Technology Conference,* IEEE, Piscataway, N.J., 1993.
9. Manson, S. S., *Thermal Stress and Low-Cycle Fatigue,* McGraw-Hill Book Co., Inc., New York, 1966, pp. 125–192.
10. Lau, J., D. Rice, and S. Erasmus, "Thermal Fatigue Life of 256-pin, 0.4mm Pitch Plastic Quad Flat Pack (QFP) Solder Joints," *Advances in Electronic Packaging,* ASME, San Jose, 1992.
11. Guo, Y., and C. K. Lim, "Hybrid Method for Stress/Strain Analysis in Electronic Packaging Using Moiré Interferometry and FEM," *Proceedings of 1994 SEM Spring Conference on Experimental Mechanics,* Baltimore Md, June, 1994.
12. Wilson, E. A. and E. P. Anderson, "An Analytical Investigation into Geometric Influence on Integrated Circuit Bump Strain," *Proceedings of the 33rd Electronic Components Conference,* 1983, pp. 320–327.
13. Sherry, W. M., J. S. Erich, M. K. Bartschat, and F. P. Prinz, "Analytical and Experimental Analysis of LCCC Solder Joint Fatigue Life," *Proceedings of the 35th Electronic Components Conference,* 1985, pp. 81–90.
14. *CAEDS Graphics Finite Element Modeler User's Guide,* version 3, release 2, IBM Corporation, 1990.
15. *ANSYS Engineering Analysis System User's Guide,* revision 4.4, Swanson Analysis Systems, Inc., Houston, Pa., 1989.
16. Corbin, J. S., "Finite Element Analysis for SBC Structural Design Optimization," *IBM Journal of Research and Development,* vol. 37, no. 5, Sept. 1993.
17. Post, D., B. Han, and P. Ifju, "High Sensitivity Moiré," Springer-Verlag, New York, 1994.
18. Guo, Y., C. K. Lim, W. T. Chen, and C. G. Woychik, "Solder Ball Connect (SBC) Assemblies Under Thermal Loading: I. Deformation Measurement via Moiré Interferometry, and Its Interpretation," *IBM Journal of Research and Development,* vol. 37, no. 5, Sept. 1993.
19. Chen, W. T. and C. W. Nelson, "Thermal Stress in Bonded Joints," *IBM Journal of Research and Development,* vol. 23, 1979, pp. 179–199.
20. Guo, Y., W. Chen, and C. K. Lim, "Experimental Determinations of Thermal Strains in Semiconductor Packaging Using Moiré Interferometry," *ASME/JSME Joint Conference on Electronic Packaging,* San Jose, CA, April, 1992, pp. 779–784.
21. Guo, Y., D. Post, and R. Czarnek, "The Magic of Carrier Patterns in Moiré Interferometry," *Experimental Mechanics,* vol. 29, no. 2, June, 1989, pp. 169–173.
22. Choi, H.-C., Y. Guo, W. Lafontaine, and C. K. Lim, "Solder Ball Connect (SBC) Assemblies under Thermal Loading: II. Strain Analysis via Image Processing, and Reliability Considerations," *IBM Journal of Research and Development,* vol. 37, no. 5, Sept. 1993, pp. 649–659.
23. Robert, M. and L. M. Keer, "An Elastic Cylinder with Prescribed Displacements at the Ends—Asymmetric Case," *Quart. J. Mechan. Appl. Math.,* vol. 40, 1987, pp. 365–381.
24. Box, G. E. P., W. G. Hunter, and J. S. Hunter, *Statistics for Experimenters—An Introduction to Design, Data Analysis and Model Building,* John Wiley & Sons, Inc., New York, 1978.
25. Ohshima, M., A. Kenmotsu, and I. Ishi, "Optimization of Micro Solder Reflow Bonding for the LSI Flip Chip," *Proceedings of the International Electronics Packaging Conference,* 1982, pp. 481–488.

26. Kamei, T. and M. Nakamura, "Hybrid IC Structures Using Solder Reflow Technology," *Proceedings of the 28th Electronic Components Conference,* 1978, pp. 172–182.
27. Lin, P., J. Lee, and S. Im, "Design considerations for Flip-Chip Joining Technique," *Solid State Technology,* vol. 13, 1970, pp. 48–54.
28. Goldmann, L. S., "Geometric Optimization of Controlled Collapse Interconnections," *IBM Journal of Research and Development,* vol. 13, 1969, pp. 251–265.
29. Goldmann, L. S., and P. A. Totta, "Area Array Solder Interconnections for VSLI," *Solid State Technology* vol. 26, 1983, pp. 91–97.
30. Jeannotte, D., L. Goldmann, and R. Howard, "Package Reliability," *Microelectronics Packaging Handbook,* R. Tummala and E. Rymaszewski, eds., Van Nostrand Reinhold, New York, 1989, chap. 5, pp. 295–299.
31. T. Caulfield, J. A. Benenati, and J. Acocella, "Surface Mount Array Interconnections for High I/O MCM-C to Card Assembles," *ICEMM Proceedings '93.*

Chapter

9

Plastic Ball Grid Array Packaging Technology

Robert C. Marrs

9.1 Introduction

The continuing drive toward more complex IC devices having lower cost, higher I/Os, greater operating speeds, increased functions per chip, and smaller device geometries has pushed the package requirements far beyond the capability of traditional IC packages.

In response to this situation a family of low-cost, easy-to-use packages called Plastic Ball Grid Array (PBGA) was developed.[1]

The PBGA is a cost-effective, high-I/O surface mount package. It utilizes an array of solder balls on the underside of the package to provide

Figure 9.1 Photo of PBGA front and back.

a high-density interconnection of quality solder joints. The package is simple to mount and reflow to a printed circuit motherboard with very high yields.

JEDEC has completed family registration[38] of standardized outlines and grid pitches of 1.00, 1.27, 1.50 mm in both full- and stagger-matrix that will result in I/O counts exceeding 2400 in up to a 50-mm BGA body size.

9.2 Why Plastic BGAs?

Ball grid array packaging is rapidly gaining acceptance in the electronics industry as a low-cost alternative to fine-pitch leaded packages such as QFPs. PBGAs are currently being used in portable telecommunication and computing applications.

9.2.1 Surface mount advantages

PBGAs offer significant surface mounting advantages over conventional leaded plastic packages.[5,12] A primary advantage is that the solder-bumped PBGAs can be attached with extremely low solder joint defect levels. Solder joint assembly defect levels under 20 parts per million solder joints are reported by Motorola, Compaq Computer, and others in the electronics industry.[8] In contrast, solder defect levels for high pin-count, fine-pitch PQFPs, such as the 208 lead, 0.5-mm PQFP are reported to be 500–2000 parts per million solder joints. IBM recently reported defect levels of less than 1 ppm on a per-joint basis for volume production assembly of BGAs, compared with 20 to 100 ppm for 0.5-mm-pitch QFPs.[35]

Other surface mount advantages include the elimination of lead inspection, lead straightening, and costly rework cycles at the board level. Additionally, PBGAs allow the use of less sophisticated solder printing and pick-and-place equipment to attach the package on the motherboard.

The high surface mount assembly yields of PBGAs is primarily due to the solder land design and use of eutectic tin/lead solder bump technology, which reduces the sensitivity to the three primary factors responsible for yield loss in conventional leaded plastic packages: (1) opens due to inadequate lead coplanarity or skew, (2) shorts between terminations due to solder bridging, and (3) opens or shorts due to package misplacement or movement.[8] With PBGAs, the lead coplanarity and skew problems are eliminated through elimination of fragile leads and the addition of a collapsible solder bump technology. Solder balls on the PBGA package are designed to collapse approximately 0.15 to 0.20 mm during the solder assembly process—eliminating yield loss due to motherboard flatness/warpage, package flatness/warpage, or solder ball size variations. Lead-tip or pad shorting is significantly reduced because of the larger solder pad pitch of PBGAs compared to fine-pitch PQFPs, and the resulting ease of solder screening. Finally, PBGA placement tolerances are not as critical as fine-pitch PQFPs because of the large pad pitch and self-aligning ability. Like IBM's C4 process, PBGAs are self-aligning during solder reflow, due to the capillary action and surface tension forces acting to align the package to the solder lands of the motherboard. Because of the self-aligning properties of the PBGA, the package solder balls can be misaligned up to 40 percent relative to the board pads and the package will align itself during reflow. This phenomenon significantly increases the allowable placement tolerance of the package on the motherboard.

This ability to use standard processes and equipment to surface-mount the PBGA packages, rather than investing in new state-of-the-art fine-pitch soldering equipment and materials, represents a significant savings in capital equipment and manpower as compared to alternative options in high I/O packaging. Several major United States computer and communications companies have implemented the high-density PBGA solder assembly technology into their production lines with little capital investment in equipment and a relatively low investment in manpower.

9.2.2 Advantages of PBGA for MCM applications

The changing market has severely pushed the limits of traditional IC packaging of the 1970s and 1980s. The products of the 1990s require cost-effective solutions with greatly improved electrical performance, lighter weight, flip-chip or high-density wire bonding (frequently mixed together in one module), and novel methods of power dissipation, all while using IC devices sourced from various manufacturers. These requirements are easily served by the low-cost plastic MCM PBGA package format.[2,6,7] Many are developing low-cost MCM-L technologies to service this rapidly expanding market. Amkor/Anam has developed plastic MCM PBGA technology called PMCM®.[13]

The plastic BGA package offers many added benefits and capabilities to the MCM market, including[3]:

Increased interconnect density. Getting more ICs into a smaller footprint is possible with the PBGA because area-array solder ball connections are used rather than peripheral leads. Also, as compared to other MCM configurations, the size of a finished package is considerably smaller. This is because with conventional MCM packages, the MCM substrate is frequently inserted into a larger second-level package—which acts to increase the effective footprint area. Users should carefully study the board-level footprint of alternative MCM packages before selecting the package for their application.

Figure 9.2 Flip chip MCM PBGA.

Lower profiles. With the exterior (bottom layer) of the MCM-L substrate exposed, the package height is about 30 to 50% less than QFP-type packages.

Lighter weight. Along with smaller footprints and lower profiles comes lighter weight.

Lower cost. By eliminating leadframe and the the interconnection from substrate to QFP leadframe, the cost is substantially reduced and interconnection simplified. This improves, to a small extent, yields and reliability. The PBGA substrates also are lower cost than competing multilayer technologies.

Larger routing areas and ease of routing. Because the printed circuit board (PCB) now forms the entire base of the PBGA package, the effective MCM routing area is increased. Also, fewer through-holes and interconnect vias are required to connect from the IC chip to the external package contacts (balls).

Quicker time to market. By eliminating the substrate-to-leadframe lamination step, several critical steps in the design, review, and manufacturing process are eliminated, which in turn reduces manufacturing time by several days or more.

Improved electrical performance. The primary improvements in electrical performance result from two key factors. The obvious is that smaller footprints provide shorter electrical paths for reduced inductance and signal delays, and more importantly, the ability to create very short connections to power and ground planes/buses. These short connections more directly couple internal power/ground connections to those on the motherboard, providing very short length current return paths, reduced voltage drops across planes, and very low power/ground inductance.

The shorter signal and power/ground paths also are significantly lower in resistance. This enables many MCM circuits to operate at reduced power levels.

Improved capability for cost-effective high-I/O packages. In leaded IC packages, manufacturing 300–500 I/O packages results in very large footprints, with higher associated package costs and reduced electrical performance. PBGAs are quite cost-effective in these lead counts, and offer the additional benefit of easy depopulation for custom MCM applications (which frequently require larger body areas but fewer I/Os).

Ease of design and electrical modeling. The PBGA format is easier to design, and has only one package structure to design and model. Thus, the design and analysis time is shortened. As a result, cost is reduced.

Lower nonrecurring expenses. By utilizing standard PBGA formats, users benefit from the established high-volume infrastructure which comes with standard outlines. The only nonrecurring expenses are for design and phototooling, which is low-cost.

Ease of handling through testing and burn-in. Most multichip modules require several extra testing and handling steps. The PBGA package eliminates concerns of high package lead yield loss and lead inspection costs.

Availability of test and burn-in sockets. Standard test and burn-in sockets are also now available from many manufacturers.[35] Families of BGA test and burn-in sockets are now available from Enplas, Plastronics, and Textool, 3M Electronic Products Division. Cinch Connectors has a variety of test sockets and will custom-tool for special requirements. Production sockets for socketing of BGAs on the motherboard are available. Wells Electronics also makes a family of low-force clamshell production sockets for BGAs.

9.3 PBGA Weaknesses

With any new technology there are some initial limitations in use. PBGAs are no exception. The major limitations, at this time, appear to be

1. The installed base of qualified PBGA suppliers is limited at this time (but expected to increase quickly).

2. The QFP package has been thoroughly tested and its limits are well understood. As the PBGA is newer, at this time less data is generally available. The PBGA does appear to be somewhat more corrosion-sensitive due to its added interconnect complexity.

9.4 PBGA versus CBGA Considerations

The principle drawbacks of ceramic BGAs (over PBGAs) are the added complexities in soldering ceramic BGAs directly to PCBs, and their higher cost.

Unlike plastic packages, the CBGAs can be hermetically sealed. A metal lid may be soldered to a plated metal surface, or a ceramic lid may be glass-sealed directly to the package surface. Alternate methods of sealing include brazing a seal ring of ASTM F-15 material to the metal surface. The lid may also be attached by conventional parallel seam welding methods. Some of the processes for hermetic lid sealing takes place at quite elevated temperatures which is a problem for some IC's. Other advantages of ceramic BGAs over PBGAs is the lack of moisture absorption and generally better thermal performance. Also, for IC's which are surface sensitive, the cavity structure of CBGAs is preferrable.

During the design of an electronics system, designers must make many decisions regarding performance, cost, and reliability. For example, a hermetically sealed package is more reliable than a plastic package for some applications, but is the added reliability justified by the added cost? A plastic package can be expected to survive long past the expected product lifetime for all but a few applications, such as military and invasive medical. In today's competitive market, the cost of ceramic packaging alternatives is too high for many situations. In many cases, the added reliability of a ceramic hermetic package is partly or completely negated by the failure mechanism created by the difference in thermal coefficient of expansion (TCE) between the CBGA and the parent PCB (although IBM appears to have solved this[34]). This problem is generally not a concern in plastic packages.[10]

9.5 Using PBGAs

9.5.1 Design of motherboards for PBGA use

The design of a typical motherboard PBGA solder pad is shown in Fig. 9.3. The solder pad on the motherboard is generally defined by an opening in the solder resist. The diameter of the opening in the solder resist will determine the amount of collapse of the package during the reflow process.

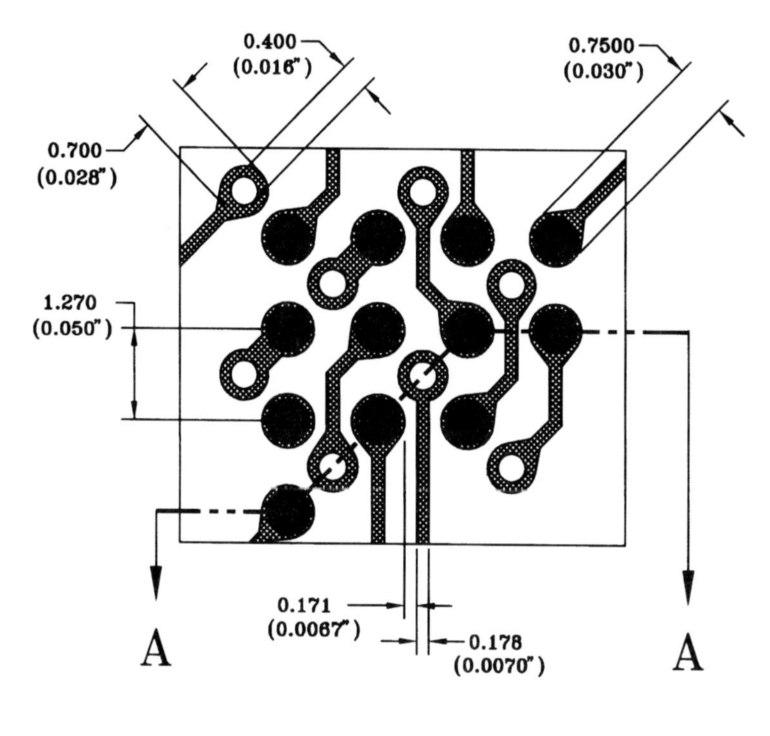

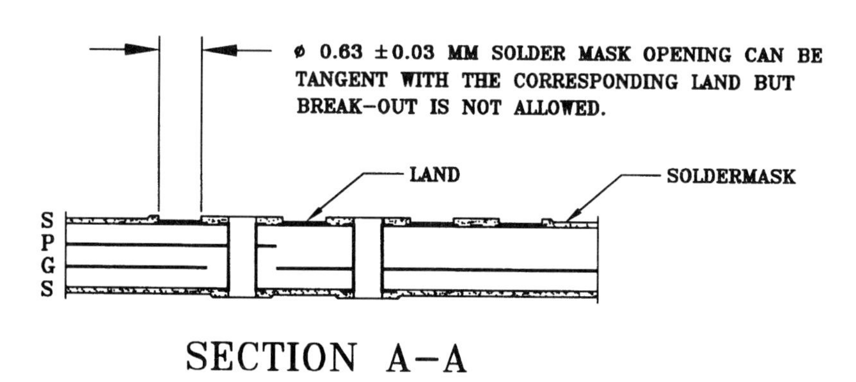

Figure 9.3 Typical motherboard solderpads.

The solder paste screen opening should match the solder mask opening on the motherboard (which is 0.63 to 0.65 mm for ball sizes of 0.71 to 0.76 mm). The recommended copper pad size under the solder mask opening is 0.80 to 1.0 mm depending on solder mask registration tolerance. The package pad pitch is either 1.0, 1.27-, 1.50-, mm array, or 1.27-mm stagger format (parts are not designed in inches).

The motherboard solder pad metal finish is dependent on the requirements of the end user. Several metallizations have been used successfully, including bare copper with an antioxidation coating, hot-air solder-leveled solder, electroplated gold, and electroless gold.

9.5.2 Surface mount attach process

As noted previously, a key advantage of PBGAs is the compatibility with existing surface mount processes and equipment. The standard PBGAs offered by Amkor/Anam, Citizen, and Motorola are all capable of being placed by any surface mount process which is characterized for placement of 0.80- to 1.0-mm pitch plastic QFPs. Fine-pitch placement equipment is not necessary.

In surface mounting of PBGAs, solder paste screening is done first with a commercial solder paste of 63/37 Sn/Pb, 60/40 Sn/Pb, or 62/36/2 Sn/Pb/Ag (the solder paste contains the desired flux type). Either a no-clean flux or a water-clean base flux can be successfully used with PBGAs. The thickness of the screened paste is generally governed by the other components on the motherboard, as PBGAs are relatively insensitive to solder-paste thickness. For example, if the board has other fine-pitch QFP packages to be screened at the same time, the screened paste thickness would probably be in the range of 0.10 to 0.15 mm. If no other fine-pitch products are to be attached at the same time, a screened-paste thickness of 0.20 mm would be more typical. Solder particle size in the paste is again dependent on whether or not other fine-pitch parts are to be placed in the same pass.

PBGA parts are then picked from standard JEDEC[37] or from tape and reel trays, registered, and placed. Registration can be accomplished by either mechanical centering, package edge recognition, or other features on the package. Reflow conditions are typical of an IR or hot-air convection furnace. If a no-clean flux system is used, a hot nitrogen gas convection furnace with oxygen control provides the best results. Next, the motherboard and all solderjoints are cleaned using a standard CFC-free cleaning process.

Amkor/Anam uses a 0.76-mm solder ball which collapses to about 0.71 mm after attachment to the PCB. The parts are designed to collapse about another 0.15 to 0.20 mm during final assembly to the motherboard. The actual amount of collapse depends somewhat on the amount of solder paste, solder-paste thickness, and the reflow temperature profile. The final standoff is then about 0.51 to 0.56 mm, which ensures excellent under-package cleaning. BGAs can be assembled in double-sided PCB configurations using the conventional two-pass method. Some users have found that the high surface tension of the solder joints is adequate to hold the previously placed units during subsequent reflows or during rework processes. Many newer PBGA designs are depopulated at the center (no solder ball connections in the center region of the package). This allows epoxy adhesive to be dotted under the part prior to placement on the board. The epoxy then cures during reflow and holds the part in position during flip-side processing or rework.

Once the process is established for a particular configuration, for process control, many users simply monitor solder ball attach process with electrical test yields. When inspection of finished solder joints is required, commercially available x-ray systems are utilized. Mechanical shear testing can also be employed to monitor joint strength and joint shape, and to analyze failures modes.

When defectives solder joints (or components) are found, solder joint rework is not easy; therefore part replacement is generally required. Rework is accomplished by one of a number of methods which have been developed. Most involve localized heating and removal of component, wicking away of the excess solder, clean-up, and reattachment of a new component.

9.6 Plastic BGA Package Structure and Manufacturing Process

The typical plastic BGA utilizes a BT laminate PCB as a substrate, which is configured in a leadframe format. Chips are assembled to the PCB using nearly the same manufacturing processes employed for high-volume plastic QFP packages. Assembly processes such as chip attachment, wire bonding, and molding are done using conventional automated plastic IC assembly equipment, with a few modifications to fixtures, temperatures, and materials.

An example of one type of 225 PBGA bottom side package design is shown in Fig. 9.5. Many variations of the patterns and through-hole locations are possible.

The PBGA package substrate is composed of two or more metal layers formed on an organic substrate which has glass reinforcement. Like its QFP cousins, PBGAs are processed in strips. The PBGA strips are punched from the PCB panel in which the PBGA substrate fabrication was done and then each site is electrically tested and visually inspected. This processing is generally done at a PCB fabrication com-

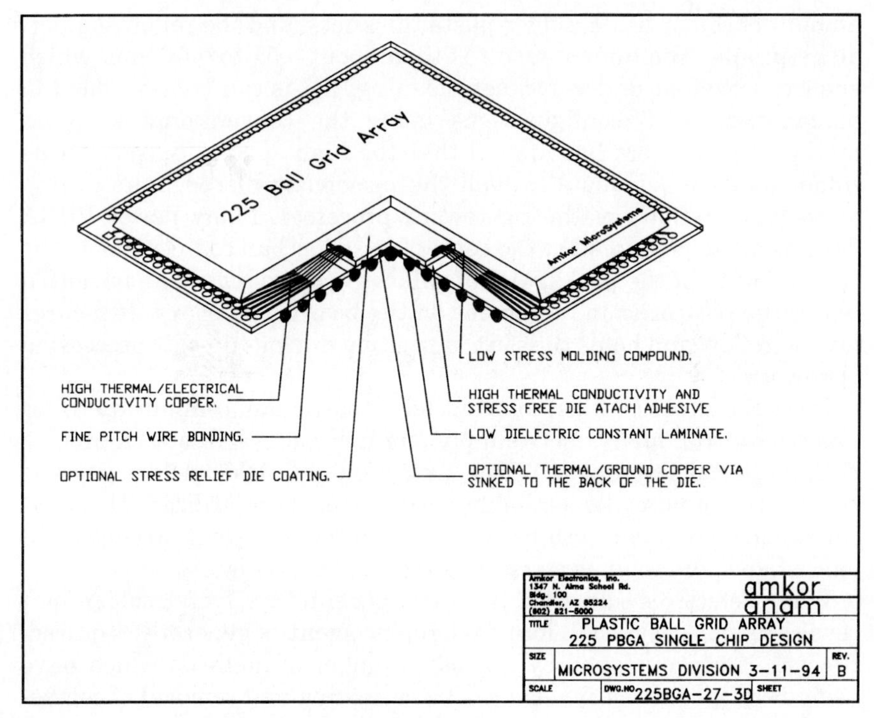

Figure 9.4 Cross section view of standard PBGA.

pany. Next, the strips of PBGAs are packed and shipped to the IC assembly site. At the IC assembly site, IC wafers are saw cut into individual chips using a high-speed diamond-impregnated cutting blade. The individual chips are then mounted to the sites on the PBGA strip using a silver filled epoxy. The attachment is done on the same kind of chip attacher used for plastic QFP assembly. The various workholders and carrying magazines are usually modified slightly to handle the thicker PBGA strips. Next, cure of the epoxy attach material is done. The epoxy cure is done typically at 175°C for one hour in an air-circulating oven.

Wire bonding is performed with standard gold bonding wire which is thermosonically bonded from the aluminum bond pads on the IC chip to gold-plated pads on the PBGA strip. Wire bond temperatures are generally lower than QFPs, and as a result the process is somewhat more critical to control. The bond pad metallurgy on the PCB is typically copper-nickel-gold. The critical aspects of the PCB bond pad to control are gold purity, gold thickness, nickel hardness and nickel thickness, along with pad size. Unlike plastic QFPs, PBGAs have no

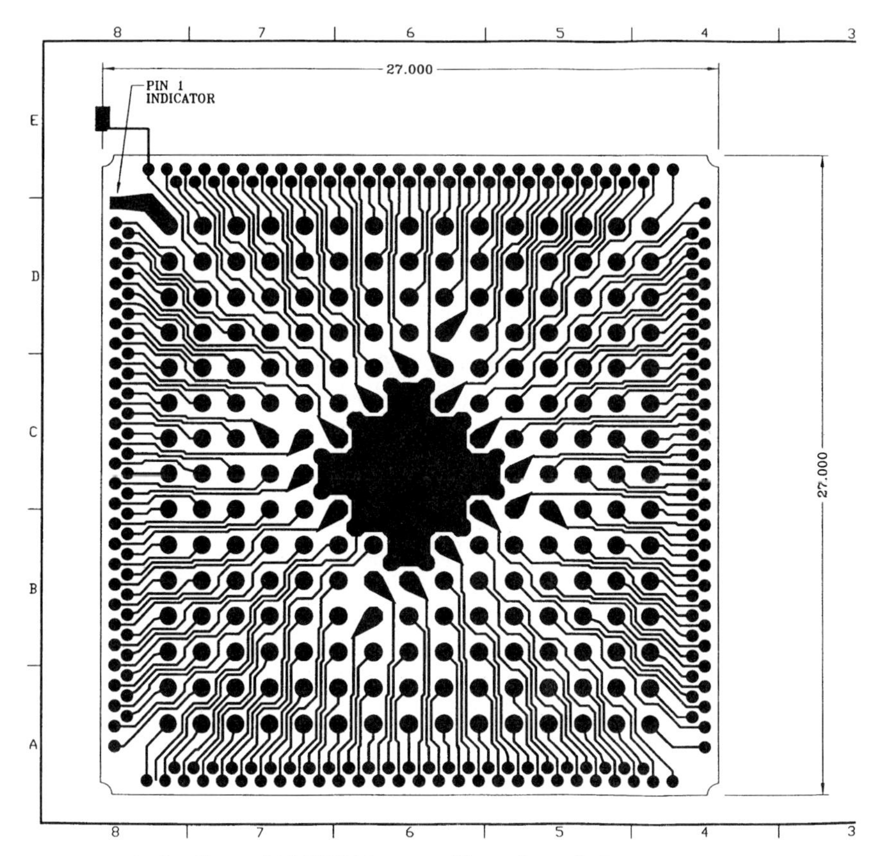

Figure 9.5 Amkor/Anam 225 PBGA bottom side package design.

"floating leads" to deal with during die attach, wire bond, and molding. As a result, yields on high-pin-count devices are frequently higher in PBGAs.

After wire bonding, the strips are transfer-molded with a Novolac based epoxy, with the outer surface (bottom) of the PCB remaining exposed. The mold compound is cured in an air-circulating oven, typically for 6 hours at 175°C. Unit marking (symbolization) is performed on the molded surfaces using either a UV curable, heat-cure-type ink, or laser marking.

Eutectic lead/tin solder balls are then attached to the pads on the bottom side of the PCB substrate. In attaching the balls, first flux is applied to the units and balls are placed on the pads. Next the solder balls are reflowed in an IR furnace to form metallurgical solder joints. The solder ball height is carefully engineered to produce the required height and planarity for final surface mount attach by the end user. The variables affecting this process are solder ball diameter, temperature/time of reflow, and the solder mask opening over the pad on the PCB substrate. Next, the units are cleaned in a CFC-free process to remove any solder flux residue. After cleaning, individual units are punched from the strip and placed in industry-standard shipping trays.

9.7 Thermal Considerations

The primary mechanism of heat dissipation in the standard PBGA structure is conduction to the PCB to which the part is mounted. The PBGAs have reasonably good heat conduction to the motherboard, and as a result, the motherboard acts like a heat sink attached to the package. The thermal performance of the PBGA is therefore highly dependent on the design of the motherboard to which the part is attached and to any airflow across the PCB. Properly configured, the motherboard design can contribute to a 50 percent or greater reduction in the thermal resistance of the package.[4,31,32]

To enhanced thermal performance with standard PBGAs, thermal analysis shows the following general guidelines can be used for optimization of the motherboard design:

1. Utilize thick copper ground and power cores in the motherboard—2 ounce preferred, 1 ounce minimum.
2. The ground plane is most effective as a thermal conductor if it can be located as close as possible to the surface on the side to which the PBGAs are mounted—use as thin as possible prepreg core to attach the outer layer.
3. Make connections from PBGA to motherboard for both the power and ground plane with as many through-holes as possible.
4. Create as large a ground and power bus or heat-spreading area on the mounting surface as the design will permit.
5. Signal leads on the outer layer should also be as wide as possible for a distance covering about 10× the package area (2–3 times the width of the package in each direction).
6. The design objective is to "fill" the area under and around the BGA with copper everywhere possible (on all layers possible) and to couple heat from layer to layer with redundant through-holes. Much of the heat will be conducted across the outer mounting layer and be radiated or conducted away.
7. Filled or plated-up through-holes can further enhance conductivity.

8. Don't forget to balance the metal on both sides and in the inner layers so that excessive stress areas and nonflat regions are not inadvertently created.

Thermal enhancements are frequently made to the PBGA package through the addition of thermal vias located under the chip on the PCB. The thermal vias are formed from plated-through-holes which are created during the PCB fabrication process. These vias provide an enhanced thermal conduction path from the backside of the chip to the solder balls and then into the motherboard. Takubo, et al.[33] of Toshiba have demonstrated that the thermal via structure in a PCB packaging application can decrease the thermal resistance by 50 percent compared to the same structure without thermal vias.

Additionally, the solder balls in the middle of the 225-PBGA package are shorted together with a solid copper at the center of the package that is formed from the PCB artwork pattern. This copper pad is the functional equivalent of a bottom side heat spreader.

Thermal resistance of a via can be calculated by conventional means.[9] The thermal resistance of the cylinder of thickness (T), length (L), and diameter (D) is given by:

$$R_{\text{th}} = \frac{4L}{K_c\,\pi\,[(4\text{DT}) - ((4)T^2)]}$$

For example, consider a 0.0270-inch-thick board, with a plated-through-hole having a diameter of 0.0118 inch and a copper wall thickness of 0.0010 inch; the thermal resistance of the via is as follows. Where *thermal conductivity of copper* = K_c = 9.6 W/°C/inch:

$$R_{\text{th}} = \frac{4\,(0.027)}{(9.6)\,(3.14159)\,[(4(0.0118)(0.010)) - (4(0.010^2))]} = 49.736\text{°C/W}$$

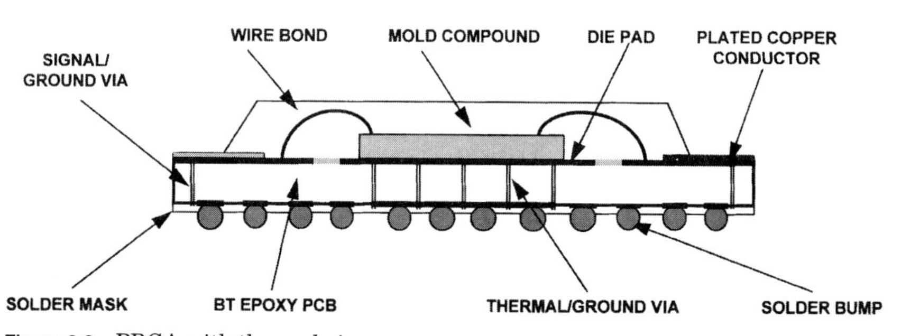

Figure 9.6 PBGA with thermal vias.

While this may seem high, a considerable number of these vias can be placed under a chip. The net thermal resistance may be calculated by dividing the resistance of one via by the number of vias. By contrast, the thermal resistance of an equivalent section of PCB resin the same size as the via is about 61,000°C/W, so the plated-through-hole represents a substantial improvement.

9.8 Electrical Considerations

One of the major advantages of the PCB-based PBGA is the ease with which the package can be customized to provide enhanced electrical performance. Integrated power and ground bus structures can be added to the package design. However, customized piece parts are higher cost due to the smaller purchase quantities and the requirement to manage specialized inventories (procurement, incoming inspection, obsolescence, and inventory costs). Thus, the use of "standard" designs is recommended unless the extra cost is warranted.

Typically, standard PBGA designs have two layers with some provision for ground management, usually in the form of an extended chip attach pad. Standard "pad sizes" are tooled to fit most square die applications (excessively rectangular die shapes frequently require customized designs). These standard designs perform quite well electrically—generally twice as good as their plastic QFP counterparts, and are very suitable for most applications. In many case the performance gains are more significant. Motorola has reported that the performance of PBGAs is superior to both plastic PFP and TAB packages in terms of inductance, switching noise and crosstalk tests.[28]

As an example, the typical inductance of a signal connection in a 169 PBGA is about 3 to 5 nH. In comparison, the typical signal connection in 160L plastic QFP is about 12 to 15 nH. For some applications, ground inductance is a more important parameter. The standard PBGAs being designed today by Amkor/Anam have very low ground inductance. A value of 2 nH or less can be expected in many applications.

The reduced ground inductance is a result from the common bus design used for grounds (or power) which is located directly under the chip, the short chip-to-bus bond wire lengths, and the short connection from the PBGA ground bus to the external PBGA ground structure on the underside of the PBGA package. From the underside of the package to the ground plane on the motherboard, the connections are the low inductance solder balls. Frequently 20 or more solder ball connections are used—all of which are bused together at several points. These busing structures act to reduce current crowding effect which further reduces the potential for creation of unwanted delta-I noise.

The excellent performance in standard and customized PBGAs is also due to changes in design of the PBGA. Original designs[8] had connections which were routed from the inside of the package to the outer edge, down a through-hole, and then back under the part to the solder ball pad. Newer generations of plastic BGAs feature some connections which go directly from the bond wire pad through the PCB and over to the closest solder ball pad. These connections are many times shorter than early design PBGAs, with the frequency response being correspondingly higher. Further, routing of the leads on both the PBGA and the associated parent board is greatly simplified, minimizing cross-coupling between signal lines on adjacent layers.

If greater performance gains are required, an additional power bus structure can be created. Further gains can be realized with power and ground planes built into the PBGA and optional decoupling structures. Critical signal isolation can also be provided for in custom designs.

Electrically, MCM PBGAs can effectively transmit signals at a frequency approaching 2 GHz—sometimes higher. One consideration is the effect of different trace lengths due to the location of the traces and package leads with respect to the package side. Consider a 0.250-inch chip in a 1-inch package. A lead in the center of the package is approximately 0.300 inch long, while a lead going from corner to corner may be as long as 0.424 inch. Depending on the characteristic impedance of the line compared to that of the corresponding terminal on the chip, the propagation delay between a signal in the center of the package and a signal in the corner may be sufficient at very high frequencies to cause synchronization problems. It should be noted, however, that for frequencies below 200 MHz, this is not a problem, and may be considered and compensated for during the layout step.

The limitation of the electrical performance of a package depends primarily on the degree to which the distributed inductance and capacitance of the leads limits the frequency response. Normally, the electrical resistance of the leads does not come into play unless the lead must carry considerable current. Currents large enough to be significant are limited to single-device packages which require a custom design.[9]

The overall frequency response of a device/package unit is the combination of the response of the package interconnect system and the response of the active device. It is clearly impossible to design a package with no distributed electrical parameters, and the goal is to design it such that the overall response is due to the device alone. The equivalent circuit of a phase angle of a typical package is shown in Fig. 9.7. Referring to Fig. 9.7,

L_s = equivalent series inductance of a package lead
L_p = equivalent parallel inductance of a package lead
C_p = equivalent parallel capacitance of a package lead
R_p = equivalent parallel resistance of a device terminal

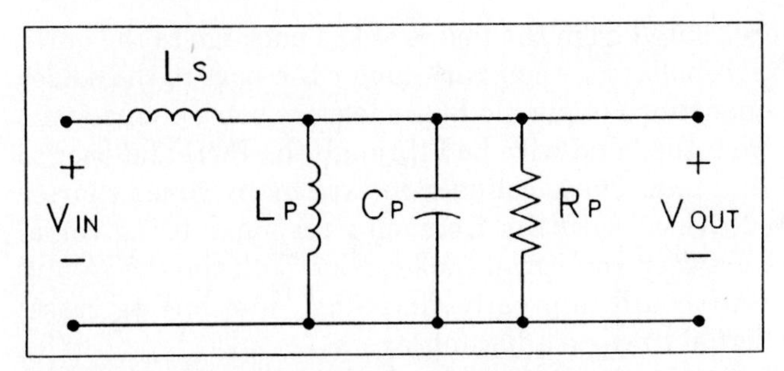

Figure 9.7 Equivalent circuit of a package connection.

The transfer function of this circuit is given by Eq. (9.1).

$$\frac{V_O}{V_1} = \frac{1}{\frac{L_p + L_s}{L_p} - \omega^2 + \frac{j\omega}{R_p C_p}} \tag{9.1}$$

where w = angular frequency = 2 nf
f = frequency in Hz

The cutoff frequency of the filter occurs at the point where the magnitude of the signal is equal to 0.707. This corresponds to a phase angle of 45 degrees, which in turn corresponds to the frequency where the real part of Eq. (9.1) is equal to the imaginary part, as shown in Eq. (9.2).[9]

$$\frac{L_s + L_p}{L_s L_p C_p} - \omega^2 = \frac{\omega}{R_p C_p} \tag{9.2}$$

For most practical cases,

$$\frac{L_s + L_p}{L_s L_p C_p} > \frac{1}{(R_p C_p)^2}$$

and Eq. (9.2) simply reduces to

$$\omega = \sqrt{\frac{L_s + L_p}{L_s L_p C_p}}$$

A typical set of parameters for a QFP package is

$$L_s = 13 \text{ nh}, L_p = 15 \text{ nh, and } C_p = 1 \text{ pf}$$

For these figures, the total cutoff frequency of the leads is:

$$f = \frac{\sqrt{\frac{15x10^{-9} + 13x10^{-9}}{15x10^{-9}x13x10^{-9}x1x10^{-12}}}}{2\pi} = 1.91x10^9\text{Hz}$$

which is adequate for all but microwave-integrated circuits.[8]

9.9 Reliability Considerations

The reliability of any new package technology is of major importance. Many tests are required to characterize a new package completely. The more important tests are those relating to moisture, mechanical stress, and operating life.

Moisture. A major reliability concern in plastic packages is minimizing the moisture content. Moisture in the liquid form is one of the worst enemies of an integrated circuit. Even pure distilled water can absorb phosphorous from a passivation oxide such as P_2O_5, creating phosphoric acid which can readily corrode the aluminum bond pads. While moisture will evenly pass through the plastic molding compound, a key reliability concern is moisture that penetrates through other interfaces.

In PBGAs the possible moisture paths are:

1. Along the glass fibers, then up through the PCB resin
2. Between solder mask and the PCB
3. Between mold compound and solder mask
4. From the bottom through the bulk PCB materials
5. Along any plated-through-hole walls which are located under the molded body regions
6. Along copper traces which extend to the edge of the PCB

The amount of moisture that penetrates is primarily due to two factors: the path length and the degree of adhesion of the interfaces. By proper materials selection, processing conditions, and substrate design, the amount of moisture intrusion can be minimized, thereby dramatically extending the life of the device as well as improving resistance to delamination or package cracking during surface mount attachment.

Package crack and delamination. "Popcorn" is defined as a crack in a plastic package that has reached the outside of that package. It is caused by the rapid expansion of moisture to water vapor inside the plastic package. Delamination is a reduced form of the popcorn mechanism and is very important to check in PBGAs. Delamination is simply the separation of interior layers. The possible delamination modes in PBGAs are listed as follows:

1. Separation of the mold compound from the PCB or soldermask
2. Mold compound to die surface separation
3. Soldermask to PCB separation
4. Delamination between the underside of the chip and the PCB surface

In the PBGA, the mold compound to die surface separation is not any greater risk than in standard plastic QFPs. However, delamination of the mold compound from other surfaces or under the IC chip is a more critical issue because the delamination could propagate along the mold compound to PCB interface to the stitch bond region and possibly lift bonds at the PCB surface. As a result, considerable effort is spent in improving the adhesion of these interfaces. Package crack and delamination resistance of PBGA is good. It is usually tested through a preconditioning of 5× temperature cycle, 168 hours of 30°C/60% relative humidity exposure, followed by three times through a typical IR or vapor phase solder reflow process. Parts should pass functional electrical test and C-Sam or T-Sam testing for delamination in functional areas.

Temperature cycle. Temperature cycle performance is another important factor in PBGAs. Today's PBGAs are not yet as rugged as plastic QFP devices.[13] This is due to the decreased compliance of the solder bump connections as compared to a conventional IC package lead. Typically, PBGAs today are tested to and pass test conditions of 1000 cycles from –55 to +125°C. Extended testing to failure shows that most commonly the failure mechanisms are broken traces, cracked plated-through-holes, lifted stitch bonds, and solder mask separation from copper surfaces. On units attached to PCB motherboards, failure is solder ball fatigue, generally at the balls located under the edges of the IC chip. Improvements in thermal design, and proper selection of materials as well as substrate design and control of processing conditions assures consistent quality and reliability. Future designs and materials improvements will soon result in PBGAs which can pass in excess of 2000 cycles of –65 to +150°C.

Pressure cooker. Pressure cooker (autoclave) testing is another concern in any plastic package, and PBGAs are no exception. Typical designs today pass 168 to 336 hours of 121°C at 15 psi, which is less than a typical plastic QFP package, but is adequate for most applications. The reduced performance is due to the more complex materials system. The predominate failure mechanism is aluminum bond pad corrosion resulting from the presence of excessive water plus ionics. The moisture ingress paths were discussed in section on moisture. The possible sources of ionics are bromine from flame retardant levels in the PCB (flame-retardant levels are not as tightly controlled in PCBs as compared to semiconductor grade molding compounds), poor cleaning during PCB manufacturing, unclean PCB manufacturing conditions, plating bath or chemical etching residues on the surface, plating bath or etching bath residues trapped in irregularities of the plated-through-hole or other surface, and impurities in the solder mask material. The glass fibers in the PCB can also be a source of ionics and other contamination.

The PCB fabrication industry and PCB materials suppliers are working to upgrade the cleanliness levels to semiconductor standards, which will provide ongoing improvement in pressure cooker performance. A few suppliers have already made the investment in new clean-room processing facilities, advanced materials, and advanced cleaning systems. Careful selection and evaluation of materials and suppliers is recommended to achieve the best performance.

Other tests. Hast testing and operating life (burn-in) can also be affected by many of the same mechanisms which affect pressure cooker performance.

9.10 Mechanical Outline Information

See Fig. 9.8.

9.11 Multichip PBGA Market and Applications

MCM market. Many industry analysts have published the current and forecast market of MCMs.[29] The forecasts for the late 1990s range from $9B to $13B (including silicon). Applications such as pocket computers, portable phones, portable fax/modems/answering systems, and integrations of these functions in various forms will drive unprecedented demands for new products which demand low-cost, quick-time-to-market, high-density interconnect technologies. Additionally, laptop/portable workstation level systems will drive high-performance applications requiring high density and speed. Other popular applications will include the PCMCIA cards which have quickly transformed from merely memory cards to the subsystem building blocks of the future.

MCM PBGA applications. The PBGA is an excellent medium for cost effective MCMs.[2,3,6,7,11,14,15,25,26] Given the high degree of thermal and electrical performance, a number of options will exist for the MCM user. Amkor/Anam's PMCM PBGA's can be designed to be tested before encapsulation to maximize final test yields. The high-speed performance of the PBGA facilitates a number of applications combining microprocessor and cache memory in close proximity. With high clock speeds in microprocessors, the PBGA offers a number of possibilities for combining personal-computer-related chips into one package. In MCM applications, the application of applying flip-chip technology to PBGA packaging increases chip densities within the package, as well further improvement in the electrical performance.

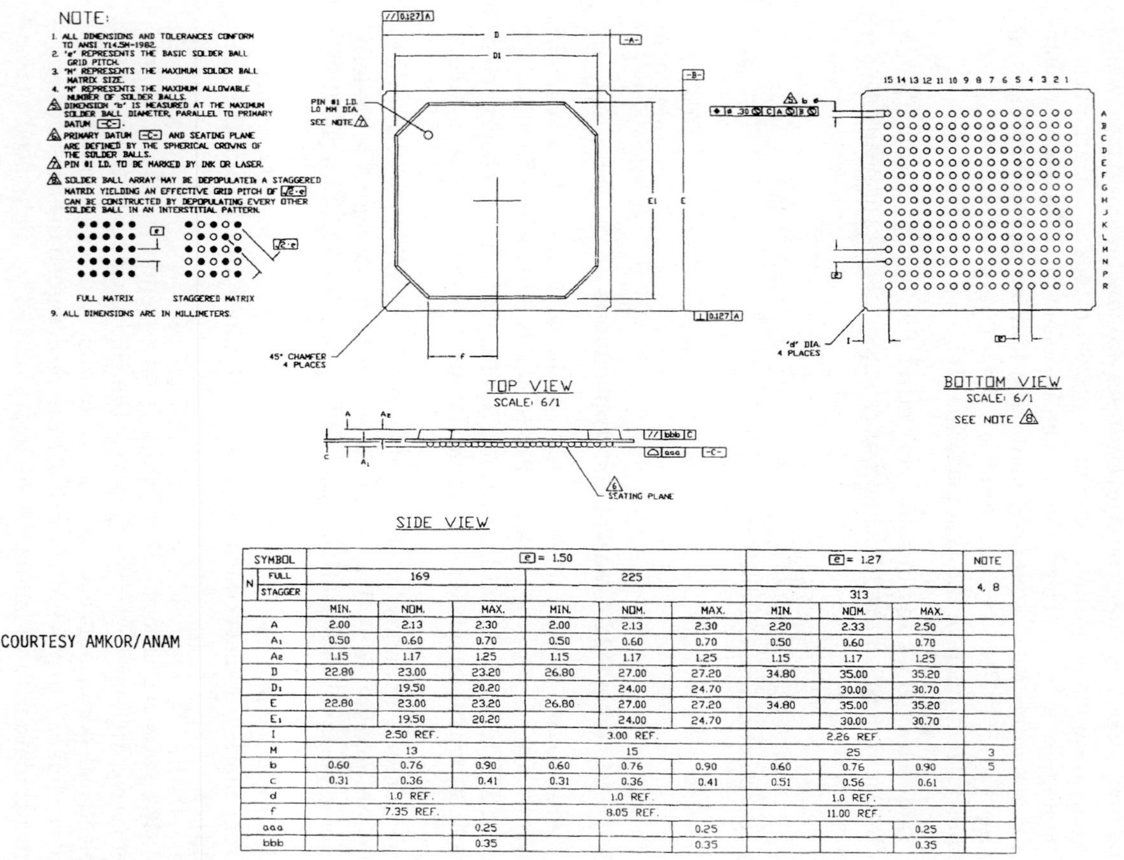

SYMBOL	e = 1.50						e = 1.27			NOTE
N FULL	169			225						4, 8
N STAGGER							313			
	MIN.	NOM.	MAX.	MIN.	NOM.	MAX.	MIN.	NOM.	MAX.	
A	2.00	2.13	2.30	2.00	2.13	2.30	2.20	2.33	2.50	
A_1	0.50	0.60	0.70	0.50	0.60	0.70	0.50	0.60	0.70	
A_2	1.15	1.17	1.25	1.15	1.17	1.25	1.15	1.17	1.25	
D	22.80	23.00	23.20	26.80	27.00	27.20	34.80	35.00	35.20	
D_1		19.50	20.20		24.00	24.70		30.00	30.70	
E	22.80	23.00	23.20	26.80	27.00	27.20	34.80	35.00	35.20	
E_1		19.50	20.20		24.00	24.70		30.00	30.70	
I		2.50 REF.			3.00 REF.			2.26 REF.		
M		13			15			25		3
b	0.60	0.76	0.90	0.60	0.76	0.90	0.60	0.76	0.90	5
c	0.31	0.36	0.41	0.31	0.36	0.41	0.51	0.56	0.61	
d		1.0 REF.			1.0 REF.			1.0 REF.		
f		7.35 REF.			8.05 REF.			11.00 REF.		
aaa			0.25			0.25			0.25	
bbb			0.35			0.35			0.35	

Figure 9.8 Mechanical outline table.

Figure 9.9 Typical MCM PBGA assembly.

MCM PBGAs can also be used to enable mixed IC chip technologies inside one package. This provides product and subsystem designers low-cost, easy-to-implement solutions for many situations. The MCM PBGA technology provides a simple alternative to the difficulties faced in combining mixed fab processes or mixed IC technologies into one chip. It also enables the ability to easily integrate proprietary chip products (such as microprocessor or others which are not readily available to integrate into a single custom chip) into one package, and it eliminates the problems in testing highly integrated chips with mixed technologies.

Historically, solving such chip integration problems lead to delayed product development/introduction and resulting loss of market share or reduced selling prices of new products due to late entry. MCM PBGAs provide a simple and low-cost alternative.

9.12 Technology Trends and Future Developments

I/O counts, memory PBGA modules. I/O counts in PBGA packages are increasing quickly. Volume packages are currently in the 169–313 lead-count ranges. Products are under development with 350–600 pins. Future product requirements will result in packages with 600–1000 I/Os. At the other end of the spectrum are lower I/O PBGAs for memory devices and simple processors.

PBGA board technology. PBGA substrate technology benefits greatly from the large infrastructure and ongoing investment in the PCB industry. Hundreds of large PCB companies exist worldwide which are potential substrate suppliers.

Fine line and space capability is increasing rapidly in the PCB industry. This will have tremendous benefit for applications requiring fine-pitch bonding, dense MCMs, or flip chip. Many MCM-D and MCM-C suppliers do not want to believe that laminates will soon be produced with 25 micron lines and spaces and proportionally smaller vias. However, 75 micron line/space products are in production today, and development work is already ongoing to qualify 50-micron line/spaces and 50-micron vias.

Other development work is concentrating on lower moisture absorption PCB materials, cleaner resin systems, improved cleaning processes, and improved adhesion of dielectric layers. High-speed substrates and PBGA package designs with improved thermal dissipation will also be seen.

Number of ICs. Plastic QFP MCMs and PBGA MCMs typically contain two to four I/ICs and few passive components. Today's capability in test technology limits most configurations to four to nine chips plus passive components. Longer term, 10- to 20-chip plastic MCM PBGAs may be seen, but the author believes that two- to nine-chip solutions will predominate, because chip integration will bring 10- to 20-chip modules down to mostly two- to four-chip solutions.

Wire bonding. Interconnection of ICs today is 96 to 97% wire bonds.[15,27] It is expected that this trend will continue due to the flexibility and low cost of wire bonding. Continued wire-bond technology improvements will allow sub-100-micron-pitch bonding in the near future.

Thinner packages. The market's endless push for thinner and lighter-weight components drives the requirement for thin PBGAs. The capability to produce PBGAs with 1.75-mm overall height is well within the reach of the package technology. As flip-chip PBGAs are introduced, even thinner PBGAs will be seen. Amkor/Anam recently announced its high performance *Super*BGA™ technology which has configurations developed with 1.40-mm height (package top to bottom of ball before board attach and solder ball collapse). After board attach, the parts will be about 1.20-mm height. Thinner versions are in development.

Smaller footprints. The need for smaller footprints will also drive the market to develop finer ball pitch products with smaller footprints. An example of this is the SLICC package recently announced by Banerji of Motorola.[39]

Flip chip. The drive toward higher density combined with electrical performance gains provides the market a tremendous incentive to invest in flip chip.[11,21] In the past, issues such as high initial investment costs, wafer-bumping costs, bumping logistics, and testing of bumped chip has limited the market's jump into flip-chip technology. This appears to be changing. Many companies are now working on low-cost flip-chip technologies.[11,16–26] Many of these processes will be compatible with MCM-L PBGAs, and will be seen in commercial use in PBGAs in the near future.

***Super*BGA™.** For high-performance applications Amkor/Anam has recently developed the new *Super*BGA™ (also called SBGA™)[30] technology which can be utilized to help the designer meet the most stringent thermal and electrical requirements.

The key benefits of the SBGA technology are its simple design, low cost, very thin profile, light weight, and good reliability. It is suitable for high and low I/O applications. Additional internal power and ground planes can also easily be incorporated into the package. Because of the inherent construction technology of the *Super*BGA, new designs can achieve a very fast design-to-market cycle time. Analysis of the heat flow and heat flux patterns show that the SBGA packages can transmit tremendous heat into the PCB to which it is mounted. The PCB to which the SBGA is mounted can therefore act as an additional heat sink for the package. The package is constructed much like a standard PBGA except that the chip faces down (cavity-down) and the entire top surface of the package is covered with a thin copper heat sink/spreader. The heat sink layer is applied during the PCB lamination and is therefore an integral part of the substrate for improved reliability. With the heat-sink layer fully covering the top surface of the SBGA part, the SBGA also provides reduced EMI radiation (and resistance to external EMI sources).

A typical SBGA part is only 1.40 mm thick, with the heat-sink layer and balls included. In comparison, the standard PBGA is 2.30 mm maximum thickness, and a conventional cavity PBGA with heat sink is typically 3.65 mm thick.

Thermal resistance is dependent on the PCB to which the SBGA is mounted, but theta J-A ranges of 12 to 18°C/W can be expected with the 27 × 27 mm SBGA package at zero airflow. Theta junction-to-case values are very low and vary directly with chip size. A chip size of 10 mm will have a theta J-C value of about 0.4°C/W.

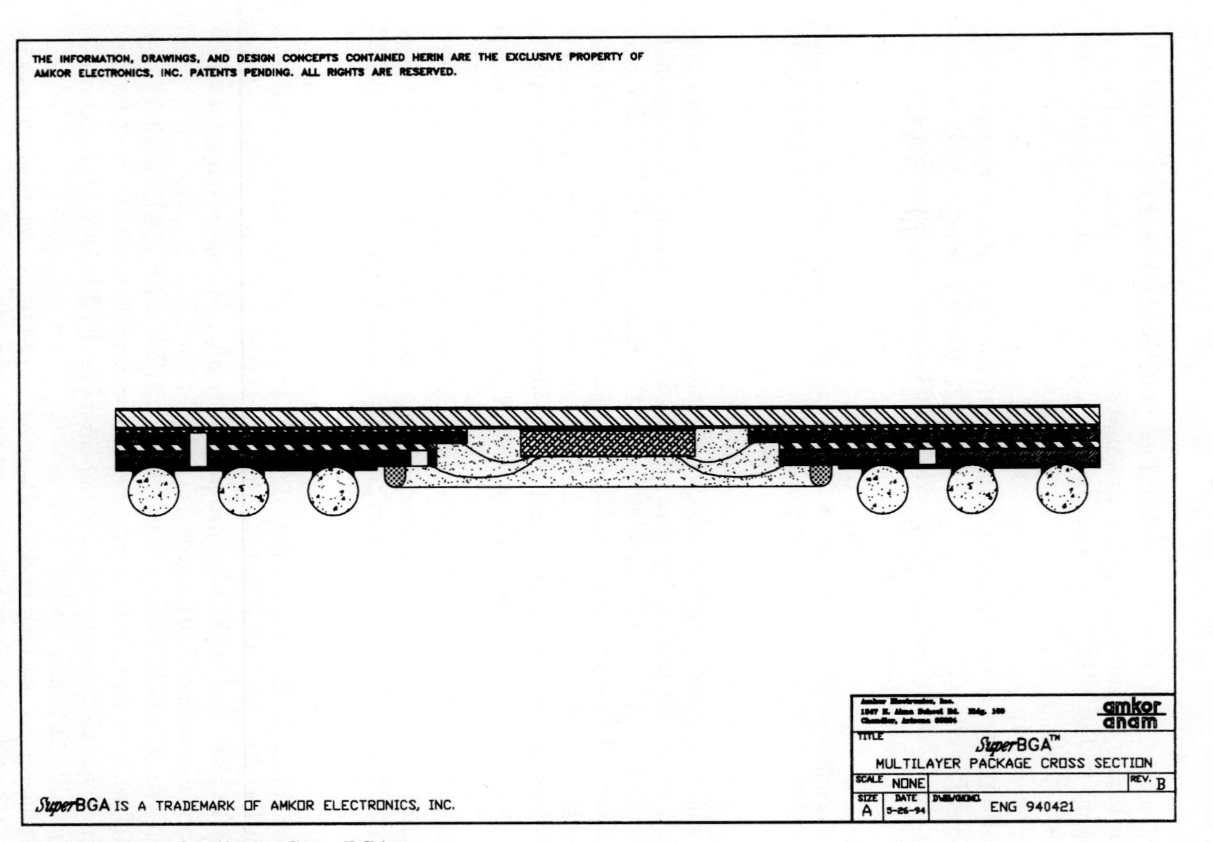

Figure 9.10*a* Amkor/Anam *Super*BGA™.

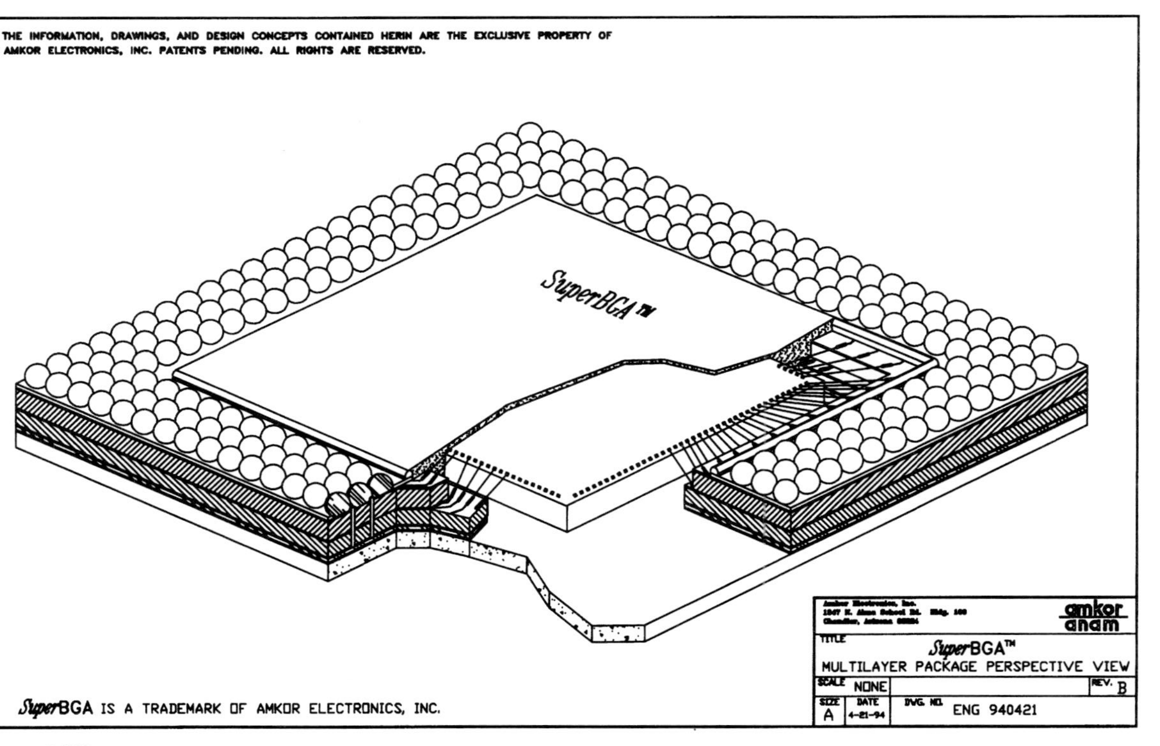
THE INFORMATION, DRAWINGS, AND DESIGN CONCEPTS CONTAINED HERIN ARE THE EXCLUSIVE PROPERTY OF AMKOR ELECTRONICS, INC. PATENTS PENDING. ALL RIGHTS ARE RESERVED.
SuperBGA™
SuperBGA IS A TRADEMARK OF AMKOR ELECTRONICS, INC.
amkor anam
TITLE
SuperBGA™
MULTILAYER PACKAGE PERSPECTIVE VIEW
SCALE NONE
REV. B
SIZE A
DATE 4-21-94
DWG. NO. ENG 940421

Figure 9.10*b*

9.13 Conclusions

The continued market demand for smaller products with more features and higher speed continues to push packaging barriers. In response the plastic PBGA was developed.

The plastic BGA represents an exciting solution to both IC and system designers, as well as the system manufacturer. Its simple solder ball interconnection technology eliminates the coplanarity and fine-pitch problems found in QFPs. By assembling MCMs in PBGA packages, greatly increased densities can also be achieved. Additionally, the PBGA provides a very high degree of board-level integration into individual packages. Improved electrical performance also occurs as a result of the shortened interconnect lengths and ability to incorporate multiple layers into power/ground planes, as well as shorter bond wires.

Ongoing research in high-density substrates, thermal enhancements, and low-cost flip-chip developments will fuel the excitement and continued rapid growth of the PBGA market.

9.14 Acknowledgments

The author would like to thank Amkor/Anam for permission to publish the information on PBGA and MCM PBGAs. Also the author would like to thank B. Freyman and D. S. Youm, as well as J. Houghten, J. Briar, Y. W. Heo, C. N. Kim, J. Lee, and J. I. Kim for their endless energy and support during the development of Amkor/Anam's PBGA technology. Also thanks are due to R. Molnar, D. Cooley, and R. Wenzel for their superior work in the development of the PMCM®[12] MCM PBGAs and *Super*BGA™.[30] Thanks also to J. Boruch and I. K. Hwang for their support and encouragement in advanced packaging technology projects.

9.15 References and Notations

1. Freyman, B. and Pennisi, R., "Overmolded Plastic Pad Array Carries (OMPAC): A Low Cost, High Interconnect Density IC Packaging Solution for Consumer and Industrial Electronics," *Proceedings of the Technical Conference, 1991 ECTC,* pp. 176–182.
2. Marrs, R. C., "Recent Developments in Low Cost Plastic MCM's," *1993 Proceedings International Conference on Multi-Chip Modules,* April 14–16, 1993, Denver, Colo., pp. 220–229.
3. Marrs, R. C., "Advanced BGA Technology for MCM Applications," *Nepcon West 1994,* May 1–3, 1994, Anaheim, Calif., pp. 2006–2111.
4. Marrs, R. C., "Thermal Performance of the 28×28 mm QFP PowerQuad®," *1994 ITAP & Flip Chip Proceedings,* February 16–18, 1994, San Jose, Calif., pp. 150–164.
5. Marrs, R. C., Freyman, B. J., et al., "High Density BGA Technology," *1993 Proceedings International Conference on Multi-Chip Modules,* April 14–16, 1993, Denver, Colo., pp. 326–329.
6. Marrs, R. C., "Trends & Drivers in Technology of Low Cost MCM's," Workshop Speech, *January 1993 Semi-ISS Conference,* Monterey, Calif.

7. Marrs, R. C., "Low Cost Plastic MCM Technology Trends," Luncheon Speech, *5th International TAB & Advanced Packaging Conference,* February 3, 1993, San Jose, Calif.
8. Freyman, B. J. and Marrs, R. C., "Ball Grid Array (BGA): The New Standard for High I/O Surface Mount Packages," *1993 Japan IEMT Symposium,* June 9–11, 1993, Kanazawa, Japan, pp. 41–45.
9. Sergent, J. E., Marrs, R. C., et al. "Designing a High Lead Count Surface Mount Package," *Inside ISHM,* May/June 1993, pp. 5–8.
10. Marrs, R. C. and Sergent, J. E., "Selecting a Package for High Lead Count MCM Applications," *Electronic Packaging & Production,* November 1993, pp. 56–59.
11. Marrs, R. C., "Technology Trends in Low Cost Flip Chip," Luncheon Speech, *First International Flip Chip & 6th International TAB & Advanced Packaging Conference,* February 16, 1994, San Jose, Calif.
12. Freyman, B. F., Briar, J., and Marrs, R. C., "Surface Mount Process Technology for Ball Grid Array Packaging," *Surface Mount International,* August 29–September 2, 1993, San Jose, Calif., pp. 81–85.
13. Mawer, A., Darvaeaux, R., and Petrucci, A., "Calculation of Thermal Cycling and Application Fatigue Life of the Plastic Ball Grid Array (BGA) Package," *1993 I.E.P.S. Conference,* September 12–15, 1993, pp. 718–727.
13. PMCM is a registered trademark of Amkor Electronics, Inc. Patents pending. All rights reserved.
14. Tummula R., "Multi-Chip Technologies from Personal Computers to Mainframes and Supercomputers," *Nepcon West 1993,* February 7–11, 1993, pp. 637–643.
15. Khadpe S., "A Global View of Technology and Market Trends in TAB/Advanced Packaging," *5th International TAB/Advanced Packaging Symposium,* February 2–5, 1993, pp. 1–8.
16. Basavanhally N., D. Chang, B. Cranston, and S. Segar, "Direct Chip Interconnect with Adhesive Conductor Films," *IEEE Transactions on Components, Hybrids & Manufacturing Technology,* vol. 15, no. 6, December 1992, pp. 972–976.
17. Y. Tsukada, Y. Maeda, and K. Yamanaka, "A Novel Solution for MCM-L Utilizing Surface Laminar Circuit and Flip Chip Attach Technology," *1993 Proceedings International Conference on Multi-Chip Modules,* April 14–16, 1993, Denver, Colo., pp. 252–259.
18. G. Hill, J. Clement, and J. Palomaki, "Epoxy Encapsulation Improves Flip Chip Bonding," *Electronic Packaging & Production,* August 1993, pp. 46–49.
19. J. Eldring, E. Zakel, and H. Reichl, "Flip Chip Attach of Silicon and Gas Fine Pitch Devices as Well as Inner Lead Tab Attach Using Ball Bump Technology," *1993 I.E.P.S. Conference,* San Diego, Calif., September 12–15, 1993, pp. 304–320.
20. C. Becher, et al., "Direct Chip Attach, The Introduction of a New Packaging Concept for Portable Electronics," *1993 I.E.P.S. Conference,* San Diego, Calif., September 12–15, 1993, pp. 519–533.
21. E. Vardaman and R. Crowley, "Emerging Flip Chip Use," *Advanced Packaging,* Fall 1993, pp. 32–34.
22. Y. Tsukada, Y. Mashimoto, and N. Wantanuki, "The Design and Reliability of Flip Chip Attach Joint on Surface Laminar Circuit," *1993 ISHM Proceedings,* November 9–11, 1993, pp. 349–354.
23. C. Montgomery, "Flip Chip Assemblies Using Conventional Wire Bonding Apparatus and Commercially Available Dies," *1993 ISHM Proceedings,* November 9–11, 1993, pp. 451–456.
24. J. Simon, A. Ostmann, and H. Reichl, "Electroless Bumping for TAB and Flip Chip," *1993 ISHM Proceedings,* November 9–11, 1993, pp. 439–444.
25. P. A. Sandborn, H. Hashemi, and L. Bai, "Design of MCM's for Insertion into Standard Surface Mount Packages," *Nepcon West 1993,* February 7–11, 1993, Anaheim, Calif., pp. 651–660.
26. A. Lin, "Low Cost Multi-Chip Modules," *Nepcon West 1993,* February 7–11, 1993, pp. 644–649.
27. S. Khadpe, "A Global View of Technology and Market Trends in TAB/Advanced Packaging," *5th International TAB/Advanced Packaging Symposium,* February 2–5, 1993, pp. 1–8.

28. Yip, W. and Tsai, C., "Electrical Performance of an Overmolded Pad Array Carrier (OMPAC)," *1993 I.E.P.S. Conference,* September 12–15 1993, San Diego CA, pp. 731–739.
29. R. Iscoff, "Just How Big is the MCM Market Anyway?" *Semiconductor International,* December 1992, p. 28.
30. *Super*BGA and SBGA are trademarks of Amkor Electronics, Inc. The package and its design concepts have patents pending. All rights are reserved.
31. A. Malhammer, "Heat Dissipation Limits For Components Cooled By The PCB Surface," *1991 I.E.P.S. Conference,* San Diego, Calif., pp. 304–311.
32. V. Manno, N. Kurita, K. Azar, "Experimental Characterization of Board Conduction Effects," *1993 Ninth IEEE Semi-Therm Symposium,* pp. 127–135.
33. Takubo, C., et al., "A Remarkable Thermal Resistance Reduction in a Tape Carrier Package on a Printed Circuit Board," *5th International TAB/Advanced Packaging Symposium,* February 2–5, 1993, San Jose, Calif., pp. 44–51.
34. Cole, M., Caulfield, T., "BGA's Are Extending Their Connections," *Electronic Engineering Times,* February 28, 1994, pp. 48 and 63.
35. Derman, G., "IC Attachments, Interconnects & Packaging," *Electronic Engineering Times,* February 28, 1994, pp. 45, 60, 61.
36. Houghten, J., "Capturing Design Advantages of BGA's" *Surface Mount Technology,* March 1994, pp. 36–43.
37. JEDEC, Solid State Products Engineering Council, Committee Ballot JC-11.5-93-104, item 11.5-375.
38. JEDEC, Solid State Product Outline, SPXGA-X/PBGA PLASTIC BALL GRID ARRAY FAMILY REGISTRATION, number MO-151.
39. Banerji, K., "Development of the Slightly Larger Than Integrated Circuit Carrier (SLICC)," *Nepcon West 1994,* February 27–March 4, 1994, Anaheim, Calif., pp. 1249–1256.

Chapter

10

Plastic Ball Grid Array Assembly

Bill Mullen, Allen Hertz, Barry Miles, and Robert Darveaux

10.1 Introduction

Utilization of pad array carrier (PAC) technology has been commonplace in Motorola's portable products since the mid-eighties. Selection of this packaging technology was driven primarily by the need for packaging density which was not commercially available. With the evolution of larger integrated circuit devices, both dimensionally and with higher pin counts, the need for additional process and interconnect compliance became essential. With PAC technology firmly embedded into Motorola product and commercially available packages still imposing product volume and process compliance penalties, the widely publicized technology of solder ball attach for additional interconnect ductility was the logical path to follow. One of the not-so-obvious benefits of implementing the BGA and PBGA technologies was in the degree of process compliance to be realized in the manufacturing environment. The first PBGA (OMPAC™) to be implemented in a production product occurred in late 1990 into a Motorola pager, pictured in Fig. 10.1, and has been in production in various products since. Subsequently OMPACs have been incorporated into many portable hand-held radios and mobile radios. Figure 10.2 shows the most recent production product to utilize OMPAC™s.

The board in Figure 10.2 contains three OMPAC™s on the side pictured. One vacant site is reserved for option implementation. The edge runners visible around the periphery of each carrier are used for solder connection verification. This feature is especially useful in the development phases of new products such that solder connections can be eliminated from the nonfunctional aspects of the new product.

Today and shortly after production introduction the assembly yields experienced with this packaging technology was and is an order of magnitude better than substantially more mature processes with other

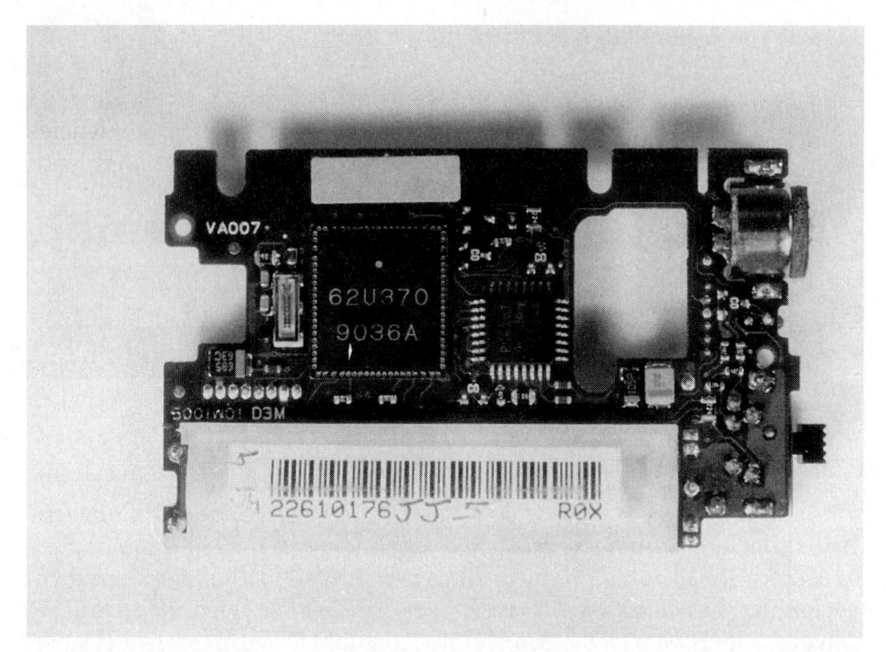

Figure 10.1 The first implementation of PBGA (OMPAC™) technology into Motorola product. The larger package pictured is an 86-pin 8 × 9 array OMPAC™ containing an 8-bit microprocessor. Solder interconnects are arrayed on 0.55-in centers.

Figure 10.2 One of Motorola's most recent products to utilize OMPAC™ technology.

packages. Some cost tradeoffs are currently necessary to achieve the packaging efficiency needed for small portable products. Packages with pin counts less than 140 cost more than their equivalent leaded plastic packages, but as volumes increase and packaging efficiencies are optimized the pin count at which OMPAC™ reaches parity with leaded devices will come down dramatically.

Within the OMPAC™ family itself substantial gains are being realized from optimization around the packaging process. These gains are reflected in increased packaging density, lower manufacturing cost, improved performance, and stability in its less than three PPMJ defect performance in the manufacturing environment. As illustrated in Fig. 10.3 the packaging density of the original Motorola HC11 microprocessor has been dramatically changed from its original (14 × 15 mm) area to an optimized (9 × 11 mm) area. The resultant interconnect density moves from the original 1.5-mm array to the illustrated 1-mm array. In the case of the optimized package the size is limited by the integrated circuit area and requirements for the die and wire-bonding processes; whereas the initial version of this same integrated circuit is size-limited by the BGA interconnect array area required for 1.5-mm grid.

The packages pictured in Fig. 10.3 are of the interconnect-limited and the packaging-technology-limited OMPAC™s. Both packages contain HC11 microprocessor devices with 68 interconnects and were photographed at the same magnification.

Figure 10.3 1.55 mm/1.0 mm comparison.

Motorola's newest products make extensive use of PBGA technology. The newest products, Jedi and Visar radios, contain as many as nine PBGAs each. With this packaging technology Motorola has been able to realize substantial gains in packaging density, product reliability, and assembly quality.

Historically, increases in interconnect density have carried with them some substantial penalties in yields and quality as well as increased equipment costs (e.g., TAB and fine-pitched QFP); but to the contrary, with the exception of cost, PBGA technology has replaced penalties with benefits and created one of the most appealing transitions for manufacturing and engineering.

The phenomenon of "popcorning," wherein the plastic package violently delaminates from the die and/or lead frame, cannot be ignored as a potential showstopper. All plastic packages suffer from this phenomenon to one degree or another. The delamination is caused when moisture absorbed into the package during improper storage is converted to steam during the heating process used for soldering. As with any plastic package the impact of "popcorning" can vary substantially. Popcorning and how to prevent it is treated in more detail in subsequent passages.

This chapter focuses on the considerations of using the PBGA in the assembly environment and covers the following issues:

1. Placement considerations associated with pad design and placement accuracy requirements, including details of the self-aligning pad concept as utilized with Motorola OMPAC™s
2. Compliance in the manufacturing environment, key to the success of the PBGA packaging technology and described and quantified throughout the chapter
3. Assembly methodologies
4. Yield issues
5. Factory handling
6. Repair
7. Failure analysis and the constraints imposed by this packaging technology

10.2 Placement Considerations

Conductor routing from the integrated circuit (IC) pinouts to the product circuitry requires a new philosophy. The solder pads are arranged in an array beneath the component resulting in a maze which disperses the conductors before exiting from the array, compared to a typical fanout of a peripheral-style carrier. In addition to the standard conductor distribution, the use of multilayer substrates enhances the ability to direct the conductor distribution. Plated through-holes (vias) prepare a vertical distribution channel to additional power, ground, and signal layers, thereby multiplexing the exit paths for conductors.

For single-layer distribution, the maximum pinouts can be calculated by:

$$\text{Internal interconnects} < \text{available exit paths}$$
$$(N - 2)^2 < 4\,A^*\,(N - 1)$$

where N is defined as the number of columns and rows of square array and A is defined as the number of lines between pads (see table below). For rectangular arrays, N = SQRT (length * width)

The number of layers required can be determined by:

$$\text{Required PCB layers} \geq \text{available exit paths}$$
$$L \geq (N - 2)^2/\{4\,A^*\,(N - 1)\}$$

where L is defined as the number of layers.

The optimum pad design for pad array carriers would be circular. This geometry minimizes any notching effects and evenly distributes all wetting forces. Maintaining a circular pad with a junction to a conductor introduces some difficulties. The industry standard is to define a solder pad by etched metallization and secure a keepout for resist (etchback). Strategically designing the resist etchback can isolate the solder pad from the conductor. Being consistent with an imperfect world, manufacturers of PCBs require some tolerance. To design out those tolerances, one can define the receiving pad by enlarging the metallization and reducing the resist opening. If the overlap in any one direction is greater than the vendor tolerance window, the design insures a perfect circular pad.

In addition to creating a perfect circle, the design in Fig. 10.4 enhances reliability. First it provides a larger interface area between the pad and the substrate material. Second, it creates a sandwich between the pad and the substrate.

The use of resist over the pad also introduces some new challenges to the PCB manufacturer. The Hot Air Solder Level (HASL) process has been characterized for fully exposed solder pads. This design creates a "bowl" which can be compared to the wind blowing over a small pond. The water would be blown from the leading edge and center to be accumulated at the far side. The electroplated and fused solder application process introduces a third level for registration. It also provides a path where solder may be applied under the resist or copper to be exposed, potentially lifting the resist after reflow. Lifted resist has not been found to cause any concerns, unless rework is required. Bare copper

A: Number of lines between pads

Pitch	Ball diam.	Pad diam.*	0.010	0.009	0.008	0.007	0.006	0.005	0.004	0.003	0.002
0.100	0.030	0.030	3	3	3	4	5	6	8	11	17
0.080	0.030	0.030	2	2	2	3	3	4	5	7	12
0.070	0.030	0.030	1	1	2	2	2	3	4	6	9
0.060	0.030	0.030	1	1	1	1	2	2	3	4	7
0.055	0.030	0.030	0	0	1	1	1	2	2	3	5
0.050	0.030	0.030	0	0	0	0	1	1	2	2	4
0.045	0.025	0.025	0	0	0	0	1	1	2	2	4
0.040	0.025	0.025	0	0	0	0	0	1	1	2	3
0.035	0.020	0.020	0	0	0	0	0	1	1	2	3
0.030	0.020	0.020	0	0	0	0	0	0	0	1	2

* Pad diameter is based upon metallization area with resist over copper (worst case).

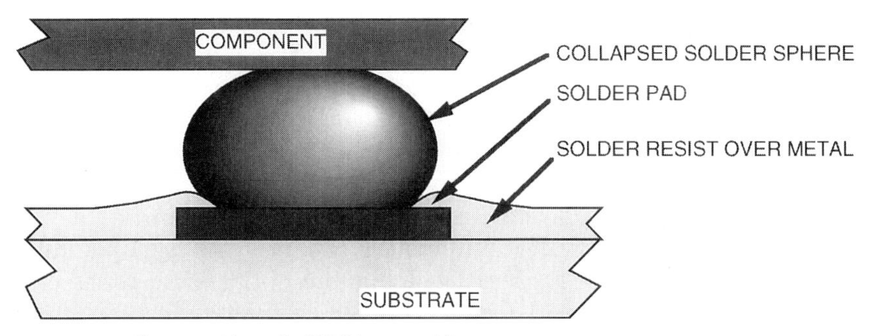

Figure 10.4 Cross section of a PBGA assembly.

protectants (such as Enthone's Entek® series or Protecto®) can replace the tin lead finish. Implementation of this technology has relieved the vendors of their issues and demonstrated increased yields in PBGA assembly.

Alternatives to the resist defined pad can be used. Typical pad designs with a standard resist etchback have demonstrated similar reliability expectations. One advantage this design brought forth was the ability to determine wetting. After reflow, the assembly can be examined by x-ray. Proper wetting would be exhibited by the solder flowing onto the conductor. As in Fig. 10.5, a nonsymmetrical solder fillet is obtained. The lack of symmetry will cause x-ray attenuation variations and produce an indication that wetting took place.

Automated placement routines should also be considered when determining the pad design. Where pads are defined by resist, the vision recognition marking (fiducial) would be preferred to be defined by

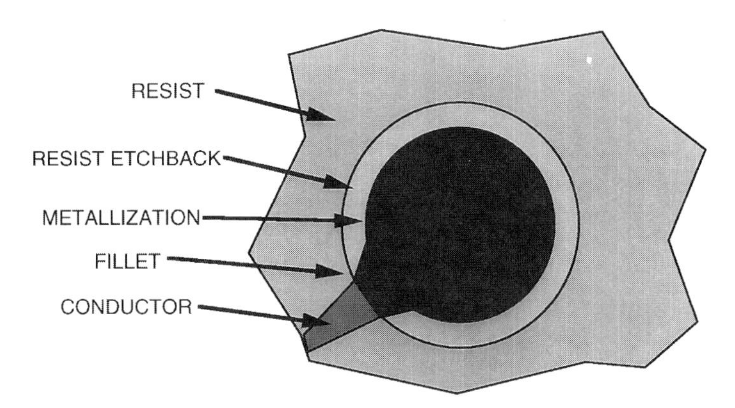

Figure 10.5 A nonsymmetrical solder mask.

resist. A different identification mark should be used for pads defined by metal etch. If the two defining features differ, the tolerance between them should be included when completing the tolerance analysis.

Reliability at the conductor (or via)/solder pad junction should be considered. The PCB and PBGA separately are somewhat flexible, but when attached the interconnect system becomes considerably rigid. Beyond the interconnects remains flexible, creating a cantilever. The formation of the interconnect stiffens the copper, creating a potential stress concentration point where the solder meets the bare copper, particularly on the peripheral joints. The use of fillets can reduce the focused stress concentrations by distributing the stresses over a larger surface. A via interconnected by its annular ring to the solder pad would demonstrate the worst case (see Fig. 10.6).

Placement inspection must be considered when designing the receiving substrate. Peripheral carrier packages (TSOP, QFP, etc.) can be visually inspected by comparing the lead alignment to the solder pads. With pad array carriers, the lead-to-pad alignment is obscured by component package; therefore a second means to verify alignment is required. One method found effective is to add registration markings to the receiving substrate artwork (Fig. 10.7). This can be accomplished utilizing the conductor layout, separate metallization etch, or even resist etchback. The preferred layout would be two L-shaped markings of either exposed metal (isolated from the circuitry) or resist etchback located at opposing corners of the component body. Numerous options can provide the same feature giving the designer the freedom to incorporate what is available. The self-centering phenomenon of the controlled collapse solder spheres naturally produces a large placement window where this alignment inspection method is acceptable. (See *X-Y* alignment.)

Combining the thoughts noted previously regarding via placement and the loss of the interconnect surface, one can conclude to locate the via in the pad center (Motorola patent applied for). A portion of the sol-

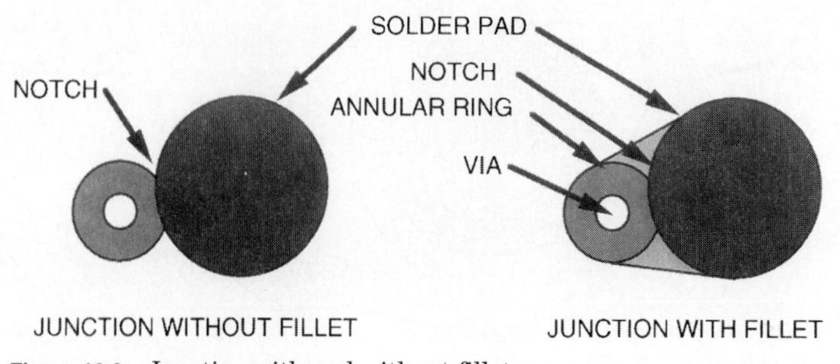

Figure 10.6 Junction with and without fillet.

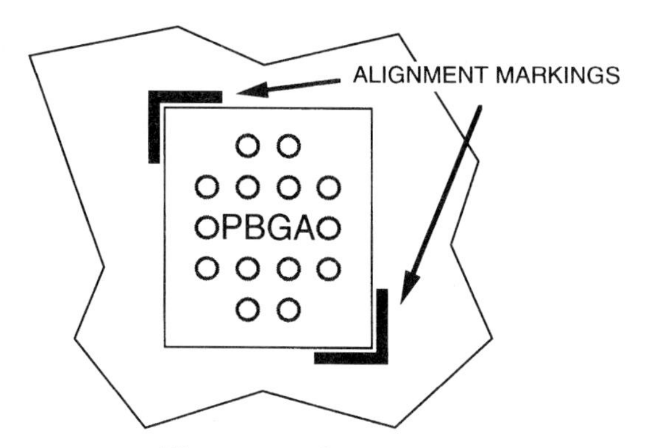

Figure 10.7 Alignment markings on substrate.

der sphere volume reflows into the via creating a rivet-like structure, increasing the field reliability by forcing crack propagation away from the joint interface and into the bulk solder, and also providing an additional path for conductor routing.

Component-to-component distances are determined by the process capabilities of the placement equipment. The PBGA self-centering phenomenon ensures the final location of the component after reflow, simplifying mechanical arrangements of adjacent assemblies.

The low weight, similar coefficients of thermal expansion, manufacturing robustness, and tolerances of PBGA devices provides a multitude of options for the printed circuit board designer. Working with the design team and process engineer can best optimize the layout for the particular application.

X-Y placement. Controlled collapse solder spheres utilize the physics of surface tension. The solder becomes liquidous and flows outward until covering the entire solder pad. Once the solder is liquidous, the component is essentially floating. During reflow, the component readjusts its location until centered on the grid array. This can be demonstrated by dipping a component into a thin film of flux (approximately ½ the ball diameter), placing it slightly misaligned onto the grid array, then placing the assembly onto a hot plate (>200°C) and observing the centering action after the part reaches liquidous.

Placement characterization is a function of the placement process. The large pitch (relative to QFPs), self-centering characteristics, and option for package-based fiducials provide a variety of options for automated placement, such as vision alignment (either the silhouette or fiducials), mechanical alignment, or even manual placement. Most placement equipment manufacturers offer BGA and PBGA placement capabilities on their standard equipment today.

The flexibility of PBGA packages allows the option for fiducials on the underside of the component substrate. Although including fiducials on the underside of the chip carrier may increase the required PCB real estate. The placement process would govern the characterization and component requirements. If the package is aligned using the component silhouette (mechanical or backlit vision), the variation from the sheared edge to the solder spheres needs to be included. If fiducials are used (frontlit reflective vision), the artwork registration is not an issue. Mechanical alignment can be accomplished via edge alignment or direct alignment to the solder spheres.

Presentation for automated placement of the PBGA can be accomplished by several means. The PBGA component is robust, allowing for tray, tape and reel packaging, tube, etc. The variety of options maximizes the real estate utilization and component-exhaust issues while not limiting the opportunities for equipment selection. The defined assembly process influences the placement equipment requirements. For example, if the receiving substrate has the tacking media or solder paste applied (as subsequently described) prior to component placement, the equipment requirements are only to pick, align, and place the component. If the assembly process requires dipping the component in flux, the equipment must be capable of completing the additional process step.

Calculations based upon variable limits in the function of the assembly process can determine the assembly process yield. The process capability index (Cp) is defined as:

$$\text{Cp} = \frac{\text{design tolerance}}{\text{total process variation}}$$

where Cp is desired to be greater than one, or where with the mean of all processes centered, the process yield would be 99.73 percent. If the process means were to shift 1.5 sigma, the yield would be reduced to 93.32 percent. The optimum Cp would be 2, where with the mean of all processes shifted 1.5 sigma, the process yield would be 99.99966 percent:

Desired (Cp = 1) Design tolerance ≥ total process variation
Optimum (Cp = 2) Design tolerance ≥ 2* (total process variation)

The process design tolerance would be defined by the placement window. The self-centering characteristics of controlled collapse carriers provides a window of placement defined as anywhere the solder sphere contacts the pad metallization. Mathematically, the placement window would be defined as ± half the pad diameter when looking at a single axis of error.

The process variations are the sum of the limits of all features of the process, including the following:

1. Placement equipment capabilities (provides limits)
2. Placement process (defines part/PCB tolerances)
   a. Tooling
      (1) Tooling alignment to PCB
      (2) PCB drill to artwork tolerance (pad definition: resist or etch)
      (3) Mechanical alignment process for part
      (4) Alignment means to solder sphere centers

or

   b. Vision
      (1) Component fiducials: zero offset to solder spheres
      (2) Resist defined receiving substrate pads: resist tolerance
      (3) Etched defined receiving substrate pads: zero offset

Example: The placement of a .060-in pitch component on a .025-in diameter pad using silhouette vision for component recognition, placed on pads defined by resist. The automated placement accuracy is ±.005 in.

The design tolerance is defined as ± half the pad diameter (±.0125 in).

The process tolerance is defined by the sum of: resist tolerance (±.003 in), solder sphere to sheared edge registration (±.0026 in), and automated placement accuracy (±.005 in). Since a vision recognition process is used, no tooling tolerances would be included. The accumulated process tolerance is ±.0106 in. The process capability index (Cp) is 1.18 where the design tolerance is 1.18 times the process tolerance.

To achieve a Cp = 2, the process tolerance would have to include tighter limits or the design tolerance limits would have to be relaxed. With a delta of only .0019 in, the potential for shift in any of the contributors would greatly affect the process capability. The process tolerances are generally defined by the manufacturers' capabilities. Restricting the specifications would handicap the manufacturers, resulting in increases to the direct materials costs. The PCB layout is defined by the assembly house providing a simpler means to modify the design tolerances. One example is the pad design illustrated in Fig. 10.8, developed at Motorola extending the placement window from ½ pad diameter to 1 entire pad diameter (patent pending). For this example the Cp would increase from 1.18 to 2.36, taking the process beyond six sigma!

The theory is to utilize large-edge pads arrayed in such a manner as to allow for corner pad wetting even when the devices were grossly misplaced as far as .025 in from center. The surface tension forces which are affected by these edge pads then cause the device to migrate to the approximate center of the internal pads. This technique has

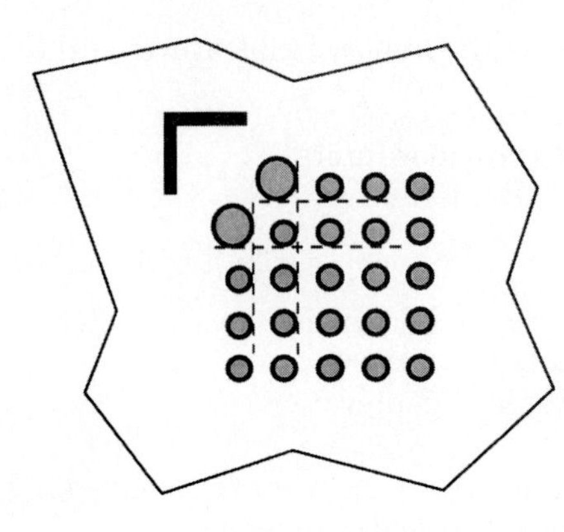

Figure 10.8 A pad design on PCB.

been implemented into production and proven successful, and is one of the key elements in taking the PBGA attachment process into the performance levels expected in a six-sigma environment. Figure 10.9 demonstrates the process of self-aligning corner pads as practiced in our manufacturing environment.

The self-centering and robust nature of the PBGA™ package opens many avenues for the assembly process. With the minimum of expense, the package can provide six-sigma placement tolerances. With placement equipment being capable of placing QFP devices within .003 in of the mean target, placement of PBGA™s with their lower accuracy requirements allows for those elements which extend beyond our three-sigma world to perform successfully.

10.3 Process Compliance

One of the major assets contained in the PBGA interconnection process is its capability to accommodate variations in substrate topography and camber. This is accomplished by designing in a collapse process which occurs during the reflow process. To better understand the collapse process we will follow one specific size package through its assembly process. In this example we will use a process which does not utilize printed paste and therefore represents a worse-case scenario for the process compliance.

In Fig. 10.10 we see the solder spheres attached to the PBGA in its as-received form. The solder sphere's mean height is .027 in with a standard deviation of .0003 in. In this case we will be placing the package onto a PWB with .030-in-diameter uniform bare copper pads defined by photoimaged solder resist. Flux will be applied to the spheres, and the device will be placed onto the receiving substrate

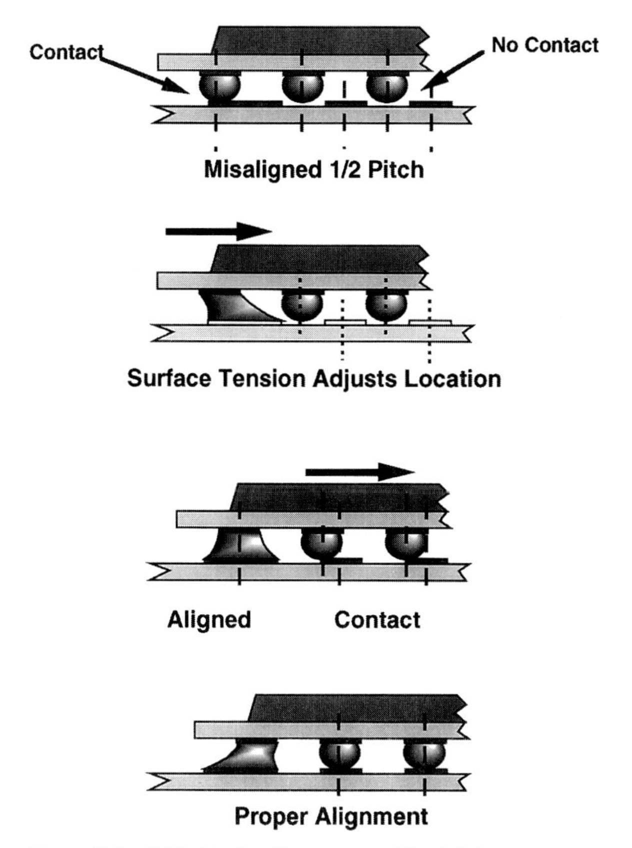

Figure 10.9 Self-aligning the corner solder joints.

and reflowed in a nitrogen atmosphere. To predict the process compliance we will assume that if a .001-in interference fit is effected between the liquidous solder and the PWB pad, wetting will take place. As the reflow process takes place the solder spheres melt and they wet to the PWB pads. This wetting action allows the surface tension of the liquidous solder to pull the PBGA toward the PWB. On conclusion of the reflow cycle the assembled units can be sectioned and measured, at multiple points, to determine the distance between the PBGA surface, from which the original sphere height was determined, and the PWB pad surface. In this example the resultant dimension has a mean of .019 in and a standard deviation of .00023 in. In all of the devices studied with this methodology no failures to effect a robust connection have been observed. To look at the worst case imagined, by taking the six-sigma extrapolated value of the minimum sphere height, .0252 in, and comparing it to the six-sigma extrapolated value of the greatest distance, .02038 in, between the two measured surfaces, we conclude that there will always be an interference fit of .00462 in. One might expect that the compliance of this process allows for more, possibly .003 in, of topography and camber variation before an impact on the joint quality can be realized. Fundamentally the robustness can be determined for any given package and board layout using this method. With the die sizes currently used in OMPAC packages there is little to no jeopardy to yields to design in collapse factors which are larger than the example cited. Having conducted the same evaluation for a pad array carrier (non-BGA) using the solder paste printing and reflow process, and having found the predicted process yield losses to be consistent with yields then currently being experienced in our production lines, we felt the procedure to be valid.

When using solder paste to attach PBGAs to a substrate, additional compliance is realized. A suitable means of characterizing the impact of the additional metal volume might be as follows: (1) to print the PBGA sites with solder paste, and (2) to reflow the printed sites and measure the resultant solder bump height. This data can then be combined with the sphere height data and compared to the assembled height. Suffice to say that if there is sufficient compliance with the non solder-paste assembly, the utilization of printed paste will only enhance the compliance value.

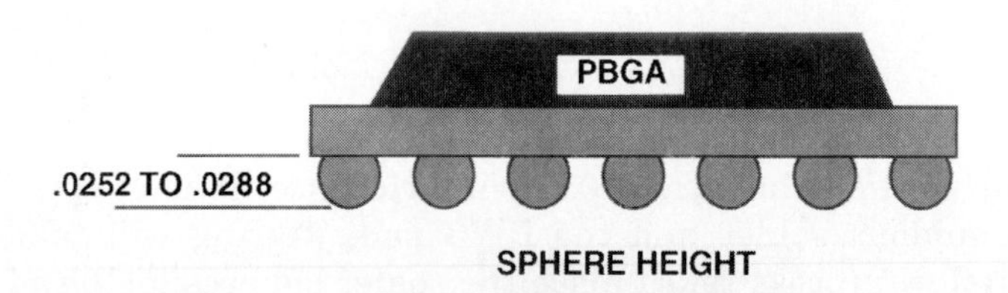

Figure 10.10 PBGA compliance collapse measurement.

10.4 Assembly Methods

Implementation of BGA technology into the manufacturing environment offers a degree of flexibility unseen in other high-density integrated circuit packages. In the typical SMT factory of today solder paste printing followed by component placement and reflow is probably the most common process in use. Other less-popular but nonetheless beneficial technologies include: (1) assembly using solid solder deposition (SSD) whereby boards bearing solid (fused) solder are brought to the assembly line, flux and components are applied and the assembly is then reflowed; and (2) preform assembly whereby a combination of solder preforms and flux precedes the placement and reflow of components.

The implemented BGA assembly technologies the author has observed in recent years include two fundamentally different methods which accommodate a variety of in-place processes and board finishes. The first of these processes is the solder paste printing, subsequent component placement and reflow method. Utilizing this process, whereby solder paste is applied using printing or dispensing techniques, to the metallized pads, as described in Sec. 10.2 and the BGA devices are then placed into the applied paste; allows for process compatibility with the most widely used SMT assembly process in the electronics industry. With this process the substrate finish can include all of the common finishes available today. Specifically finishes currently in use include organically protected bare copper, various solder-leveled finishes, and gold-plated and solder-plated finishes. Since the solder needed to effect the interconnect is carried by the BGA itself, the volume of solder which is printed is of lesser importance than when non-solder-bearing components are used. As discussed in Sec. 10.3, the use of solder paste will enhance the process compliance of the PBGA process as discussed. Areas of possible concern associated with solder paste use would include the formation of solder balls, which could in turn cause shorting or peripheral damage to adjacent component functions, or the introduction of solder voids into the solder sphere.

Several solutions to a solder ball problem can be implemented:

1. Since the typical standoff height of a PBGA is greater than .25 mm, a solvent-cleaning process can be adequately designed to remove solder balls from under the device.
2. The solder paste can be designed to eliminate production of solder balls.
3. No-clean solder pastes can be used here as solder balls are typically contained by the vehicle residue and are therefore passivated from interaction with other components.

Item 2 provides the least problematic solution to a potential solder ball problem, and if combined with the use of no-clean solder paste presents the best of both worlds.

The introduction of voids into the solder sphere is a concern primarily from a reliability standpoint as voids contained within the sphere tend to migrate to one interface surface and provide a less stress-resistant interface. Although it is known that small levels of void content do not present a reliability problem and are currently tolerated in several manufacturing processes it is the experience of the author that large (>50 percent) void contents may lead to earlier fatigue-life failures and will sustain less bending stresses.

The second commonly used methodology for attaching PBGAs is to apply flux to either the PWB pads or the PBGA spheres, place the PBGA onto the PWB, and reflow the solder spheres. Application of flux to the PWB can be accomplished by any number of methods such as dispensing with a programmable system, transferring using pins in an array matching that of the PBGA, spraying, or brushing. Application of flux to the spheres can be accomplished by dipping the PBGA into a controlled film of flux, brushing or spraying flux onto the bottom surface of the package. As with the printing of solder paste this method places no constraints on the commonly available board finishes. In this case, unless SSD is used, all the solder available for forming a solder joint will be provided by the package. Of the two methods for applying the flux, dipping the package into a controlled film of flux will provide the most certainty that all spheres and pads receive fluxing action. Whether applied to the PWB or the PBGA the flux selected must have sufficient retention properties to force the PBGA to remain in place through any intermediate processes which precede reflow soldering. Solder joints will be reformed, once in the soldering environment, given sufficient flux, a sufficiently active flux, solderable pads, sufficient heat, and adequate placement. If the emerging assembly carries a device, which is now seated substantially closer to the board than it was on entry to the reflow environment, it is a rational conclusion that all connections have been effected. A mature, characterized, PBGA process requires no inspection.

10.5 Soldering Yields

With production lines currently running in excess of 10,000 PBGAs per day and defect rates at or below 4 PPMJ there is not a lot of engineering focus on improving the yields of PBGA package soldering. Historically documents discussing process yield related issues for new packaging technology start with defect categories and proceed to discuss the activity focused on reducing the yield defect numbers. The typical list of defect sources might include solder shorting, solder opens, misoriented part, missing part, wrong part, insufficient solder, and debris (such as a misplaced chip capacitor) under the package. Of these elements the process-related issues would be opens, shorts, and insufficient solder. As mentioned previously, a typical PBGA assembly process utilizes no inspection and as a result defects, if there are any, are caught at electrical test. As such, opens and shorts remain as the only two categories to be recorded which are assembly process-related. Shorts have been observed, when solder paste printing is used as the method of assembly. Two sources of shorts have been identified: (1) paste bridging between the pads when the part is placed one-half pitch off pads such that a solder bridge is formed from bump to bump, or (2) when the part is moved, intentionally or unintentionally, subsequent to placement. Although it may be possible to effect a short at the package pitches available today, shorts have not been found as a result of assembly processing without solder paste. In one memorable instance shorts were observed where an out-of-control repair process allowed for substantial downward force to be effected on the top of the PBGA during solder and solidification. Once corrected this has become a nonexistent problem. Ultimately we are left with opens. Opens are detected at electrical test and are included along with all yield loss sources in the yield learning curve graphs (Fig. 10.11). The migration from low-density interconnects to high-density interconnects has classically meant finer pitches and more fragility associated with the packages and the processes; this is supported by the much more mature QFP process curves which still demand the dedicated attention of a manufacturing engineering staff. The learning curves chart (Fig. 10.11) compares QFP with the PBGA. In the case of the QFP, curves data is included from many sources on .5-mm and .65-mm packages. The area between the learning curves for the QFP is wide such that it encompasses the best, the average, and the less-than-average performers all between the two lines. The end point of the QFP curve would indicate a best-in-class performance of 80 PPMJ while the not-so-yield-focused organizations would perform at 800 or more PPMJ. As of this writing the author does not have sufficient data to generate a pair of curves that would differ sufficiently to identify "best in class" and "typical" for PBGA assembly. One observation might be that the PBGA yields have started where the QFP thirteen-year-evolved yields have matured to.

The dominant sources of yield loss are unsolderable PWB pads, unfluxed PWB pads, misoriented components, and the product engineer's imagination (the culture of accepting an invisible solder joint as being good). The most difficult of these to solve is the last, as a hidden solder connection is often assumed to be the reason for a nonfunctional circuit. It is altogether too easy to take an analysis through a defective

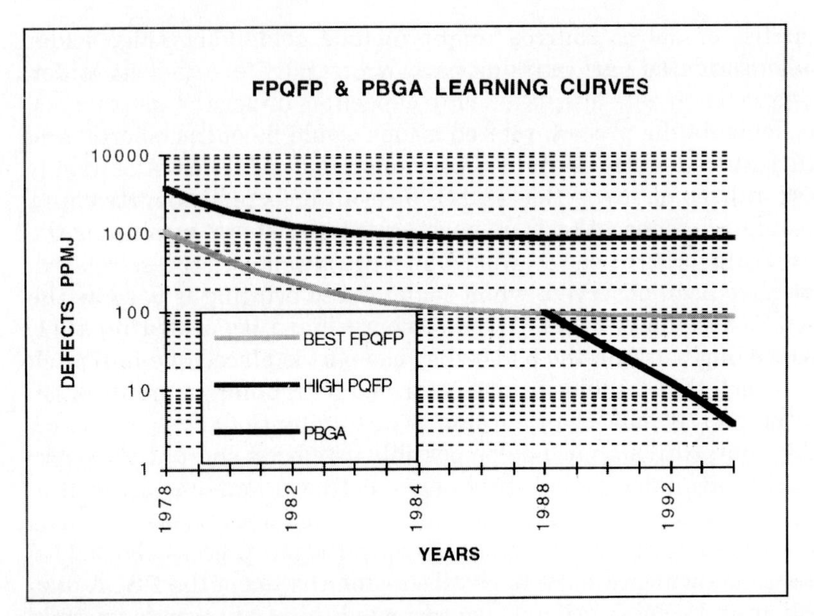

Figure 10.11 FPQFP and PGBA learning curves.

assembly until the PBGA is encountered and then assume the PBGA to be the problem. Developing a culture where you push on and find the real problem, which in the authors' experience seldom is the PBGA, is the difficult part. Elsewhere in this chapter a methodology for identifying the real nonsoldered connections is described. Addressing the other three sources of defects is well within the level of expectations of the manufacturing engineering staff and presents no unique technological challenges.

10.6 Factory Handling Requirements

10.6.1 Popcorn/delamination issues

The PBGA package is the latest in the succession of plastic semiconductor packages to exhibit susceptibility to moisture-induced package cracking. As semiconductor devices continue to be designed into smaller, lighter, and thinner packages, the inherent ability to absorb moisture of the materials used to fabricate plastic packages has become a serious issue. Package miniaturization, coupled with the continuing introduction of larger semiconductor devices, has exposed interfacial adhesion as a weakness of surface mount plastic packages. This interfacial adhesion weakness is magnified by the fact that materials with significantly different coefficients of thermal expansion (CTE) are incorporated into a final product which will need to survive multiple thermal excursions.

Plastic packages, including the PBGA, are susceptible to the popcorn cracking phenomenon. Several papers have described the occurrence and the mechanisms of surface mount package popcorn cracking.[1,2,3] Popcorning is defined as a delamination or a crack in a plastic package, which is caused by the rapid expansion of moisture (i.e., water vapor) inside the package during solder reflow. This problem is particularly noted in surface mount packages, which experience higher thermal and mechanical stresses due to the exposure of the entire package to reflow temperatures. These cracks can either be internal or external, and those which extend from the surface of the package to the semiconductor device can provide a path for contaminants to reach the die pad. The device can then experience a long-term reliability failure due to bond pad corrosion.[4]

Fukuzawa et al.[1] first reported popcorn cracking of a leaded, surface mount, plastic package in 1985. The cracking mechanism is shown in Fig. 10.12. Upon exposure to reflow temperatures, either vapor phase or IR reflow, the plastic mold compound releases from the backside of the lead frame and initiates cracks at the edge of the die paddle. When a critical combination of die size and plastic thickness is reached, this bulging of the plastic away from the paddle can be severe enough to propagate the cracks through the entire molded body at its thinnest point.

The failure mechanism for the PBGA package can be seen in the cross section in Fig. 10.13. The crack or delamination is initiated in the die attach region. The failure mode can be cohesive to the die attach material, adhesive between the die attach and the printed circuit board (PCB) substrate, or adhesive between the die attach and the back of the die. Typically, a combination of these three modes is the result in a popcorn failure. As the crack propagates beyond the die attach region, it runs between the resin butter coat and the glass fabric in the PCB. Excellent adhesion between the mold compound and

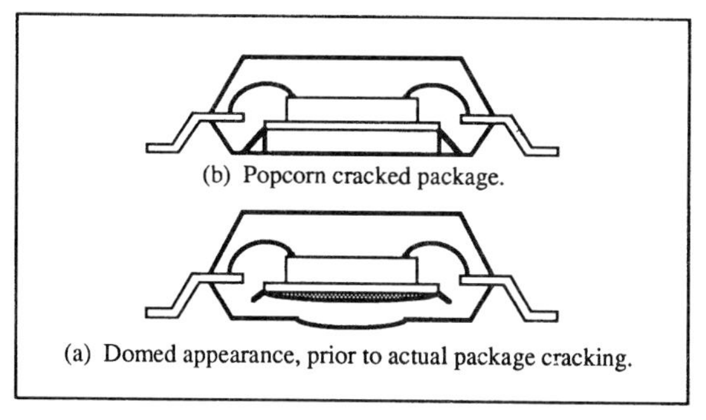

Figure 10.12 Typical popcorn crack mechanism in a leaded plastic package.

the PCB surface causes the crack to move inside the PCB construction. Finally, as the crack propagates further, it moves to the PCB/solder mask interface. Once again, excellent adhesion between the mold compound and the solder mask forces the crack to propagate along this weaker interface. The crack continues to follow the weakest adhesive interfaces until it exits the side of the package.[5]

Upon reflowing the PBGA package to the board, popcorn cracking or package delamination can cause a number of different types of failure. The stresses on the integrated circuit may be so severe that the device cannot electrically function as it was originally designed. This type of failure can occur immediately after reflow or after longer term operation of the device. A second type of failure can be observed if popcorn cracking of the PBGA package occurs due to the path created within the package for contaminants to reach the die pad, wire bond, and subsequently the die surface. As stated above, the bond pads may then become corroded, and the device may experience a long-term reliability failure.

One other type of moisture-related failure mode of the PBGA package is solder shorts. This failure mode has been observed on large packages (>144 I/O) and is due to the larger IC device size and corresponding die attach region. The pressure exerted by the vaporization of the moisture on the package is quite high immediately prior to the actual popcorn crack. The printed circuit board substrate actually bows out from the back of the package, particularly centered over the die attach region. This combination of events may result in solder shorts at the board level because if a popcorn failure occurs, it occurs at typical solder reflow temperatures.

10.6.2 Dry pack

Standard methods have been adopted as an alternative to baking plastic semiconductor packages prior to board mounting. The method of dry packing these types of components is the preferred one. A standard spec-

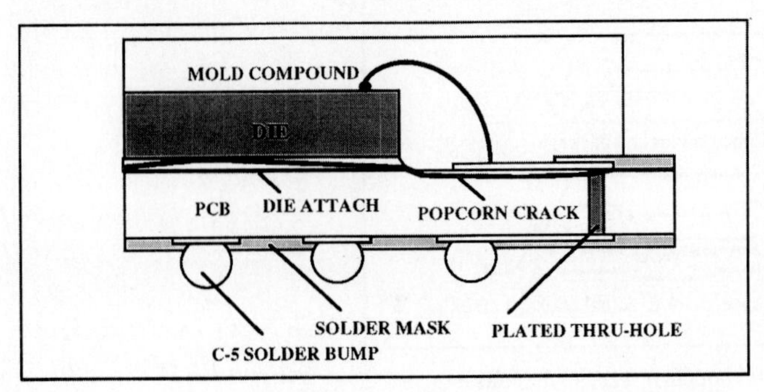

Figure 10.13 Typical popcorn crack mechanism in a PBGA package.

ification has been developed by the Institute for Interconnecting and Packaging Electronic Circuits, IPC-SM-786.[4] This document explores the impact of moisture on plastic semiconductor package cracking. The document also defines standards for dry packing.

A dry pack consists of desiccant material and a humidity indicator card being sealed with the packages inside a moisture barrier bag. This type of dry pack system provides a minimum of six months shelf life for typical plastic packages, and the PBGA package is no exception.

The materials (i.e., moisture barrier bag and desiccant) have specifications which define their respective effectiveness at controlling moisture. These specifications can be found in IPC-SM-786. The humidity indicator card is placed inside the bag as an aid in determining device exposure to moisture. The indicator contains a series of blue dots designating various levels of relative humidity exposure. Typical indicator cards have a resolution of 10 percent RH and range from either 10 to 60 percent RH or 30 to 50 percent RH. The blue dots turn pink when the moisture exposure has reached the corresponding humidity level. The PBGA package is considered to have reached an unacceptable level of moisture if the 30 percent RH dot is pink.

The final element of the dry pack process is the bag label. The packages contained in the dry pack need to be labeled as moisture-sensitive devices. The label should provide the user with a factory seal date. The user can then determine the remaining exposure limits based on the specific ambient conditions to which the packages will be exposed. For specific information on PBGA exposure limits, see Sec. 10.6.3.

10.6.3 Factory limits

10.6.3.1 Introduction. Numerous studies have been conducted to characterize the PBGA in terms of post-soldering package integrity as a function of moisture content. The objective of these studies was to evaluate the moisture performance of PBGAs in various ambient conditions for the purpose of providing storage and usage guidelines for assembly factories. Standard designs of 68-pin and 80-pin PBGA packages were used for these studies. Moisture absorption curves have been generated, and usage zones have been defined to denote the onset of package delamination and the ultimate occurrence of the popcorn failure.

The test procedure to define the moisture absorption limits for delamination and popcorn cracking is given in Fig. 10.14. The procedure begins by completely drying the package, followed by moisture conditioning at various ambient conditions for specified periods of time. Much work has been done on moisture sensitivity classifications. In order for a package to be considered "not moisture sensitive," it must survive typical reflow temperature excursions without degradation, after being conditioned at 85°C/85 percent RH for 168 hours.[6] The standard design of the PBGA package cannot survive this severe moisture conditioning step without popcorning when subjected to reflow operations. Therefore, for the purposes of establishing handling procedures specific to known factory ambient conditions, the following ambient conditions were used to define the PBGA package's moisture sensitivity:

1. 85°C, 85 percent RH
2. 30°C, 85 percent RH (a worst-case factory scenario)
3. 23°C, 55 percent RH (a clean room ambient)
4. 22–24°C, <60 percent RH (a typical factory environment)

10.6.3.2 Analysis methods. Scanning laser acoustic microscopy (SLAM) and C-mode scanning acoustic microscopy (C-SAM) analyses were used to evaluate the PBGA packages for delamination and/or "popcorning." For SLAM analyses, the resulting picture is much like an x-ray photograph, with delamination shown by large black areas. These black areas indicate an air gap in the package that attenuates the signal as it passes

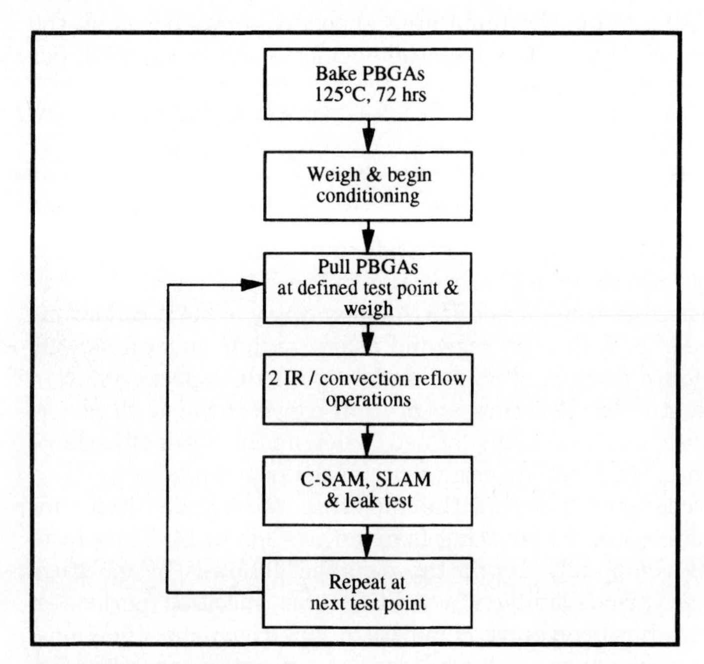

Figure 10.14 Test procedure for moisture characterization of PBGA package.

through the sample. However, the interface at which the delamination is occurring cannot be determined with SLAM analysis, since it operates by through-transmission of the acoustic signal. C-SAM analysis allows the user to focus the signal at a specific depth in the package. Using an oscilloscope, the signal can be gated to provide detailed information on specific depths or interfaces in the package. (Detailed descriptions of SLAM and C-SAM analyses are given later in this chapter.) Interfacial delamination will typically range from very slight amounts to total release of the die and finally to a separation or crack extending to the outside edge of the package.

The second analysis method was the leak or bubble test. The samples are immersed in a perfluorinated liquid heated to 130°C. A popcorned sample will release a stream of bubbles (the emission of water vapor) at the point of failure in the package. Delaminated, but not popcorned, packages will not release any bubbles.

10.6.3.3 Moisture characterization results. The initial moisture performance benchmarking was performed using an 80-pin, 1.4-mm pitch PBGA package as the test vehicle. (Physical package dimensions are given in Fig. 10.15.) Earlier testing had shown gross adhesion problems at the die flag or paddle. This was particularly evident when a conventional solid die flag pattern was used in the design. The PBGA, used in this study, was designed with a nonsolid die flag configuration (see Fig. 10.16), which eliminates the adhesion problems presented at the plated gold flag, die attach interface. Referring to Fig. 10.17, the popcorn zone is defined by 4, 24, and 120 hours of conditioning at 85°C/85 percent RH, 30°C/85 percent RH, and 23°C/55 percent RH, respectively.

A second round of testing was performed after optimizing many of the assembly process steps. Bakeout steps were implemented into the assembly process, and the PBGAs were shipped in vacuum-sealed bags. The die attach material was changed to a material which possessed better mechanical properties, especially at elevated temperatures. Referring to Fig. 10.18, an improvement in the popcorn performance of the same 80-pin PBGA was observed. The popcorn zone is defined by 8 and

	80-pin PBGA	68-pin PBGA	106-pin PBGA
Die size	0.262″ × 0.268″	0.253″ × 0.302″	0.295″ × 0.318″
Solder pad pitch	0.055″ (1.4 mm)	0.060″ (1.5 mm)	0.060″ (1.5 mm)
PCB thickness	0.014″	0.016″	0.016″
Total package thickness	0.050″	0.060″	0.060″
Overall package size	0.530 × 0.590″	0.530″ × 0.590″	0.685″ × 0.750″

Figure 10.15 Physical package dimensions.

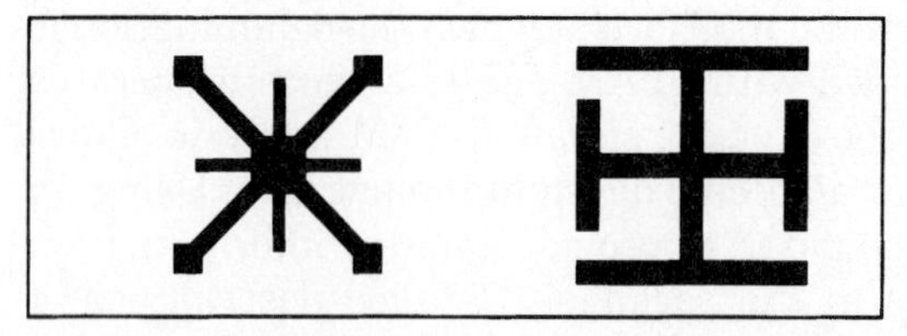

Figure 10.16 Examples of nonsolid die flag or paddle designs.

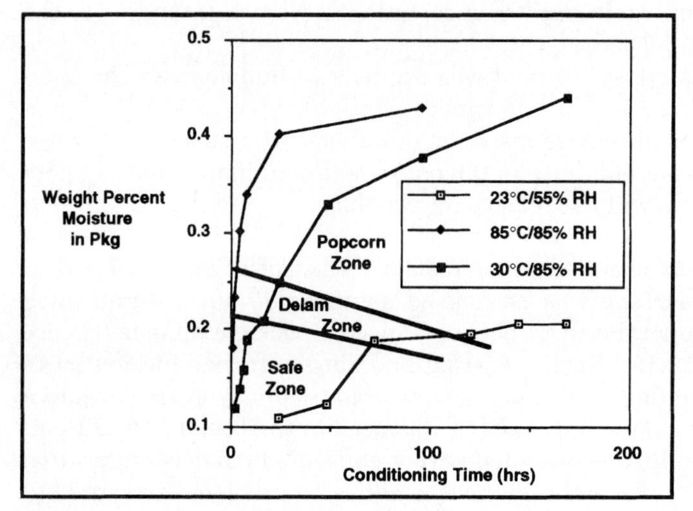

Figure 10.17 80-pin PBGA moisture characterization chart—initial study.

72 hours of conditioning at 85/85 and 30/85, respectively. The delamination zone also was improved and is currently defined at 4 and 48 hours of conditioning at 85/85 and 30/85, respectively. It should be noted that the zones are difficult to define for the 85/85 conditioned samples because the moisture absorption curve is quite steep in the region in question. However, for the 30/85 study, the first delaminated samples were observed after 72 hours of conditioning, and the first popcorned samples were observed after 96 hours. Therefore, the zones were denoted by the last test point with no delamination or popcorn degradation of the PBGA package.

An additional moisture characterization study was performed using some typical factory ambient conditions (22–24°C/ <60 percent RH). For this study, 68-, 80-, and 106-pin PBGA packages were characterized. (See Fig. 10.15 for physical package dimensions.) The moisture absorption curves can be seen in Fig. 10.19. The work from the previous studies indicated that a moisture content of 0.2 percent is a critical one and denotes the onset of package degradation. For this study, the

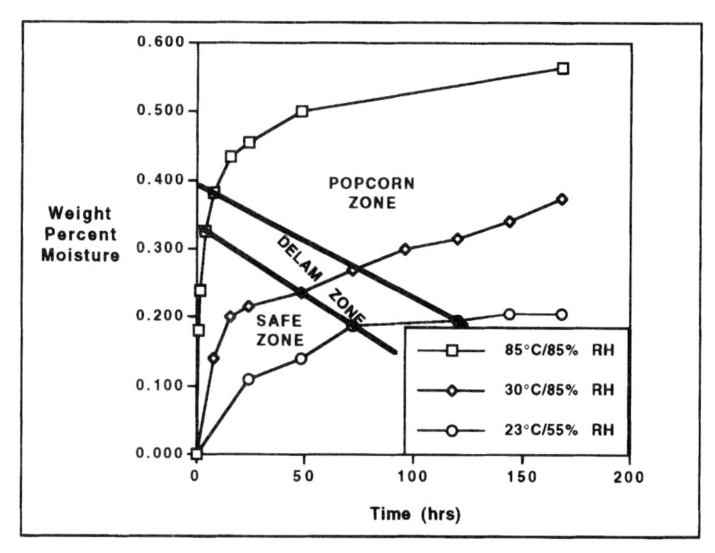

Figure 10.18 80-pin PBGA moisture characterization chart—enhanced process.

moisture level approaches this critical level after approximately 168 hours of conditioning. SLAM and C-SAM analyses of the samples prove this fact to be true for these different package sizes at the ambient condition given above. Varying amounts of delamination were observed on the 168-hour samples, with little to no delamination seen on samples conditioned for 144 hours or less. The first popcorned samples occurred after conditioning for 240 hours.

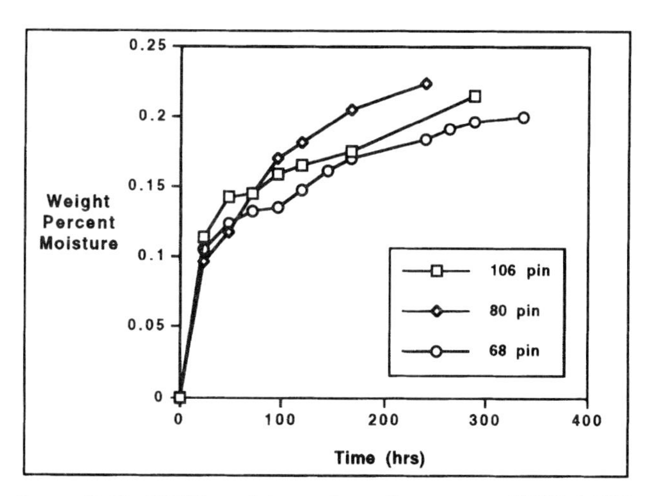

Figure 10.19 PBGA moisture absorption curves (22°C–24°C/ <60% RH ambient conditions).

Based on the results of these studies, it has been determined that the critical moisture level is approximately 0.2 percent by weight. As the moisture content approaches 0.2 percent, package delamination begins to occur when the PBGA is subjected to reflow temperatures. It should be noted that the critical moisture level increases with temperature and humidity, but the time to achieve that critical level decreases dramatically for the PBGA package. This effect is due to the fact that moisture is being forced into the package, and specifically into the die attach region. When the die attach region becomes saturated with moisture, package degradation is most likely to occur.

For the purpose of generating storage and usage guidelines for assembly factories, the results of the 30°C/85 percent RH and the typical factory ambient (22–24°C/<60 percent RH) studies provide the most information. The critical exposure times were defined at 48 hours and 144 hours for these two sets of ambient conditions. Therefore, the factory exposure limits can be set somewhere between 48 and 144 hours, with the actual factory conditions defining the time.

10.6.4 PBGA bakeout

The best means of ensuring that no degradation of the PBGA package occurs during reflow operations is to limit the amount of moisture that the package absorbs. However, even the best controls cannot totally eliminate situations where the package exposure limits are exceeded. In these situations, the PBGA must be dried.

Bake time must be specified such that the moisture is removed from the package, and it is especially critical to completely remove all of the moisture from the die attach region. The bake temperature is the other important variable. Lower bake temperatures will result in unreasonably long bake times. Higher temperatures can accelerate the growth of intermetallics or allow oxides to form on the solder balls. Both of these conditions can diminish the solderability and the reliability of the final solder joint.

The typical bake cycle should reduce the moisture level in the PBGA package to less than 0.05 percent. A common bake cycle is 125°C (<5 percent RH atmosphere) for 24 hours. This cycle is common for baking plastic semiconductor packages before dry-packing and shipping. Figs. 10.20 and 10.21 show typical moisture desorption or bakeout curves for 68-pin and 80-pin PBGA packages. As is evident, the saturation point and the conditions at which the moisture was absorbed affects the rate at which moisture is removed from the package. For 125°C bakeout, the minimum time to dry the PBGA package is 8 hours for packages with at least 0.3 percent moisture content. A bakeout at 140°C can reduce the bake time to the 3- to 4-hour range; however, intermetallic growth at the PBGA/solder ball interface increases significantly at this temperature. Study of this intermetallic growth and its effect on the final solder joint is continuing.

The final thickness of the PBGA package will determine the optimum bake time and temperature. Thinner package designs will require shorter bake times, and thicker packages may require lower temperature bakes for longer times. Another consideration is the packing materials used for the PBGA. Packing trays and tape-and-reel materials may not be able to withstand a high-temperature bake. One solution is a low-temperature bakeout, possibly under vacuum. The other alternative is to remove the packages from the packing material, bake out, and then repack. The risk of removing the PBGA package from its packing media is low because solder balls are quite robust and not easily damaged.

10.6.5 Moisture sensitivity summary

Many methods are being pursued with the end goal of improving the moisture sensitivity of the PBGA package. The materials manufacturers are focusing on developing mold compounds and die attach materials with improved moisture absorption characteristics along with excellent adhesive properties. As was discovered in the studies described in the previous sections, the selection of materials is critical to the overall popcorn resistance of the package. In the end analysis, the proper combination of materials and package assembly processes may provide the solution to the popcorn issues with the PBGA package.

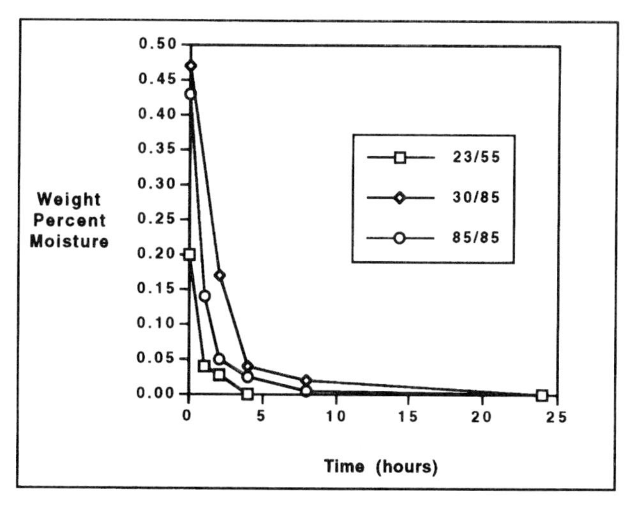

Figure 10.20 PBGA moisture desorption or bakeout curves for 80-pin PBGA at 125°C in nitrogen atmosphere.

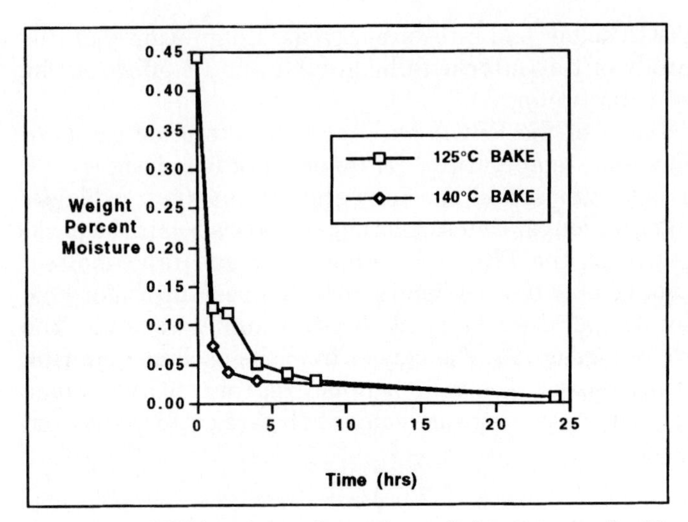

Figure 10.21 PBGA moisture desorption or bakeout curves for 68-pin PBGA at 125 and 140°C in nitrogen atmosphere.

Package design may provide a means to improve the moisture sensitivity of the package. A complete understanding of the multitude of interfaces in the PBGA package may create opportunities to enhance popcorn performance through simple design features. One such feature is a vent hole located under the die attach region of the package. Fukuzawa et al.[1] describes a vent hole placed at the bottom of the die flag in a thin, leaded plastic package. Such a vent hole allowed the package to pass moisture resistance tests without failure. Similar testing has been performed at Motorola on PBGA packages with a vent hole. The vent hole is drilled through the PCB substrate to provide a moisture relief path from the die attach to the bottom side of the package. Preliminary studies have shown that the 68-pin PBGA package (see Fig. 10.15 for package dimensions) can withstand solder reflow temperatures without package delamination after conditioning at 85°C/85 percent RH for 168 hours. A detailed description of these vent hole studies can be found in the 1992 IEPC Proceedings.[5]

As overall PBGA package sizes continue to increase and as new materials (e.g., mold compound, die attach, PCB laminates, etc.) are used, additional characterization studies must be performed. Moisture absorption and desorption curves must be generated to determine factory handling specifications. Package design features must be evaluated for effectiveness at improving or eliminating moisture-sensitivity problems. In general, moisture-sensitivity issues must remain a focus of all involved in the manufacturing process, from the design and assembly of the PBGA package, to the packing and shipping of the device, and finally to the handling of the package on the factory floor.

10.7 Repair

Both factory and field repair depots are required to replace defective devices on a daily basis. A common defect might be static damage to the silicon die. Hot air repair stations are typically used, and placement and fluxing operations are generally done manually. Plastic BGA packages do not present any significant repair challenges relative to other surface mount components. In fact, they are probably easier to deal with than fine-pitch leaded parts because they are more robust to handling, and the placement tolerance is much wider. In the following two sections, a repair process for small PBGAs (less than 0.75 in) will be described, and considerations when repairing neighboring components will be discussed. Issues such as board bowing, which become important for larger packages, will not be addressed here.

10.7.1 Removal, site preparation, and replacement process

Motorola facilities generally use standard hot air repair stations with no special modifications. Repair profiles are set up based on component size and board density. The following procedure has been found to work quite well for small PBGAs (less than 0.75 in):

1. Removal
   - Dispense liquid flux underneath the carrier
   - Apply heat to reflow solder (220°C max.)
   - Remove carrier with a vacuum tip
2. Site Preparation
   - Remove excess solder on pads using solder wick and soldering iron
   - Clean pads with alcohol and small brush
   - Dry and inspect
3. Replacement
   - Use "dry" PBGA
   - Apply flux to solder balls using flux block with 0.014-in-deep trench
   - Place part on PC board and align corners to marks on board
   - Apply heat to reflow solder (220°C max.)

When the part reflows, the solder balls collapse and the part drops about 0.006 to 0.008 in.

10.7.2 Neighboring component considerations

As with other plastic packages, PBGAs are prone to die bond delamination and popcorn failure during solder reflow. Motorola repair procedures were originally written to require an 8-hour bakeout (@ 125°C) of the PC board before package replacement to prevent delamination or popcorn cracking of neighboring PBGAs. Such a bakeout procedure met with strong resistance from both the factory and the field repair depots. In order to develop an improved procedure, studies were run to determine the maximum temperature that saturated PBGAs could withstand before experiencing any package delamination or popcorn cracking. The knowledge gained from this study was then used to establish repair profiles for replacing components without heating the neighboring PBGAs to this maximum temperature.

A hot air repair station was used to perform the experiments, with a PBGA taped to a typical FR-4 motherboard. A thermocouple was used to monitor the package temperature and to set the repair profile. The maximum temperatures ranged from 155 to 230°C, in 15° increments, and the ramp time ranged from 30 to 60 seconds. Scanning Laser Acoustic Microscopy (SLAM) analysis was used to detect package delamination.

Saturation conditioning was done at 30°C/85 percent RH for 168 hours. A total of six PBGAs were incorporated into the study. The following general trends were observed:

1. Popcorn cracking only occurred on the larger die PBGAs at temperatures of 200°C or greater.
2. Some individual samples of the larger die OMPAC™s did experience major delamination (>50 percent of the die flag area) at a temperature above 185°C.
3. If repair temperatures are kept below 185°C, the jeopardy of significant delamination and/or popcorn cracking of small PBGAs is greatly reduced.

After experiments with various repair stations, it was determined that three guidelines can be followed to replace a component with minimal heating of neighboring PBGAs:

1. Apply heat primarily from the top side of the component.
2. Use a nozzle which is the same size as the component.
3. Position the nozzle ⅛ to ¼ in above the component.

Some representative repair profiles are shown in Fig. 10.22 to illustrate the temperature differential between devices in close proximity on the board. Two different hot air repair stations were used. Temperature was monitored on the device being repaired, an adjacent device, and a device on the opposite side of the board. It is obvious that heating of neighboring parts is minimized when only top heat is used, even when the maximum repair temperature is allowed to rise above the nominal value of 220°C.

10.8 Failure Analysis

10.8.1 Acoustic microscopy

Ultrasonic frequencies ranging from 10 to 100 MHz can be used to nondestructively characterize the quality of materials and the integrity of the interfaces within a plastic semiconductor package. At these frequencies, ultrasound is extremely sensitive to elastic properties and will not transmit through air. Therefore, ultrasound can be used to locate internal defects, such as package cracks, package voids, and delamination. It can also be used to measure the quality of adhesion between two or more interfaces. Acoustic microscopy methods include through-transmission and reflection techniques which are highly sensitive to material or interfacial defects smaller than one micron.[7,8]

For failure analysis of PBGA packages, two acoustic imaging methods are most widely used. The first method, scanning laser acoustic microscopy (SLAM), is a through-transmission mode of analysis. This method can be used to evaluate packages which are not connected to

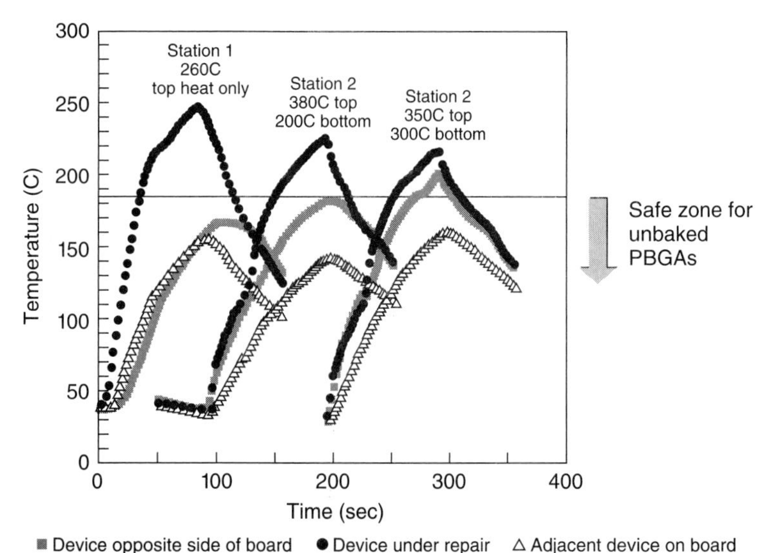

Figure 10.22

any type of printed circuit board. Scanning laser acoustic microscopy was used to evaluate most of the samples in the studies (described earlier in this chapter) to determine acceptable factory storage limits. This technique allows for real time feedback on the integrity of package interfaces.

After the PBGA package has been attached to the printed circuit board, C-mode scanning acoustic microscopy (C-SAM) can be used to evaluate the package for internal defects. Reflection or pulse echo techniques allow this form of microscopy to nondestructively analyze the package without removal from the board. Also, detailed information can be obtained about specific interfaces when employing this technique. C-SAM analysis is typically the first step in the failure analysis process after detection of a PBGA package problem. If no package delamination is evident, the board is then routed to the next step in the failure analysis procedure for solder joint analysis.

Both of the acoustic techniques described above require that the package or printed circuit board be immersed in deionized water or another type of fluid couplant. This medium aids in the transmission of the ultrasound without impeding the signal. Neither the package nor the circuit board is adversely affected by immersion in deionized water or another type of fluid couplant for the short period of time required for analysis. However, proper drying practices must be performed before heating the board or the PBGA package for further processing or analyses.

10.8.1.1 Scanning Laser Acoustic Microscopy (SLAM). In scanning laser acoustic microscopy analysis, a continuous plane wave of ultrasound is introduced at the lower surface of the sample and travels through the entire thickness of the material or component. The pattern of transmitted sound is then detected by a rapidly scanning, finely focused laser beam which acts like an ultrasensitive detector (see Fig. 10.23). The images are produced in real time and reveal the bond integrity of all interfaces throughout the entire sample. The transmission of the ultrasound is affected by internal features and defects, discontinuities, and material properties, as any air gaps block the ultrasound from the detector and render them visible with high contrast in the acoustic image. Any air gap acts as a zero acoustic impedance, terminating the transmission of an acoustic signal.[9]

The real time imaging capability of scanning laser acoustic microscopy provides the benefit of employing SLAM as a quality assurance tool in the PBGA package assembly process. Immediate feedback on interfacial integrity can be obtained after molding the packages. The SLAM can be used on the assembly factory floor to check PBGA packages suspected of containing defects or an unacceptably high

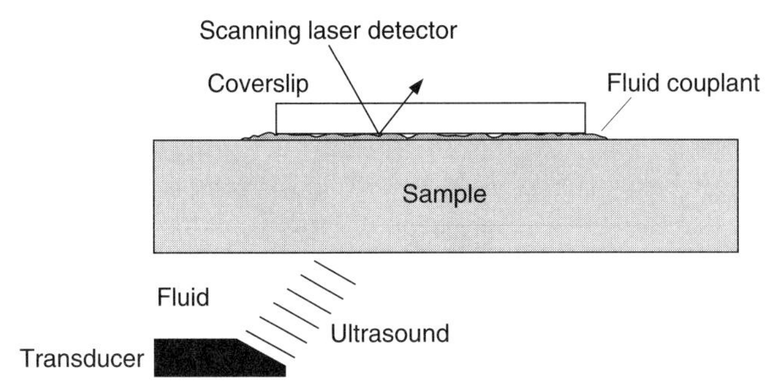

Figure 10.23 Scanning Laser Acoustic Microscope (SLAM)—through transmission mode. *Courtesy of Sonoscan, Inc.*

moisture content. The packages can be subjected to a simulated solder reflow operation and immediately analyzed for package delamination.

10.8.1.2 C-mode Scanning Acoustic Microscopy (C-SAM). For layer-by-layer discrimination of plastic semiconductor packages, C-mode scanning acoustic microscopy is an extremely useful failure analysis tool. Detailed information regarding the internal construction and integrity of adhesive interfaces can be obtained with this reflection mode technique.[10] C-SAM analysis utilizes reflection mode or pulse echo technology in which a single, focused acoustic lens mechanically raster-scans a tiny "dot" of ultrasound over the sample. As the ultrasound is introduced into the sample in a pulsed mode, a reflection is generated at each interface and returned to the transducer for processing (see Fig. 10.24). Images can then be generated from specific depths, cross sections, or through the entire sample thickness. The images are typically produced in ten to thirty seconds.

The C-SAM equipment is equipped with acoustic impedance polarity detectors. These detectors will compare the acoustic impedance of the returned signal. If the ultrasound travels from a plastic mold compound material, which has high acoustic impedance, into an air gap, which has a low acoustic impedance, the echo exhibits a very negative

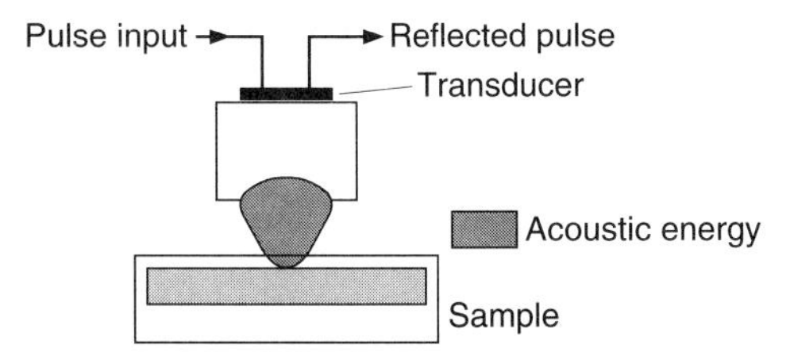

Figure 10.24 C-mode Scanning Acoustic Microscope (C-SAM). *Courtesy of Sonoscan, Inc.*

polarity. Likewise, if the ultrasound travels from a lower to higher impedance material, the returning echo will have a positive polarity. The polarity measurements will accentuate any discontinuities or defects in the packaging materials or interfaces.

Specific to PBGA analysis, the C-SAM is used to inspect the package for delamination at any of three interfaces: (1) mold compound to die surface, (2) mold compound to printed circuit board surface, and (3) die attach to printed circuit board surface. The scan response is gated to provide information only at the specific interface in questions. The adhesion of the mold compound to the die surface is evaluated to determine if proper cleaning and surface preparation procedures have been followed in the assembly of the PBGA package. Improper processing will be manifested in poor adhesion or delamination of the mold compound from the surface of the integrated circuit.

The various printed circuit board interfaces are scanned to determine if any moisture-related or popcorning problems have occurred while soldering the PBGA package to the motherboard. As described earlier in this chapter, the die attach region is the first area affected by excess moisture in the package. The die attach material will delaminate from the die flag, and the C-SAM can detect this separation. The mold compound to printed circuit board interface is analyzed to determine if gross package delamination or popcorn cracking has occurred. The crack typically exists within the printed circuit board laminate along the fiberglass layer near the surface. The C-SAM can easily and effectively detect the delamination in this area around the outside of the die.

10.8.2 Solder joint failure analysis

Solder joint failures can occur due to a poor manufacturing process or due to excessive thermal or mechanical loading. When performing analysis on suspected BGA solder joint failures, conventional metallurgical cross-sectioning techniques work well. However, these techniques are time consuming, destructive, and only give a two-dimensional view of the joint array. There are also nondestructive techniques such as acoustic microscopy and fine focus x-ray analysis. Acoustic microscopy does not work well through a printed circuit board, so cracks or nonbonded areas in solder joints are difficult to resolve. X-ray radiographs can resolve voids quite easily, but more sophisticated techniques are required to inspect for cracks or nonbonded surfaces (these issues are discussed in detail in Chap. 16).

As an alternative to sophisticated x-ray techniques, a simple mechanical removal procedure was developed to inspect ball grid array solder joints. The procedure utilizes dye penetrant to stain cracked or nonbonded surfaces, then the package is mechanically torqued off or pried off the product circuit board to expose the solder joint array. Before this step, however, the package is generally inspected for internal delamination using C-mode scanning acoustic microscopy (as discussed in Sec. 10.8.1). Also, it is wise to identify exactly which joints are suspected to be failed by measuring resistance between the exposed runners on the top surface of the PBGA substrate and the product circuit board. If the package is not designed with exposed runners on the top surface, failure analysis can be greatly hindered.

Once the suspected joint failures have been identified, the following procedure is used to remove the package from the product board:

1. Use solvent to remove all flux residue.
2. Apply generous amount of dye penetrant around periphery of package, allowing surface tension to draw it underneath (Dykem #80496 steel red layout fluid works well).
3. Place in vacuum of 25-in mercury for 1 minute, then bring back to atmospheric pressure.
4. Bake at 100°C for 15 minutes to dry dye penetrant.
5. Rough PBGA mold compound top surface with sand paper.
6. Glue fixture to mold compound (Loctite Black Max works well with 1-hour cure at 25°C)
7. Torque or pry package from board.

Once the package is removed, the joints can be readily inspected. Any previously failed surfaces are stained with the dye. During the removal process, copper pads can be torn from the package substrate or the product board. If this happens, the joints were obviously good, but further testing of the package or board is not possible. This type of damage can be avoided by torquing the package off slowly (over a period of several hours), because the strength of the solder decreases with strain rate. Hence, at very low strain rates, the stresses are low enough that the array of solder joints always fractures before the copper pads delaminate, and both the board and package are undamaged and available for further electrical analysis. A torque of about 0.2 to 0.4 in-lb/joint is required for a 12-hour chip carrier removal time.

10.9 References

1. I. Fukuzawa, et al., "Moisture Resistance Degradation of Plastic LSIs by Reflow Soldering," *Proceedings of the 23rd International Reliability Physics Symposium,* 1985, pp. 192–197.
2. R. Lin, et al., "Control of Package Cracking in Plastic Surface Mount Devices During Solder Reflow Process," *Proceedings of the 7th Annual Conference of the International Electronics Packaging Society,* 1987, pp. 995–1010.

3. R. Lin, et al., "Moisture Induced Package Cracking in Plastic Encapsulated Surface Mount Components During Solder Reflow Process," *Proceedings of the IEEE International Reliability Physics Symposium,* 1988, pp. 83–89.
4. IPC-SM-786, "Impact of Moisture on Plastic I/C Package Cracking," Institute for Interconnecting and Packaging Electronic Circuits, 1990.
5. B. Miles and B. Freyman, "The Elimination of the Popcorn Phenomenon in Overmolded Plastic Pad Array Carriers (OMPAC™s)," *Proceedings of the 1992 International Electronics Packaging Conference,* pp. 605–614.
6. T. Moore and R. McKenna, *Characterization of Integrated Circuit Packaging Materials,* Butterworth-Heinemann and Manning Publications, Co. 1993, p. 81.
7. "Nondestructive Acoustic Micro Imaging," Sonoscan, Inc. product brochure.
8. L. W. Kessler and S. R. Martell, "Acoustic Microscopy Technology (AMT) Analysis of Advanced Materials for Internal Defects and Discontinuities," *Proceedings of the International Symposium for Testing and Failure Analysis,* 1990, pp. 491–504.
9. J. Semmens and L. Kessler, "Nondestructive Evaluation of Thermally Shocked Plastic Integrated Circuit Packages Using Acoustic Microscopy," *Proceedings of the International Symposium for Testing and Failure Analysis,* 1988, pp. 211–215.
10. L. Santangelo and L. Kessler, "Acoustic Microscopy: A Key Inspection Tool for Improving the Reliability of Surface Mount Capacitors and Plastic IC Packages," *Surface Mount Technology,* September 1989.

Chapter

11

Thermal and Electrical Performance Management in Plastic Ball Grid Array Packages from the Vendor's Perspective

Phil Rogren

11.1 Introduction

The focus of this chapter is a discussion of the electrical and thermal issues that we as manufacturers see around the BGA format. Before jumping straight into the technical issues, however, I think that it is relevant to talk a bit about the author's perspective on the development of the BGA market and infrastructure, and the position that BGAs will assume in the semiconductor packaging hierarchy. While this discussion is from the perspective of the vendor of BGA packages, it is also from the unique perspective of the author.

11.1.1 Package suppliers perspective

The author's experience is in supplying high-performance, cost-effective packages for semiconductors. The company has been working with packages incorporating a multilayer laminate substrate and with molded package bodies for more than six years. The early work was with Plastic Pin Grid Arrays (PPGA), Plastic Leaded Chip Carriers (PLCC) and Plastic Quad Flat Packs (PQFP). As a natural extension of the technology, the company began working with area array packages, other than PGAs, about 4 years ago. Early work with what were called Land Grid Arrays (LGA) led directly to what we have called Solder Grid Arrays (SGA), which are essentially the current Ball Grid Array (BGA) package.

The author's perspective is unique in several ways. First, the company is among the smallest of the BGA suppliers and as a result, is focused on the niche markets, particularly those applications that require high lead count with high electrical and thermal performance. We also thrive on applications that require relatively low production volumes of customized packages that take advantage of our flexibility. Last, we are ardent supporters of the concept of laminate-based molded packaging for the optimum combination of performance and cost. Because of the emphasis on high-performance and highly customized packages, the company probably has exposure to a wider range of packages than the other suppliers of plastic ball grid array packages.

From our work to date, we are convinced that the BGA will become a viable semiconductor package. We feel that the format is particularly well-suited to the portion of the market that we serve because there is no other format that is as cost-effective or easy to use in the over-208-lead range. Particularly above 300 leads, there is really no other viable format for surface mount, and the PGA becomes an expensive and cumbersome format for through-hole assembly.

The thermal and electrical performance of the format, as we will discuss here, is on a par with anything else available on the market. Probably the most universally important performance measurement is assembly yields, and that is the feature that is most striking about BGAs. Standard surface mount packages have taken us through a progression of tighter lead pitches and more delicate leads. BGAs promise a rare opportunity to make the task of board assembly easier and improve their yields at the same time. Not only can BGAs increase yield while eliminating rework, but they can do it with equipment and processes that have much wider process windows than the industry has seen in many years.

The author does not believe that BGAs will become a major factor for devices with fewer that 208 leads because of the additional cost of the laminate substrate. However, it is possible that the cost advantages of small QFP will be washed out by higher assembly yields and lower costs for equipment used to place BGAs. Our experience to date suggests that there is very little difference in yield expectations between the smallest BGAs and those with 400 or more leads. The current expectation is that a process yielding less than 50 DPM on a package basis is a good starting point for a new process. Normal levels of process refinement will push those levels to less than 5 DPM.

One of the features that makes the BGA format so attractive to a supplier of custom packaging solutions is the relatively low expense involved in tooling for a custom package. While high-volume production tooling is still quite expensive, the tooling costs required to produce anything over a few thousand parts can be amortized over the production run with minimal pain.

11.1.2 Thermal and electrical performance management in plastic BGAs

Performance, like beauty, is in the eye of the beholder. As designers and suppliers of semiconductor packages, we tend to focus on the properties of the package alone, in isolation from the environment in which it will be used. That is not to say that the package manufacturer is totally insensitive to the needs of the end user. We are simply more preoccupied with the things that we can have a direct influence on than with performance influences of the final application environment. That is also not to say that what we perceive as improved performance does not yield a performance improvement to the end user. The real issue is the relative magnitude and importance of the performance features of a given package at the system level. Unfortunately, every system benefits differently from the built-in performance of the package. Consequently, the system designer/manufacturer has to evaluate the performance of a package in the light of his own system environment. The objective of this chapter is to give the user of PBGAs the information they need to evaluate the package performance in their applications.

Thermal. The good news with respect to thermal management is that semiconductor devices are becoming more efficient every day. The thermal energy dissipated by each junction is being pushed lower and lower, despite increases in clock rates. The bad news is that higher levels of integration and smaller design rule geometries are increasing the number of "transistors" per chip and squeezing them into the same size, or even smaller chips. On top of generally increasing power densities, bigger chips are being squeezed into smaller packages (ball grid arrays are a perfect example) and the volume of the systems is being reduced (subnotebook computers for example). All the while, active cooling systems like fans are virtually being legislated out of existence. The bottom line is increased demand for thermal management at all levels.

The challenge for plastic ball grid arrays is the same as for any other package type. The package must move the heat to the outside of the case where it can be moved out of the system environment. By virtue of their second-level interconnect method, PBGAs have some advantages over other surface mount packages in terms of thermal transfer from the surface of the die to the outside of the package. The problem is that the primary thermal transfer mechanism from a PLCC or QFP is a combination of conduction and convection. In the standard PBGA, the primary thermal transfer mechanism is pure conduction. Depending on the construction of the PCB on which the packages are mounted and the general environment of the system, the leaded packages, with higher thermal resistance from junction to case (θ_{JC}) values, may exhibit lower thermal resistance from junction to ambient.

As with most package types, there are enhancements that can be incorporated in PBGA packages to improve the thermal conductivity between the junction and case. In virtually all cases, reducing thermal resistance to the outside of the package also reduced thermal resistance to ambient. Again, the degree to which θ_{JA} is improved when θ_{JC} is improved is largely dependent upon system design. As package suppliers, our objective is to minimize the thermal resistance of the package and to do whatever is possible to maximize the thermal coupling between the outside of the package and the next level of thermal transport.

Electrical. Electrical performance of the semiconductor packages translates more directly to performance at the system level than does thermal performance. Generally, electrical performance of the PC board can be designed to match or exceed the performance of the package without going to extreme measures. The miniaturization that increases power density and causes major thermal problems for systems designers can actually have a positive effect on electrical performance. As the components are brought closer together, propagation delay and parasitic affects are reduced. Increased clock rates places more demand on the PCB design but the design and manufacturing technologies are fully capable of matching the package performance.

Electrical performance in PBGAs is highly dependent upon the design and construction of the specific package. Perhaps more than any other packaging type, PBGA designs address a wide range of price and performance objectives. When compared with the other available surface mount package types, PBGAs have the dubious honor of both the worst and best electrical performance. As with thermal performance, the second-level interconnect method (the solder balls) have electrical performance advantages over the leads of PLCCs or QFPs. The area array of the PBGA leads also reduces the straight line distance from the bond pad in the die to the interconnect to the PC board. The problem is that there are no straight lines in the package and as the conductor length deviates from the minimum possible, the inductance of the signal traces increases and the general electrical performance decreases. The conductor length is greatly influenced by the basic construction of a specific PBGA design. The design and basic construction options are the predominant factors affecting electrical performance in PBGAs and those issues will be addressed in detail.

11.2 Thermal Performance Management

11.2.1 Thermal transfer basics

From the package supplier's perspective, thermal management is strictly a matter of moving heat from the surface of the die to the outside of the package. Thermal transfer within conventional semiconductor packages is almost exclusively via conduction. Convection is a factor only in cavity-style packages and typically plays an insignificant role. Thermal transfer via radiation does occur in all packages but the contribution under practical conditions is totally insignificant to any practical application.

Consequently, our success in moving heat from the die to the package surface is dependent upon the thermal conductivity of the materials that go into the package and upon how those materials are used in the package. In the case of PBGAs, the basic materials and designs are well-defined. In general, those basic materials and the ways in which they are combined to produce PBGA packages do not lend themselves to high thermal performance. Fortunately, there are avenues for increasing the thermal performance significantly by altering the designs slightly and using additional, more thermally conductive materials.

To say that the materials are key to thermal performance is an understatement. The materials that affect the thermal performance of a PBGA package are listed in Table 11.1 along with their relative thermal conductivities. The actual thermal conductivities of these and other common materials are listed in Table 11.2 for your reference. The point is that there is a difference of more than four orders of magnitude among the materials that affect the thermal performance of the package.

In most electronic systems, moving heat away from the outside of the packages is effected via convection. Whether forced or natural convection, and regardless of the fluid medium, modeling and predicting convection cooling is an extremely complex problem involving fluid dynamics, thermal conductivities, and thermal capacities. The fluid dynamics part of the equation is typically so complex that it can only be predicted by comparison to measurements of very similar systems. Fortunately for the package supplier, that problem is out of our hands, left to the system designer to solve.

Conduction mode thermal resistance is an almost perfect analog to electrical resistance, right down to changes in resistivity with changes

TABLE 11.1 Relative Thermal Conductivity of Package Materials

Material	Relative conductivity
Copper	15,000
Gold	12,000
Silicon	3,000
Sn 63 Solder	1,900
Die Attach Epoxy	190
Mold Compound	22
BT Laminate	7
Air	1

TABLE 11.2 Thermal Conductivity of Various Materials (Watt/Cm°C)

Material	Conductivity
Diamond	5.9
Silver	4.3
Copper	4.0
Gold	3.2
Beryllium oxide	2.5
Aluminum	2.4
Tungsten	1.8
W85/Cu15	1.7
Lead frame	1.6
Silicon carbide	.90
Iron	.84
Silicon	.80
Platinum	.73
Tin	.66
96% aluminum oxide	.35
Die attach epoxy	.05
Mold compound	.006
Laminate substrate	.002
Hydrogen	.002
Air	.0003
Steam	.0002

in temperature of the conducting medium. As such, models are relatively easy to build. The basic equation that applies to conductive heat transfer across a material of uniform composition and cross section is

$$q = -kA \frac{dt}{dx}$$

where q = heat conduction in the x direction
A = cross-sectional area normal to heat flow
dt/dx = temperature gradient in x direction
k = thermal conductivity of conducting medium

In any practical application, the situation is more complex than this simple case, but most cases can be estimated closely enough for practical purposes. In the case of semiconductor packages, there are three issues that make the problem more complex. First is the stack-up of several different materials between the heat source and the outside of the package. Second, the heat source is confined to the surface of the die and will be conducted in all directions. This problem is exacerbated by the fact that in many cases, the die surface is not a uniform heat source but that most of the thermal energy will come from one or two hot spots. In addition to layers of different materials normal to the heat flow, there are also areas of highly different thermal conductivity that lay parallel to the heat flow. The problem of layers of different material in the heat path is a relatively simple one. The illustration in Fig. 11.1

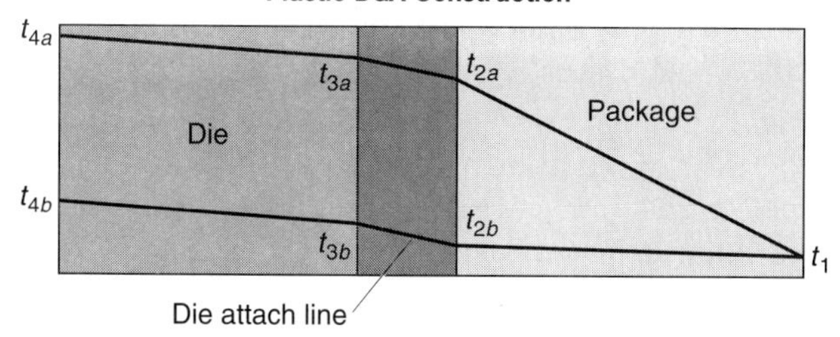

Figure 11.1 The delta t across the material drives the heat flow out of the package. Unfortunately, the objective is to keep the temperature of the die, and thus, the delta t low. The only way to have both is to keep the thermal resistance as low as possible in all materials.

shows schematically the stack-up of die, die attach, and package. t_1 is the temperature of the outside of the package. t_{2a} and t_{2b} are the temperatures at the interfaces between the die-attach material and the package. The difference in those two temperatures illustrates the problem of high thermal resistance in the package. The thermal resistance of the die is represented by the temperature increase between t_3 and t_4. The delta t across the die is virtually the same along both lines a and b. The same is true across the die attach line, from t_3 to t_2.

In the multilayer case, the heat flow across all of the layers is equal; therefore we can write

$$q = k_a A \frac{t_1 - t_2}{\Delta x_a} = k_b A \frac{t_2 - t_3}{\Delta x_b} = k_c A \frac{t_3 - t_4}{\Delta x_c}$$

These equalities can be rewritten to express the temperature across each layer.

$$t_1 - t_2 = q \frac{\Delta x_a}{k_a A}$$

$$t_2 - t_3 = q \frac{\Delta x_b}{k_b A}$$

$$t_3 - t_4 = q \frac{\Delta x_c}{k_c A}$$

Unfortunately, real-world thermal transfer problems are never as simple as flat plates with the heat source on one side and an infinite heat sink on the other side. Even multiple-layer models are very simple when compared to real applications. Semiconductor packages are obviously no exception. In virtually all cases, the heat source—the junctions on the surface of the die—are completely surrounded by the materials that make up the package. If you draw a line from the junction to the outside of the package, it may cross a single, homogeneous layer, or it could cross half a dozen different material layers. The very simplest situation is that of a standard molded plastic package, illustrated in Fig. 11.2.

In the molded package, the heat source on the top of the die has a direct path to the outside through the mold compound. In addition, there is a path to the backside of the package, which goes through the die itself, the die-attach line, the die-attach flag, and then through the mold compound on the back of the package. Those are the physically shortest paths, but because of the high thermal resistance of the mold material, they do not represent the primary thermal path. Most of the thermal energy moves out of the package through the leads. It is easy to understand why the longer path along the lead is the shorter thermal path, but it is a complex task to model the coupling between the die and the leads. There are two major thermal connections between the two. One is through the wire bonds and the other is through the die and die attach line, into the die attach flag, and then across the mold compound between the flag and the lead fingers.

Modeling the thermal transmission of the bond wires is fairly simple. They really look like long, flat plates, and for all practical purposes, they are wrapped in a perfect thermal insulator since gold wire is more than 500 times more thermally conductive than the mold compound. Modeling the path through the back of the die is much more complicated. Essentially, there are a number of thermal resistances in series between the heat source(s) and the lead fingers. Each resistance can be determined and the series resistance calculated. The problem is that none of these physical regions resembles a flat plate. In all cases the cross section of the region is constantly changing as you move along the path of the heat flow. In some cases the direction of heat flow is not clear from casual observation. The only way to determine the thermal transfer with any degree of accuracy is through the application of some form of finite element analysis.

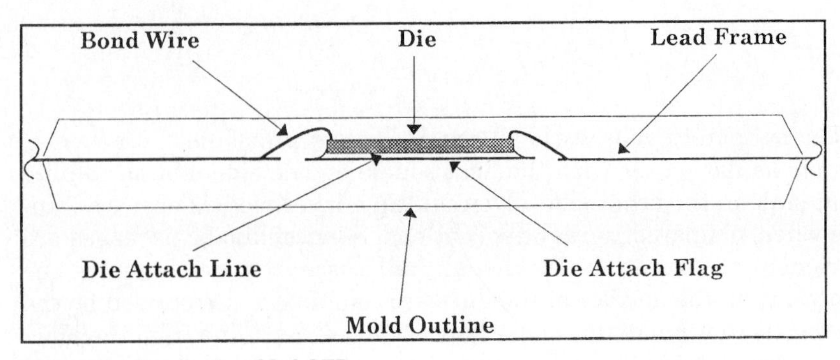

Figure 11.2 Standard molded QFP.

In cases where the heat source is well-defined and the surrounding material is uniform in composition, simple relaxation models can be constructed that yield reasonable results. In most cases the problems are complex enough that a real solution is not practical without sophisticated thermal analysis software.

The good news about thermal analysis of the package itself is that the hard parts can often be ignored. We have learned empirically that devices dumping a watt or a little more will perform well in a standard molded package. We also know that as the die and package get larger, they will transfer more power to the outside of the package. The result is an imperial set of practical power levels for each package family. The difficult situations come as the application approaches those assumed limits.

For high-power applications, those well beyond the limits of standard packages, the problem of predicting the overall thermal performance of the package is actually simplified. In order to go from one or two watts to 6 to 10 watts coming out of a plastic package, the thermal path must be radically improved. Typically those improvements are effected by replacing mold compound with high thermal conductivity material. In the case of plastic packages, that material is usually copper. For all packaging designs, adding a good thermal conductor reduces the primary thermal path to a flat plate model. While this enhanced thermal path is only one of several parallel paths to the outside of the package, it has a very low resistance relative to the others and thus dominates the thermal performance of the package. In most applications, the contribution of the secondary paths are swamped out by variations in the application environment and can be ignored.

The role of the package. Thermally, the role of the package is to move the heat dissipated by the active elements on the surface of the die to the outside of the package and then to transfer it to the next level of thermal conductor. The role of the package designer, with respect to thermal management, is to ensure that a sufficient thermal path exists to remove the power that the chip dissipates.

11.2.2 Package construction

Plastic ball grid arrays are available in two basic configurations. The most common and most basic package is a thin PC board with the die on one side and the solder balls on the other. The body of the package and the protection for the die is a transfer molded structure on the side of the substrate opposite the balls. This basic configuration goes by the Motorola trade name OMPAC®,* which stands for Over Molded Pad

---

* OMPAC is a registered trademark of Motorola, Inc.

Figure 11.3 Cut-out of an OMPAC-style package.

Array Carrier. Figure 11.3 illustrates the construction of the basic OMPAC style package. In the basic configuration, the OMPAC style package is slightly superior in terms of thermal transfer to an equivalent-sized PQFP. It is important to note that the improved performance is in thermal resistance between junction and case (θ_{jc}). The improved performance is the result of the relatively shorter thermal path between the die and the balls on the back of the package. The primary heat sink in OMPAC style of package is the PC board to which they are attached. The equivalent QFP delivers its heat to the leads at the edge of the package, which in turn pass it to the air around the package. The relative efficiency of these two thermal paths depends upon the construction of the PC board and the quality of the airflow around the packages. The result is that even though the PBGA has a lower $\theta_{jc,}$ the thermal resistance between the junction and the final heat sink may be higher because of the way it couples to that heat sink.

There are a number of variations to the OMPAC construction but the only one that has any real effect on thermal management is the addition of thermal vias through the substrate under the die. Again, this enhancement delivers the heat primarily to the PC board. If the board is not capable of moving the heat away from the package, the thermal vias will be of little value even though they reduce the θ_{jc} significantly.

The pure OMPAC style package is the PBGA standard and the equivalent of the standard, lead-frame-based molded plastic QFP. It is appropriate for the majority of semiconductor devices and is the most economical of PBGA designs. All of the other designs address higher performance applications and much smaller markets. As with enhancements to most products, the enhancement comes at a certain cost. Consequently, the higher thermal performance PBGAs are finding acceptance where there is a need to dissipate more than about 2 watts. There are essentially three approaches to improving the thermal performance of a PBGA as illustrated in Fig. 11.4. The simplest is to put thermal vias through the substrate from the back of the die to the back of the package. A thermal via is simply a hole through the substrate that is plated with copper. Electrical vias can serve as thermal vias. The only real difference between the two is that thermal vias are employed in arrays that place as much copper as possible in the area of the array. The effectiveness of thermal vias is dependent upon the area of copper in the array and is inversely proportional to the thickness of the substrate.

As mentioned previously, this approach can be effective if the PC board to which the package is attached is capable of transferring the heat away from the package. In applications where the heat can be moved away from the package, thermal vias can take PBGA thermal performance from the 1- to 2-watt range to the 3- to 4-watt range.

The next step up in thermal performance enhancements for PBGAs is copper planes in the substrate that spread the heat over the entire area of the package. Typically, thermally conductive planes are used in conjunction with thermal vias. As with thermal vias, thermal planes can and most commonly do act as electrical planes as well. The difference between thermal and electrical planes is that copper planes included for electrical reasons perform their duty very well at thicknesses of .0007 to .0014 in. While a 1.4-mil-thick copper plane is of some help, the typical thickness for a thermal plane is between 4 and 8 mils. The primary thermal path in a package with thermal planes is parallel to the surface of the plane, through the solder balls, and into the PC board. Unlike the thermal via approach, thermal planes also enhance the secondary thermal path through the mold compound to the top of the package.

While the basic construction of a package with thermal planes is the same as the basic OMPAC style, the substrate is significantly different. Basic OMPAC has a metallized layer on the top and bottom surface of the substrate only. To incorporate thermal planes, it is necessary to

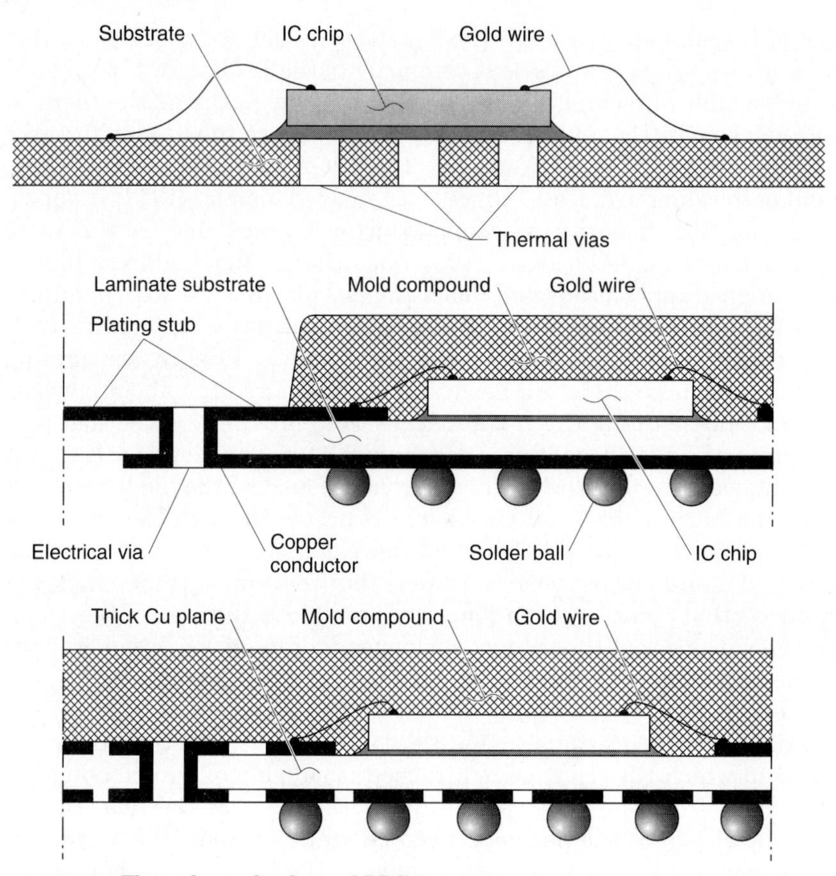

Figure 11.4 Three thermal-enhanced PBGAs.

have more than just the two layers of metal. The added complexity of a multilayer substrate typically yields electrical as well as thermal benefits, but it also increases the cost of the package.

For the maximum in thermal performance, the configuration of the PBGA package is changed considerably. The package is constructed with the die attached directly to a copper slug which forms most of the top surface of the package. The substrate is also attached to the copper slug and forms a cavity around the die. The balls are attached to the surface of the substrate opposite the copper slug and they form a depopulated array around the cavity. Figure 11.5 illustrates the typical construction of what Hestia calls a *high-performance solder grid array.* Aside from the copper heat spreader, the most interesting feature is the cavity-style substrate. In order to provide room for the wire bonds, the wire-bond ledge must be set well below the ball surface of the substrate. This construction dictated the use of a multilayer substrate and as such, electrical performance of these packages can be significantly improved over standard PBGAs. The cost also increases.

The price for the improvement in thermal performance is modest and the price/performance ratio is extremely low when compared to ceramic multilayer with Copper/Tungsten heat spreaders. The thermal performance of this style of package, as represented by $\theta_{jc,}$ is dependent upon the size of the heat source and the size of the heat spreader. In the smallest practical sizes for this construction, around 100 balls, θ_{jc} should be less than 4°C/W. With packages designed to house die more than 15 mm sq. and heat spreaders in the range of 37 mm sq., θ_{jc} can go to less than 0.15°C/W. This is really the area where the package supplier and the system manufacturer can interrupt the performance differently. While the package supplier is likely to see this package as capable of dissipating more than 300 watts with a 50°C delta above ambient, the systems designer may see it as stretching to deal with 7 watts. The difference in perspective is a function of how each sees the transfer of heat from the package to ambient. As packages suppliers, we know that the systems guys can run these things in liquid nitrogen if something gets too hot. Unfortunately, the systems designer wants to make the system as small and as quiet as possible and cooling systems are incompatible with those objectives.

The real story on laminate packages with copper heat spreaders is that their θ_{jc} is lower than a multilayer ceramic package with the same size copper tungsten heat spreader for a fraction of the cost. They do have the same problems as any thermally enhanced package. The thermal enhancement can only move the heat from the die to the back of the package. There must be some provision made to thermally couple the package to the final heat sink, and the efficiency of that path dictates the practical efficiency of the package.

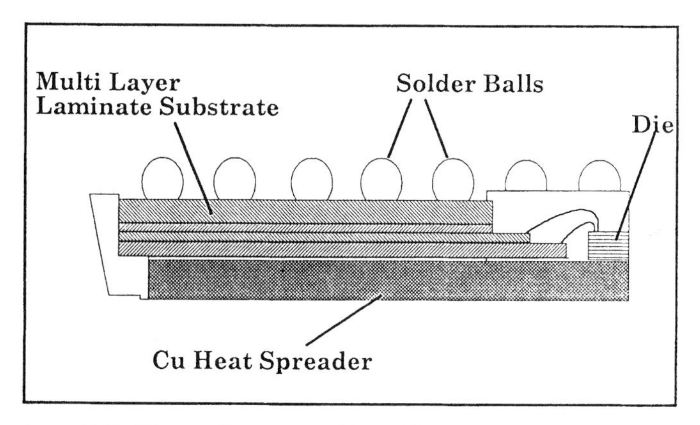

Figure 11.5 High-performance solder grid array.

External augmentation. One way to improve the transfer of thermal energy from the package to the air inside the system enclosure is to increase the area of the package. One of the simplest ways of doing that is to add a heat sink to the back of the package. These devices, which are available in many different configurations and materials, can improve the performance of any package. The real improvement is a function of the θ_j of the package, the attachment of the heat sink to the package and the efficiency of the heat sink itself. One of the most common types is illustrated in Fig. 11.6.

While external heat sinks offer a quick and easy way to increase the thermal efficiency of any package, the PBGA format offers a few challenges that are not issues in other form factors. The most notable problem is with the balls themselves. During the reflow of a PBGA, the solder ball melts completely. In most cases, the weight of the packages literally floats on the ball of molten solder. The advantage with this system is that the surface tension of the molten solder pulls the package into the alignment with the metal pattern on the PC board that produces the lowest cumulative surface tension among all the balls. By definition, this is the best possible alignment of package to board. Adding a copper heat spreader to the package can compromise its ability to float. Each ball can support just so much weight before collapsing into contact with its neighbor. External heat sinks are typically heav-

Figure 11.6 Examples of heat sinks for OMPAC® style plastic ball grid arrays. (*Photo courtesy of Wakefield Engineering.*)

ier than integral heat spreaders and in most cases would catastrophically collapse the solder balls. The solutions are fairly straight forward but do compromise some of the advantages of the PBGA form factor. First, standoffs can be molded into the package to prevent the added weight from collapsing the balls. The standoff, however, can prevent the package from being pulled to best alignment by the surface tension of the solder. The other solution is to attach the heat sink after the package has been reflowed to the PC board. Probably not the most elegant solution from a manufacturing perspective, but effective.

11.3 Electrical Performance Management

11.3.1 Vendor vs. end-user perspective

As with thermal performance, the real electrical performance contribution of a semiconductor package is influenced by its environment. The electrical environment of a package is primarily the board to which it is directly attached. Increasing the performance of the board typically involves adding wiring layers to the board and designing the interconnect to maximize performance. Along with performance, cost is also increased. However, in contrast to the measures necessary to improve the thermal environment, the monetary cost is low and there is virtually no change to the outline and components of the system when the PC board is optimized for electrical performance. In addition, electrical performance improvements in the PC board can be effected for virtually any system whereas thermal improvements require physical attributes which can be impossible to provide for some applications.

The bottom line is that high electrical performance in the package is much more likely to be translated into equivalent performance at the system level. Consequently, the perspective of the package supplied and the system designed are much more closely aligned.

11.3.2 Basic electrical requirements

The basic electrical features in semiconductor packages are conductive paths between the chip and the outside world and insulating separations between the conductive paths. The conductive paths between the chip I/O and the outside world must be of low enough resistance to maintain required voltage levels in the signal. For most packages, this in not a problem. For PBGAs with robust copper conductors, low resistance traces are unavoidable, and ultra low resistance is easy to design in.

Resistance between adjacent conductors is a slightly more complicated issue. With most high-speed digital devices, resistances in the range of a 10^6 ohm are all that are required for proper device function. No package makers worth their salt would consider shipping a package with resistances that low. The problem comes when very high isolation resistance is required. It is not uncommon to see leakage currents specified at less than 1 nA at 100 volts, or $10^{11}\Omega$ on multilayer ceramic packages. There are very few devices that really require that level of electrical isolation but there are a few. Laminate-based packages are certainly capable of that level of isolation when they are dry, but in real systems, that is probably not the case. One of the biggest problems with PBGA packages is that both the laminate substrate and the mold compound absorb moisture, which, among other things, reduces the resistivity of both those materials. The degree to which this is a problem is primarily dependent upon the device. In most applications, we do not see excessive leakage current as a failure mode until the package has been exposed to pressure cooker conditions (121°C, 100 percent relative humidity and 2 atmospheres pressure) for more than 168 hours. Even then the condition can be reversed simply by baking the component dry. The point is that semiconductor devices that operate with current levels in the pico amp range may not be candidates for PBGA packages, or plastic packages in general. Higher power devices should not experience any problem.

11.3.3 Noise problems in semiconductor packages

All commercially available packaging solutions for semiconductor devices meet the basic requirements for low resistance in the conductors and high resistivity in the dielectric materials. The more pressing issue with the electrical performance in semiconductor packages is electrical noise, and there are three major sources of noise in a semiconductor package: (1) coupling of signals between nearby conductors due to parasitic capacitance and inductance, (2) voltage bounce generated by transient currents in inductive elements, and (3) reflections from impedance discontinuities in the conductive elements.

Noise associated with capacitive and inductive coupling is more commonly known as *crosstalk*. Capacitive crosstalk is a function of the dielectric constant of the package insulating materials and of the separation between conductors. Reducing the dielectric constant and increasing the separation of conductors reduces capacitive coupling. Unfortunately, the dielectric constant of the insulating materials is defined to a narrow range by the materials applicable to a given package type. The physical separation of conductors is a design issue, but largely constrained by the die and the form factor of the package. There is, however, more latitude to influence capacitive-coupled noise with design than with materials selection. Increasing the number of conductive layers provides more design flexibility to both space conductors farther apart and to isolate traces that are particularly sensitive to capacitive coupling. Inductive crosstalk is largely a design problem in most packages but is influenced by the magnetic permeability of the conductors and the other materials in the package. Again, increasing the number of metal layers and providing flexibility to vary dielectric thicknesses provides the latitude necessary to reduce inductive coupling by design.

For packages housing high-speed digital circuits, induced voltage in the ground circuit (or ground bounce) is a problem of at least equal magnitude to crosstalk. Ground bounce—or for that matter, supply voltage bounce—is caused when many outputs switch simultaneously causing a current surge in the ground or supply circuit. The induced voltage is a function of the self-inductance of the circuit and the rate of change in the current pulse. The voltage potential, that can be calculated by the equation

$$v = L \frac{di}{dt}$$

appears as noise on the inputs and outputs of the IC and thus reduces the noise margin. The voltage "bounce" can be reduced by reducing the rate of change in the switching current, but that implies slowing down the chip. For all practical purposes, voltage bounce must be controlled by reducing the self-inductance in the power and ground circuits. This is accomplished by avoiding materials with high magnetic permeability and by design. Increasing the number of metal layers is particularly important in controlling inductance. The easiest way to reduce self-inductance is to increase the area of the current path while minimizing the length of the current path. The result is that wherever practical, power and ground circuits are actually planes that, as completely as possible, cover the entire area of a metal layer.

The problem of reflections from impedance mismatches is really two problems. The first is signals reflected from physical discontinuities in the package. The most problematic of these in most cases are the traces used to plate the exposed metalization on laminate-based and cofire ceramic packages. In virtually all cases, these traces simply end at the edge of the package and reflect a large portion of the signal incident upon them back into the circuit and to the input of the chip. The impact of these reflected signals increases as the clock rate increases and, again, appears as noise at the inputs and outputs of the chip.

The second part of the problem with reflections from impedance mismatches is probably more important in most cases. In order for the signal to be passed from one circuit element to the next, the characteristic impedance of the two elements must match as closely as possible. To the degree that the impedance of a transmission line and the inputs of the chip do not match, the signal is reflected back into the input circuitry. While this reflection appears as noise in the circuit, the more important issue is the reduced transfer of the signal onto and off of the chip. To complicate the problem, characteristic input impedances of the various semiconductor devices range from around 50 to 110 Ω. Impedance in semiconductor packages is primarily a function of capacitance and inductance, which really gets back to dielectric constant, magnetic permeability, and design. In most package types, magnetic permeability is not an issue and dielectric constant is closely dictated by the packaging type. That leaves only design as a major impedance variable. However, with design, it is possible to vary capacitance, inductance, and resistance over a wide range. This is particularly true when multiple conductive layers are available to the designer.

Figure 11.7 is the equivalent circuit for a finite length of the typical transmission lines on all semiconductor packages. The value of G, the conductance between the two legs of the transmission line and R, the resistance of each leg are insignificant in most applications and can be neglected. Figure 11.8 shows the equivalent circuit as it appears without considering R and G. This network is terminated with an impedance that represents the impedance of an infinite number of sections exactly like the section under consideration. Z_0 is the characteristic impedance of the line, Z_1 is the distributed reactive impedance of the section, and Z_2 is the capacitive reactance of the section. Without going into the mathematics, the characteristic impedance of a transmission line is primarily dependent on the ratio of the distributed inductance and capacitance in the line. That allows the line to be characterized by looking at the lumped values of capacitance and inductance for a finite section of the line according to the equation

$$Z_0 = \sqrt{\frac{L}{C}}$$

What is not obvious from any of the basic discussion of factors influencing impedance is the effect of the inherent geometries of the processes and how they interact with the basic materials parameters.

11.3.4 The role of the package in electrical performance

The role of the packages is to deliver the functionality of the chip to the rest of the system with as little degradation as possible. Having said that, the implication is that the package does damage to the intrinsic performance of the chip. The implication is accurate. The chip is capable of significantly better performance before any electrical or mechanical connections are made to it. Unfortunately, the chip is of virtually no practical value in that state.

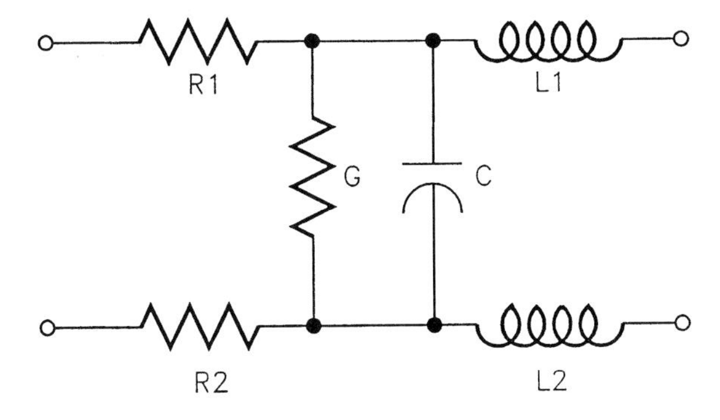

Figure 11.7 Equivalent circuit for a finite-length transmission line of an IC.

In addition to acting as the electrical interface to the rest of the system, the package provides a translation of geometries from those of the chip to those of the real world and to provide mechanical and environmental protection. In all packages, the space translation is a performance problem. The mechanical protection features may or may not impose significant performance limitations, depending upon the design.

The degree to which the performance of the chip is degraded can be reduced by the package design. In keeping with the natural laws of economics, the simplest (and therefore least expensive) designs do the most damage to performance. The performance of the chip can be preserved with careful design of the package. The problem is that these

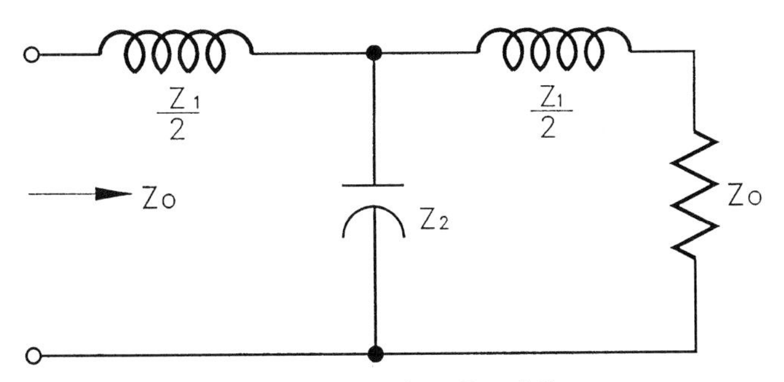

Figure 11.8 Equivalent circuit without R and G.

designs are more complex mechanically and thus more expensive. Fortunately, the majority of chips can deliver adequate performance with the simplest of packages. The question of electrical performance, then, only becomes an issue for the fastest or most sensitive devices.

11.3.5 The materials set

Plastic ball grid array packages are blessed with a good basic materials set. Table 11.3 lists the basic properties of those materials. First of all, the electrical properties of all the materials are good. The electrical conductivity of the copper conductor material is among the highest available. The dielectric properties of both the glass/epoxy substrates and the mold material are favorable with dielectric constants in the range of 4. The mechanical attributes of the materials and the processes used to produce laminate substrates lend themselves to advantageous geometries in both the conductors and the dielectric structures. The practical conductor pitches, down to around 100 μm, are well-suited to translate the fine geometries of the chip and the first-level interconnect to the courser geometries of the PC board. Those natural geometries, combined with the basic electrical properties, allow for the easy design of transmission lines with characteristic impedance that match to most semiconductor devices.

The range of geometries possible and practical in laminate manufacturing processes bring more to the party than just the ability to do the space transformation from the die to the PC board. The combination of those natural geometries and the basic electrical properties allows for the easy design of transmission lines with characteristic impedances that match to most semiconductor devices. Practical dielectric thicknesses for PBGA substrates range from .002 to .020 in. Minimum conductor width is in the range of .002 in and there is no maximum except for the width of the substrate. Using conductor designs with maximum width of .020 in, it is possible to achieve impedance values from 2 or 3 Ω to 110 Ω. While neither the range nor the absolute impedance numbers are unique to laminate substrates, there is no other packaging type that has a materials set that allows those numbers to be achieved so easily.

TABLE 11.3 Electrical Properties of Laminate Materials

Sheet resistivity	1 mΩ/Sq.
Dielectric constant	4.2
Impedance	20 to 100 Ω
Propagation delay	70 pS/Inch
Minimum conductor pitch	.004″
Dielectric thickness range	.002″ to .030″

Plastic ball grid arrays are still in the industry-acceptance portion of their life cycle, and as such, there continues to be significant evaluation and development done of materials, processes, and designs. The current basis for PBGA packages is a substrate made of Bismalimide Triazene (BT) epoxy and E glass. The primary purpose for choosing this material is its high glass transition temperature at a moderate cost premium over FR-4. The high Tg, in the range of 190°C, is necessary to withstand the thermal exposure associated with thermosonic bonding and transfer molding. As an added benefit, the BT board has a lower dielectric constant than FR-4 at around 4.1.

The current conductor is copper laminate. In standard designs, the copper is ½ oz/sq. ft, or about 17.5 μm thick. At that thickness, the conductors exhibit a sheet resistivity of around 1 mΩ/sq. The mold compound that complete current PBGA designs is one of several that are also used for standard molded packages. Typically these materials have dielectric constants from about 3.7 to 4.4.

11.3.6 Package construction

The advantages of the materials set are analogous to having good genes. You may be genetically predisposed to live to a ripe old age, but you can always find a way to squander your advantage and die young. In the case of PBGAs, the problem is that the most natural design for economical packages is close to the worst case for electrical performance. The most basic PBGA conductor design is illustrated in Fig. 11.9. Essentially, the conductor traces are routed from the edge of the die to the outer edge of the package, through vias to the backside of the package, and then to a solder ball location. This design makes it virtually impossible to avoid doubling the conductor over on itself, separated only by the thickness of the substrate. The result is very high mutual- and self-inductance in the traces. Generally, the individual conductors in PBGAs of this design can be expected to have self-inductance that is higher than in the conductors of a QFP of comparable lead count. Even with this problem, the standard OMPAC® design is adequate for most semiconductor devices and applications.

Where improved electrical performance is required, there are changes in the package construction and design that allow the package designer to take full advantage of the properties of the materials set. The biggest problem with the OMPAC® design is the fact that the conductors transition from the top to the bottom of the substrate only at the extreme outside edge of the substrate. This design eliminates the possibility of mold material leaking through vias since there are none in the molded area. It is also mechanically logical since the traces must be routed to the edges to be tied to the electrolytic plating bus. By rout-

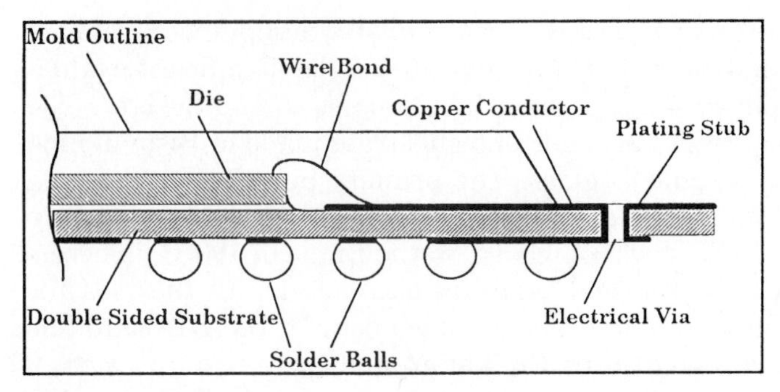

Figure 11.9 OMPAC conductor configuration.

ing the traces to vias under the molded area, and designing for the shortest possible conductor lengths, much of the self-inductance of the trace can be eliminated. With shortened traces, the inductance of the leads in a PBGA would typically be less than in the leads of a standard QFP with the same lead count. Unfortunately, the extension of the traces for plating—from the vias to the edge of the substrate—still adds capacitance and inductance and is a source of reflected noise from the stubbed-off end of the conductor. The only way to eliminate the plating stubs is to eliminate electrolytic plating of the conductors. Since the standard conductor material is essentially pure copper, either the material must be changed, or the surface of the copper must somehow be treated to be compatible with the gold thermosonic bonding processes. At this point, the most practical way to provide that surface without plating stubs is by electroless plating of gold on the copper conductors.

Once we eliminate the plating traces and provide for vias under the molded area of the package, the next step in improving the performance of the package is to increase the number of wiring layers. As the number of wiring layers increases, there are several performance-enhancing features that can be incorporated in the package. For practical purposes, wiring layers are added two at a time. Effectively, a four-metal layer board is made by sandwiching a layer of epoxy/glass laminate material, with no metallization, between two double-sided boards. With four metallized layers to work with, there is room for two layers for signals and two layers for low-inductance power and ground planes. Further increasing the number of available metal layers increases the flexibility to design precise transmission lines and to add even more complex power-distribution systems to isolate sensitive portions of the circuit.

The multilayer PBGA construction is basically equivalent to multilayer ceramic package construction in concept and in the feature geometries. The big difference is in the properties of the two materials sets. The resistivity of the tungsten conductors of cofire ceramic is an order of magnitude higher than the copper in the multilayer laminate, and the dielectric constant of the ceramic is two times that of BT epoxy/glass. The result is that laminate-based multilayer substrates can be implemented in fewer layers than is possible in cofire ceramic multilayer.

11.3.7 Electrical advantages of BGAs vs. other surface mount packages

BGAs do have one inherent electrical performance advantage over other surface mount packages. That advantage lies in the interface between the package substrate and the PC board: the solder balls themselves. When compared to the leads of a QFP or a PLCC, the solder balls are short, they are relatively large in area, and they are widely spaced. The result is an electrical connection of significantly lower inductance than with the leads of other surface mount packages.

In addition to the lower lead inductance, the average length of the package conductors is much shorter in a BGA than in a QFP. Shortening the conductor does reduce the capacitance, inductance, and the total time of flight for the signal. Unfortunately, this is one of the areas where the package supplier and the system designer are likely to see this performance advantage differently. In cases where the footprint of a BGA package is the same size or larger than an equivalent QFP, reduced package conductor length does not translate into an overall shorter distance for the signal to travel. The only real effect in those cases is to move the signal path more to the PC board. Depending upon the PC board design, that may improve or degrade the electrical performance.

For high-lead-count, multilayer BGAs, not only is the average package conductor length less than the equivalent QFP's, but the maximum conductor length is also lower. BGAs become more area efficient than QFPs at around 100 leads. As the I/O count goes up the area efficiency advantage of the BGA increases. The 304 QFP occupies 1849 sq. mm as opposed to 625 sq. mm for the 324 BGA with 1.27-mm ball pitch. That translates to a real reduction in trace length and, therefore, to propagation delay between system components.

It probably comes as no surprise to anyone that increasing the performance of PBGAs increases their cost. Since cost is also a real performance attribute, it is necessary to weigh the cost against the benefit of high-performance packages. In most cases, the price for enhanced performance in PBGAs will be lower than for other packages types with the same lead counts. The primary reason for the lower cost is the smaller size of the PBGA. To build a QFP that is electrically equivalent to a PBGA with a four-layer board, the QFP will have to have a wiring system that is very similar to the PBGA. In fact, the most practical option is to use essentially the same board in the QFP as is used in the PBGA. The only difference is the size. Looking again at the comparison of the 304 QFP to the 324 PBGA, the substrate for the QFP would be in the range of 38 mm sq. as opposed to 25 mm sq. for the PBGA. The direct area ratio of QFP to PBGA is 2.3:1 If you extend the comparison to less area-efficient formats such as PGAs, the area ratios become even larger. Again looking at the 300 lead-count range, a standard PGA would occupy at least 3.4 times the space of a PBGA, and in a more practical configuration, the ratio is closer to 5:1.

The bottom line is that increasing the electrical performance of a BGS does increase the cost, but that the incremental cost for increased performance is less in the PBGA format than it is for other formats. Additionally, most high-performance PBGAs have greater potential to increase the system performance than the other formats simply because of their smaller size and therefore shorter conductor paths. It is also important to consider that the smaller BGA footprint also reduces the size of the system board, yielding both electrical performance and cost improvements.

11.3.8 Electrical modeling and high-performance design

Having come to the point of understanding that PBGAs are endowed with a materials set that lends itself to high electrical performance, and that there is sufficient latitude in the construction of the packages to take advantage of the materials set, the focus goes to design. Intrinsically, plastic ball grid array packages have all of the advantages of other package formats based on a laminate substrate and have shed some of the limitations by growing smaller than their cousins. The general design features are similar to those used in laminate-based QFPs, PLCCs, and PGAs. To tune the designs to optimum performance, we need careful evaluation and measurement of actual packages. To this point, we do not have sufficient data to model PBGA's electrical performance to the degree that we will ultimately need.

We can, however, take what we have learned from other packages and from our early work with PBGAs and postulate a number of rather solid rules for PBGA design. The issues of crosstalk are the same as with most other package types. The one advantage that PBGAs have is that their leads are generally shorter—and thus couple less energy—both through capacitance and inductance, into their neighbors. The keys to low crosstalk are to minimize the length of parallel conductors and to space them as far apart as possible. Reducing voltage bounce in either power supply or ground circuitry is a matter of lowering the inductance of the circuits. In most cases this is accomplished by increasing the number of pins dedicated to each circuit and distributing those pins around the package to equalize current in each. All of the pins for a given circuit are shorted together to a plane that occupies as much as possible of one metal layer. In many cases, a few signal lines are more sensitive to noise than most and need special treatment. In these cases it is often desirable to split power and ground planes into quiet and noisy planes. The quiet planes are reserved for the few extra-sensitive signals, and the less-sensitive lines connect to the noisy planes. Even though these planes are connected to common power and ground sources somewhere on the PC board, moving that unification closer to the power supply can significantly reduce the noise apparent at the chip.

One of the problems with the leaded surface mount packages is that the parasitics of the leads are both unfavorable and essentially fixed for a given package configuration. The inductance of a solder ball is variable within a narrow range, but is typically less than one-third that of the gull wing lead of a QFP. More importantly, the capacitance, and therefore the impedance of the leads can be varied significantly by varying the spacing of the balls. While there are effectively six different pitches approved by JEDEC, it is not generally practical or desirable to change the ball pitch to modify the impedance of a few signal traces. Instead, JEDEC, and the configuration flexibility built into the ball grid format, allow the removal of individual balls to increase individual ball spacing and thus reduce the capacitance. By using this design approach and controlling the size of the solder ball it is possible to adjust the impedance of the solder ball connection from around 20 to 50 Ω without much difficulty. Increasing the impedance to the 100 Ω range is possible but causes significantly more disruption of the ball pattern.

11.4 Summary

Plastic ball grid array packages offer an array of options in construction and design that allows the format to be competitive with any other in terms of electrical and thermal performance. Thermally, the standard PBGA package has a shorter thermal path to the PC board than any other package type. To be useful, the PC board needs to be of a construction that will absorb and dissipate the thermal energy transferred through the balls. Moving to more complex packages, there are a number of features that can be added to PBGA packages that can move their thermal performance into competition with the very best of semiconductor packages. The beginning step is to add thermal vias that can improve the thermal performance by 20 to 40 percent depending upon the substrate and the exact configuration of the package. Internal copper planes can spread the heat out, over the full area of the package and move the heat to both the top and bottom of the package. Typically, thermal planes will improve the thermal dissipation capabilities of a package by 50 percent. The real big improvement comes when the back of the package is replaced by a copper plane. This requires making the package a cavity-down design and, necessarily, increases the size of the package. The pay-back for the larger package is a package that is a thermal transfer champion. Actual performance depends on both the package and die size, but θ_{jc} values less than 0.2 °C/W are possible in real-world packages. θ_{ja} values for these packages can be less than 1°C/W.

Electrically, the PBGA package benefits from a very good materials set. The conductor is nearly pure copper and has a design sheet resistance of 1 mΩ/sq. The dielectric constant of both the substrate dielectric and the mold compound are in the range of 4. The relationship between these basic values and the natural manufacturing geometries of the laminate production processes allow for very easy design of transmission lines that match the impedance of virtually all semiconductor logic types. The combination of these attributes and the flexibility to tailor the package for the chip to a greater degree than in any other surface mount format makes the plastic ball grid array package an unsurpassed electrical performer.

The combination of electrical and thermal performance options in plastic ball grid arrays will make them one of the most useful surface mount packages introduced to date.

11.5 References

1. J. Kennedy and S. Diamond, "High Speed Performance: Spice Modeling Helps Optimize Interconnect," *Advanced Packaging,* summer 1993, pp. 10–13.
2. R. Sigliano, "Design for Test: Eliminating the Noise From Packages," *Advanced Packaging,* spring 1993, pp. 10–12.
3. D. Hattas, "BGAs Face Production Testing," *Advanced Packaging,* Summer 1993, pp. 44–46.
4. C. Luchinger, J. Beuers, A. Fiebig, and W. Werner, "Inside Solder Connections: Examining the Metallurgy of Soft-Solder Alloys and the Influence of Intermetallic Layers," *Advanced Packaging,* Jan./Feb. 1994, pp. 14–19.
5. R. Knight, R. Johnson, J. Suhling, J. Evens, C. Romanczuk, and S. Burcham, "Thermal Analysis: Modeling the Thermal Performance of an Automotive Powertrain Controller's Components Mounted on an Insulated Metal Substrate," *Advanced Packaging,* Jan./Feb. 1994, pp. 30–34.
6. H. Markstine, "Impedances Dictate Backplane Design," *Electronic Packaging & Production,* Dec. 1993, pp. 38–40.
7. I. Bhutta, A. Elshabini, and Sedki Riad, "Measurement, Modeling, and Simulation of Electronic Packages, *The International Journal of Microcircuits & Electronic Packaging,* second quarter 1993, pp. 161–166.

8. C. Lassen, "Silicon Packaging: The Aluminum Nitride Domain," *Advanced Microelectronics,* Jan./Feb. 1994, pp. 41–42.
9. "Technical Discussion," *EG&G Wakefield Engineering, Heat Dissipation Components Catalog,* Apr. 1990, pp. 5–14.
10. D. Mallik and B. Bhattachryya, "High Performance PQFP," *Proceedings IEEE 39th Electronic Components Conference,* Houston, May 1989.

Chapter

12

Thermal and Electrical Management of Plastic BGA Packages—A User's Perspective

David B. Walshak, Jr. and Hassan Hashemi

12.1 Introduction

One of the attributes generally listed by proponents of ball grid array (BGA) packages is their perceived improved thermal performance over conventional quad flat pack (QFP) and pin grid array (PGA) packages. However, these improvements can only be achieved through a system-level approach to thermal performance. Plastic BGAs are similar in construction to plastic pin grid array (PPGA) packages except that where pins are located on a PGA, solder balls are attached on a BGA. Therefore, many of the thermal performance characteristics of BGAs are similar to PPGAs. Due to the shorter traces provided by its fine-line laminate base and to the lack of long pins, the electrical performance of a BGA is generally better than that of the best PPGA.

This chapter describes the important factors affecting the thermal and electrical performance of plastic BGA packages that have been reported in the literature. Wherever possible, comparisons with similar conventional packages are presented. Since this package type is still relatively new to the industry, only a small number of studies have been conducted and published, but the BGA package is currently being scrutinized by a large number of packaging designers and users.[1]

12.2 Thermal Management of Plastic BGA Packages

12.2.1 Introduction

Three common types of BGA packages are currently being studied: plastic, ceramic, and tape. The plastic BGAs can be manufactured more cheaply, but the cheapest of these packages inherently have the poorest thermal characteristics—which may be completely acceptable for many low-power applications. For this reason, when faced with putting a high-performance chip (or chipset) in BGA format, designers have typically pursued ceramic or tape BGAs. Several enhancements have been conceptualized for plastic BGAs to improve their thermal performance and consequently expand their power-operating window. Thermally enhanced plastic BGAs (EBGAs) are also being investigated as a lower-cost alternative in moderate- to high-performance applications. The EBGA packages will be discussed in Sec. 12.2.3, while the thermal aspects of ceramic BGAs and tape BGAs are discussed in other chapters.

In certain "borderline" situations the cost savings of plastic overmolded BGAs must be weighed against the performance attributes of both ceramic and tape BGAs. Just as different package types are compared to obtain the optimum match, the mix of enhancement options provides a large variety of selections, each option with its own cost penalty or benefit. The effects of utilizing some of these performance enhancements should be thoroughly studied to determine the impact on overall system performance and cost.

A schematic showing the major heat-flow paths in a typical BGA package is provided in Fig. 12.1 and a simple one-dimensional electrical analogy to the thermal problem is shown in Fig. 12.2. While the thermal analysis problem is a three-dimensional one, this 1-D schematic shows the associations between the significant factors in the system. The thermal resistances in Fig. 12.2 are discussed in detail later in this section.

The factors affecting thermal performance of the plastic BGA packages may be divided into three main categories, although there is some overlap: materials variables, geometric variables, and environmental variables. Each of these categories is described below, including the important factors for each area.

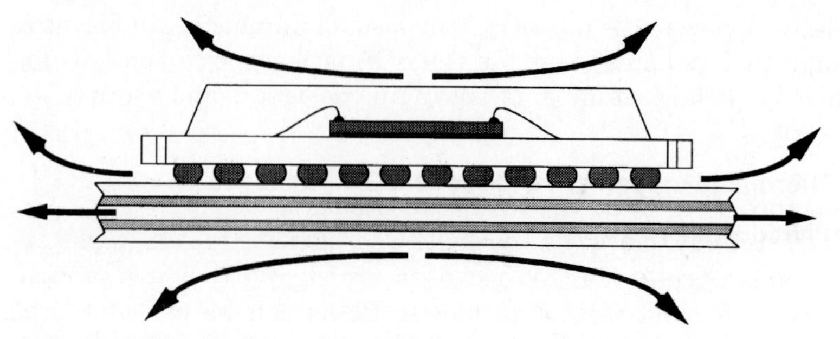

Figure 12.1 Schematic of major heat-flow paths in a typical BGA package.

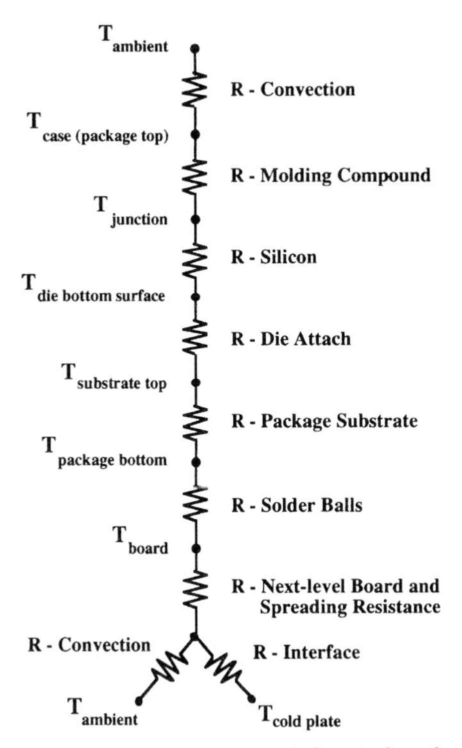

Figure 12.2 One-dimensional electrical analogy to the thermal problem.

12.2.2 Materials variables

From a thermal management perspective, the three major materials variables include the molding compound, the die-attach material, and the substrate dielectric material.

The molding compound completely surrounds the chip, providing mechanical protection as well as a surface for marking the package. Typically, the molding compound is in direct contact with the surface of the chip, although, in some cases, an encapsulation layer may be utilized prior to the overmolding operation. This encapsulation layer (typically either silicon gel or an epoxy) can act as a mechanical stress buffer and can provide additional corrosion protection. Unfortunately, from a thermal point of view, the molding compound thickness over the die surface is typically on the order of 15 to 25 mils, providing a relatively high thermal-resistance path to the top surface of the package. For this reason, most of the attention placed on thermal performance of standard plastic BGAs has been directed toward improving the thermal paths through the package to the external board. One study has been published showing that the mold compound thermal conductivity is of small importance relative to other packaging variables.[2] Of course, if an extended-surface heat exchanger were added to the top of an overmolded BGA package, the thermal conductivity of the molding compound (or more correctly, the thermal resistance from the chip surface to the top surface of the package) would be of tantamount importance.

With the current emphasis on through-the-board thermal enhancement, the selection of die-attach material is critical. When thermal vias are incorporated in the package substrate, this material directly couples the silicon chip to the die-attach pad at the top of the thermal via stacks. From a 1-D perspective, the thermal resistance across this layer is directly proportional to the thickness and inversely proportional to the thermal conductivity of the die-attach material. Typically, a silver-filled epoxy is used for the die attach with a thermal conductivity of around 1.9 W/m·K with die-attach layer thicknesses ranging from 1 to 2 mils.

The thermal conductivity of the substrate dielectric material is secondary in importance relative to several design-related variables that are discussed in Sec. 12.2.3. Most plastic BGA users are using a modified polyimide resin-glass (bismaleimide triazine, or BT) as the core for the substrate, while others use polyimide. The thermal conductivities for these materials vary from 0.21 W/m·K for BT to 0.155 W/m·K for polyimide.

12.2.3 Geometric variables

Using materials which are optimized for thermal performance is only one of the tools that package designers have at their disposal. In fact, design-related factors can have an even larger effect on thermal performance than materials-related variables. Several design enhancements have been incorporated into plastic BGAs to improve thermal performance. Some of these enhancements are related to the package substrate design, while others involve making a heat spreader integral to the package. The package designer must observe the thermal performance from a system-level point of view in making intelligent decisions related to package design optimization.

Package size. The "standard" plastic BGA utilizes a double-sided substrate with plated-through-hole vias located around the perimeter of the package for routing the I/O from the chip to the solder balls on the bottom of the package (see Fig. 12.3). This molded plastic package configuration, often referred to as the OMPAC (Over-molded Pad Array Carrier), provides a low-cost, easy-to-manufacture package which is more than adequate for low thermal-performance applications (i.e., less than one watt).

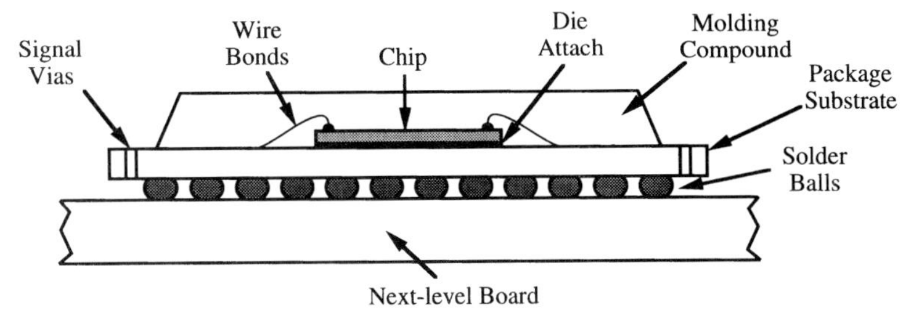

Figure 12.3 Standard plastic BGA package (or OMPAC).

Values quoted for junction-to-ambient thermal resistance, θ_{ja}, for plastic BGAs with no thermal enhancements are presented in Table 12.1. With the package mounted onto a low thermal conductivity board[3] (i.e., no thermal vias, no heat—spreading layers, and with a minimum number of traces) and with no forced airflow (i.e., in a natural convection environment) the values for θ_{ja} ranged from 52 to 74°C/W depending on package size. With 1 m/sec forced airflow, the thermal resistances ranged from 38 to 57°C/W. The relationship between θ_{ja} and package size is shown in Fig. 12.4.

Along these same lines for a given size package, a package with a larger chip will exhibit a better perceived thermal performance compared to the package containing a smaller chip. The thermal conduc-

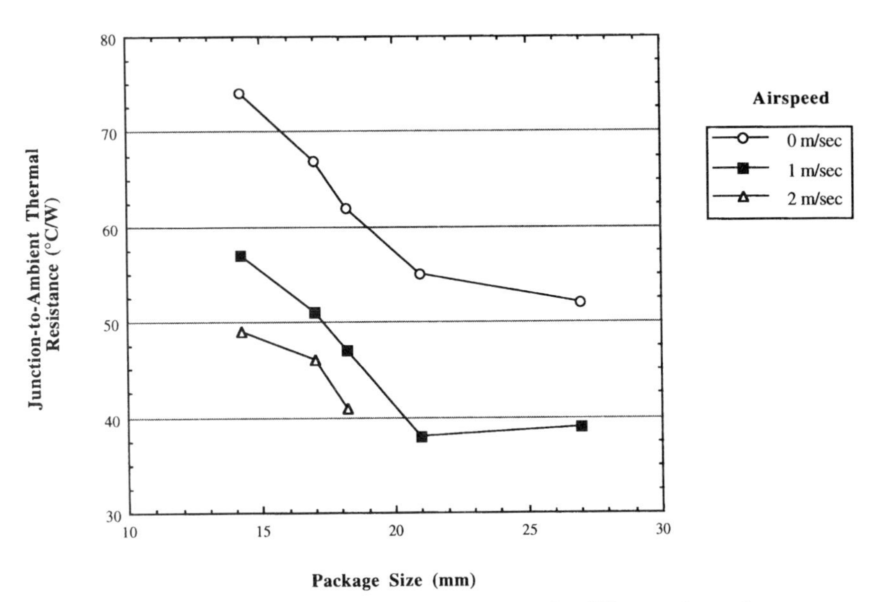

Figure 12.4 Relationship between package size and θ_{ja} for different airspeeds.

TABLE 12.1 Package Comparison of Standard Plastic BGAs with No Internal Thermal Enhancements

		Junction-to-ambient thermal resistance, θ_{ja}, (°C/W)		
		Natural convection	Forced convection	
Package description	Package size (mm)	0 m/sec	1 m/sec	2 m/sec
68 OMPAC[4]	14.2	74	57	49
86 OMPAC[4]		67	51	46
106 OMPAC[4]	18.23	62	47	41
169 BGA[5]	21	55	38	
225 BGA[6]	27	52	39	

tivity of the silicon is so much larger than that of the molding compound (149.6 versus 0.7 W/m·K, respectively) or most of the other materials in the package that, from a global perspective, the temperature of the chip will appear much higher and quite uniform relative to the temperature distribution throughout the rest of the package. For two different size chips dissipating the same amount of power, the larger chip will have a lower heat flux (in watts per unit area) than the smaller chip. The larger chip can also provide some of the benefits of a heat spreader because of the higher thermal conductivity of the silicon. In summary, a BGA containing a larger chip will thermally outperform an identical BGA containing a smaller chip.

Thermal test data may also be skewed by power dissipation, especially at low airflows where the buoyancy effect of natural convection is the major heat-transfer driver. Depending upon package design, next-level board layout, and airspeed, the thermal resistance measurements may vary by up to 28 percent simply because of the different powers being dissipated within the package.

The BGA substrate also presents a platform on which more than one chip can be interconnected, providing an overmolded multichip module. The electrical, assembly, and cost benefits of multichip modules have been well-documented. The thermal benefits with a multiple chip BGA are also pertinent due to the heat-spreading attributes inherent in the package simply by having multiple heat sources within the same package.[2]

In summary, if system-level space allows it, one method of improving thermal performance of BGAs is to increase the package size. A larger package will generally accommodate more solder balls to help spread the heat to the next-level board as well as a larger surface area for improved convective heat transfer. In many situations, the constraints are such that it is the engineer's job to minimize both the size and cost parameters. In fact, smaller package size is a primary reason why

BGAs are being considered for use. For this reason, an alternative to enlarging the package must be pursued.

Package substrate design. The major emphasis on improving thermal performance for plastic PGAs has been on improving the through-the-substrate heat transfer. Two parallel design approaches have been used to optimize through-substrate thermal conduction: using either thermal vias or heat-spreading planes.

When dealing with laminate substrates, thermal vias are simply plated through-holes in the substrate which happen to be located underneath the chip (see Fig. 12.5). In general, the thermal vias are tied together with a rather large copper land on both the top and bottom surfaces of the substrate. The copper land on the top side of the substrate is also used as the die-attach pad. Although the cross-sectional area of copper from the thermal vias is relatively small, the difference in thermal conductivities between the copper and the dielectric is quite large (i.e., 389 W/m·K for copper versus 0.21 W/m·K for BT). Thus, the effective through-substrate thermal conductivity can be dramatically improved by utilizing thermal vias.

The design of the thermal vias can obviously affect the effective through-substrate thermal conductivity underneath the substrate. Compaq has used 0.300-mm (12-mil) diameter thermal vias for two plastic BGA packages considered.[7] A modeling study was conducted at MCC on these two packages to determine the dependence of through-substrate effective thermal conductivity on several parameters: via pitch, via diameter, and plating thickness.[8] The net result of this study is that increasing the amount of copper in the cross-section of the via array increases the effective through-substrate thermal conductivity. For example, an array of 12-mil diameter vias on 60-mil pitch plated with ½-oz copper, the predicted through-substrate effective thermal conductivity in the thermal via region is approximately 1.75 W/m·K versus 0.21 W/m·K for BT without thermal vias—an improvement of 833 percent.

The thermal vias aid in getting heat from the top surface to the bottom of the substrate, but, used alone, only in the region directly underneath the chip. In this case, the solder balls directly under the chip must trans-

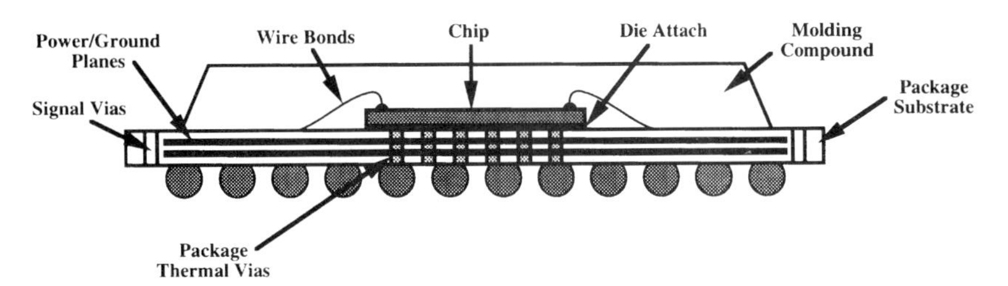

Figure 12.5 Plastic BGA with internal thermal vias and heat-spreading layers.

fer most of the heat to the next-level board. To efficiently utilize the benefits of thermal vias, some method of spreading heat throughout the package must be used.

With an internal heat-spreading layer incorporated into the package substrate, heat can be spread over a larger portion of the package area (see Fig. 12.5). This heat-spreading layer is usually a ground plane that is nearly solid and provides a low-resistance path across the package area. For best results, the thermal vias should be connected directly to the heat-spreading layer. The net effect of tying the thermal vias to this ground plane is that it improves the thermal coupling to the next-level board through the solder balls and also improves the convective heat transfer from the top surface of the package.

These additional layers can be added to the package substrate without significantly increasing the overall package cost. These layers can be used for power, ground, or additional signal planes and can provide opportunities for improved electrical, thermal, and routing characteristics.[9] The thermal vias occupy a section of the substrate that would normally be used for routing purposes. Again, the system-level tradeoffs must be analyzed to determine the appropriate direction to proceed.

The thermal management strategy currently being used by Compaq is to route all power and ground traces into the center of the package to connect with the thermal vias. This also puts performance-dependent pads on the perimeter of the array, thereby shortening the electrical paths to the next-level board and reducing electrical parasitics.[7]

Thermally enhanced plastic BGAs. Each of the above variables described varying degrees of subtle changes to package design and materials relative to overmolded plastic BGAs. Another approach that has been used to improve the thermal performance of plastic BGAs is similar to incorporating a copper slug in plastic PGAs. This package is usually considered a "cavity-down" BGA or enhanced BGA (EBGA) as opposed to the traditional overmolded plastic BGA. In this approach, the interconnect substrate has a hollow center and usually has multiple tiers to facilitate wire bonding and routing (see Fig. 12.6). Usually the copper slug is adhesively attached to the backside of the substrate, and solder or other electrical connections are made to ensure proper grounding. The chip is then placed in the cavity followed by a die-attach cure cycle. After the wire bonds are formed, the chip is either over-coated or a lid is affixed for mechanical protection. With this package, the bond pad array is not a continuous array, but has omissions in the region occupied by the chip and lid (i.e., the package only has solder balls under the package substrate).

Since the major heat path is through the slug, thermal vias in the substrate are not necessary with this package. From a thermal man-

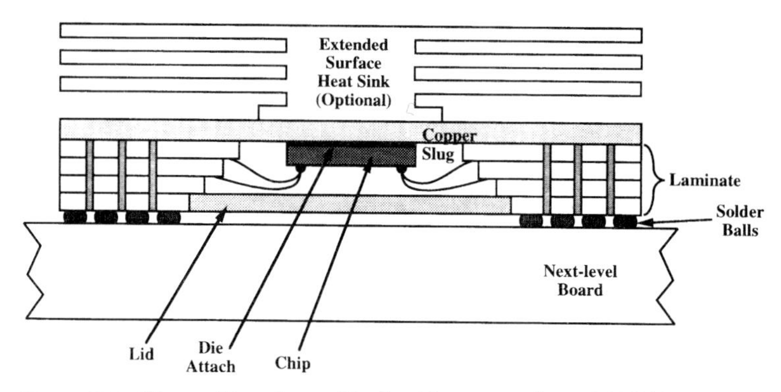

Figure 12.6 Thermally enhanced ball grid array package (EBGA).

agement perspective, if the package is to be used without a heat sink, the slug should be made as large as possible. This ensures good heat spreading for the package and also provides a larger surface area which enables more efficient convective heat transfer. The EBGA package also allows for a heat sink to be attached to the package top with a relatively low junction-to-case thermal resistance path.

Unfortunately, very little thermal work has been published regarding these enhanced BGA packages. Figure 12.7 shows some results generated by Amkor comparing a 169-I/O BGA in different package configurations. The thermal resistances show the dramatic potential of this thermally enhanced package.[5]

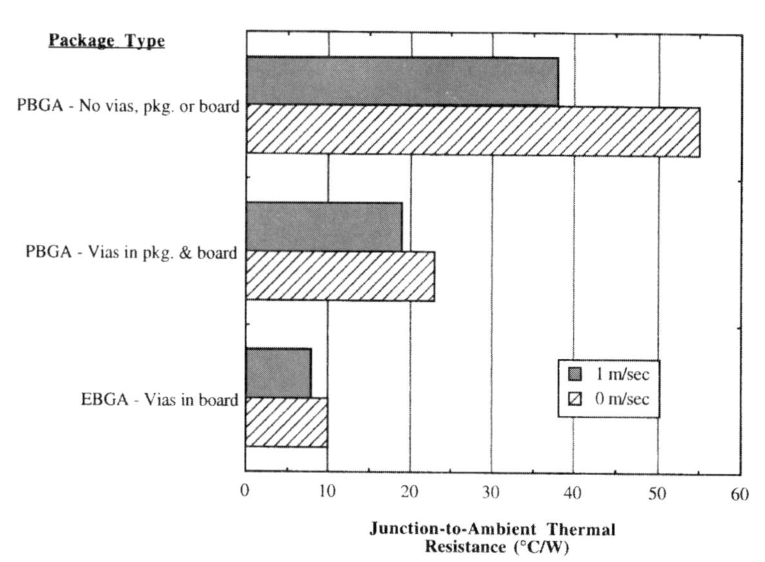

Figure 12.7 Comparison of two different PBGAs with an EBGA at 0 and 1 m/sec airflow.

12.2.4 Environmental variables

While most thermal testing is performed in controlled, specified environments, the application- or system-level environment is where the thermal performance of any package is most critical. The degree to which the operating environment can have an effect on the perceived thermal performance of BGA packages is presented and the major factors are identified in this section.

System-level cooling approach. Several options are available for the packaging engineer to remove heat from a package. Usually, these cooling decisions are made at the system level and are specified to the device engineers as requirements. The optimum design of a BGA package for any situation will depend upon which system-level cooling strategy is to be employed.

In many consumer, telecommunication, and portable products, system power, size, and/or noise restrictions prevent forced-air convection from being utilized as a cooling method. In this situation, the heat must be either removed from the package by conduction or by natural convection from the package and board surface.

Natural, or free, convection is a result of the motion of air due to density changes arising from the heating process. As the air in proximity to a hot surface heats up, its density decreases. This change in density imposes buoyancy forces on the air, resulting in upward movement of the air.[10] Even so, this movement of air is typically much less than 0.25 m/sec. Thus, the major vehicle for cooling these products requires conducting the heat to the outside surface of the product.

In contrast, the power dissipated in many workstations and other computer applications requires some form of forced convection to cool the electronics. Typical airspeeds used in these applications range from 0.25 m/sec up to 3 m/sec. To improve the heat transfer from an electronic package, a heat sink may be attached to the top surface of the package with a high thermal conductivity material, usually either a metal-filled epoxy or an adhesive film. The enhanced BGAs with their copper slugs are the most likely candidates for utilizing a heat sink.

External board design. Since the most efficient thermal path for a BGA package is through the substrate to the next-level board, another method for removing the heat from the package is by optimizing the conduction path to this next-level board.

The effective thermal conductivity of the next-level board can have a dramatic effect on the package thermal resistance.[11] The design of the next-level board can be modified in many ways to improve the thermal performance of BGAs. As mentioned previously, several designers have used thermal vias in the package substrate to provide a lower thermal resistance path through the substrate. These thermal vias are usually located underneath the chip and many times are shorted to either the power or ground planes in the substrate. The solder balls in the center of these packages are typically connected to these package thermal vias. For best results, the mating bond pads on the next-level board should be connected to thermal vias in the board, which are in turn connected to power or ground planes in the board.

Using the next-level board's ground plane is an excellent method for cooling BGAs. System-level designers may find it best to place the BGA packages close to the edge of the next-level board so that heat can be carried away easily.[12] In comparing the presence or absence of ground planes in the next-level board, Compaq has reported thermal-resistance improvements of up to a factor of two by having a ground plane. An additional ground plane tended to reduce the thermal resistance measurements by another 5°C/W.[7]

Again, the package designer must balance improvements in thermal performance with overall system benefits and costs. For example, it may be easy to aggressively assign too many solder balls to power or ground for better thermal coupling to the next-level board. Unfortunately, this action can also limit the number of solder balls available for the I/O. Another concern for the designer is to ensure that not too many closely spaced packages are dependent on the system (or next-level board) ground plane to provide heat conduction away from these packages. Although this potential problem should be manageable in most situations, it must be evaluated at the system level and could affect the overall cooling strategy.

12.3 Electrical Management of Plastic BGAs

12.3.1 Introduction

The electrical performance issues for plastic ball grid array (PBGA) packages are discussed in this section. First the electrical parameters that affect the package performance are briefly reviewed, and then the typical values of these parameters for conventional PBGAs and enhanced BGAs are listed. Some comparisons between quad flat package (QFP) and pin grid array (PGA) electrical performance specifications and BGAs are also included in these discussions.

Many concerns about BGAs have been raised regarding package reliability and thermal performance, but even less published data is available regarding the electrical performance of these packages. However, the simplicity of the conventional single-layer OMPAC-like BGAs allows one to use back-of-the-envelope closed-form equations to arrive at first-order electrical parameters. In the meanwhile, the construction of enhanced BGAs is very similar to cavity-down plastic pin grid arrays

(PPGAs) and there have been numerous publications on their electrical performance. The main difference between enhanced BGAs and PPGAs is that the 3-mm pins are replaced with 0.8-mm to 0.9-mm solder bumps.

As mentioned earlier in Sec. 12.2 there are several configurations of PBGAs. The first-level interconnection between the chip(s) and the BGA substrate can be direct-chip attached (DCA), tape automated bonded (TAB), or wire-bonded. The overall performance in this case is dependent on the overall net length, conductor geometry, and the location of current return path. For example, in the simplistic OMPAC-like layout, the conductor traces are routed to the periphery of the top layer and through a host of through-hole connections for connection to the redistribution bottom layer. A typical net from the IC chip to the next layer interconnect board (see Fig. 12.8) consists of a wire bond, substrate conductive trace (top-layer), through-hole via, redistribution conductor layer, and the solder bump. In the enhanced face-up configuration, a pair of power and ground planes could be designed into the interconnect substrate to allow for lower inductance net and more controlled signal propagation characteristics.

12.3.2 Electrical analysis methodology

The electrical analysis of a package needs to be addressed from a system's point of view to fully appreciate its performance effects. However, in this chapter, a very simplistic approach has been taken to discuss the typical electrical parameters of PBGA packages. While the methodologies may vary, these characteristics are derived from the same fundamental set of circuit parameters (i.e., resistance, inductance, capacitance—*RLC*). Separate measurements or modeling techniques may be used to indepen-

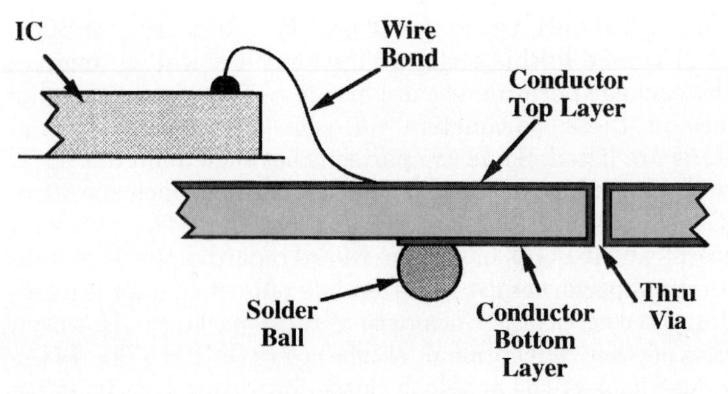

Figure 12.8 A typical net connection in a PBGA which consists of a wire-bond connection from an IC to the BGA substrate, the conductive trace along the substrate, and the solder ball.

dently confirm the listed data. The PBGA package performance needs to be studied from two points of view: signal fidelity and power distribution integrity. Signal fidelity is analyzed in the context of propagation delay and crosstalk noise, while power distribution integrity is discussed from inductance of power network and how it can be minimized in a PBGA design.

Three electrical parameters—resistance, capacitance, and inductance—are always present in packaging and interconnect systems. Resistances can cause DC drops on signal lines and contribute to *RC* charging delays, both of which are unwanted effects. Resistances, however, often help to dampen unwanted noise in systems. Capacitance is the primary cause of delays in packaging and interconnect systems. Capacitive loading of signal lines is decreased by reducing net lengths. Finally, inductance contributes to switching noise and time-of-flight delays in packaging systems. These basic circuit parameters define the signal fidelity in a BGA package and contribute to the overall power distribution integrity. The definitions for each term affecting the electrical performance of a typical plastic BGA package follow, and the predicted values of these parameters for a 225 I/O BGA are summarized in Table 12.2.

Propagation delay. Propagation delay is the time delay for a signal edge to propagate along a given length of interconnect, measured at the 50 percent signal transition level. In digital signal transmission applications, propagation delay is perhaps the most significant composite characteristic. Many factors contribute to the propagation delay of a signal along a trace. Different components to the propagation

TABLE 12.2 Typical Package Features and Electrical Parameters Comparing a 225-I/O BGA with a 208-I/O QFP Package[2]

Attribute	225 I/O BGA	208 I/O QFP
Size (mm^2)	27 × 27	31 × 31
Pitch of out lead (mm)	1.5	0.5
Shortest lead (mm)	6.8	10.2
Longest lead (mm)	22.5	13.7
Smallest L (nH)	3.3 to 5.8	6.3 to 7.1
Largest L (nH)	10 to 11.2	8.9 to 9.8
Inductive coupling (K_L)	0.12 to 0.25	0.68 to 0.70
Z_O (Ω)	60 to 85	—
Delay (psec)	5.02 to 9.07	9 to 14.5
Capacitance (pF)	1.28 to 1.31	1.61 to 2.38
R_{DC} (mΩ)	20 to 24	70 to 80
ΔI-noise (mV)	550	730
Crosstalk (mV)	90	510

NOTE: This data corresponds to the packages only and no chip-to-package interconnect data is included.

delay, ranked in decreasing order of their importance, are time of flight, *RC* charging, and delays associated with noise (including reflections, crosstalk, and switching noise). The packaging delay depends not only on the length of the nets but also the distribution of loads on the nets, the size of the loads, and the characteristics of the circuits driving the nets. Finally, the signal delay varies as a function of frequency—at slower frequencies the *RC* charging delays dominate, and at high frequencies the time of flight ($\sqrt{LC}$) dominates the delay.[13] The speed of signal propagation is inversely proportional to the square root of the dielectric constant (ε_r) of the dielectric medium. For typical laminate materials used in PBGA substrates (with $3.5 < \varepsilon_r < 4.9$) a delay of 62 to 74 psec/cm is expected.

Crosstalk. Crosstalk is defined as the coupled noise from active lines onto quiet lines due to mutual capacitive and inductive coupling. Crosstalk increases as the distance between lines (pitch) decreases, the distance between the lines and their circuit return path (ground plane, for example) increases, and the length of their close proximity (coupled length) increases. Coupled transient analysis of transmission lines can be used to simulate the crosstalk noise. Capacitive coupling ($K_C = -C_m/C_s$) and inductive coupling ($K_L = L_m/L_s$) between adjacent lines add at the driving end and subtract at the far end of the line. C_m and L_m are the mutual capacitance and mutual inductance between adjacent leads, respectively. Likewise, C_s and L_s are the self-capacitance and self-inductance of an active signal line, respectively. A low dielectric constant material is desired for reducing the effective crosstalk in a PBGA package. Alternatively, inclusion of a ground trace between adjacent signal lines reduces the crosstalk by approximately 50 percent. Critical nets can be designed in a stripline format by taking advantage of an internal power or ground plane. This allows PBGA packages to be used for very high end applications which demand low crosstalk.

Switching noise. A switching event causes current change in the supply network which delivers energy to the chip. When this transient current is passed through a power distribution network, a noise voltage is produced, which is referred to as switching, or delta-I, noise. Power distribution systems represent a source of coupling between the circuits that depend on characteristic impedance of the distribution system. The magnitude of the noise voltage depends on the current slew rate and the amount of inductance in the current path as shown in Eq. (12.1). The duration of the noise disturbance depends primarily on the impedance of the power distribution network. Switching noise limits the achievable system performance and can induce a number of significant problems in digital systems if it is not handled correctly.

The transient noise generated by simultaneous switching of n output drivers is expressed by

$$\delta v = \frac{L\, n\, \delta i}{\delta t} \tag{12.1}$$

where δv = amplitude of noise
L = inductance of interconnect
δi = current demand per driver at the time of switching
δt = rise (fall) time of the signal

Switching (or ΔI) noise for conventional PBGAs is approximately 10 percent of the swing voltage. Conventional single-layer PBGAs inherently have some long routing paths which result in highly inductive traces. In this case, the allocation of power and ground pads is essential to minimize the overall net (trace) length which would in turn allow small switching noise voltage in the power distribution network.

First-level interconnect concerns. The first-level interconnect between an IC and the PBGA substrate is generally through gold-wire bonding. However, solder flip-chip bonded parts are also used in ≤169 I/O BGA packages. Flip-chip bonding offers the shortest interconnect lengths and highest signal fidelity. The mutual coupling between the interconnects is small due to very short electrical paths. Typical solder bumps may be 150 to 200 μm in diameter. Wire bonds introduce a transmission discontinuity for the signals traversing the integrated circuit boundary. The inductance of thin wires, of diameter 0.025 to 0.032 mm (0.001 to 0.00125 inches) and length 1.9 to 2.5 mm (0.075 to 0.100 inches), is on the order of 2 nH. Wire bonding can be performed down to a wiring pitch of 0.1 mm (0.0045 inches). The inductive mutual coupling between the wires is significant (1 nH cumulative mutual inductance and 1 pF of mutual capacitance). The inductance terms scale logarithmically with distance, so their coupling values do not drop very rapidly as one moves away from a signal line. It is necessary to have ground wires placed among every few signal wires in order to keep the crosstalk within reasonable bounds.

Effect of BGA design on electrical performance. The electrical parameters of a 225 I/O BGA package are given in Table 12.2. In the single-layer implementation of a BGA (e.g., Motorola OMPAC design), a metal layer for routing is located on both the top and bottom sides of the substrate. These two metal layers are electrically connected through vias at the perimeter of the substrate and there are no metal layers inside the substrate. There usually is a die attach pad that may have a host of thermal vias connecting it to the bottom layer of the substrate.

Inductance of a package may be the single biggest factor affecting the overall performance of a BGA package, especially in a high-frequency (≥50 MHz) system. High-frequency inductance measurement of a 225 I/O OMPAC shows that the longest trace has an inductance of nearly 11 nH, and the shortest trace's inductance is 1 nH with ≈51 percent of all the traces having an inductance of ≤6 nH.

12.3.3 PBGA Performance Comparison versus QFP

The results of a comparison between a 208-lead quad flat package (QFP) and a 225-I/O PBGA are presented in Table 12.2.[2] Neither the QFP nor the PBGA in this study has any type of electrical enhancement such as built-in voltage planes in the package structure. The data in Table 12.2 shows that the PBGA has a smaller per unit inductance than the corresponding QFP package.

The BGA package, even in its simplest configuration (i.e., single-layer OMPAC), exhibits very favorable electrical performance specifications. Due to the circuit structure of the conducting traces (see Fig. 12.9), each conductive trace is partly routed on the top side and partly routed on the bottom side of the substrate. Therefore, the two segments of the same trace have close coupling.

The overall inductance of a given trace is determined by Eq. (12.2):

$$L_{tot} = L_1 + L_2 - 2M_{12} + L_{VIA} \tag{12.2}$$

where L_1 corresponds to the inductance of the top-layer conductor, L_2 is the inductance of the bottom-layer conductor, M_{12} refers to the mutual coupling between the two segments, and L_{VIA} is the inductance of the through-hole via.

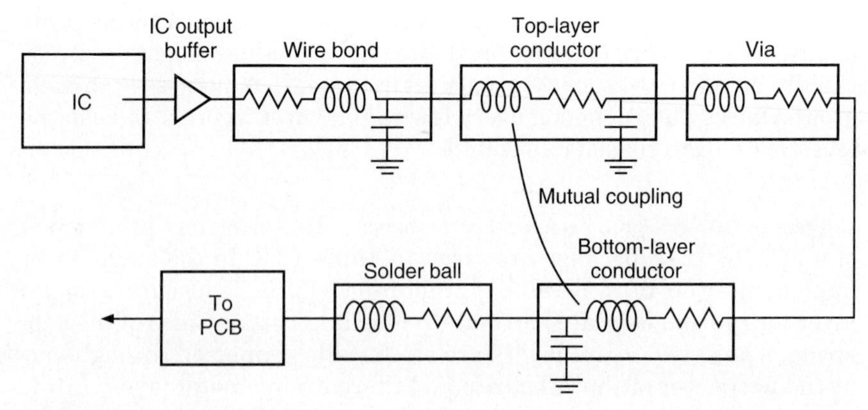

Figure 12.9 An equivalent circuit model corresponding to a typical net in the PBGA illustrated in Fig. 12.8.

In an outgoing signal, the current flows outward along the top-layer conductor, and flows inward along the bottom-layer conductor. Due to opposing magnetic fields generated by the two different current flows, the overall conductor inductance is reduced. The stronger the coupling between the top-layer conductor and the bottom-layer conductor, the larger the M_{12} term [Eq. (12.2)] and the smaller the L_{tot} term will be. Motorola has shown that if the traces were to be simply routed directly from a bonding pad on the top layer through a via to the bottom layer and routed to the solder bump, the overall net length would be one-third of the conventional routing and the inductance would be less than one-fourth the inductance in the conventional case.[15]

Mutual inductive coupling between adjacent lines is primarily a function of geometry, but it is also affected by the magnetic fields due to nearby signals. The coupling between the top-layer component of one signal trace and the bottom-layer component of another signal trace causes opposing magnetic fields that partially reduce the coupling between adjacent traces on the same layer. The manufacturable substrate design rules also dictate the routing limitations between the solder pads, thus minimizing the number of adjacent traces running in parallel with one another, which effectively reduces the overall coupling length.

In cavity-up (chip face-up) BGA configurations with built-in power and ground planes, the signal routing on the top layer to a periphery and redistribution on the bottom layer may not change. However, the electrical parasitics are substantially different due to the presence of such voltage planes. Referring to Eq. (12.2), the self-inductance terms L_1 and L_2 are much smaller (microstrip line or buried microstrip line with a solder mask) as the current return path is across a very low-inductance plane. Compaq has reported that the loop inductance of the ground or power path in a 225-I/O BGA package with built-in planes is 130 to 170 pH, respectively.[9]

In the enhanced BGA configurations, a through-hole via is placed near a bonding pad routing the signal through a very short path from the IC to the solder bump. In this case, the overall net inductance is reduced simply by shortening the net length. BGA packages with 300 to 500 I/O are under considerations that are geared for 100- to 200-MHz computing applications. The enhanced BGA packages with solder balls on 1.27-mm or 1.5-mm pitch simply allow multiple voltage and signal planes to be laminated in a donut-shaped laminate structure where all the traces are routed in a controlled-impedance manner. For the very high end processors with sub-nanosecond edge rates, the ball pitch of 1.27 mm is being used to minimize the overall substrate size. This way the microstrip or stripline structures are very short and package parasitics are minimized. When 32- or 64-bit wide drivers switch simultaneously and cause significant switching noise problems, the low inductance of the enhanced BGAs simply minimizes these effects.

12.4 Summary

Overmolded plastic BGAs are being touted as a low-cost, easy-to-manufacture, and easy-to-implement package. With no additional cost-related requirements, they fit very well in many low-power, high-I/O applications. As device engineers explore using these packages for their high-performance applications, the thermal limitations of these overmolded BGAs are quickly encountered.

With little cost penalty, users of plastic BGA packages have been able to substantially increase the thermal performance of this class of packages by optimizing the package and next-level board design. Package modifications include the intelligent design of vias and power and ground planes to serve as centralized thermal vias with heat-spreading layers. By coupling a BGA with thermal vias to a next-level board that also has thermal vias and heat-spreading layers, additional improvement is made to the overall thermal performance.

Another approach is being investigated that dramatically improves thermal performance. By incorporating a copper slug into the package and mounting the chip directly to this slug, the thermal resistance from the chip to the outside of the package is minimized. These enhanced BGAs should be able to dramatically increase the power capabilities of this class of package, especially with the addition of a heat sink.

Of course, each of these package improvements comes with a cost increase, which must be considered early in the product-design cycle.

In general, the electrical performance of PBGA packages, like any other semiconductor package type, needs to be studied from a system point of view. The PBGA construction allows the use of this family of semiconductor packaging from the low-end consumer electronics to the very high-end computing machines with 100- to 200-MHz processors. In an isolated one-to-one comparison, a single-layer PBGA has similar or better performance than a comparable leadcount QFP. The packaging technology choice has a very strong effect on delay and noise, and simulation and modeling are needed to accurately predict the performance of a PBGA in a system. Nevertheless, the PBGA electrical performance can be improved significantly by optimizing the layout of the module with the intent of shortening the overall trace length. PBGAs with built-in power and ground planes have better electrical performance than conventional single-layer PBGAs and, to maximize their performance, the signal routing should be optimized. Finally, the enhanced BGA packages with solder balls on 1-mm or 1.27-mm pitch are suitable packaging options for advanced personal computers and workstation applications.

12.5 References

1. Vardaman, J., "BGA Manufacturability & Market Trends," *High-Density Packaging User Group,* October 1993.
2. Walshak, D. and H. Hashemi, "Thermal Modeling of a Multichip BGA Package," *Proc. NEPCON West,* March 1994, pp. 1266–1276.
3. SEMI specification G42-88, "Thermal Test Board Standardization for Measuring Junction to Ambient Thermal Resistance of Semiconductor Packages."
4. Sloan, J., V. Nomi, and H. Wilson, "Over Molded Pad Array Carrier (OMPAC): A New Kid on the Block," *Proc. First VLSI Packaging Workshop of Japan,* November 1992, pp. 17–27.
5. Marrs, R., B. Freyman, and J. Martin, "High Density BGA Technology," *Proc. International Conference and Exhibition on Multichip Modules,* April 1993, pp. 326–329.
6. Groover, R., C. Huang, and A. Hamzehdoost, "BGA—Is It Really the Answer?" *1994 ITAP and Flip Chip Proceedings,* February 1994, pp. 57–64.
7. Johnson, R. and D. Cawthon, "Thermal Characterization of 140 and 225 Pin Ball Grid Array Packages," *Proc. NEPCON East,* June 1993, pp. 423–430.
8. Hashemi, H. and D. Walshak, "Enabling Technologies for Few-chip Packaging—A Study in Thermal Performance," *Proc. 1993 ASME Winter Meeting,* November 1993.
9. Johnson, R., A. Mawer, T. McGuiggan, B. Nelson, M. Petrucci, and D. Rosckes, "A Feasibility Study of Ball Grid Array Packaging," *Proc. NEPCON East,* June 1993, pp. 413–422.
10. Holman, J., *Heat Transfer,* McGraw-Hill, 4th ed., 1976, p. 235.
11. Wyland, C., "The Effect of PCB Thermal Conductivity on Package Thermal Resistance," *Proc. 1993 International Electronics Packaging Conference,* September 1993, pp. 90–94.
12. Costlow, T., "Ball grid arrays: the hot new package," *Electonic Engineering Times,* March 15, 1993, p. 35.
13. Doane, D. and P. Franzon, *Multichip Module Technologies and Alternatives, The Basics,* Van Nostrand Reinhold, 1993, pp. 525–567.
14. Katopis, G., "Delta-I Noise Specification for a High-performance Computing Machine," *Proc. IEEE,* vol. 75, no. 9, Sept. 1985, pp. 1405–1415.
15. Yip, W. and C. Tsia, "Electrical Performance of an Overmolded Pad Array Carrier (OMPAC)", *Proc. of IEPS,* Sept. 1993, pp. 731–739.

Chapter

13

Reliability of Plastic Ball Grid Array Assembly

Robert Darveaux, Kingshuk Banerji, Andrew Mawer, and Glenn Dody

13.1 Introduction

One of the most important aspects of PBGA technology is the controlled collapse solder joint. Historically, leadless packages have had serious limitations with respect to size and interconnection count because of solder joint reliability concerns.[1–8] However, it is becoming clear that PBGA packages have attachment reliability that rivals their leaded counterparts. This is primarily due to relatively large joint height and to package materials which are nearly expansion-matched to the product's printed circuit board. This chapter is devoted exclusively to PBGA solder joint reliability. Delamination and popcorn failure is another reliability issue related to PBGA technology, but it is discussed separately in Chap. 10 on PBGA assembly.

This chapter provides extensive data, analytical techniques, and a generalized methodology for BGA solder joint reliability assessment. The failure modes discussed are creep rupture, interface failure, and thermal fatigue. With the tools provided in this chapter, an engineer will be able to predict solder joint failure under a wide range of conditions.

Mechanical testing was used to characterize the solder behavior and develop constitutive relations which predict strain as a function of stress, temperature, and time. All creep tests were conducted on actual solder joints to ensure that the exact microstructure in real products was duplicated. Both shear and tension tests were conducted in the temperature range from 25 to 134°C, and at strain rates from 10^{-8} sec^{-1} to 10^{-1} sec^{-1}. The alloys studied were 62Sn36Pb2Ag, 60Sn40Pb, 96.5Sn3.5Ag, 97.5Pb2.5Sn, 100In, and 50In50Pb. A single set of constitutive relations fits the data for all of these alloys; only the constants are alloy-dependent. Creep rupture design guidelines were also established for 62Sn36Pb2Ag solder joints. It was found that dispersed intermetallics and constraint at interfaces make actual joints significantly more creep-resistant than bulk solder specimens.

Crack growth and thermal fatigue studies were conducted on BGA assemblies under several different conditions. The temperature range was varied from –55°C <-> 125°C to 25°C <-> 75°C, and the cyclic frequency was varied from 144 cycles/day to 2 cycles/day. Both ceramic and plastic BGA carriers were studied, but only near-eutectic tin-lead balls were utilized (no CBGA data with 90Pb10Sn balls are reported in this chapter). Crack growth was measured by removing three chip carriers from the sample population at regular intervals during thermal cycling. The number of cycles to crack initiation and the crack growth rate per cycle were both correlated with viscoplastic strain energy density by using nonlinear finite element analysis. A fatigue life model was formulated based on these correlations, and it was successfully applied to a wide range of data from BGA assemblies. Finally, the model was used for field reliability assessment.

13.2 Constitutive Relations

Since solder is above half of its melting point at room temperature, creep processes are expected to dominate the deformation kinetics. Steady-state creep is generally expressed by a relationship of the form[9,10,11]

$$\frac{d\gamma_s}{dt} = C_4 \frac{G}{T}\left[\sinh\left(\alpha \frac{\tau}{G}\right)\right]^n \exp\left(\frac{-Q}{kT}\right) \tag{13.1}$$

where $d\gamma_s/dt$ is the steady-state strain rate, G is the shear modulus, k is Boltzmann's constant, T is the absolute temperature, τ is the applied stress, Q is the activation energy for the deformation process, n is the stress exponent, α prescribes the stress level at which the power law dependence breaks down, and C_4 is a constant characteristic of the underlying micromechanism. The stress exponent, n, is dependent on the rate-controlling mechanism. At low stresses, $n = 1$ for diffusional creep, and $n = 2$ for grain boundary sliding (superplasticity). At intermediate stresses, $n = 3$ to 4 for dislocation glide-controlled kinetics, and $n = 5$ to 7 for dislocation climb processes. To obtain the true activation energy, the temperature dependence of the shear modulus must be incorporated

$$G = G_o - G_1(T - 273) \tag{13.2}$$

where G_o is the modulus at 0°C (273K), and G_1 gives the temperature dependence.

A simplification of Eq. (13.1) is often used to describe steady-state creep

$$\frac{d\gamma_s}{dt} = C_5[\sinh(\alpha_1\tau)]^n \exp\left(\frac{-Q_a}{kT}\right) \tag{13.3}$$

where Q_a is the apparent activation energy.

Steady-state creep is not generally achieved immediately when stress is applied. A certain amount of transient (or primary) creep occurs before attaining steady state. For normal decelerating transient creep, the strain rate starts high and decreases to the steady state value as the material work hardens. Transient creep at constant stress and temperature can be described by[12]

$$\gamma_c = \frac{d\gamma_s}{dt}\,t + \gamma_T\left(1 - \exp\left(-B\frac{d\gamma_s}{dt}t\right)\right) \tag{13.4}$$

where γ_c is the creep strain, $d\gamma_s/dt$ is the steady state creep rate, γ_T is the transient creep strain, and B is the transient creep coefficient. Taking the time-derivative of both sides yields

$$\frac{d\gamma_c}{dt} = \frac{d\gamma_s}{dt}\left(1 + \gamma_T B \exp\left(-B\frac{d\gamma_s}{dt}t\right)\right) \tag{13.5}$$

where $d\gamma_c/dt$ is the instantaneous creep rate and $d\gamma_s/dt$ is the steady-state creep rate. Hence at $t = 0$, the instantaneous creep rate is $(1 + \gamma_T B)$ times greater than the steady-state creep rate, and after a long time, the instantaneous rate is equal to the steady-state rate. At high stresses, $\tau/G > 10^{-3}$, there is also a time-independent plastic strain component to the deformation. The following strain-hardening law can be used to describe high-stress deformation

$$\gamma_p = C_6\left(\frac{\tau}{G}\right)^m \tag{13.6}$$

where γ_p is the time-independent plastic strain, and C_6 and m are constants. The total inelastic strain is given by the sum of creep strain and plastic strain

$$\gamma_{in} = \gamma_c + \gamma_p \tag{13.7}$$

where γ_{in} is the total inelastic strain. Equations (13.1) and (13.3) to (13.7) can also be written in their tensile forms

$$\frac{d\varepsilon_s}{dt} = C_{4t}\frac{G}{T}\left[\sinh\left(\alpha_t\frac{\sigma}{G}\right)\right]^n \exp\left(\frac{-Q}{kT}\right) \tag{13.8}$$

$$\frac{d\varepsilon_s}{dt} = C_{5t}[\sinh(\alpha_{1t}\sigma)]^n \exp\left(\frac{-Q_a}{kT}\right) \tag{13.9}$$

$$\varepsilon_c = \frac{d\varepsilon_s}{dt} t + \varepsilon_T \left(1 - \exp\left(-B_t \frac{d\varepsilon_s}{dt} t \right) \right) \tag{13.10}$$

$$\frac{d\varepsilon_c}{dt} = \frac{d\varepsilon_s}{dt} \left(1 + \varepsilon_T B_t \exp\left(-B_t \frac{d\varepsilon_s}{dt} t \right) \right) \tag{13.11}$$

$$\varepsilon_p = C_{6t} \left(\frac{\sigma}{G} \right)^m \tag{13.12}$$

$$\varepsilon_{in} = \varepsilon_c + \varepsilon_p \tag{13.13}$$

Note that the shear modulus is used in Eqs. (13.8) and (13.12), not Young's modulus. The following relations can be derived based on application of von Mises yield criteria for effective strain and effective stress[13]

$$\sigma = \tau\sqrt{3} \tag{13.14}$$

$$\varepsilon = \frac{1}{\sqrt{3}} \gamma \tag{13.15}$$

Based on Eqs. (13.14) and (13.15), the conversion factors between shear and tensile deformation constants are given below:

$$C_{4t} = \frac{1}{\sqrt{3}} C_4 \tag{13.16}$$

$$\alpha_t = \frac{1}{\sqrt{3}} \alpha \tag{13.17}$$

$$C_{5t} = \frac{1}{\sqrt{3}} C_5 \tag{13.18}$$

$$\alpha_{1t} = \frac{1}{\sqrt{3}} \alpha_1 \tag{13.19}$$

$$\varepsilon_T = \frac{1}{\sqrt{3}} \gamma_T \tag{13.20}$$

$$B_t = \sqrt{3} B \tag{13.21}$$

$$C_{6t} = \left(\frac{1}{\sqrt{3}} \right)^{m+1} C_6 \tag{13.22}$$

13.3 Mechanical Characterization of BGA Solder Joints

All of the data were collected on actual soldered assemblies to properly account for the effects of grain size and intermetallic compound distribution. Six solder alloys have been characterized to date: 62Sn36Pb2Ag, 60Sn40Pb, 97.5Pb2.5Sn, 96.5Sn3.5Ag, 100In, and 50In50Pb. The Controlled Collapse Chip Carrier Connection (C5) test samples were formed by soldering two identical ceramic chip carriers together using an array of 0.030-in-diameter solder spheres. The chip carriers were metallized with sputtered Cr/Cu followed by plated Cu(1000 μin)/Ni(50 μin)/Au(20 μin), with a limited number of samples being plated Cu with an Entek Plus coating. For all C5 samples, the soldering conditions were as follows: (1) RMA flux, (2) peak temperature 25–40°C above liquidus, (3) time above liquidus <60 sec, and (4) 2 reflows, bumping, and joining. The indium die bond samples were made by soldering together two metallized silicon chips 0.394 in in size, and the metallizations were Cr/Ni, Cr/Ni/Cu, Cr/Ni/Pd, and Cr/Ni/Au.[4,14]

Once the solder joint samples were fabricated, they were subjected to thermal aging. The aging conditions were either several months at room temperature, 100 hrs at 100°C, or 24 hrs at 125°C. After aging, the samples were bonded to steel rods using a cyanoacrylate adhesive for tests up to 100°C and a one-component epoxy for tests above 100°C. The C5 test samples and fixtures are shown schematically in Fig. 13.1. A double lap shear configuration was used to minimize bending. A Schaevitz 025 MHR LVDT was used to measure displacement with a resolution of 1 μin. All tests were conducted in a screw-driven INSTRON 4501 with either an 1100-lb or a 22-lb load cell. LabVIEW data acquisition software was used to control the INSTRON and to record load and displacement readings. The temperature chamber was capable of maintaining a constant temperature within ± 1°C or better. The indium die bond fixtures and test procedures are described in Refs. 4 and 14.

Constant crosshead displacement rate tests were used for most of the data above 10^{-4} sec^{-1}, and constant load creep tests were used below 10^{-4} sec^{-1}. Typical plots from these two types of tests are shown in Figs. 13.2 and 13.3. A limited number of load relaxation tests were run in which the crosshead was stopped after a steady-state condition had been reached. However, this test was not very reproducible because of temperature fluctuations and the relatively long load train (approximately 2 ft). Short, stiff load trains and tight temperature control are generally required to minimize thermally induced displacements during load relaxation tests.

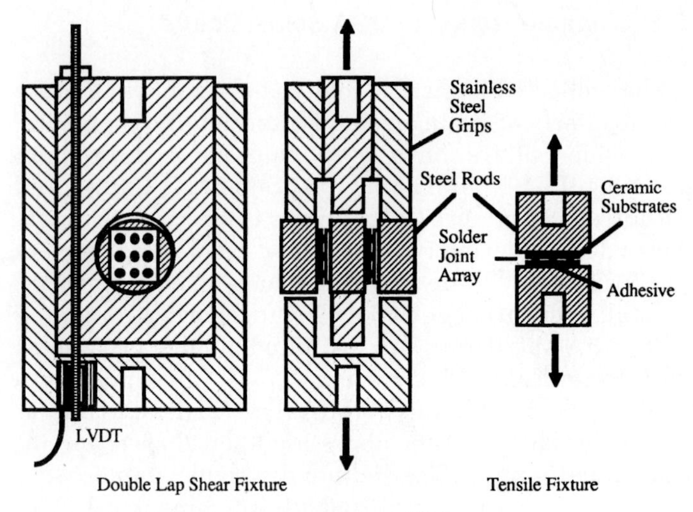

Figure 13.1 Mechanical test samples and fixtures.

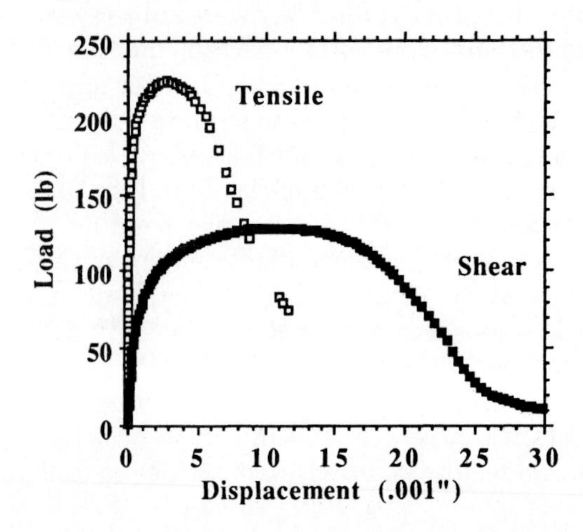

Figure 13.2 Typical load-displacement plot for constant displacement rate tests.

Testing actual solder joints is the best way to account for microstructural effects in the solder. However, it does introduce some experimental error compared to testing bulk samples. First, the cross-sectional area of a controlled collapse joint varies slightly across its thickness, so the stress and strain distribution in the material will not be perfectly uniform. Second, the solder material is constrained at the interfaces with the substrate and/or IC chip, which also affects the stress and strain distribution. Nevertheless, an average shear stress and shear strain can be defined by the following relations

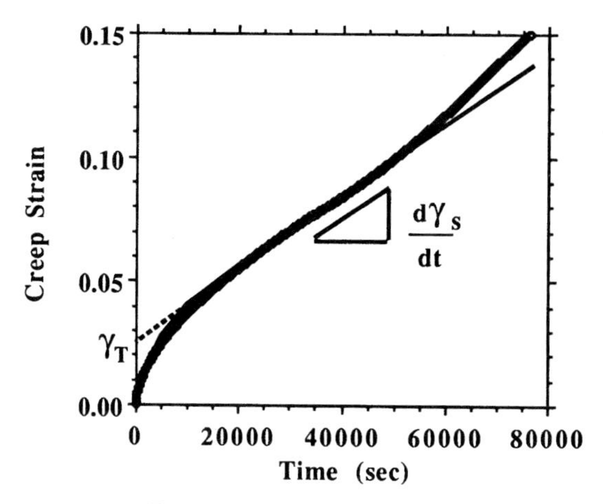

Figure 13.3 Typical creep-strain plot for constant load test.

$$\tau = \frac{P}{A} \tag{13.23}$$

$$\gamma = \frac{\delta}{h} \tag{13.24}$$

where P is the load, A is the pad area, δ is the displacement, and h is the joint height. It is very difficult to measure only displacement across a solder joint; typically displacements in the grip assembly are measured as well. In this study, all elastic displacements, both in the grips and in the solder, were subtracted out of the data. Hence, only inelastic solder displacements are used to calculate strain.

Under tensile loading, the true stress and strain are given by

$$\sigma = \frac{P}{A_o}\left(1 + \frac{\delta}{h_o}\right) \tag{13.25}$$

$$\varepsilon = ln\left(1 + \frac{\delta}{h_o}\right) \tag{13.26}$$

where σ is the true stress, ε is the true strain, A_o is the initial area, and h_o is the initial height. Equation (13.25) assumes that the solder volume remains constant during inelastic deformation as the height increases and the area decreases. Clearly, the cross-sectional area at the middle of the joint decreases during deformation. However, the area at the interfaces changes very little due to the constraining effect of the ceramic, which is not deforming inelastically. Hence, engineering stress might be a more representative quantity for solder joints under tensile loading

$$s = \frac{P}{A_o} \tag{13.27}$$

where s is engineering stress. All tensile data reported herein are given as engineering stress.

A trial-and-error curve-fitting procedure was used to determine the steady-state constants Q, Q_a, C_4, C_5, α, and α_1. The stress exponent, n, is prescribed by the lowest stress data. The steady state creep rate, γ_s, and the transient creep strain, γ_T, were determined from the slope and intercept of the linear part of a creep strain–time plot, as shown in Fig. 13.3. The transient creep coefficient, B, was determined by trial and error from the creep strain–time plot. The time-independent plastic strain constants, C_6 and m, were determined from the low strain portion of the highest strain rate, lowest temperature test.

13.3.1 Steady-state creep

A plot of temperature-normalized strain rate versus modulus-normalized stress is shown in Fig. 13.4 for 62Sn36Pb2Ag solder joints. Tests were run at approximately 27, 67, 100, and 132°C. Load relaxation data are indicated by open symbols. It is apparent that Eq. (13.1) gives a good fit to the data over 9 orders of magnitude in normalized strain rate. The power law breakdown is at approximately $\tau/G = 10^{-3}$, which is consistent with all of the other alloys tested and with most other metals.

The raw steady-state data for each alloy are plotted in Figs. 13.5 to 13.10. In each figure, the best fit curves using Eq. (13.1) are also plotted. It is seen that Eq. (13.1) gives a good fit to all of the data, only the

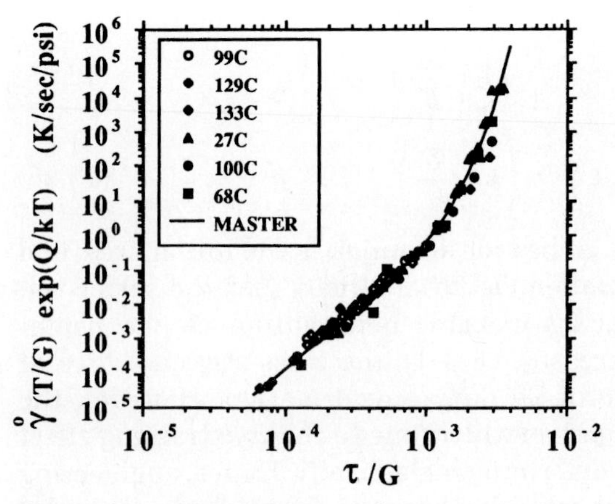

Figure 13.4 Normalized steady-state creep of 62Sn36Pb2Ag solder joints.

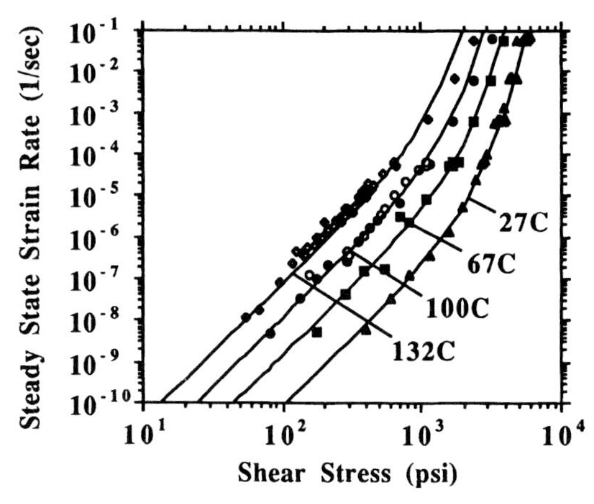

Figure 13.5 Steady-state creep of 62Sn36Pb2Ag solder joints with Cr/Ni/Au pad metallization.

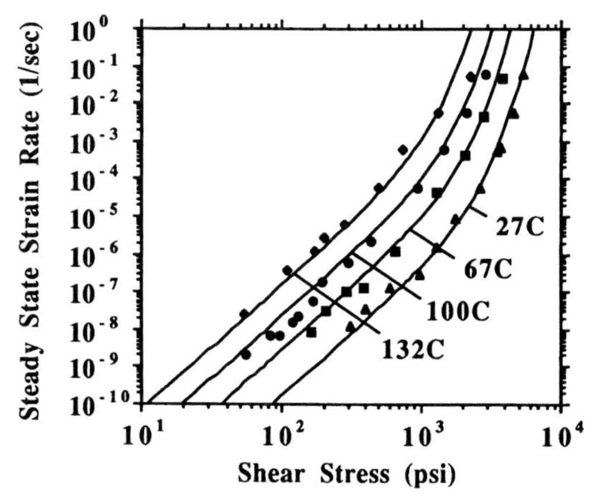

Figure 13.6 Steady-state creep of 60Sn40Pb solder joints with Cr/Ni/Au pad metallization.

constants are alloy-dependent. There is a slight discrepancy between the die bond data and the C5 joint data for 100In in Fig. 13.10. The C5 joints appear to be more creep-resistant, but they probably are not in reality because the die bond data was not taken under steady-state conditions. Instead, the die bond data was taken at a strain of 0.05. The steady-state curves would all be shifted to the right on the stress axis, especially for the highest stress conditions in which strains of 0.50 are needed to reach steady state. Therefore, the model fit is only strictly valid for strains up to 0.05 for the 100In data. Fortunately, the vast majority of applications fall into this regime.

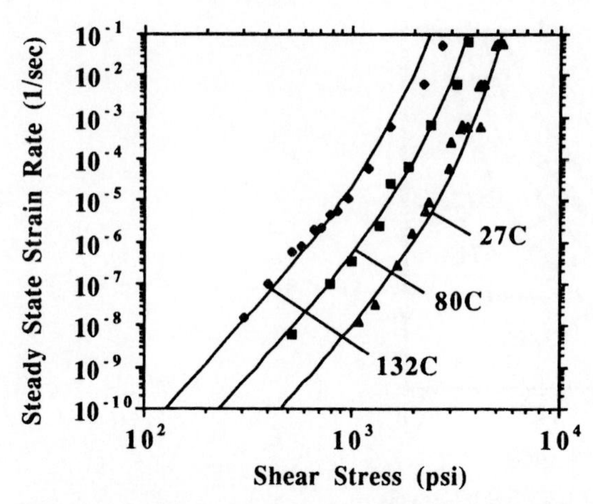

Figure 13.7 Steady-state creep of 96.5Sn3.5Ag solder joints with Cr/Ni/Au pad metallization.

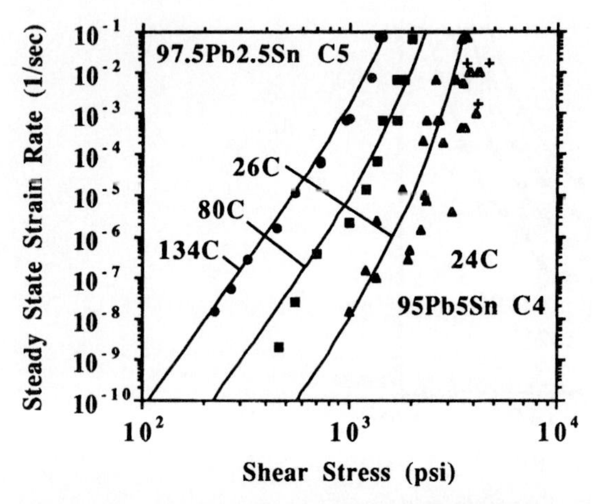

Figure 13.8 Steady-state creep of 97.5Pb2.5Sn with W/Ni/Au pads and 95Pb5Sn with Cu/Ni/Au pads.

Pad metallization can also affect creep behavior of solder joints due to precipitation strengthening from dispersed intermetallics[4,14–16] and due to altering the alloy grain structure.[17] Shown in Fig. 13.11 are creep data for indium die bond joints with various pad metallizations. The Cr/Ni/thick Au samples (which had the highest volume fraction of dispersed intermetallics) were the most creep-resistant, and the Cr/Ni samples (which had the lowest volume fraction) were the least creep-resistant. The effect is also seen for 62Sn36Pb2Ag C5 joints as shown in Fig. 13.12, where the samples with Cr/Cu/Ni/Au on one side and Cr/Cu on the other side are slightly more creep-resistant than the samples with Cr/Cu/Ni/Au or Cr/Cu on both sides.

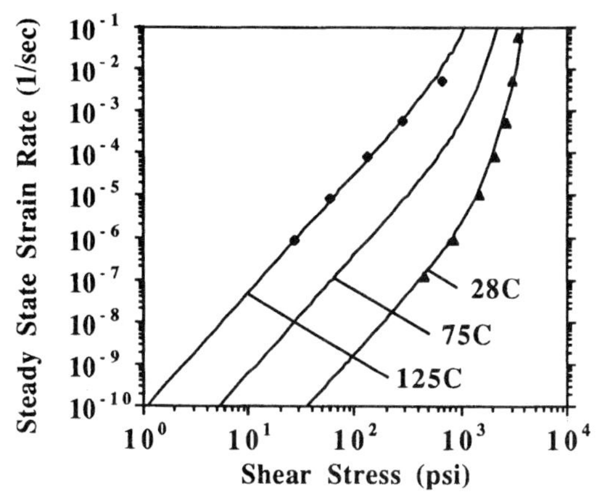

Figure 13.9 Steady-state creep of 50In50Pb solder joints with Cr/Ni/Au pad metallization.

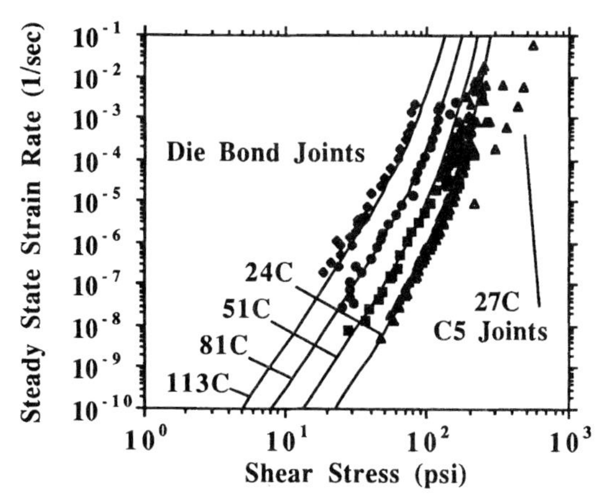

Figure 13.10 Creep of 100In solder joints with Cr/Ni, Cr/Ni/Cu, Cr/Ni/Pd, and Cr/Ni/Au pad metallization. Tests on die bond joints were taken to 0.05 strain, which was not steady state.

There is typically a correlation between high ductility and long life under low cycle fatigue conditions.[18] In the present study, strain at the onset of failure is defined as a figure of merit to compare solder alloy ductility, and it was taken at the saturation stress level during a constant displacement rate test, or at the departure from steady-state creep during a constant-load creep test. Strictly speaking, this is not exactly the same as "ductility," but it is similar. Furthermore, strain at the onset of failure probably gives a better figure of merit for shear loading because a joint can be totally failed and still provide resistance to deformation due to the fractured surfaces sliding across each other.

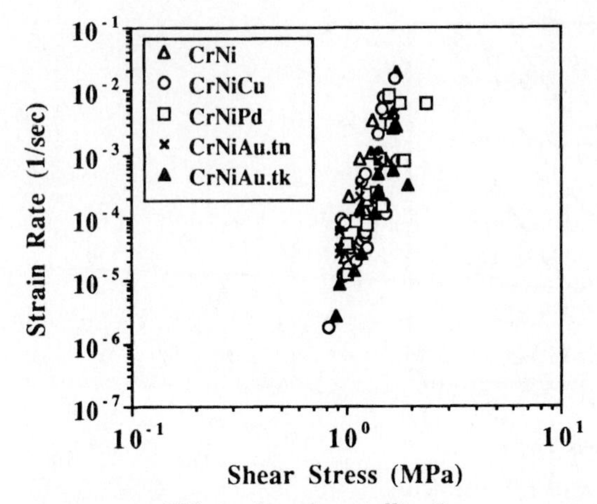

Figure 13.11 Effect of pad metallization on creep behavior of indium die bond joints, 24°C.

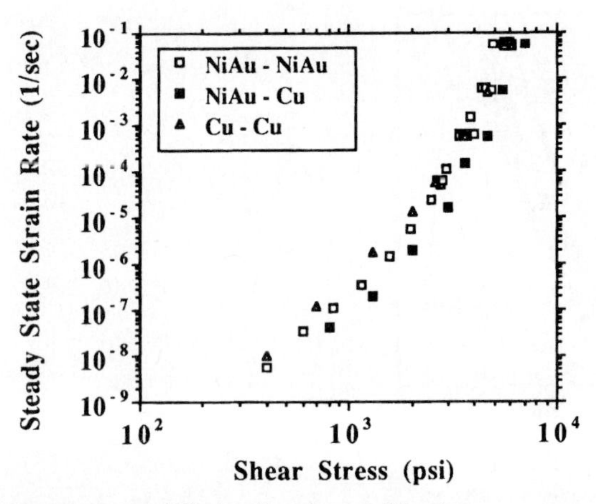

Figure 13.12 Effect of pad metallization on creep behavior of 62Sn36Pb2Ag C5 joints, 27°C.

However, a problem with this figure of merit could arise if a drop in load is caused by microstructural softening instead of a decrease in the load-bearing area. Fortunately, this was not the case in the present study. Cross-sectional analysis of samples after mechanical testing revealed that there was a decrease in load-bearing area immediately after departure from steady-state deformation had begun.

A plot comparing strain at the onset of failure is shown in Fig. 13.13. It is seen that the dislocation climb-controlled alloys, 97.5Pb2.5Sn and 96.5Sn3.5Ag, can absorb much more strain than the dislocation glide-controlled alloys, 60Sn40Pb and 62Sn36Pb2Ag. However, in all cases, the alloys absorb less strain at lower stress levels (higher temperature or lower strain rate). This observation is consistent with a strain range partitioning approach to cyclic fatigue damage, where the intent is to quantify the damage due to plastic flow versus the damage due to creep.[18,19] Typically, creep is considered more damaging. The 95Pb5Sn alloy C4 joints were the least ductile. There are two possible explanations for this: (1) extensive tin precipitation during the long room temperature aging before testing, and (2) the constraining effect at the interfaces for the small joint geometry (height ~ 0.002 in). Frost et al.[20] reported increased strength with tin precipitation in bulk lead tin alloys, but it is not clear whether a decrease in ductility was also observed. To make a true comparison with the other alloys, 95Pb5Sn should be tested in the C5 joint configuration.

13.3.2 Deformation constants

The best fit deformation constants for all of the alloys tested are shown in Table 13.1. It should be noted that some of these constants might be slightly different from those previously published,[21–23] because more data has been obtained since the earlier publications. Also, in Ref. 22 the shear stress was calculated based on the average joint cross-sectional area, instead of the pad area. It was later determined that the pad area gave a better correlation between different-shape joints, so the pad area is used now.

First, the elastic constants for Eq. (13.2) are given based on data from Refs. 2, 15, and 24 to 26. Next, the steady-state creep constants for Eqs. (13.1) and (13.3) are given from the data on joints under shear loading. The value of α is fairly consistent between all of the alloys to be near 1000. This results in a power law breakdown at $\tau/G = 10^{-3}$,

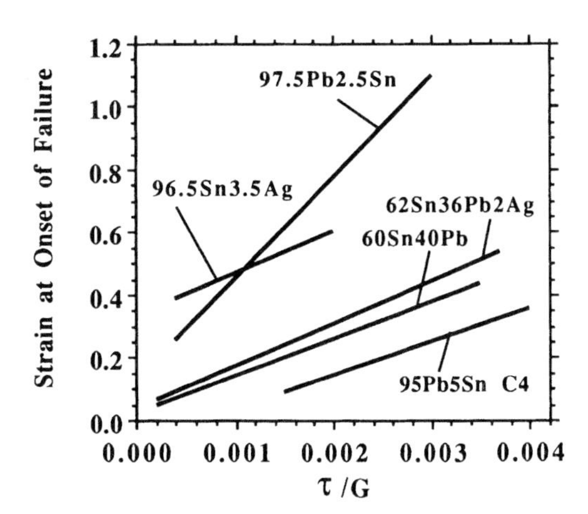

Figure 13.13 Strain at the onset of failure for C4 and C5 solder joints.

which is also consistent with most other metals. The only exception in the present study was the die bond data for 100In, that had $\alpha = 7500$. However, this data was taken at a strain level of 0.05, which is not steady state in the high stress regime, so the true value of α is probably less than 7500. This is evident in Fig. 13.10, where the C5 data (that was taken at steady state) does not correlate exactly with the die bond data.

The next set of constants under the steady-state creep heading are the derived tensile constants for Eqs. (13.8) and (13.9). These constants were derived from the shear constants using Eqs. (13.16) to (13.19), which assume a von Mises yield relation is obeyed.

Shown in Fig. 13.14 are steady-state creep data for 62Sn36Pb2Ag solder joints under shear or tensile loading at room temperature. The shear data have been converted to equivalent tensile data using Eqs. (13.14) and (13.15). It is seen that the von Mises conversions work well at strain rates above 10^{-4} sec^{-1}, but at lower strain rates the joints are more creep-resistant under tensile loading. Nonlinear finite element modeling was conducted to determine if this discrepancy is related to the constraining effect at the interfaces. Models of both the shear and tensile loading configuration were built using ANSYS 5.0, and the derived tensile constants in Table 13.1 were used in the simulations (Q_a, n, C_{5t}, and α_{1t}). In the shear configuration, the predicted displacement rate was slightly greater than the measured value for a given applied load. In the tensile configuration, the predicted displacement rate was less than the value derived from the shear data, but greater than the true measured value. Hence, the model indicates that there is a small amount of interface constraint under tensile loading, but not enough to explain the discrepancy between the converted shear and tensile data in Fig. 13.14. Hence, Eqs. (13.14) and (13.15) are not completely valid in the low stress regime.

The only alloy for which enough tensile data was taken to fit Eq. (13.9) was 62Sn36Pb2Ag. The measured tensile constants for this alloy are given in Table 13.1. As expected from the above discussion, the measured tensile constants do not agree with the tensile constants derived from the shear data. Once more tensile data is obtained, it should be possible to develop an alternative yield relation to the von Mises criteria in order to better correlate shear and tensile deformation behavior.

The last two sections of Table 13.1 give the constants for transient creep and time-independent plastic flow. In general, the alloys with a higher stress exponent, $n = 5$ to 7, also undergo more transient creep (higher γ_T), which is consistent with conventional creep theory. Alloys with a lower stress exponent, $n = 3$ to 4, exhibit much less transient creep, especially at low stress levels. Time-independent plastic flow

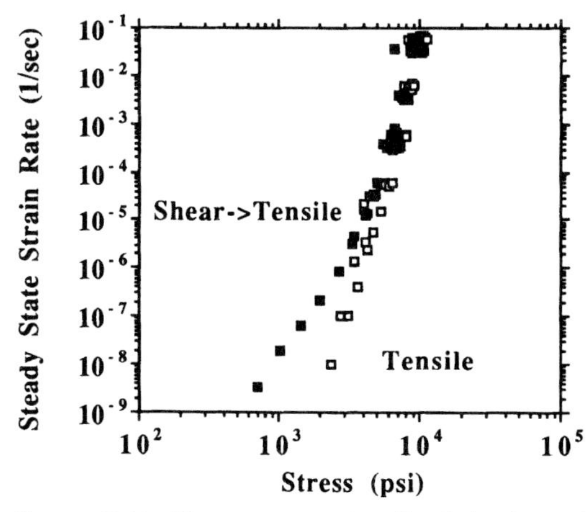

Figure 13.14 Shear versus tensile behavior of 62Sn36Pb2Ag C5 solder joints at 27°C. Shear data converted to equivalent tensile using Eqs. (13.14) and (13.15).

was observed in all of the alloys at high stresses, $\tau/G > 10^{-3}$. Also given in Table 13.1 are the derived tensile constants based on a von Mises flow criteria using Eqs. (13.20) to (13.22).

Most finite element codes have the capability to model both time-independent and time-dependent nonlinear deformation. There are standard input equations built into the code, as well as user-defined options where the code can be customized to accommodate any equation. Typically, the standard inputs are more convenient to use. Many finite element codes offer a sinh steady state creep relation such as Eq. (13.9) for time-dependent deformation, and stress-strain input tables for time-independent deformation. In general, the input data is required in terms of uniaxial tensile deformation, and a von Mises relation is used to calculate the effective stress and strain in the material. For time-independent deformation, various work-hardening options are available. Transient creep behavior can also be incorporated with most codes.

Based on the data presented in this chapter, several recommendations can be made with respect to finite element model deformation constant input. ABAQUS and ANSYS will be used as examples, since they are most familiar to the author. Neither ABAQUS nor ANSYS offers a sinh steady-state creep equation simultaneously with time-independent stress-strain input as a standard input. ANSYS offers a sinh equation for viscoplastic elements, but it cannot be used simultaneously with time-independent plastic flow with these elements. For SOLID45 3-D brick elements in ANSYS, creep and time independent

TABLE 13.1 Solder Alloy Deformation Constants

	60Sn40Pb	62Sn36Pb2Ag	96.5Sn3.5Ag	97.5Pb2.5Sn	100In	50In50Pb
Elastic						
G_o (Mpsi)	1.9	1.9	2.8	1.3	0.64	0.84
G_1 (kpsi/K)	8.1	8.1	10.0	1.5	2.0	1.8
Steady-state creep						
Shear						
Eq. (13.1)						
C_4 (K/sec/psi)						
α	0.198	0.0989	3.13E-3	1.62E7	35.2	3.18E7
n	1300	1300	1500	1000	7500	1000
Q (eV)	3.3	3.3	5.5	7.0	5.0	2.8
Eq. (13.3)	0.548	0.548	0.50	1.10	0.72	1.03
C_5 (1/sec)						
α_1 (1/psi)	2.78E5	1.39E5	2.46E5	4.60E11	1.00E8	5.19E11
Q_a (eV)	8.0E-4	8.0E-4	6.3E-4	8.0E-4	1.2E-2	1.25E-3
Derived tensile	0.70	0.70	0.75	1.15	0.90	1.08
Eq. (13.8)						
C_{4t} (K/sec/psi)						
α_t	0.114	0.0571	1.81E-3	9.35E6	20.3	1.84E7
Eq. (13.9)	751	751	866	577	4330	577
C_{5t} (1/sec)						
α_{1t} (1/psi)	1.61E5	8.03E4	1.42E5	2.66E11	5.77E7	3.00E11
Measured tensile	4.62E-4	4.62E-4	3.64E-4	4.62E-4	6.93E-3	7.22E-4
Eq. (13.9)						
C_{5t} (1/sec)						
α_{1t} (1/psi)		4.11E6				
Q_a (eV)		5.0E-4				
		0.90				
Transient creep						
Shear						
Eq. (13.4)						
γ_T	0.026	0.040	0.167	0.115	0.013	0
B	403	152	131	137	163	—
Derived tensile						
Eq. (13.10)						
ε_T	0.015	0.023	0.096	0.066	0.0075	0
B_t	698	263	227	237	282	—
Time-independent plastic flow						
Shear						
Eq. (13.6)						
C_6	2.34E13	1.21E13	2.04E11	6.36E7	1.40E19	2.31E9
m	5.58	5.53	4.39	3.10	5.92	4.13
Derived tensile						
Eq. (13.12)	6.30E11	3.35E11	1.06E10	6.69E6	3.13E17	1.38E8
C_{6t}						

flow can be used simultaneously, but a sinh expression is not offered as one of the creep equations. Hence, one is forced to invoke the user-defined option to create a sinh-based creep equation, or to switch the element type back and forth between VISCO107 and SOLID45 throughout the simulation. Similarly, in ABAQUS, time-independent plastic flow can be used for the loading part of a thermal cycle and sinh creep can be used for the dwell time, but they cannot be used together in the same time step.

With respect to time-independent flow, ABAQUS is slightly superior to ANSYS because it offers a standard option with no cyclic work hardening (which is the observed behavior in most solders under thermal cycling conditions). ANSYS offers only kinematic or isotropic work hardening as standard options. Both codes offer transient creep behavior, but not in the form of Eq. (13.10) as a standard input.

In an effort to simplify finite element modeling and use standard input relations, one-dimensional nonlinear simulations used to determine an approximate stress-strain rate relation which incorporates both creep and time-independent plastic flow. A FORTRAN code named HLOOP2 has been written which uses the complete set of relations, Eqs. (13.2) and (13.8) to (13.13). The code was used to simulate constant rate loading over a range of temperatures and displacement rates. Next, the stress and inelastic strain rate was recorded at a strain level of 2 percent. An example of the simulation results at 25°C is shown in Fig. 13.15 where the initial creep rate, steady-state creep rate, and total inelastic strain rate at 2 percent strain are plotted as a function of applied tensile stress. It is seen that the total inelastic strain rate is considerably higher than the initial creep rate in the high stress regime, but is between the initial and steady-state creep rates in the low stress regime.

A series of simulations were run in the temperature range between −40 and 125°C, and the results were fit to Eq. (13.28)

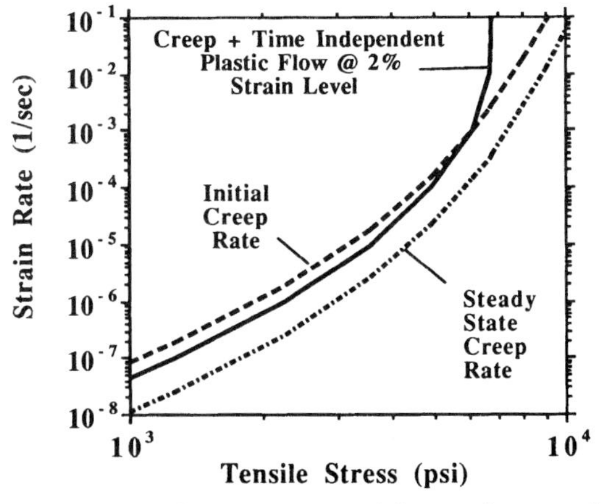

Figure 13.15 Constitutive model simulation of 62Sn36Pb2Ag solder joints at 25°C.

$$\frac{d\varepsilon}{dt} = C_{7t}[\sinh(\alpha_{2t}\sigma)]^n \exp\left(\frac{-Q_a}{kT}\right) \tag{13.28}$$

where $d\varepsilon/dt$ is the total inelastic strain rate at a strain level of 2 percent. Equation (13.28) is a form which is available as standard input in both ANSYS and ABAQUS. Only the constants C and α had to be modified from Eq. (13.9). The recommended deformation constants for a simplified nonlinear finite element analysis using standard input are given in Table 13.2. It is interesting to note that the stress-strain rate behavior at a 2-percent strain level is essentially equal between 62Sn36Pb2Ag and 60Sn40Pb, even though 62Sn36Pb2Ag has a 2× lower steady state creep rate.

13.3.3 Bulk versus joint behavior

Standard mechanical tests utilize bulk samples which are typically cast into ingots, then extruded and machined into final form. There are

TABLE 13.2 Recommended Constants for Simplified Nonlinear Finite Element Modeling

	60Sn40Pb	62Sn36Pb2Ag	97.5Pb2.5Sn	96.5Sn3.5Ag
Eq. (13.28) C_{7t}(sec^{-1}) *ANSYS** A(sec^{-1}) *ABAQUS* A(sec^{-1})	9.62E4	9.62E4	2.20E8	9.00E5
Eq. (13.28) α_{2t}(psi^{-1}) *ANSYS** φ(psi^{-1}) *ABAQUS* B(psi^{-1})	6.0E-4	6.0E-4	1.4E-3	4.5E-4
Eq. (13.28) n *ANSYS** 1/m *ABAQUS* n	3.3	3.3	7.0	5.5
Eq. (13.28) Q_a/k (K) *ANSYS** Q/R (K) *ABAQUS* ΔH/R (K)	8110	8110	13330	8690

* ANSYS uses the Anand model. Set $S_o = 1$, $h_o = 1.0e\text{-}9$, $s\hat{} = 1$, $n = 1.0e\text{-}9$, and $a = 1$, to negate the evolution equation.

several reasons why data obtained from bulk solder samples might not be directly applicable to solder joints: (1) the constraining effect of the solder/substrate interfaces, (2) precipitation strengthening from dispersed intermetallics, and (3) difference in grain structure, grain size, or grain/specimen size ratio. In order to show the magnitude of these effects, room-temperature creep data obtained from solder joints are plotted along with data obtained from bulk solder specimens in Fig. 13.16. All of the bulk data and two sets of solder joint data were obtained under tensile loading. The other three sets of the solder joint data have been converted from shear to tensile (S → T) using Eqs. (13.14) and (13.15).

In comparing the bulk data to the joint data, it is seen that there is a huge discrepancy in creep behavior. Schmidt's[27] solder wire samples had two orders of magnitude higher strain rate at a given stress level than the solder joints tested by Darveaux or Subrahmanyan.[3] Skipor's[28] and Kashyap's[29] bulk samples showed closer agreement to the solder joints at stresses above 5000 psi, but the 9.7 μm material deviated significantly at lower stress levels. Only the 28.4 μm bulk material showed similar creep behavior to actual solder joints. However, at lower stresses, even the 28.4 μm bulk material deviates, showing a stress exponent of $n = 2.0$, versus $n = 3.3$ for solder joints.

These differences between joint and bulk behavior might be partially attributed to microstructure. Kashyap's bulk material had little, if any, lamellar structure, and it showed increasing creep resistance with grain size. In addition, Kashyap's stress exponent of 2.0 indicates grain boundary sliding. Hence, the larger grain material was more resistant

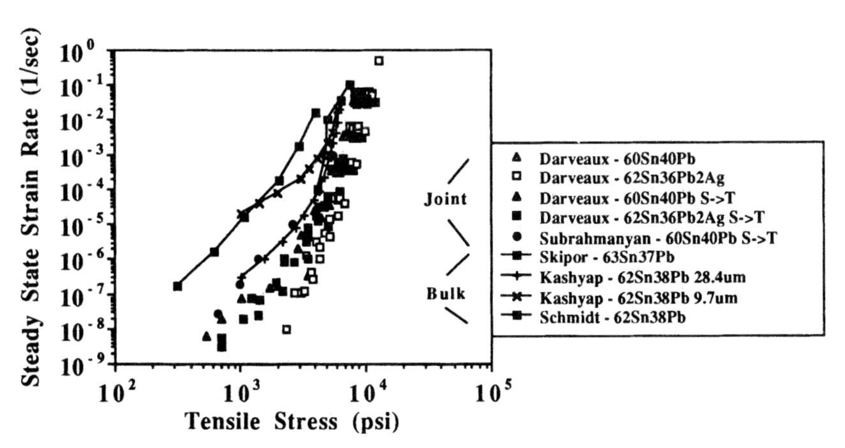

Figure 13.16 Comparison of bulk specimen to solder joint creep behavior at 25°C for eutectic lead-tin. S → T indicates shear data converted to tensile using Eqs. (13.14) and (13.15).

to grain boundary sliding than the smaller grain material. On the other hand, Darveaux's solder joints had some regions with lamellar structure, although most of the lamellae were broken up because the samples were aged at 100°C for 50 to 100 hrs. The average grain size was on the order of 1 μm (details on the microstructure and the effects of aging are given in Ref. 30). Based on Kashyap's data, one would expect the joints to be less creep-resistant and show more grain boundary sliding due to the smaller grain size. But the opposite was true. The solder joints were more creep-resistant than the bulk material, and did not show complete grain boundary sliding. The stress exponent was $n = 3.3$, which indicates dislocation glide-controlled deformation kinetics, or a combination of grain-boundary sliding and dislocation glide or climb.

It is possible that the regions of lamellar structure in the joints resulted in some strengthening. However, one would still expect grain-boundary sliding at the colony boundaries and at the equiaxed grain boundaries. A second possibility is that there is an effect from gold-tin or copper-tin intermetallics dispersed in the solder joints. If the intermetallics precipitate near colony boundaries, one would expect a reduction in grain-boundary sliding. Based on the 100°C aged microstructures shown in Ref. 30, it does appear that the gold-tin intermetallics preferentially segregate near colony boundaries, but it is not completely obvious. One would have to test solder joints without dispersed intermetallics to verify this theory. As discussed in the previous section, finite element modeling indicates that there is some interface constraint in the C5 joints under tensile loading, but not nearly enough to explain the discrepancy between joint and bulk data shown in Fig. 13.16. Therefore, it is concluded that most of the discrepancy is due to microstructural effects such as grain structure and dispersed intermetallics.

The effect of interface constraint is more clearly shown in Fig. 13.17, where the maximum stress is plotted versus height/radius ratio for 100In. All of the data in Fig. 13.17 have been normalized to a strain rate of 10^{-3} sec^{-1}. Although there is considerable scatter in the die bond data, it is seen that decreasing the height-to-radius ratio from 1/1 to 1/10 increases the peak stress by almost 2×. This is consistent with the work of Ashby et al.[31] on lead wires constrained by glass. They also saw a 2× difference in peak stress between constrained and unconstrained samples. Further evidence of interface constraint is seen in the fracture surfaces of the failed samples. The C5 joints draw down to a point under tensile loading, whereas the fracture surface of a die bond joint is dimpled. A dimpled surface is indicative of internal cavities which have nucleated and grown in the highly constrained solder. Such cavitation was also observed by Ashby et al.[31] and Skipor et al.[32]

13.3.4 Creep rupture

Solder joints on leadless chip carriers can fail by creep rupture due to bending of the PC board during product assembly. In most creep rupture situations, the joints will be under both shear and tensile loading, but the tensile component will be the most significant. It was decided to define creep rupture guidelines based on tensile data, because tensile loading is probably more representative of an actual application. Shown in Fig. 13.18 are creep rupture results for 62Sn36Pb2Ag C5 joints under tensile loading. The best fit curves are based on a variation of Eq. (13.9) shown below

$$t_{\text{rupture}} = \frac{1}{C_{5t}[\sinh(\alpha_{1t}\sigma)]^n \exp\left(\frac{-Q_a}{kT}\right)} \tag{13.29}$$

where $C_{5t} = 1.47\text{E}8\ \text{sec}^{-1}$, $\alpha_{1t} = 5.0\text{E-}4\ \text{psi}^{-1}$, $n = 3.3$, and $Q_a = 0.90$ eV.

To make an up-front prediction of creep rupture failure, an engineer needs to: (1) calculate the equivalent tensile stress using FEA or other techniques, then (2) read off or calculate the time to failure for the field use temperature of interest. For a room temperature field condition, tensile stress must be kept below 630 psi to ensure a 10-year use life. However, if the use temperature is raised to 50°C, the maximum allowable stress is only 300 psi.

Once prototypes are built, the entire assembly can be creep-rupture tested by aging it at elevated temperature. Plotted in Fig. 13.19 is the time an assembly must survive at 100°C to ensure a failure-free life at various use temperatures. For example, an assembly must survive about 48 days at 100°C to ensure a 10-year use life at 50°C.

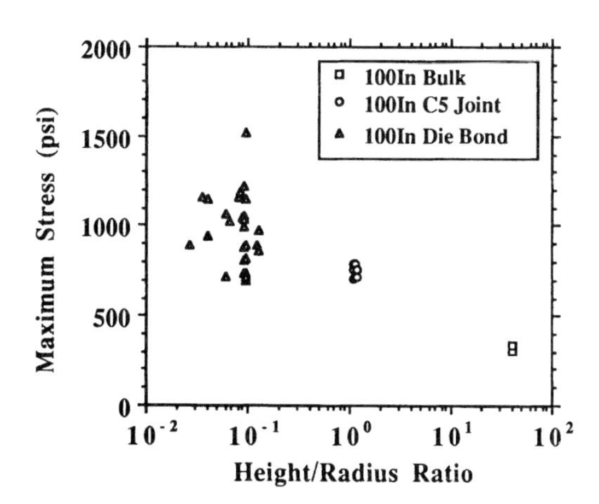

Figure 13.17 Increase in solder joint strength with interface constraint. Indium, 25°C, 1.0E-3 sec^{-1} strain rate.

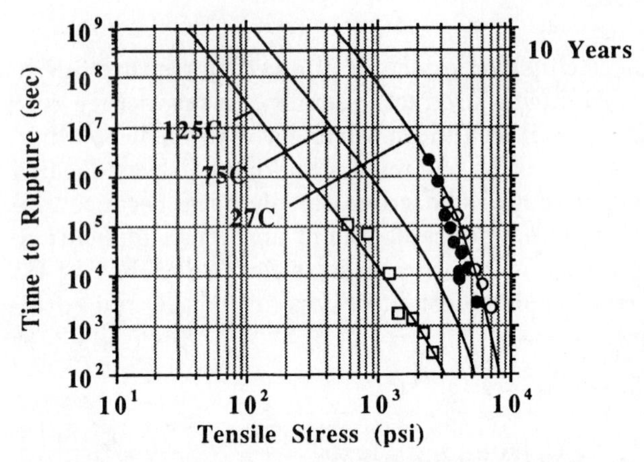

Figure 13.18 Creep rupture of 62Sn36Pb2Ag C5 solder joints. Open symbols indicate Cr/Cu/Ni/Au – Cr/Cu metallization, closed symbols Cr/Cu/Ni/Au – Cr/Cu/Ni/Au metallization.

13.3.5 Gold embrittlement

Gold-plated terminations are extensively used in the electronics industry to preserve solderability of surfaces. Ductility loss of tin-lead eutectic solder due to presence of gold is often quoted as "gold embrittlement" and cited as a reason for failure of solder joints. Typically, it is reported that gold embrittlement occurs in solder joints containing in excess of 5 percent gold by weight.[33–35] However, experience has shown that brittle joints can occur at nominal gold concentrations much less than 5 percent.

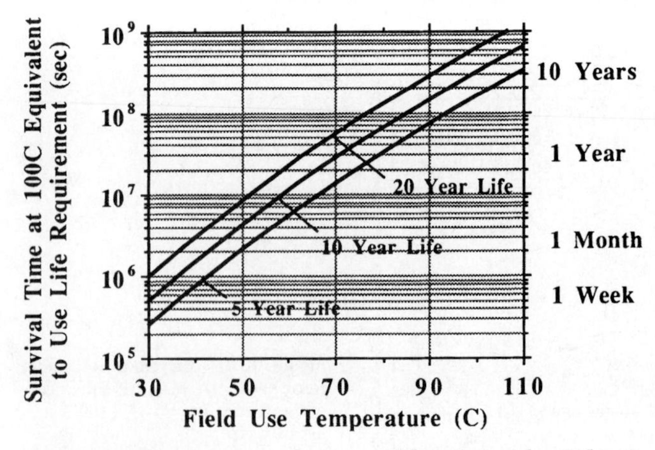

Figure 13.19 Creep rupture design guideline for accelerated testing of assembled products.

A study was conducted to determine the impact of gold embrittlement on BGA solder joints. Samples identical to that shown in Fig. 13.1 were fabricated with gold thicknesses of 4.2, 12.1, 20.1, and 50.7 microinches. Hence, the resulting weight fraction of gold in the joints was 0.17, 0.49, 0.82, and 2.03 percent, assuming gold from an entire 0.040-in pad was dissolved into a joint which was defined by a 0.030-in opening in the solder mask. After fabrication, the samples were aged at room temperature, 100 and 150°C for various times for microstructural characterization and mechanical testing. Mechanical testing consisted of monotonic tensile or shear loading at strain rates of approximately 10^{-2} and 10^{-4} sec^{-1}.

Shown in Fig. 13.20 are micrographs of etched solder joints revealing the distribution of the gold tin intermetallics. These particles have been identified in the SEM-EDAX as $AuSn_4$. In Fig. 13.20*a* are the needlelike particles for the 51-microinch-thick gold layer without any aging. Fine needles are uniformly distributed throughout the tin-lead matrix. Compare this with Fig. 13.20*b* which is for the substrate with only 4 microinches of gold. Only a small amount of the particles are seen near the termination interface. Shown in Fig. 13.20*c* and *d* are the distribution and morphology of the $AuSn_4$ for 500 hours at 100°C and 200 hours at 150°C respectively for the 51 µ-inch gold samples. It is obvious that with high temperature aging, the needles are transformed to a lath or blocky morphology, and there appears to be some growth at the interfaces. In particular, comparing Fig. 13.20*a* and *d,* it is obvious that there is thick growth of some compound at the interface after 200 hours at 150°C, but not in the unaged sample.

Figure 13.20*a* 51 µin gold, unaged.

Figure 13.20*b* 4 μin gold, unaged.

Figure 13.20*c* 51 μin gold, 500 hours at 100°C.

Shown in Fig. 13.21*a* to *d* are the substrate metallization–solder interface for the conditions (*a*) 51 microinch, no aging, (*b*) 4 microinch, no aging, (*c*) 51 microinch, 500 hours at 100°C, and (*d*) 51 microinch, 200 hours at 150°C. The bright layer at the bottom is copper. The sam-

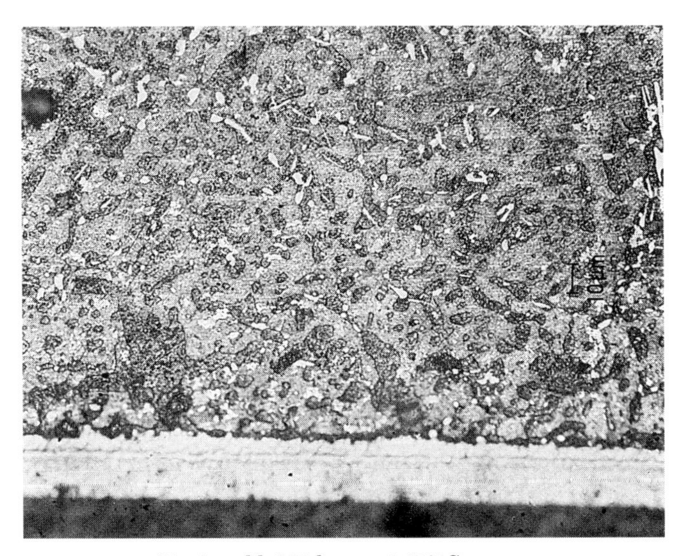

Figure 13.20*d* 51 μin gold, 200 hours at 150°C.

ples were etched for copper and solder to reveal the intermetallics. The two distinct intermetallic layers in Fig. 13.21*d* have been identified with SEM-EDAX as $AuSn_4$ adjoining the solder and the gray layer in between is nickel and the nickel tin intermetallic, Ni_3Sn_4. The boundary between the original nickel barrier layer and the nickel-tin intermetallic layer is not easily resolved. It is apparent from Fig. 13.21 that a gold-tin intermetallic layer grows at the interface at the expense of the same compound in the bulk solder. In Fig. 13.21*a* and *b* it is seen that the intermetallic layer is similar for the solder joints formed with 51 and 4 microinches of gold. This indicates that even at 51 microinches, all the gold is leached away from the interface during reflow to form the fine needlelike $AuSn_4$. Gold has the highest dissolution rate in solder among common terminal metallurgies.[36] However, with aging, the needles transform to lathlike structures, dissolve, and reprecipitate as $AuSn_4$ at the interfaces. Although the exact mechanism has not been identified, it is speculated that due to its high mobility, the gold diffuses into the tin and starts to grow on the interface because of favorable energetics due to the crystallography of nickel-tin and gold-tin compounds. Comparing Figs. 13.21*c* and *d* it is seen that 500 hours at 100°C is an intermediate stage in the growth process. Initially, long columns grow into the solder at discrete locations along the interface. At higher temperatures or longer times (Fig. 13.21*d*), there appears to lateral growth of these columns to form a continuous layer.

Figure 13.21*a* 51 µin gold, unaged.

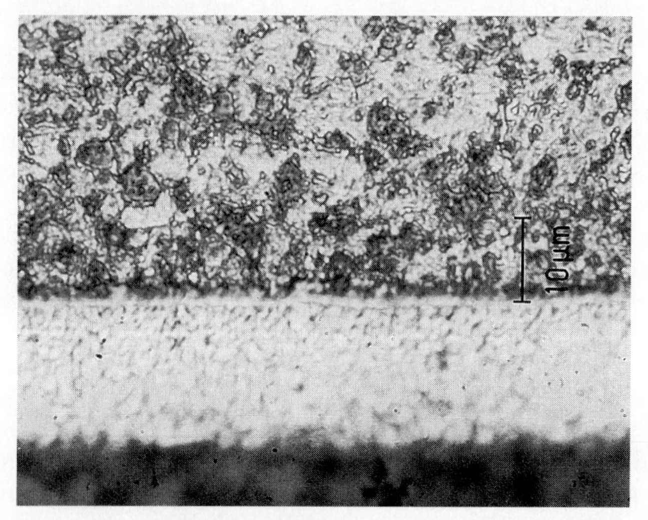

Figure 13.21*b* 4 µin gold, unaged.

Plotted in Fig. 13.22 is the stress corresponding to the peak load sustained by the solder joints as a function of the gold thickness. The fracture stress is relatively insensitive to gold thickness for aging at 100°C (Figs. 13.22*a* and *b*). There is a slight decrease in the stress values going from 50 to 500 hours. For aging at 150°C, there is a monotonic decrease of the maximum stress as a function of gold thickness at the higher strain rate of $(6.0)10^{-2}$ sec^{-1} both in tension and shear.

Figure 13.21*c* 51 μin gold, 500 hours at 100°C.

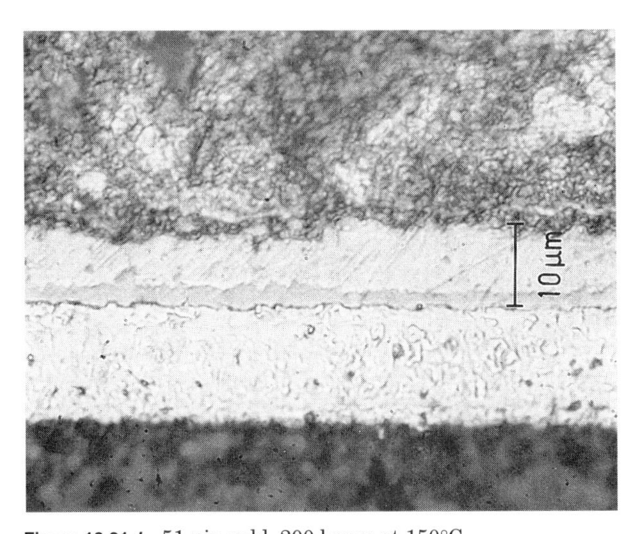

Figure 13.21*d* 51 μin gold, 200 hours at 150°C.

Plotted in Fig. 13.23 is the true strain corresponding to the maximum stress data plotted in Fig. 13.22. The strain is insensitive to gold thickness at the lower strain rate for aging at 100°C. At higher strain rates, the fracture strains decrease with increasing gold thickness. The only data point which exhibited fracture at the interface at 100°C aging was the one with the 51-microinch gold, aged for 500 hours, tested at the higher strain rate in tension. All the other samples fractured near the interface, but through the solder. Aging at 150°C causes brittle, interfacial failures for lower gold thicknesses, aging times, and strain rates (Figs. 13.23*c* and *d*) compared to aging at 100°C. For gold thicknesses of 4 and 12 microinches, the fracture was through the solder for all the aging times at 150°C that were studied. For gold thicknesses of 20 and 51 microinches, aging beyond 50 hours causes brittle interfacial failures at both the high and low strain rates in tension.

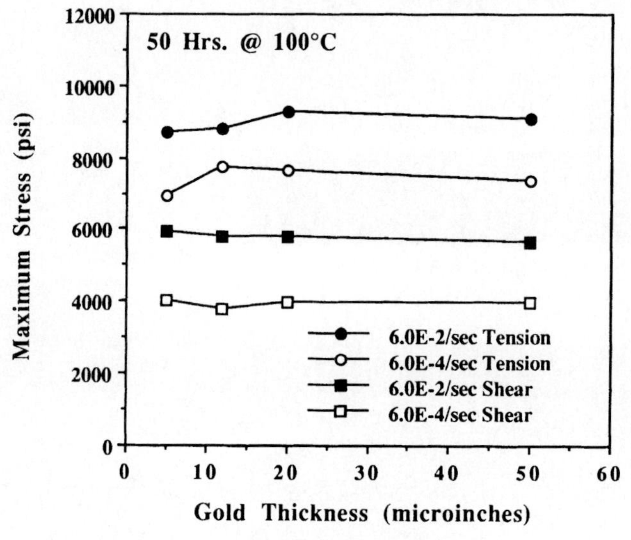

Figure 13.22*a* Maximum stress vs. gold thickness.

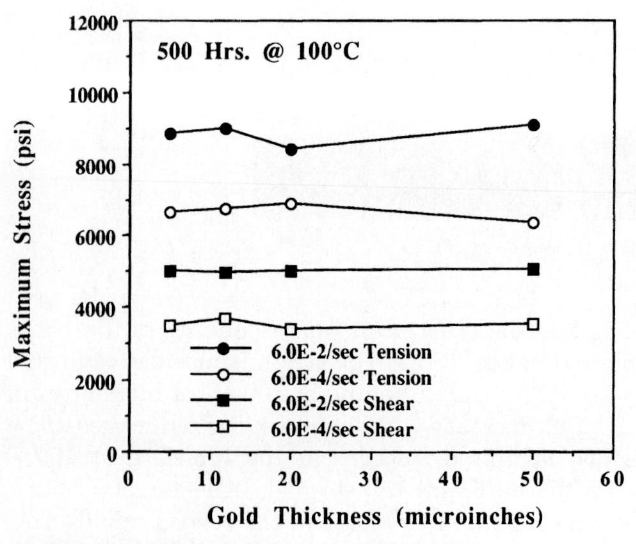

Figure 13.22*b* Maximum stress vs. gold thickness.

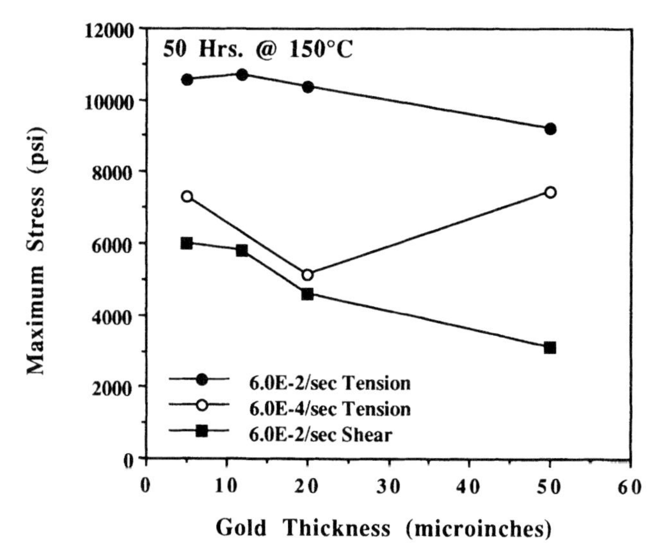

Figure 13.22*c* Maximum stress vs. gold thickness.

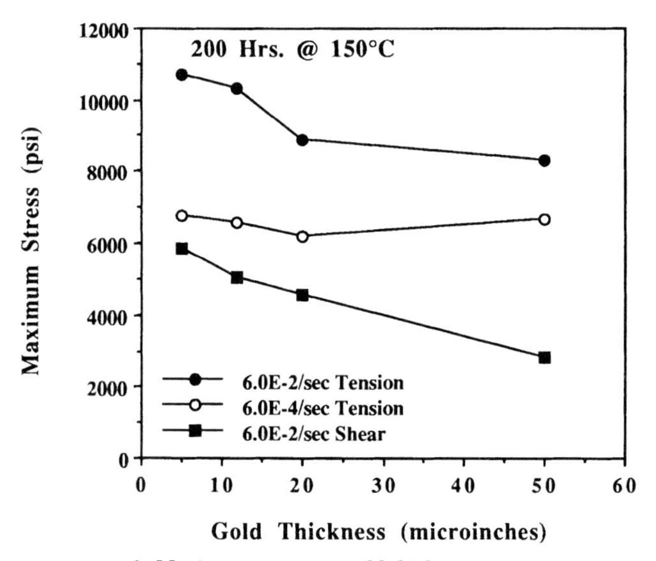

Figure 13.22*d* Maximum stress vs. gold thickness.

The drop in ductility for gold thicknesses in excess of 20 microinches is significant. The typical specification for gold plating thickness is between 20 and 30 microinches for many device terminations. The propensity for interfacial failure under tensile loading can be explained by the concept of fracture of brittle materials. The fracture instability occurs by a critical combination of imposed stress and defect size (crack size). The interfacial failure through the "intermetallic layer" is easily affected by a higher component of the tensile or opening mode stress for the same distribution of defect or crack sizes. This stress is definitely higher for the tensile than for the shear loading. Another possibility for the susceptibility to interfacial brittle failure under tension may be higher stress concentration at the interface due to the shape of the joint under tensile as opposed to shear loading. The probability of interface failure is also higher with a higher strain rate simply due to the strain rate sensitivity of the bulk solder creating higher stresses at the interface.

Shown in Fig. 13.24 are the complementary fracture surfaces of the interfacial failure for a joint with 51 microinches of termination gold and aged at 150°C for 200 hours, tested in tension at $(6.0)10^{-2}$ sec^{-1} strain rate. The morphology of the two surfaces is different. On the solder side, the material was identified by EDAX to be $AuSn_4$. On the substrate side, Ni_3Sn_4 was the predominant phase. So it is obvious that the interfacial fracture occurs by the separation between the nickel-tin and the regrown gold-tin intermetallic layers. This is a newly discovered mode of failure in this system. It can be speculated that the nucleation and growth of the gold-tin intermetallic over the nickel-tin intermetallic is discontinuous, resulting in inherent cracklike defects at the interface between the two phases. This is in contrast to the accepted premise that if the thickness of gold is high enough to be not dissolved in the bulk, the resulting "brittle" gold-tin intermetallic at

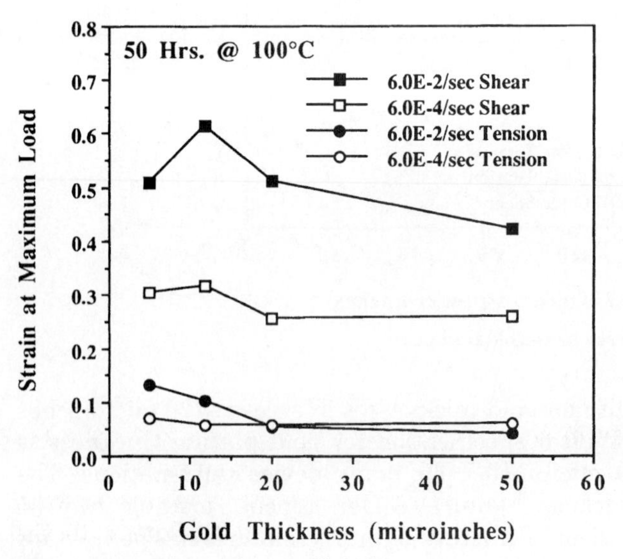

Figure 13.23a True strain vs. gold thickness.

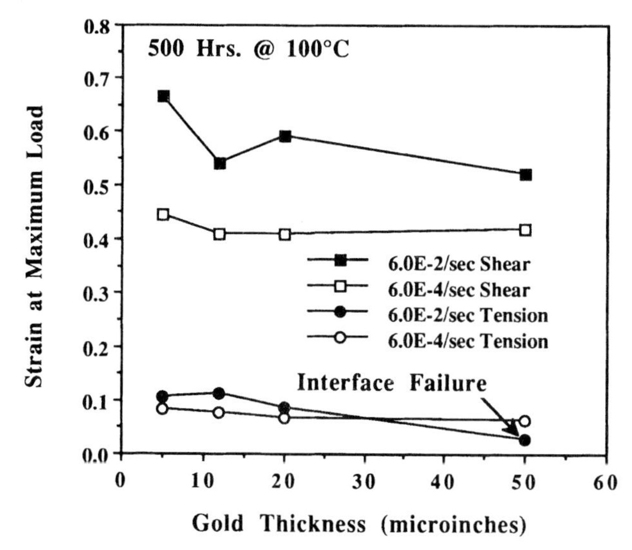

Figure 13.23*b* True strain vs. gold thickness.

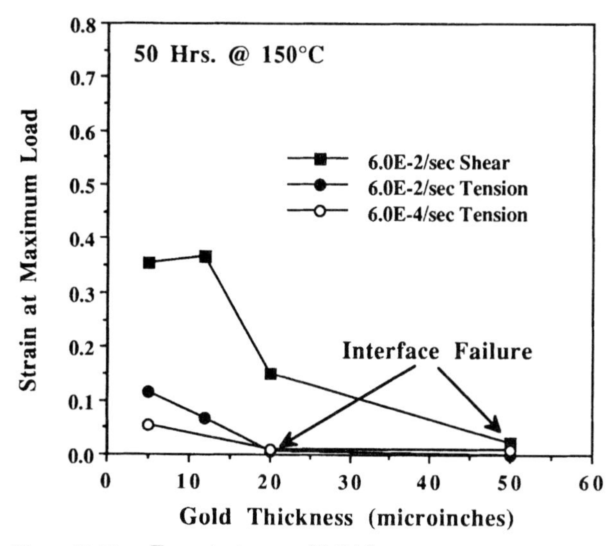

Figure 13.23*c* True strain vs. gold thickness.

the interface results in catastrophic failure of the joint.[37] From the microstructural observations presented here, it is seen that as soon as the whole surface of the primary nickel-tin interface is covered with the reprecipitating gold-tin intermetallic (minimum thickness >0), the susceptibility to brittle failure is increased.

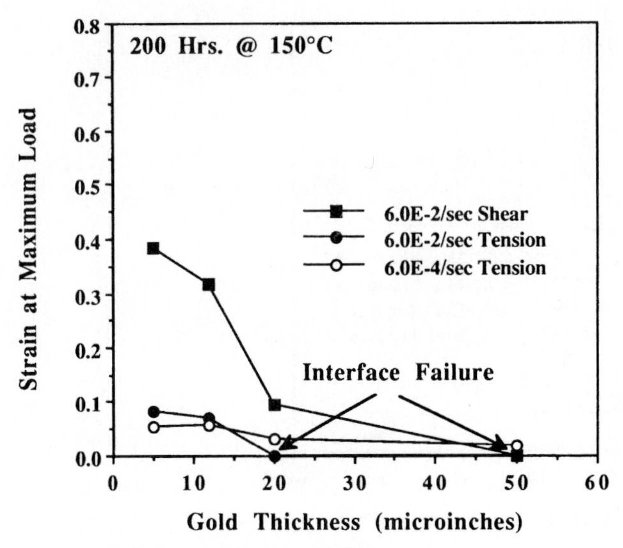

Figure 13.23*d* True strain vs. gold thickness.

The activation energy for this interfacial growth was calculated to be 0.77 eV based on the thickness data at 150 and 100°C. This approximately correlates with published values for bulk diffusion of gold in tin which is about 0.65 eV.[38] Based on the data that 500 hours at 100°C causes brittle failures in this system, it will take 19 years for the solder joint to reach a brittle state at 27°C. However, it will take only 1 year to reach an embrittled state at 60°C. Shown in Fig. 13.25 is the time to embrittlement for a BGA solder joint formed with a 30-mil-diameter sphere between two identical terminations with 51 microinches of gold (which results in 2.03 weight percent gold in the joint). Clearly, this amount of gold would be excessive for elevated temperature applications such as high power devices or automotive under the hood.

13.4 Thermal Fatigue

The actual mechanism by which a solder joint fails is dependent on several factors. In all cases, the failure is due to crack initiation and propagation through a joint, but the location and nature of the cracks depend on the joint configuration, intermetallic structure, strain rate, and temperature regime. Several fatigue studies from the literature were tabulated in Chap. 2 of Ref. 4, and the following trends in the data were noted:

Figure 13.24*a* Solder side.

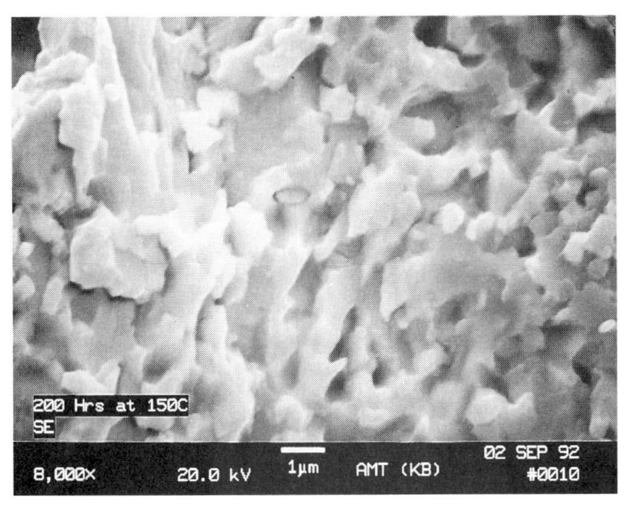

Figure 13.24*b* Substrate side.

1. Cracks usually initiate at high stress concentration sites. The direction of propagation depends on the shape of the joint and the relative degree of tensile versus shear loading.
2. Higher joint stresses due to higher strain rates or creep-resistant solders result in more fractures near solder/component interfaces.

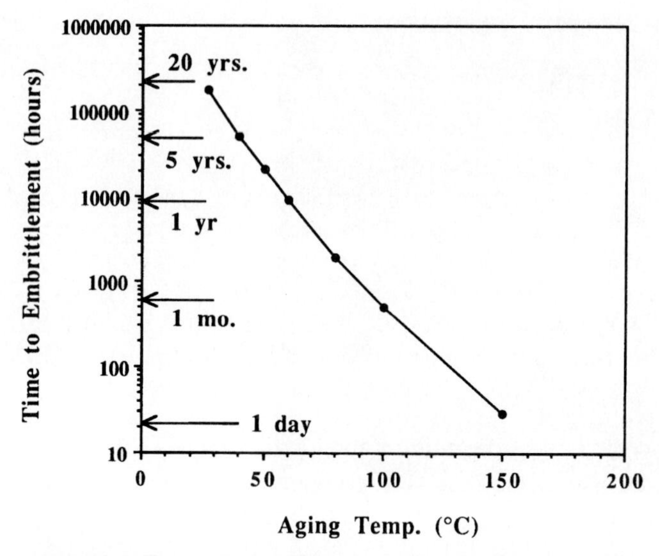

Figure 13.25 Time to embrittlement of a 0.030-in-diameter solder joint formed with two identical terminations with 51 μin of gold over copper/nickel.

3. Interfacial fracture is more likely with poor intermetallic/solder adhesion, brittle intermetallics, or depletion of a solder constituent near the intermetallic layer.
4. For bulk solder failures, higher strain rates or lower temperatures result in transgranular cracking. Lower strain rates, longer hold times, and higher temperature lead to intergranular cracking. Oxidation also leads to more intergranular fracture.
5. A higher strain range results in more transgranular fracture, and a lower strain range results in more intergranular fracture.
6. Thinner joints or single-phase alloys result in more uniform strain distribution across the solder layer.
7. For near-eutectic Pb/Sn solders, localized coarsening in high stress areas precedes cracking.

It is apparent that the observed solder joint failure mechanism is dependent on the accelerated test conditions. Furthermore, applicability of a particular lifetime prediction model depends on the activated failure mechanism. Therefore, accelerated tests should be designed based on expected field used conditions as much as possible. The only conflicting requirement is that accelerated tests need to be as rapid as possible to reduce new product development cycle time.

13.4.1 Failure statistics

All failure processes should be described in terms of a statistical distribution. The distribution function is a mathematical expression which relates the probability of failure to a variable such as number of cycles or time. Two of the most commonly used distributions to describe wearout data (e.g., solder joint fatigue) are lognormal and Weibul.[39–41] Clech et al.[42] recently compared 2-parameter (2P) Weibull, 3-parameter (3P) Wiebull, and lognormal distributions over a large database with 26 accelerated test logs and found the 3P Weibull to give the best fit. Nicewarner[43] also reached the same conclusion that a 3P Weibull distribution is the most appropriate to describe solder fatigue data. The difference between 2P and 3P Weibull distributions becomes significant below about 5 percent cumulative failures, where the data are seen to deviate downward on a 2P plot. One must test a large population to observe the difference. More importantly, predicted field failure rates at the 100-ppm level are overly conservative with the 2P Weibull distribution by as much as 3×.[42] The cumulative distribution of failures for the 3P Weibull distribution is given by

$$F_j = 0 \qquad \text{for } N < N_{ff}$$

$$F_j = 1 - \exp\left[-\left(\frac{N - N_{ff}}{\alpha_w - N_{ff}}\right)^{\beta_w}\right] \text{for } N > N_{ff} \tag{13.30}$$

where N is the number of cycles, N_{ff} is the failure free life, α_w is the characteristic life at which 63.2 percent of the population has failed, and β_w is the shape parameter which indicates the amount of scatter in the data. The reliability, or fraction of survivors, is given by

$$R_j = 1 \qquad \text{for } N < N_{ff}$$

$$R_j = \exp\left[-\left(\frac{N - N_{ff}}{\alpha_w - N_{ff}}\right)^{\beta_w}\right] \text{for } N > N_{ff} \tag{13.31}$$

To determine the number of cycles to 50 percent failure (or 50 percent survivors), R_j is set equal to 0.5 for known values of α_w, β_w, and N_{ff}. When applying Eqs. (13.30) and (13.31) to solder fatigue problems, each group of joints under the same loading conditions will be described by the same failure distribution. However, it must be noted that electronic components have anywhere from one to hundreds of joints which could fail and cause component failure. Hence, to predict component failure based on joint failure, the number of joints per component must be incorporated. Mathematically, the joints are considered to be in series, so the overall component reliability is given by[44]

$$R_c = \Pi R_j \tag{13.32}$$

For a given set of joints under identical loading, Eqs. (13.31) and (13.32) can be combined to give the component reliability at a given number of cycles

$$R_c = \exp\left[-q\left(\frac{N - N_{ff}}{\alpha_w - N_{ff}}\right)^{\beta_w}\right] \tag{13.33}$$

where q is the number of joints. Rearranging Eq. (13.34) gives the number of cycles that R_c of the components survive

$$N_c = N_{ff} + (\alpha_w - N_{ff})\left(\frac{-ln(R_c)}{q}\right)^{1/\beta_w} \tag{13.34}$$

For components with sets of joints under different loadings, Eq. (13.33) can be applied to each set of joints, and Eq. (13.32) can be applied to the reliability of the sets to give the net component reliability. It is interesting to note that the number of failure-free cycles for a component does not depend on the number of joints, but the failure distribution does. These are important issues when developing fatigue life prediction models for components based on stresses and strains in individual joints.

13.4.2 Finite element modeling procedure

Three-dimensional nonlinear finite element modeling was used in all of the following sections. The material properties, methodology, and boundary conditions were consistent throughout the various models. ANSYS 5.0 was used for all aspects: preprocessing, solution, and postprocessing. The solder material was modeled as a viscoplastic solid, the printed circuit board as an orthotropic linear elastic solid, and the rest of the materials as linear elastic solids. The nonlinear properties are given in Table 13.2, and the linear properties are given below in Table 13.3. The objective of the simulations was to calculate the maximum plastic work per unit volume in the solder. It was found that the elemental values were more dependent on the mesh size than the nodal values. Therefore, the nodal values were used in all correlations and predictions.

In order to reduce computation time, only a slice of the assembly was modeled, as shown in Figs. 13.26 and 13.27. A symmetry boundary condition was imposed on the surface through the center of the joints ($dy = 0$), and at the center of the package ($dx = 0$). A coupled boundary condition was imposed on the surface at the centerline between rows of joints, and at the centerline between packages on the mother board. The nodes on the coupled surfaces had to remain as a plane. For single sided boards, one node on the bottom surface was constrained ($dz = 0$) to prevent free body rotations. For double sided boards, a symmetry boundary condition was imposed on the entire bottom surface ($dz = 0$). A closer view of a BGA joint is shown in Fig. 13.28.

TABLE 13.3 Linear Elastic Constants

Material	Young's modulus (10^6 psi)	Poisson ratio	Expansivity (ppm/°C)
Si	23.6	0.278	–5.88 + 6.26E-2T-1.6E-4T^2 + 1.51E-7T^3
Al_2O_3	40.0	0.22	1.32 + 1.278E-2T
Cu	18.7	0.344	13.8 + 9.44E-3T
62Sn36Pb2Ag 60Sn40Pb 63Sn37Pb	11.0 – 2.2E-2T	0.35	24.5
Die attach adhesive	1.07	0.40	52.0
Mold compound	2.25	0.25	15.0
PC boards			
Data sets 1C-4C, 9C-11C 6 layer polyimide	*x,y* 4.12 – 4.12E-3T *z* 1.80 – 1.80E-3T	*xz, yz* 0.39 *xy* 0.11	*x,y* 17.6 *z* 64.1
Data sets 5C,6C,8C 2 layer polyimide	*x,y* 3.35 – 3.64E-3T *z* 1.46 – 1.59E-3T	*xz, yz* 0.39 *xy* 0.11	*x,y* 14.5 *z* 63.5
Data set 7C,1P,15P,17P 2 layer FR4	*x,y* 4.05 – 5.39E-3T *z* 1.77 – 2.35E-3T	*xz, yz* 0.39 *xy* 0.11	*x,y* 14.5 *z* 67.2
Data sets 2P-14P 6 layer Tetra 2	*x,y* 3.53 – 3.23E-3T *z* 1.54 – 1.41E-3T	*xz, yz* 0.39 *xy* 0.11	*x,y* 17.6 *z* 64.1
Data sets 18P-20P 2 layer FR4	*x,y* 4.05 – 5.39E-3T *z* 1.77 – 2.35E-3T	*xz, yz* 0.39 *xy* 0.11	*x,y* 16.0 *z* 84.0
Data sets 21P-22P 6 layer FR4	*x,y* 3.53 – 3.23E-3T *z* 1.54 – 1.41E-3T	*xz, yz* 0.39 *xy* 0.11	*x,y* 16.0 *z* 63.5
Data sets 1P-22P 2 layer BT/epoxy	*x,y* 4.30 – 5.72E-3T *z* 1.88 – 2.49E-3T	*xz, yz* 0.39 *xy* 0.11	*x,y* 15.0 *z* 57.0

In each calculation, five complete thermal cycles were simulated. The temperature ramps were divided into 10°C to 20°C substeps, and each substep was iterated to convergence. Automatic time stepping was used during the dwell periods, which resulted in anywhere from 6 to 20 substeps, depending on the thermal cycle profile. A typical simulation would have 300 to 400 iterations in total, with a run time of about 1 day. However, it was found that substructuring could be used to group all of the linear portion of the model into one superelement. This reduced run times by 4× to 14×. The only disadvantage is that temperature dependent properties cannot be used for the materials in the superelement. However, this only results in an error of a couple percent, so substructuring is recommended when possible.

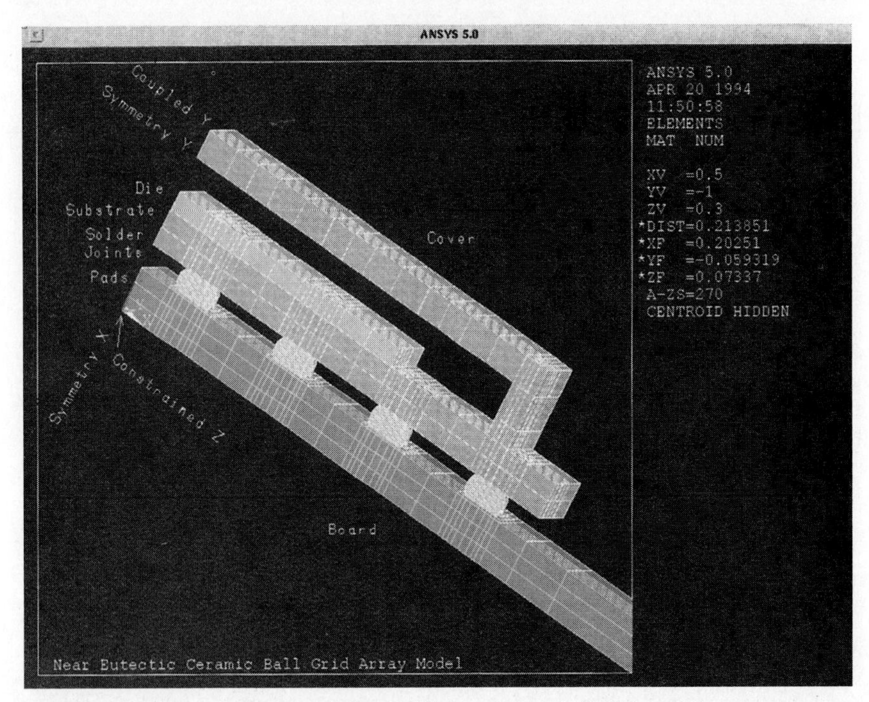

Figure 13.26 Finite element mesh for 68 I/O near eutectic CBGA model.

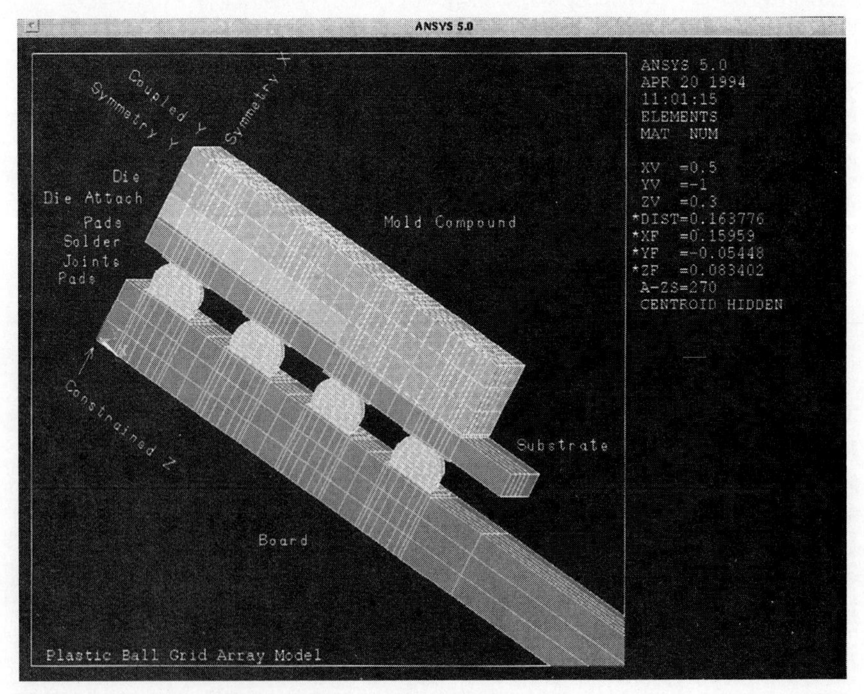

Figure 13.27 Finite element mesh for 72 I/O PBGA model.

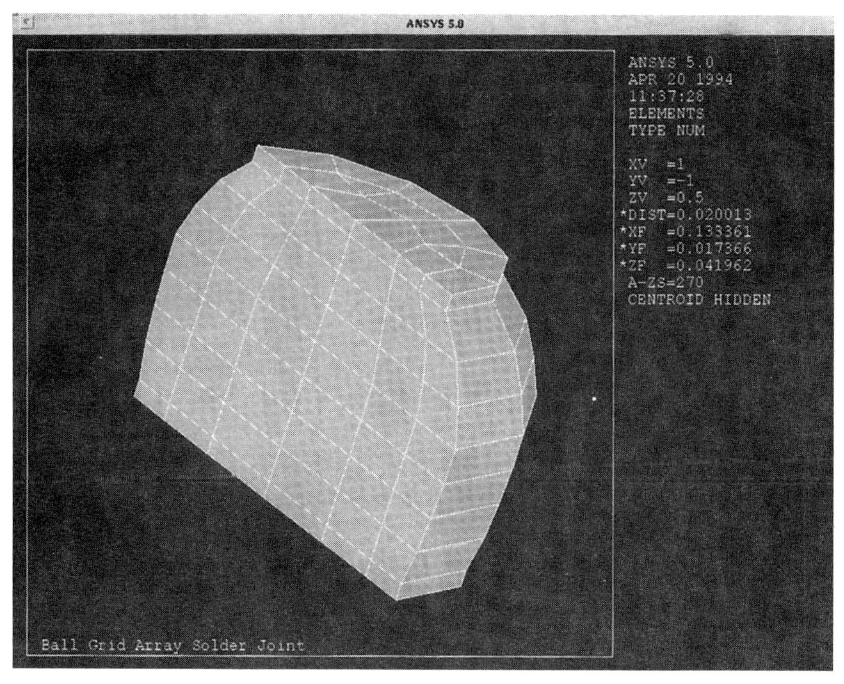

Figure 13.28 Finite element mesh for BGA solder joint.

13.4.3 Crack initiation and growth

Crack initiation and growth paths for several typical surface mount solder joint configurations are shown in Fig. 13.29. Peripheral leadless joints generally initiate a major crack at the corner of the chip carrier or underneath the carrier.[45,49–53] Propagation under the carrier is quite fast due to the high stresses/strains in this region. Propagation through the fillet is up to 10 times slower. Gull-wing leaded joints generally initiate cracks in the heal fillet, and possibly the toe fillet as well.[48,54] The two cracks propagate inward and meet in the center. Ball grid array joints initiate a primary crack on the outer edge of the joint, and a secondary crack later initiates on the inner edge. These cracks are most often near the substrate interface. Bulbous C4 joints show crack initiation on opposite corners near the interfaces, and a single crack will propagate across the joint.[6,7,48]

Several investigators have studied crack growth in solder joints. Solomon[55] measured cracked area in scaled-up joints under isothermal shear loading using ultrasonic microscopy, and approximately correlated the cracked area with load drop. Solomon[56] later used the load drop data to predict failure in leadless surface mount joints under isothermal shear loading. Subrahmanyan[3] also derived crack growth

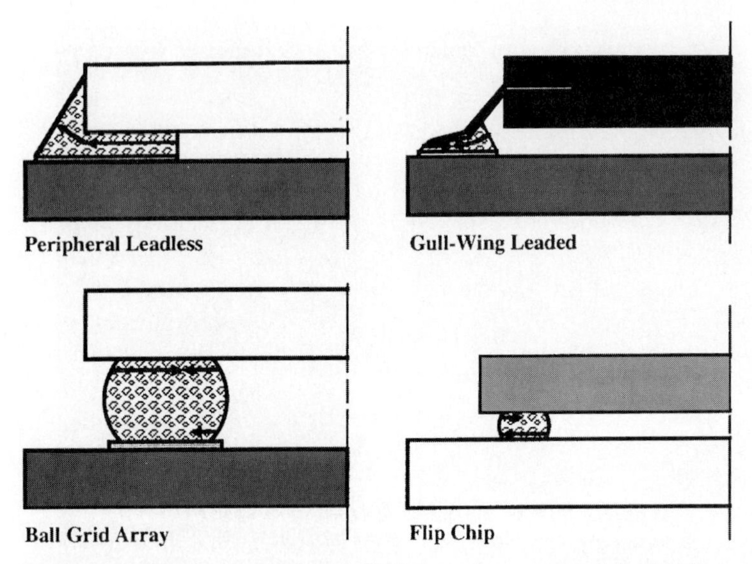

Figure 13.29 Crack initiation and growth in surface mount solder joints.

rates from load drop data, and he incorporated the data into a damage integral function for fatigue life prediction. However, when applying the approach to thermal fatigue data, he used only the damage integral to correlate different thermal cycle profiles, instead of calculating the cycles to failure directly. Satoh et al.[48] and Attarwala et al.[57] estimated crack growth rates in solder joints after thermal cycling by measuring striations using high-resolution scanning electron microscopy. Satoh et al.[48] developed fatigue life expressions which accounted for crack length, but they did not derive explicit equations to predict crack growth rate as a function of calculated stress or strain in the solder joints. Similarly, Clech et al.[42,58] developed a methodology to predict fatigue life based on the average viscoplastic strain energy density in the solder joint, and they normalized data from different joint configurations using the fracture area. Clech's predictions work over a wide range on technologies within a factor of 2 or 3.

A series of tests were run using four assemblies, two alloys, and eight thermal cycle conditions. All of the assemblies used ceramic chip carriers and eutectic ball grid array joints, but there were variations in the PC boards, chip carrier structures, and joint array patterns. The temperature range was varied from –55°C <-> 125°C to 25°C <-> 75°C, and the cyclic frequency was varied from 144 cycles/day to 2 cycles/day. All of the tests were conducted using 62Sn36Pb2Ag, and three additional tests were done using 60Sn40Pb. The test assemblies and thermal cycle conditions are described in detail in Ref. 59.

Cross sections of eutectic BGA joints after four different thermal cycle conditions are shown in Fig. 13.30. In all cases, the cracks initiate near the ceramic interface and propagate through the solder, but stay near the interface. The failure mode was consistent across the wide range of conditions tested. The stresses and strains are highest near the ceramic interface due to the large local expansivity mismatch between the solder (24.5 ppm/°C) and the ceramic (5.3 ppm/°C). There are also cracks near the PC board interface, but to a lesser degree. The solder grain size has coarsened locally around the cracked areas. Heterogeneous coarsening is typically a precursor to cracking.[60] However, the sample that experienced the smallest temperature excursion in Fig. 13.30 (15°C <-> 75°C) seemed to coarsen more uniformly. More samples need to be cross-sectioned to determine if there is a correlation between the degree of heterogeneous coarsening and the maximum temperature or the temperature range.

A simple technique was developed to measure crack growth in the joints. Dye penetrant was applied under three chip carriers from the sample population at regular intervals. The carriers were then pried off the boards revealing the fracture surfaces of the C5 joints. All cracked areas are stained by the dye, so the initiation sites, size, and shape of the cracks were readily seen.

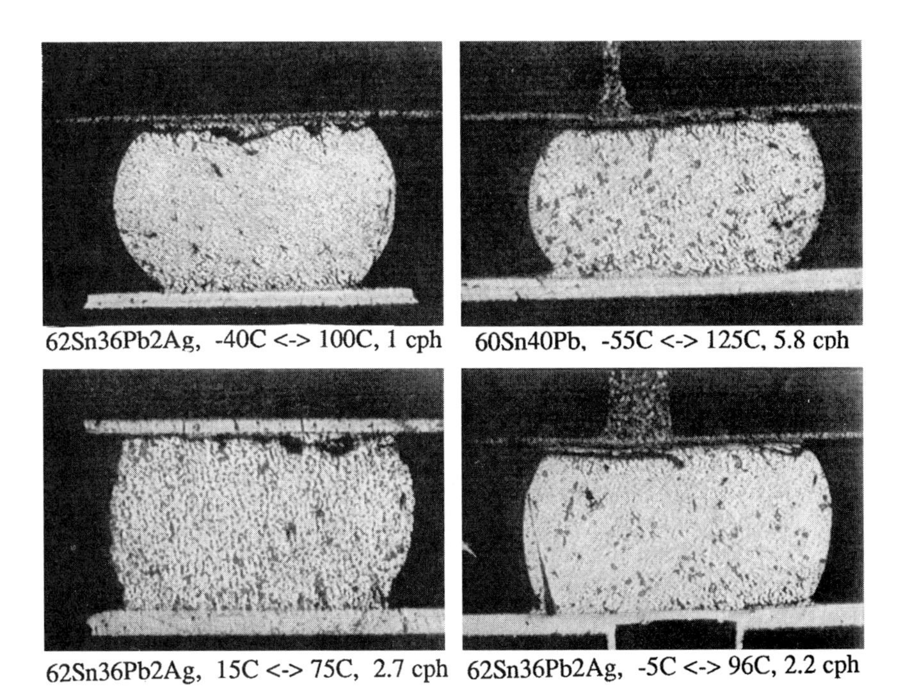

Figure 13.30 Cross sections of solder joints after thermal cycling. Near eutectic CBGA on pc boards. Failure typically near ceramic interface.

Fracture surfaces of outermost BGA joints from test 2 in Ref. 59 are shown in Fig. 13.31, where the areas that were cracked are stained by dye. As thermal cycling progresses, a primary crack initiates at the outside edge of the joint (farthest from package neutral point), and later a secondary crack initiates at the inside edge. The crack fronts move in a fairly uniform fashion toward the center of the joint, and eventually meet up to cause complete failure. Photos like that shown in Fig. 13.31 were taken periodically throughout all of the thermal cycle tests. Each test was divided into six to eight intervals, and three chip carriers were pried off at each interval. The primary and secondary crack lengths were measured along lines extending from the package neutral point to the eight outermost joints. Hence, a total of 24 joints were measured to determine each crack length data point.

In view of Clech's[42] and Nicewarner's[43] work, the crack length data were analyzed using a 3P Weibull distribution. In order to stay consistent with the concept of a failure-free time, the reciprocal crack length was used in curve fitting. A typical plot of $1/a$ versus N is shown in Fig. 13.32. Based on such plots, the characteristic crack length, maximum crack length, and shape parameter were obtained for all of the data

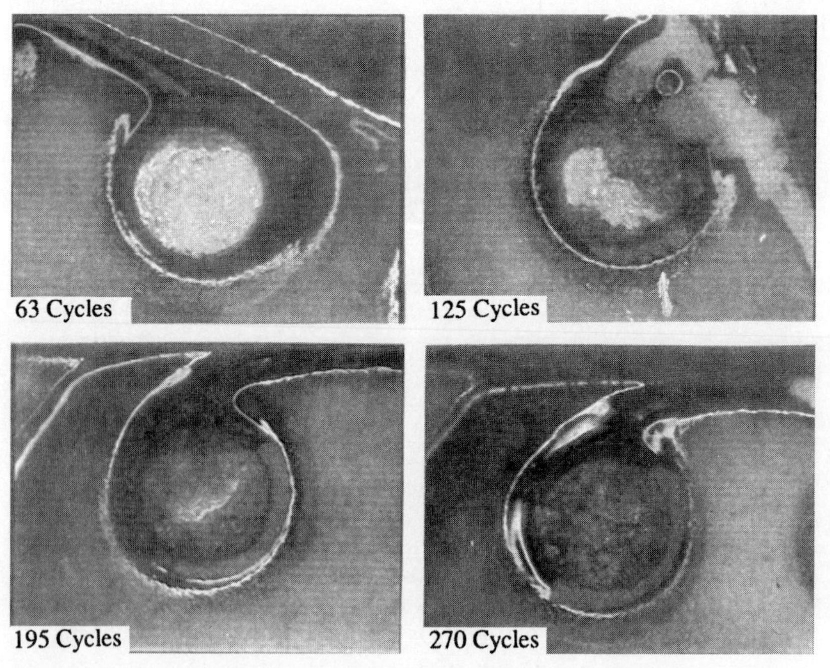

Figure 13.31 Cracked area on outermost BGA joints stained by dye penetrant. Primary crack grows from outside edge, secondary crack grows from inside edge.

sets in the study. This is an improvement over the previous work in which only the mean crack length was determined.[59]

In comparing the maximum crack length to the characteristic crack length, it was observed that there is a large variation for populations of short cracks, but much less variation for long cracks. Hence, as the cracks get longer, their length distribution gets tighter. This trend was also observed by Agarwala.[7] Shown in Fig. 13.33 is the ratio of maximum to characteristic crack length. It is seen that this ratio can exceed 10 for cracks less than 0.001 in, but reduces to about 2 for cracks longer than 0.008 in. If one considers that the fatigue life will be inversely proportional to the crack growth rate, it is expected that the minimum fatigue life will be out half as long as the characteristic fatigue life. Indeed, this is exactly what Clech et al.[42] found in analyzing a very large database where the ratio of failure-free life to characteristic life was about 0.46.

Under isothermal mechanical cycling in a constant inelastic strain range shear-shear mode test, solder joint crack growth rate typically decreases as crack length increases.[4,55] In a displacement-controlled test, where the inelastic strain range increases as the joint cracks, crack growth rate can remain constant or increase.[61] During thermal cycling, Satoh et al.[48] reported that crack growth rate increased slightly with crack length, as determined by striations on the fracture surface. On the other hand, Agarwala[7] found that crack growth rate decreased with crack length, based on cross-sectional analysis. The

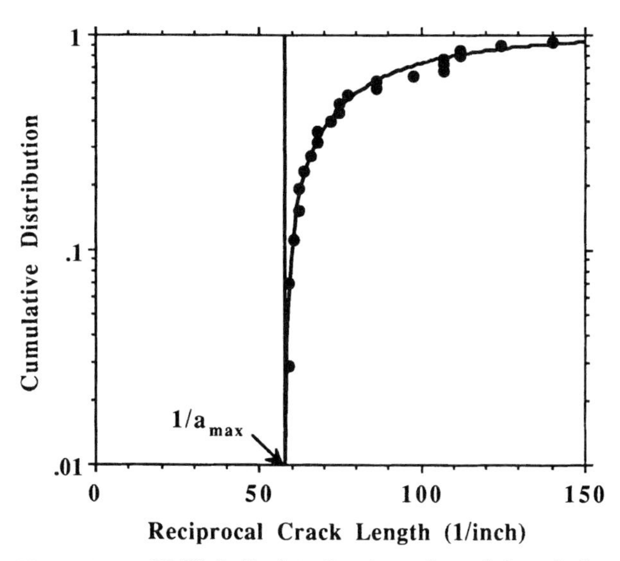

Figure 13.32 3P Weibull plot of reciprocal crack length for 55°C <-> 125°C test at 390 cycle mark.

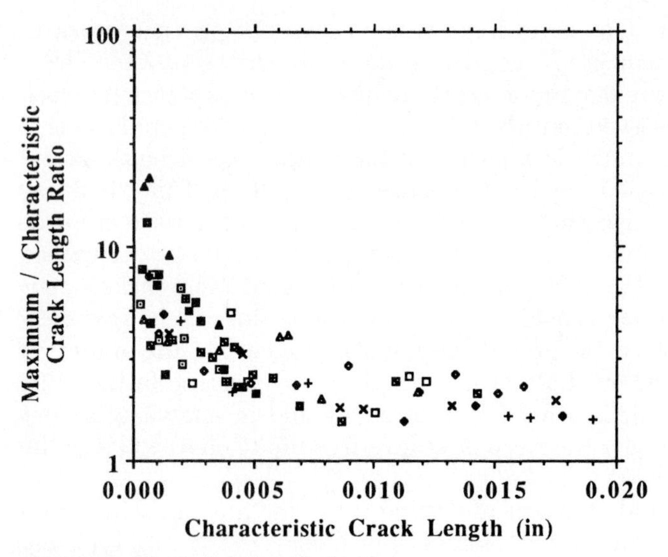

Figure 13.33 Crack length variation.

data for the present study indicates that the characteristic crack growth rate stays essentially constant during the thermal cycle tests. This is shown in Fig. 13.34 where the crack length is plotted versus cycle number for two test conditions from Ref. 59. The number of cycles to crack initiation can be determined by extrapolating the curves back to the x axis (crack length = 0). It is seen that crack initiation only accounts for about 10 percent of the fatigue life. Similar observations were made by Uegai et al.[62] in testing alloy 42 lead frame PQFP solder joints. It is expected that smaller solder joints would have a larger portion of their fatigue life spent in crack initiation. As pointed out by Clech et al.,[42] a constant crack growth rate supports the use of a Weibull failure distribution over a lognormal distribution, because the lognormal distribution is derived based on an accelerating rate of damage accumulation. Clearly, the crack growth data indicates that the rate of damage accumulation is essentially constant throughout the life of a joint.

Under identical conditions, 60Sn40Pb solder has about a 6 to 25 percent higher crack growth rate than 62Sn36Pb2Ag solder.[59] Hence, the use of 62Sn36Pb2Ag will result in slightly more reliable assemblies. However, the largest benefit is under highly accelerated conditions. For long-term fatigue tests or for field use conditions, the advantage of 62Sn36Pb2Ag solder is almost negligible.

In order to develop a predictive model, the crack initiation and growth results must be correlated to the stress and/or strain in the solder joints for all test conditions. Several failure indicators for thermal

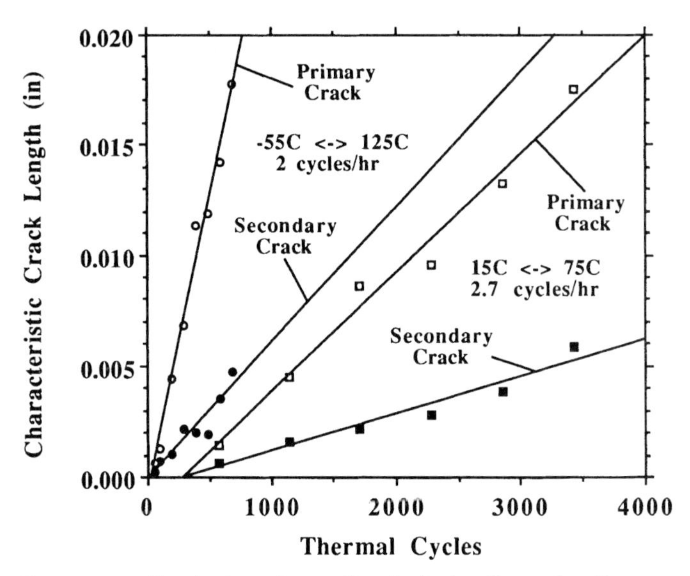

Figure 13.34 Constant crack growth rate during thermal cycling.

fatigue life correlation have been proposed: total strain range,[7,48,62,63] inelastic strain range,[47,64,65] matrix creep strain range,[66,67] viscoplastic strain energy density increment (plastic work per unit volume),[8,42,58,68] J-integral and stress intensity factor,[69] and the damage integral.[3] In the previous study,[59] both the inelastic strain range, $\Delta\varepsilon_i$, and the viscoplastic strain energy density per cycle, ΔW, were used. Strain energy density was found to give the best correlation to the crack growth data, so it will be used here.

ANSYS 5.0 was used to simulate the crack growth experiments as described in Sec. 13.4.2. The viscoplastic strain energy density is defined as the summation of the product between stress and inelastic strain increment vectors over the number of converged subsets.[70] The strain energy density per cycle was calculated using the fifth thermal cycle.

Shown in Fig. 13.35 is the correlation for the number of cycles to crack initiation in 62Sn36Pb2Ag joints. The curve can be represented by

$$N_o = 7860\ \Delta W^{-1.00} \tag{13.35}$$

where ΔW is in psi. There is considerable scatter in the plot, but this is not a great concern because initiation represents a small fraction of the total life in most assemblies. The correlation for linear crack growth rate in 62Sn36Pb2Ag joints is shown in Fig. 13.36, and it is described by the following relation

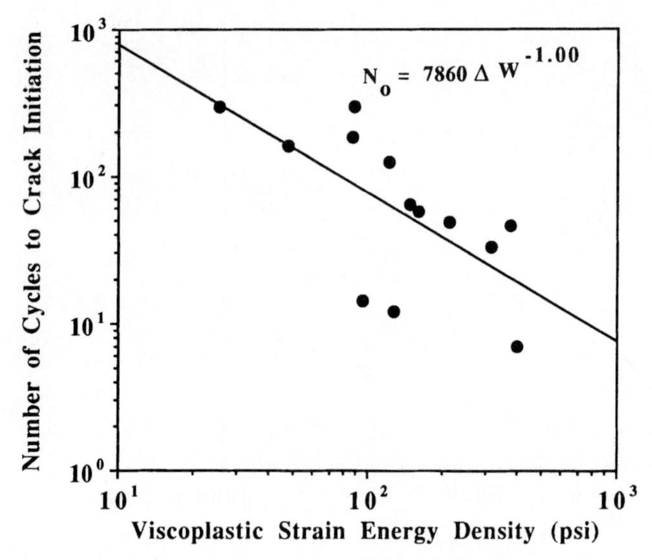

Figure 13.35 Crack initiation correlation for 62Sn36Pb2Ag.

$$\frac{da}{dN} = 4.96E\text{-}8\ \Delta W^{1.13} \tag{13.36}$$

where da/dN is in inches per cycle and ΔW is in psi. It is seen that the crack growth data are better correlated than the crack initiation data (correlation coefficient = 0.919 versus 0.433). It is significant that a single relation can be used to describe data from such a wide range of thermal cycle conditions. It is not completely surprising, however, since the failure mode was consistent between the tests (see Fig. 13.30). This finding dispels the notion set forth by Englemeir[71] that ramp rates should be kept below 20°C/min, because two of the tests in this study were liquid-to-liquid thermal shock, which imposes ramp rates up to 400°C/min.

Since Clech et al.[42,58] established a fatigue correlation using pad area, a geometric model was derived in Ref. 59 to calculate crack area from crack length. Based on this model, the areal crack growth rate correlation for 62Sn36Pb2Ag joints is plotted in Fig. 13.37 and given by

$$\frac{dA}{dN} = 1.00E\text{-}9\ \Delta W^{1.19} \tag{13.37}$$

where dA/dN is in square inches per cycle and ΔW is in psi. Only three tests were run on 60Sn40Pb solder, so there is much less data to correlate. The crack growth rate was slightly higher than 62Sn36Pb2Ag. The difference was 6 percent at 48 psi strain energy density per cycle, and 25 percent at 372 psi.

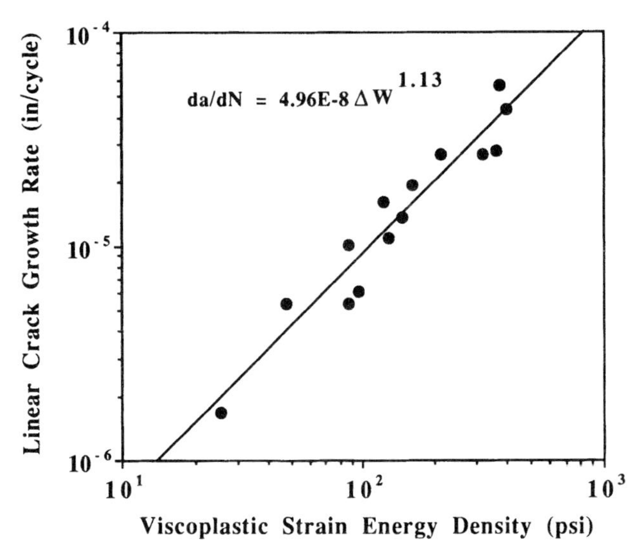

Figure 13.36 Linear crack growth rate correlation for 62Sn36Pb2Ag.

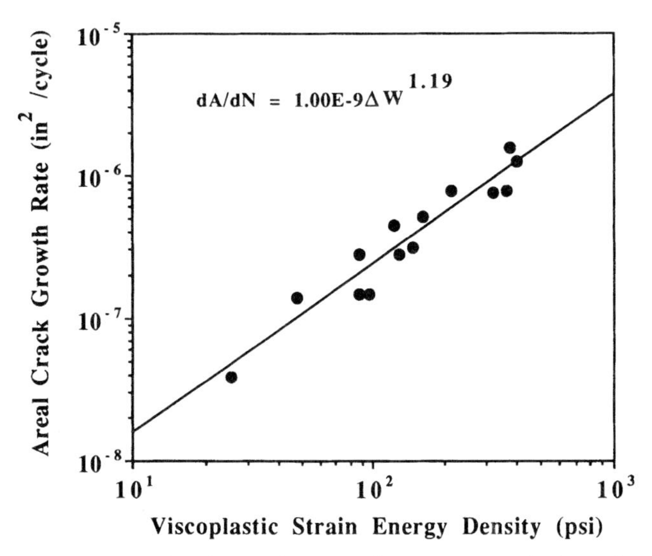

Figure 13.37 Areal crack growth rate correlation for 62Sn36Pb2Ag.

The correlations in Figs. 13.35 to 13.37 have been modified since the earlier publication[59] for two reasons: (1) the finite element models have been significantly improved by using measured board properties and finer mesh densities, and (2) the crack data were analyzed by using a

3P Weibull distribution, so the correlations are now in terms of the characteristic crack length (37 percentile) instead of the mean crack length. Hence, the constants in Eqs. (13.35) to (13.37) have changed, but the trends are still the same. Furthermore, the earlier published correlations for 60Sn40Pb solder are not recommended. Instead, Eqs. (13.35) to (13.37) should be used, with a small correction factor depending on the strain energy range as discussed above.

13.4.4 Thermal cycle reliability studies

Several studies have been conducted to characterize the thermal cycle performance of BGA solder joints. Data from four laboratories are presented in this section. In the first study, BGA assemblies with ceramic substrates were tested. The solder joints were 62Sn36Pb2Ag (they did not have 90Pb10Sn balls for standoff). Four types of assemblies were tested as outlined in Table 13.4. All boards had a 10 × 10 array of pads, but half of the chip carriers had a full array of joints, and the other half only had perimeter joints. Within those two groups half had corner joints and half did not. All chip carriers were mounted on 2-layer polyimide PC boards and subjected to air-to-air thermal shock (–55C <->125C, 2 cycles/hr). Continuity nets were continuously monitored using LabView, and failure was defined as an open circuit on 3 consecutive cycles. Each package had up to three nets: corner joints, outer row, and all inner rows.

In the other three studies, plastic BGA assemblies were tested. A wide range of test conditions and assemblies were used as outlined in Table 13.5. The package types were 72, 169, and 225 I/O. The ball diameters were .025, .030, and .038 in, and both 62Sn36Pb2Ag and 63Sn37Pb were used. The pad diameters ranged from 0.019 to 0.033″, and the pad pitch was 0.070 or 0.0591 in (1.5 mm). The joint height ranged from 0.017 to 0.029 in. The die size ranged up to 0.437 in on an edge. The packages were mounted on 0.038- and 0.062-in PC boards. Four thermal cycle conditions were used: –40C<->125C, 1cph; –25C <->100C, 2cph; 0C<->100C, 2cph, and 0C<->100C, 3cph. Continuity was continuously monitored using Anatech systems. Typically, the packages were designed with continuity nets progressing outward from the center.

Selected data are shown in Fig. 13.38 where the cumulative distribution failed is plotted versus thermal cycles. For each data set, a 3P Weibull fit is also plotted. In comparing the CBGA data sets, it is seen that a full array improves the fatigue life about 1.8× over the perimeter array, except for the first failures where they were equal. Finite element modeling indicates that the main advantage of the full array is that the inner rows of joints pick up much of the load as stress relaxes

TABLE 13.4 Thermal Cycle Test Assemblies for Near-Eutectic Ceramic BGA Reliability Study

Data set	12C	13C	14C	15C
Solder joint array	10 × 10 .030 dia .060 pitch .0179 high 62Sn36Pb2Ag	10 × 10, no corners .030 dia .060 pitch .0179 high 62Sn36Pb2Ag	10 × 10, perimeter .030 dia .060 pitch .0172 high 62Sn36Pb2Ag	10 × 10, perimeter no corners .030 dia .060 pitch .0172 high 62Sn36Pb2Ag

NOTE: All dimensions in inches.
PC boards and chip carriers same as assembly 4 in Ref. 59.
Thermal cycle profile same as Data Set 5 in Ref. 59.
For each data set, loop 1 is corner joints, loop 2 is rest of outer row, and loop 3 is inner joints.

during the dwell times. In addition, stresses are much higher on the inner edge of the outermost joints for the perimeter array compared to the full array.

The 72 I/O PBGA has about 2.5× the fatigue life of the 100 I/O CBGA under similar thermal cycle conditions. This is primarily due to package materials which are nearly expansion matched to the product's printed circuit board. For the 72 I/O PBGA, the –40C<->125C, 1cph condition is about 3.3× more damaging than the 0C<->100C, 3cph condition. In comparing the 225 I/O and 72 I/O data, it is seen that the 225 I/O package had about 1.2× to 2× lower fatigue life due to a larger die, and it was mounted on a thicker product circuit board.

A summary of all the fatigue results are given in Tables 13.6 and 13.7, where the 3P Weibull constants are tabulated (failure free life, N_{ff}, characteristic life, α_w, and shape parameter, β_w). For each data set, the three independent parameters were determined by trial and error. For the CBGA package, fatigue life increases with a full array, as discussed above. In general, fatigue life decreases as the distance from package neutral point increases. However, the nets with only four corner joints had about the same performance as the nets with the rest of the outer row joints because the probability of failure depends on the number of joints in the net [see Eq. (13.34)]. For the PBGA package, fatigue life increases with (1) taller solder joints because stress and strain are reduced, (2) larger pad size because the cracks must propagate farther to create a failure, and (3) smaller die size because the die dominates the effective stiffness and expansivity of the package. If one computes the ratio of N_{ff}/α_w, the range is from 0.21 to 0.73 with an average of 0.49. This is very similar to that reported by Clech et al.[42] over 26 data sets with a wide range of package technology and test conditions. For the shape parameter, β_w, the range is from 1.3 to 4.2 with an average of 2.6 (slightly greater than Clech et al.'s average value of 2.2).

TABLE 13.5 Thermal Cycle Test Assemblies for Plastic BGA Reliability Study

Data set	Thermal cycle profile	PC board	Package	Die	Board pad diameter	Package pad diameter	Solder joint array
1P	–40C<->125C 1 cycle/hr 5 min ramps 25 min dwells	2 layer FR4 .062 thick	.630 × .560 ×.050 .030 ϕ balls	.291 × .248 ×.015	.0300	.0280	8 × 9 .070 pitch .0181 high 62Sn36Pb2Ag
2P 11P	–40C<->125C 1 cycle/hr 15 min ramps 15 min dwells	6 layer FR4 .0378 thick	.590 × .530 ×.060 .030 ϕ balls	.266 × .261 ×.015	.0302	.0218	8 × 9 .0591 pitch .0205 high 62Sn36Pb2Ag
3P 12P	–40C<->125C 1 cycle/hr 15 min ramps 15 min dwells	6 layer FR4 .0378 thick	.590 × .530 ×.060 .030 ϕ balls	.266 × .261 ×.015	.0302	.0218	8 × 9 .0591 pitch .0205 high 63Sn37Pb
4P	–40C<->125C 1 cycle/hr 15 min ramps 15 min dwells	6 layer FR4 .0418 thick	.590 × .530 ×.060 .025 ϕ balls	.266 × .261 ×.015	.0247	.0189	8 × 9 .0591 pitch .0174 high 62Sn36Pb2Ag
7P	–40C<->125C 1 cycle/hr 15 min ramps 15 min dwells	6 layer FR4 .0381 thick	.590 × .530 ×.060 .030 ϕ balls	0 × 0 × 0	.0303	.0192	8 × 9 .0591 pitch .0212 high 62Sn36Pb2Ag
9P	–40C<->125C 1 cycle/hr 15 min ramps 15 min dwells	6 layer FR4 .0384 thick	.590 × .530 ×.060 .038 ϕ balls	.266 × .261 ×.015	.0298	.0223	8 × 9 .0591 pitch .0293 high 62Sn36Pb2Ag

TABLE 13.5 ***(Continued)***

10P	–40C<->125C 1 cycle/hr 15 min ramps 15 min dwells	6 layer FR4 .0384 thick	.590 × .530 ×.060 .038 ϕ balls	.266 × .261 ×.015	.0298	.0223	8 × 9 .0591 pitch .0293 high 63Sn37Pb
13P	0C<->100C 3 cycles/hr 5 min ramps 5 min dwells	6 layer FR4 .0389 thick	.590 × .530 ×.060 .030 ϕ balls	.266 × .261 ×.015	.0302	.0220	8 × 9 .0591 pitch .0205 high 62Sn36Pb2Ag
14P	0C<->100C 3 cycles/hr 5 min ramps 5 min dwells	6 layer FR4 .0389 thick	.590 × .530 ×.060 .030 ϕ balls	.266 × .261 ×.015	.0302	.0220	8 × 9 .0591 pitch .0205 high 63Sn37Pb
15P	0C<->100C 3 cycles/hr 5 min ramps 5 min dwells	2 layer FR4 .062 thick	1.07 × 1.07 ×.060 .030 ϕ balls	.398 × .389 ×.0189	.024	.022	15 × 15 .0591 pitch .0225 high 62Sn36Pb2Ag
17P	0C<->100C 3 cycles/hr 5 min ramps 5 min dwells	2 layer FR4 .062 thick	1.07 × 1.07 ×.060 .030 ϕ balls	.398 × .389 ×.0189	.024 Bobtail	.022	15 × 15 .0591 pitch .021 high 63Sn37Pb
18P	–25C<->100C 2 cycles/hr 3.0, 6.4 min ramps 12, 8.6 min dwells	2 layer FR4 .062 thick	.590 × .530 ×.060 .030 ϕ balls	.266 × .261 ×.015	.033	.0226	8 × 9 .0591 pitch .018 high 62Sn36Pb2Ag

TABLE 13.5 ***(Continued)***

19P	–25C<->100C 2 cycles/hr 3.0, 6.4 min ramps 12, 8.6 min dwells	2 layer FR4 .062 thick	.780 × .780 ×.060 .030 ϕ balls	.437 × .437 ×.021	.0241	.022	13 × 13 .0591 pitch .0223 high 62Sn36Pb2Ag
20P	–25C<->100C 2 cycles/hr 3.0, 6.4 min ramps 12, 8.6 min dwells	2 layer FR4 .062 thick	1.07 × 1.07 ×.060 .030 ϕ balls	.398 × .389 ×.0187	.0281	.022	15 × 15 .0591 pitch .0214 high 62Sn36Pb2Ag
21P	0C<->100C 2 cycles/hr 10 min ramps 5 min dwells	6 layer FR4 .062 thick	.95 × .95 ×.060 .030 ϕ balls	.364 × .359 ×.018	.024	.022	13 × 13 .0591 pitch .019 high 62Sn36Pb2Ag
22P	0C<->100C 2 cycles/hr 10 min ramps 5 min dwells	6 layer FR4 .062 thick	1.02 × 1.02 ×.060 .030 ϕ balls	.398 × .389 ×.018	.024	.022	15 × 15 .0591 pitch .019 high 62Sn36Pb2Ag

NOTE: All dimensions in inches.
Package substrate thickness was 0.015 inches.
Die-attach thickness was .002 inches.
Data sets 18P–20P from Refs. 44 and 72.
Data sets 21P and 22P from Ref. 73.

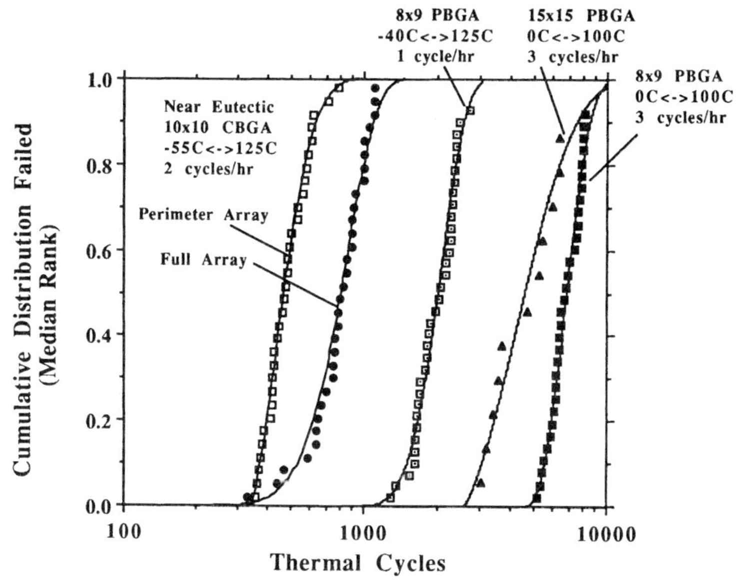

Figure 13.38 BGA thermal fatigue data.

13.4.5 Fatigue model correlation

There are at least two good reasons to develop a fatigue life model: (1) to optimize the design of an assembly, and (2) to predict field reliability based on accelerated test results. In terms of the assembly design, it is important to quantify which design features have the largest impact on reliability (e.g., joint height, pad size, die size, or substrate thick-

TABLE 13.6 Thermal Cycle Results for Near-Eutectic Ceramic BGA Packages

Data set	Sample size	Failure-free cycles N_{ff}	Characteristic life α_w	Shape parameter β_w
12C				
Corners	27	250	989	2.3
Outer row	27	220	1038	2.8
Inner rows	25	450	1289	2.9
13C				
Outer row	32	240	887	2.9
Inner rows	32	450	1153	3.1
14C				
Corners	32	340	526	1.6
Outer row	32	340	508	1.4
15C				
Outer row	29	300	491	2.6

TABLE 13.7 Thermal Cycle Results for Plastic BGA Packages

Data set	Sample size N_{ff}	Failure-free cycles α_w	Characteristic life β_w	Shape parameter
1P	36	1200	2655	3.0
2P	36	1000	2174	2.7
3P	18	1800	2456	3.0
4P	18	600	1152	3.1
7P	18	3000	4751	3.0
9P	34	1200	2835	2.8
10P	32	800	2727	3.4
11P	35	1600	2260	2.0
12P	34	1450	1987	2.0
13P	34	4700	7308	2.1
14P	32	3800	6944	1.5
15P	12	2600	5263	1.3
17P	10	1600	6747	1.7
18P	336	1700	3648	3.8
19P	336	950	2237	4.1
20P	32	2100	2899	2.0
21P	56	1500	4963	4.2
22P	56	2000	3779	3.0

NOTE: Data sets 18P–20P from Refs. 44 and 72.
Data sets 21P and 22P from Ref. 73.

ness). Once the model is developed and correlated, it is faster and less expensive to optimize a design analytically than to use a large matrix of experiments. Obviously experimental work is still necessary, but far fewer experiments need to be conducted, resulting in reduced product development cycle time.

As far as field reliability is concerned, a fatigue life model is the only way to make valid predictions. Field failure data is very difficult to come by, and if it does exist, it is likely to be on very mature technology. As technology keeps changing more rapidly from one generation to the next, field data on older technology is less useful (nevertheless, if anyone has some field data he or she would like to share, please contact the author).

In the present section, the data from the previous two sections will be tied together. Finite element analysis was conducted on all of the tests outlined in Tables 13.4 and 13.5. The results of the analysis were then used with the crack initiation and growth correlations to predict fatigue life at the component level. It was assumed that a primary and a secondary crack form in each joint, and the joint is failed when these two cracks grow together (dye penetrant analysis confirms this assumption). Finally, the measured and predicted fatigue performances were compared.

The procedure to estimate component reliability consists of seven main steps:

1. Calculate the strain energy density distribution in the solder joints using finite element analysis as described in Sec. 13.4.2.
2. Use Eq. (13.35) to calculate the number of cycles to crack initiation for both the primary and secondary region in each joint.
3. Use Eq. (13.36) to calculate the characteristic crack growth rate for both the primary and secondary region in each joint.
4. Calculate the characteristic life of each joint based on the pad diameter using Eq. (13.38) below

$$\alpha_w = N_{os} + \frac{a - (N_{os} - N_{op})\dfrac{da_p}{dN}}{\dfrac{da_s}{dN} + \dfrac{da_p}{dN}} \tag{13.38}$$

   where N_{op} and N_{os} are the number of cycles to initiate the primary and secondary cracks, respectively, da_p/dN and da_s/dN are the primary and secondary crack growth rates, respectively, and a is the total crack length (pad diameter).
5. Based on Fig. 13.33, it is estimated that for a given distribution of crack lengths, the maximum crack length will be 2× the characteristic crack length (for characteristic crack lengths greater than 0.008 in). Hence the failure-free life, N_{ff}, is assumed to be 0.5× the characteristic life calculated in step 4 above.
6. Calculate the row reliability based on the number of joints, q, using Eq. (13.33).
7. Calculate the component reliability based on the row reliabilities using Eq. (13.32).

Finite element analysis was used for step 1, then a spreadsheet was set up for steps 2 to 6. Based on the Weibull constants for each row, a second spreadsheet was set up to calculate the component reliability as a function of thermal cycles for step 7. An example of this procedure is outlined in Table 13.8. In this example, the 72 I/O PBGA was analyzed, the pad diameter, a, at the substrate was 0.0218 in, and the Weibull shape parameter, β_w, was assumed to be 2.6. It is seen that row 2, under the die edge, is the first to fail. This was also determined by Nagaraj[74,75] and Ejim.[73] The characteristic life of the component is essentially equal to that of row 2 (1860 cycles versus 1866 cycles). This is also observed in actual testing of PBGAs, the row of joints under the die edge is nearly always the first to fail. The failure-free life of the component is always equal to the failure-free life of the worst case joint, because it does not depend on the number of joints.

TABLE 13.8 Thermal Fatigue Life Calculation—72 I/O PBGA

	Row 1 6 joints	Row 2 14 joints	Row 3 22 joints	Row 4 30 joints
ΔW primary (psi)	49.7	75.9	40.0	34.9
ΔW secondary (psi)	33.2	24.2	40.0	34.9
N_{op}	158	104	196	225
N_{os}	237	325	196	225
da_p/dN (μin/cycle)	4.10	6.61	3.21	2.75
da_s/dN (μin/cycle)	2.60	1.82	3.21	2.75
α_w joint	3445	2739	3596	4190
N_{ff}	1722	1370	1798	2095
α_w row	2587	1866	2345	2694
α_w component		1860		

The failure prediction methodology was applied to all of the data sets in Tables 13.4 to 13.7. A plot of the predicted failure-free life versus the measured failure-free life is shown in Fig. 13.39. All of the data points fall in the ±2× band. This is similar to the amount of scatter Clech et al.[42] found with their predictive model over a large database. The comparison between predicted and measured characteristic life is shown in Fig. 13.40. There is a good relative prediction within the data, but on average the predicted life is less than the measured life. The average ratio of measured to predicted characteristic life was 1.53. If this factor of 1.53× is applied to all of the predicted cycles, a modified correlation can be established. The correlation between measured and modified predicted characteristic life is shown in Fig. 13.41. Once again, the data are seen to fall within the ±2× band.

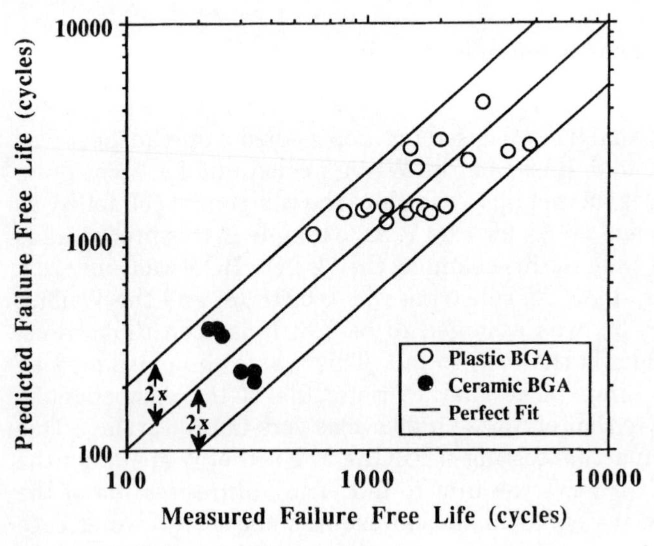

Figure 13.39 Failure-free life correlation.

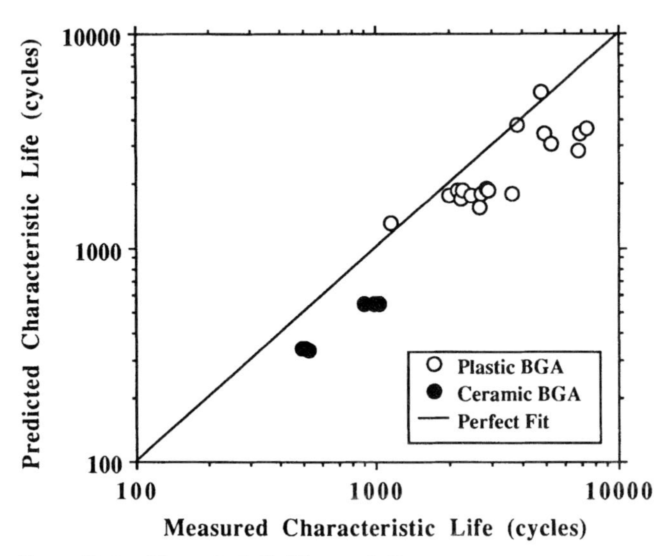

Figure 13.40 Characteristic life correlation.

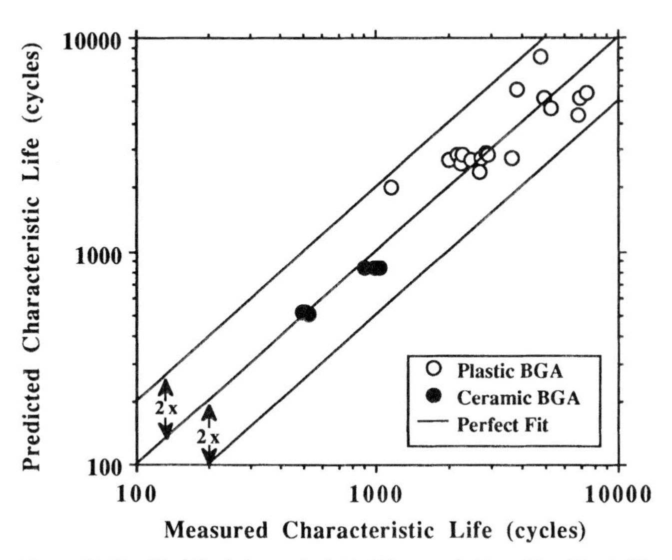

Figure 13.41 Modified characteristic life correlation. $N_{mod}/N = 1.53$.

There are two possible interpretations of these results: (1) the cycles to electrical failure is greater than the cycles to mechanical failure because the fractured surfaces are making point contact long after the joints are essentially failed, or (2) the assumption that the characteristic life is dependent on the number of joints is incorrect. If steps 6 and 7 are omitted in the above procedure, the resulting fatigue life prediction on a joint basis correlates about as well as the modified correlation in Fig. 13.41. On the other hand, Clech et al.[42] showed that it is possible to correlate the failure distribution of nets which contain ⅛ of a component's joints (octals) to the distribution of failures in the component population by using Eq. (13.34). This supports the assumption that the component failure rate would depend on the number of joints. Nevertheless, Clech et al.[42] do not currently incorporate the number of joints into their characteristic life correlation. It is possible that the distribution of failures in a population of joints on a single component is not the same as the distribution of failures of joints across a population of components. This could be true if the factors which cause variation in fatigue life are due to variations between components and boards, not between joints on a given component.

An experiment is currently underway to sort this out. A population of components is being cycled until about 30 percent have failed electrically. Then the test will be terminated and dye penetrant will be used to measure how many mechanical joint failures have occurred. At that point it will be determined if the number of mechanical failures is greater than the number of electrical failures, and if the relative distribution of failures between joints is the same as that between components or between boards.

Since the fatigue life prediction model correlates with measured data over a wide range of conditions, it can be used to estimate field use reliability. To illustrate, a typical 225 I/O assembly was simulated under representative conditions for both a desktop and a portable computer application. The package, board, and use condition assumptions are given in Table 13.9. Assuming a 5-year use requirement and a 2× safety margin, it is estimated that the package is good for more than 3W in desktop application, but only 1.5W in portable application. These simulations assumed a uniform temperature distribution. Under actual conditions there will be temperature gradients in the assembly which will effect the fatigue life somewhat. These gradients will be incorporated into future simulations.

Based on the data and analysis presented in this chapter, design changes which will improve fatigue life under thermal cycling include increasing joint height, pad diameter, and substrate thickness. Nagaraj[75] came to similar conclusions in a separate analytical study. However, it is not clear that all of these changes will improve power cycling performance as well, because increases in joint height or substrate thickness will also increase package-to-board thermal resistance. More simulations and tests are required to answer these questions.

TABLE 13.9 Field Use Reliability Predictions

Package			
0.022″ pads, 0.0223″ height, 0.0591″ pitch, 0.389″ × 0.398″ die, 0.0144″ thick substrate			
Board			
single-sided, 0.062″ thick, 19 ppm/°C TCE, 2.31E6 psi modulus			
Use conditions			
	1W	*2W*	*3W*
Desktop	23C <-> 55C	23C <-> 77C	23C <-> 99C
1 cycle/day, 15 min ramps, 525 min dwell high, 885 min dwell low			
250 cycles/year			
Portable	23C <-> 65C	23C <-> 87C	23C <-> 109C
3 cycles/day, 20 min ramps, 100 min dwells, 12 hr dwell overnight			
750 cycles/year			
Failure-free years			
	1W	*2W*	*3W*
Desktop	37.2	16.9	10.7
Portable	10.5	6.1	4.2
Years to 1% cumulative failure			
	1W	*2W*	*3W*
Desktop	43.7	19.8	12.5
Portable	12.4	7.1	5.0

13.5 Summary

1. A generalized solder joint failure prediction methodology has been outlined for creep rupture and thermal fatigue. Analytical techniques and extensive basic data have been presented.
2. A set of constitutive relations have been defined which describe solder alloy deformation under a wide range of conditions. The deformation behavior of six different alloys can be described by the same set of relations, only the constants are alloy-dependent.
3. Both time-dependent creep and time-independent plastic flow occur simultaneously. A power law sinh expression is used to describe steady-state creep, and the power law breakdown regime is approximately $\tau/G = 10^{-3}$ for all alloys tested. Time-independent plastic flow was observed at stress levels greater than $\tau/G = 10^{-3}$.
4. Solder joints are more creep-resistant than bulk solder specimens due to precipitation strengthening from dispersed intermetallics, and due to the constraining effect of the solder/substrate interfaces. Interface constraint increases as the height/radius aspect ratio decreases.
5. Gold embrittlement in BGA solder joints was characterized. A novel mechanism was identified whereby gold-tin intermetallics precipitate onto interfaces from the bulk. The activation energy for gold embrittlement due to this mechanism was 0.77 eV.

6. The number of cycles to crack initiation and the crack growth rate per cycle were both correlated with inelastic strain energy density per cycle by using nonlinear finite element analysis.
7. Fatigue data was presented for both PBGA and near-eutectic CBGA packages with various assembly and test conditions. For the CBGA, a full array had 1.8× the characteristic life of a perimeter array. For the PBGA, fatigue life increased with smaller die size, larger pad diameter, and greater joint height.
8. The crack growth correlations were used to calculate fatigue life for a wide range of test conditions, and a good correlation was obtained for both the failure-free life and the characteristic life. The methodology was then used to predict field reliability under two representative use environments.

13.6 Nomenclature

a = crack length (in)
a_s = secondary crack length (in)
a_p = primary crack length (in)
A = solder joint pad area (in^2)
A_o = initial solder joint pad area (in^2)
B, B_t = transient creep coefficient
C_4, C_{4t} = prefactor in hyperbolic sin law steady-state creep equation (K $psi^{-1}sec^{-1}$)
C_5, C_{5t} = prefactor in hyperbolic sin law steady-state creep equation (sec^{-1})
C_6, C_{6t} = prefactor in time-independent plastic flow equation
C_{7t} = prefactor in hyperbolic sin law stress-strain rate equation (sec^{-1})
F_j = cumulative fraction of joints failed
G = shear modulus (psi)
G_o = shear Modulus at 273K (psi)
G_1 = temperature dependence of shear modulus (psi K^{-1})
h = solder joint height (in)
h_o = initial solder joint height (in)
k = Boltzmann's constant (8.63E-5 eVK^{-1})
m = exponent in time-independent plastic flow equation
n = stress exponent in steady-state creep equations
N = number of cycles
N_c = number of cycles to component failure
N_{ff} = failure free life in Weibull distribution
N_o = number of cycles to crack initiation
N_{os} = number of cycles to secondary crack initiation
N_{op} = number of cycles to primary crack initiation

P = load (lb)
q = number of solder joints under identical loading
Q = true activation energy (eV)
Q_a = apparent activation energy (eV)
R_j = cumulative fraction of joints surviving
R_c = cumulative fraction of components surviving
s = engineering tensile stress (psi)
t = time (sec)
T = absolute temperature (K)
ΔW = viscoplastic strain energy density per cycle (psi)
α_w = characteristic life (or scale parameter) in Weibull distribution
α, α_t = power law breakdown coefficient in hyperbolic sin law steady-state creep equation
α_1, α_{1t} = power law breakdown coefficient in hyperbolic sin law steady-state creep eq. (psi^{-1})
α_{2t} = power law breakdown coefficient in hyperbolic sin law stress-strain rate eq. (psi^{-1})
β_w = shape parameter in Weibull distribution
δ = displacement across solder joint (in)
ε = tensile strain
γ = engineering shear strain
γ_c, ε_c = creep component of strain
γ_p, ε_p = time-independent plastic flow component of strain
γ_i, ε_i = total inelastic strain
γ_T, ε_T = transient creep strain
σ = tensile stress (psi)

13.7 Acknowledgments

The authors would like to thank Dr. Jean-Paul Clech for many enlightening discussions, Theo Ejim for providing raw data, and Motorola managers Bill Mullen, Glenn Urbish, Mike McShane, and Jim Sloan for continual support of the effort.

13.8 References

1. D. A. Jeannotte, L. S. Goldmann, and R. T. Howard, "Package Reliability," in *Microelectronics Packaging Handbook,* R. R. Tummala and E. J. Rymaszewski, eds., Van Nostrand Reinhold, 1989, pp. 225–359.
2. E. E. de Kluizenaar, "Reliability of Soldered Joints: A Description of the State of the Art, Part 1," *Soldering and Surface Mount Technology,* no. 4, Feb. 1990, pp. 27–38.
3. R. Subrahmanyan, "A Damage Integral Approach for Low-Cycle Isothermal and Thermal Fatigue," Ph.D. thesis, Cornell University, 1990.
4. R. Darveaux, "Mechanical Evaluation of Indium for Die Attachment in a Multichip Package," Ph.D. thesis, North Carolina State University, 1990.
5. J. H. Lau and D. W. Rice, "Solder Joint Fatigue in Surface Mount Technology: State of the Art," *Solid State Technology,* October 1985, pp. 91–104.

6. K. C. Norris and A. H. Landzberg, "Reliability of Controlled Collapse Interconnections," *IBM Journal of Research and Development,* May 1969, pp. 266–271.
7. B. N. Agarwala, "Thermal Fatigue Damage in Pb-In Solder Interconnections," *Proceedings International Reliability Physics Symposium,* 1985, pp. 198–205.
8. W. Engelmaier, "Functional Cycles and Surface Mounting Attachment Reliability," *ISHM Technical Monograph Series 6984-002,* International Society for Hybrid Microelectronics, Silver Springs, MD, October 1984, pp. 87–114.
9. F. Garfalo, *Fundamentals of Creep and Creep-Rupture in Metals,* The Macmillan Company, New York, N.Y., 1965.
10. K. L. Murty and I. Turlik, "Deformation Mechanisms in Lead-Tin Alloys, Application to Solder Reliability in Electronic Packages," *Proceedings 1st Joint Conference on Electronic Packaging,* ASME/JSME, 1992, pp. 309–318.
11. H. J. Frost and M. F. Ashby, *Deformation Mechanism Maps,* Pergamon Press, 1982, chap. 2.
12. K. L. Murty, G. S. Clevinger, and T. P. Papazoglou, "Thermal Creep of Zircaloy-4 Cladding," *Proceedings 4th International Conference on Structural Mechanics in Reactor Technology,* C 3/4, 1977.
13. M. A. Meyers and K. K. Chawla, *Mechanical Metallurgy, Principles and Applications,* Prentice-Hall, 1984.
14. R. Darveaux and I. Turlik, "Shear Deformation of Indium Solder Joints," *IEEE Transactions on Components, Hybrids, and Manufacturing Technology,* vol. 13, no. 4, December 1990, pp. 929–939.
15. M. Harada and R. Satoh, "Mechanical Characteristics of 96.5Sn/3.5Ag Solder in Micro-Bonding," *Proceedings 40th Electronic Components and Technology Conference,* 1990, pp. 510–517.
16. F. G. Foster, "Embrittlement of Solder by Gold from Plated Surfaces," *Papers on Soldering,* 65th ASTM Meeting, 1962.
17. J. L. Freer and J. W. Morris, Jr. "Microstructure and Creep of Eutectic Indium/Tin on Copper and Nickel Substrates," *Journal of Electronic Materials,* vol. 21, no. 6, June 1992, pp. 647–652.
18. S. S. Manson, *Thermal Stress and Low Cycle Fatigue,* McGraw-Hill, New York, 1966.
19. H. D. Solomon, "Low-Frequency, High-Temperature Low Cycle Fatigue of 60Sn-40Pb Solder," in *Low Cycle Fatigue, ASTM STP 942,* H. D. Solomon, G. R. Halford, L. R. Kaisand, and B. N. Leis, eds., American Society for Testing and Materials, Philadelphia, 1988, pp. 342–370.
20. H. J. Frost, R. T. Howard, P. R. Lavery, and S. D. Lutender, "Creep and Tensile Behavior of Lead-Rich, Lead-Tin Solder Alloys," *Proceedings IEEE 38th Electronic Components Conference,* 1988, pp. 13–22.
21. R. Darveaux and K. Banerji, "Constitutive Relations for Tin Based Solder Joints," *Proceedings IEEE 42nd ECTC,* 1992, pp. 538–551, and *IEEE Trans. on CHMT,* vol. 15, no. 6, December 1992, pp. 1013–1024.
22. R. Darveaux and K. L. Murty, "Effect of Deformation Behavior on Solder Joint Reliability Prediction," *Proc. TMS-AIME Symposium on Microstructures and Mechanical Properties of Aging Materials,* November 1992.
23. R. Darveaux, K. L. Murty, and I. Turlik, "Predictive Thermal and Mechanical Modeling of a Developmental MCM," *JOM,* July, 1992, pp. 36–41.
24. CINDAS Report 93, February, 1989, SRC Document C89107.
25. J. Weertman, "Creep of Indium, Lead, and Some of Their Alloys with Various Metals," *Trans. TMS-AIME,* vol. 218, April, 1960, pp. 207–218.
26. J. C. Wei, "High-Temperature Creep of Lead-Indium Alloys," Ph.D. thesis, Stanford University, 1981.
27. C. G. Schmidt, "A Phenomenological Constitutive Model for Type Sn62 Solder Wire," *Proceedings IEEE 39th Electronic Components Conference,* 1989, pp. 253–258.
28. A. Skipor, S. Harren, and J. Botsis, "Constitutive Characterization of 63/37 Sn/Pb Eutectic Solder Using the Bodner-Partom Unified Creep-Plasticity Model," *Proceedings 1st Joint Conference on Electronic Packaging,* ASME/JSME, 1992, pp. 661–672.
29. B. P. Kashyap and G. S. Murty, "Experimental Constitutive Relations for the High Temperature Deformation of a Pb-Sn Eutectic Alloy," *Materials Science and Engineering,* 50, 1981, pp. 205–213.

30. K. Banerji and R. Darveaux, "Effect of Aging on the Strength and Ductility of Controlled Collapse Solder Joints," *Proc. TMS-AIME First International Conference on Microstructures and Mechanical Properties of Aging Materials,* November, 1992, pp. 431–442.
31. M. F. Ashby, F. J. Blunt, and M. Bannister, "Flow Characteristics of Highly Constrained Metal Wires," *Acta Metallurgica,* vol. 37, no. 7, 1989, pp. 1847–1857.
32. A. Skipor, S. Harren, and J. Botsis, "On the Plasticity of the Solder Joint," *Proc. AMT Symposium,* July, 1993.
33. R. N. Wild, "Effects of Gold on Solder Properties," *Proc. Internepcon,* Brighton, 1968, pp. 27–32.
34. F. G. Foster, "Embrittlement of Solder by Gold from Plated Surfaces," in *Papers on Soldering,* ASTM STP 319, pp. 13–19.
35. Donald H. Daebler, "An Overview of Gold Intermetallics in Solder Joints," *Surface Mount Technology,* 1991, pp. 43–46.
36. W. G. Bader, "Dissolution and Formation of Intermetallics in the Soldering Process," Physical Metallurgy of Metal Joining, *Proc. Symp. at Fall Meeting of the Met. Soc. of AIME,* St. Louis, Mo. 1980, pp. 257–268.
37. J. L. Marshall, "Scanning Electron Microscopy and Energy Dispersive X-ray (SEM/EDX) Characterization of Solder, Solderability and Reliability," in *Solder Joint Reliability—Theory and Applications,* J. H. Lau (ed.), Van Nostrand Reinhold, 1991, pp. 173–224.
38. *CRC Handbook of Chemistry and Physics,* R. C. Weast, ed., CRC Press, Boca Raton, Fla., 1986.
39. W. Weibull, "A Statistical Distribution Function of Wide Applicability," *Journal of Applied Mechanics,* September 1951, pp. 293–297.
40. R. B. Abernethy, *The New Weibull Handbook,* Publisher: R. B. Abernethy, North Palm Beach, Fla., October 1993.
41. C. Lipson and N. J. Sheth, *Statistical Design and Analysis of Engineering Experiments,* McGraw-Hill, New York, 1973.
42. J.-P. Clech, D. M. Noctor, J. C. Manock, G. W. Lynott, and F. E. Bader, "Surface Mount Assembly Failure Statistics and Failure Free Time," *Proc. 44th IEEE ECTC,* 1994.
43. E. Nicewarner, "Historical Failure Distribution and Significant Factors Affecting Surface Mount Solder Joint Fatigue Life," *Proc. IEPS,* September, 1993, pp. 553–563.
44. A. Mawer and R. Darveaux, "Calculation of Thermal Cycling and Application Fatigue Life of Plastic Ball Grid Array (PBGA) Package," *Proc. IEPS,* 1993.
45. B. Ozmat, "A Nonlinear Thermal Stress Analysis of Surface Mount Solder Joints," *Proc. 40th IEEE ECTC,* vol. II, 1990, pp. 959–972.
46. B. Mirman and S. Knecht, "Creep Strains in an Elongated Bond Layer," *Proceedings InterSociety Conference on Thermal Phenomena,* 1990, pp. 21–32.
47. R. Subrahmanyan, D. Stone, and C.-Y. Li, "Deformation Behavior of Leadless 60Sn40Pb Solder Joints," *Proceedings MRS Symposium,* vol. 108, 1988, pp. 381–384.
48. R. Satoh, K. Arakawa, M. Harada, and K. Matsui, "Thermal Fatigue Life of Pb-Sn Alloy Interconnections," *IEEE Trans. CHMT,* vol. 14, no. 1, March 1991, pp. 224–232.
49. R. N. Wild, "Some Factors Affecting Leadless Chip Carrier Solder Joint Fatigue Life II," *Circuit World,* vol. 14, no. 4, 1988, pp. 29–36, 41.
50. W. M. Wolverton, "The Mechanisms and Kinetics of Solder Joint Degradation," *Brazing & Soldering,* no. 13, autumn 1987, pp. 33–38.
51. B. Wong and D. E. Helling, "A Mechanistic Model for Solder Joint Failure Prediction Under Thermal Cycling," *Journal of Electronic Packaging,* vol. 112, June 1990, pp. 104–109.
52. J. M. Smeby, "Solder Joint Behavior in HCC/PWB Interconnections," *IEEE Trans. CHMT,* vol. CHMT-8, no. 3, September 1985, pp. 391–396.
53. C. J. Brierley and D. J. Pedder, "Surface Mounted IC Packages—Their Attachment and Reliability on PWBs," *Circuit World,* vol. 10, no. 2, 1984, pp. 28–31.
54. M. McShane, P. Lin, G. Dody, and J. Bigler, "Lead Configuration and Performance for Fine Pitch SMT Reliability," *Proc. NEPCON West,* 1990, pp. 238–257.

55. H. D. Solomon, "Low Cycle Fatigue of 60/40 Solder—Plastic Strain Limited vs. Displacement Limited Testing," *Proc. ASM 2nd Electronic Packaging Conference,* 1985, pp. 29–47.
56. H. K. Solomon, "Low Cycle Fatigue of Surface Mounted Chip Carrier/Printed Wiring Board Joints," *Proc. IEEE 39th ECC,* 1989, pp. 277–292.
57. A. I. Attarwala, J. K. Tien, G. Y. Masada, and G. Dody, "Confirmation of Creep and Fatigue Damage in Pb/Sn Solder Joints," *Journal of Electronic Packaging,* vol. 114, June 1992, pp. 109–111.
58. J.-P. Clech, J. C. Manock, D. M. Noctor, F. E. Bader, and J. A. Augis, "A Comprehensive Surface Mount Reliability (CSMR) Model Covering Several Generations of Assembly Technology," *Proc. IEEE 43rd ECTC,* 1993, pp. 62–70.
59. R. Darveaux, "Crack Initiation and Growth in Surface Mount Solder Joints," *Proc. ISHM International Symposium on Microelectronics,* 1993, pp. 86–97.
60. D. Frear, D. Grivas, and J. W. Morris, Jr., "Parameters Affecting Thermal Fatigue Behavior of 60Sn-40Pb Solder Joints," *Journal of Electronic Materials,* vol. 18, no. 6, 1989, pp. 671–680.
61. Z. Guo, A. F. Sprecher, and H. Conrad, "Crack Initiation and Growth During Low Cycle Fatigue of Pb-Sn Solder Joints," *Proc. IEEE 41st ECTC,* 1991, pp. 658–666.
62. Y. Uegai, S. Tani, A. Inoue, S. Yoshioka, and K. Tamura, "A Method of Fatigue Life Prediction for Surface-Mount Solder Joints of Electronic Devices by Mechanical Fatigue Test," *Proc. ASME International Electronic Packaging Conference,* 1993, vol. 1, pp. 493–498.
63. J. P. Clech and J. A. Augis, "Engineering Analysis of Thermal Cycling Accelerated Tests for Surface-Mount Attachment Reliability Evaluation," *Proc. 7th IEPS,* 1987, pp. 385–410.
64. J. Sauber and J. Seyyedi, "Predicting Thermal Fatigue Lifetimes for SMT Solder Joints," *Trans. ASME Journal of Electronic Packaging,* vol. 114, December 1992, pp. 473–476.
65. R. Darveaux and K. Banerji, "Fatigue Analysis of Flip Chip Assemblies Using Thermal Stress Simulations and a Coffin-Manson Relation," *Proc. 41st IEEE ECTC,* 1991, pp. 797–805.
66. M. C. Shine and L. R. Fox, "Fatigue of Solder Joints in Surface Mount Devices," *Low Cycle Fatigue,* ASTM STP 942, H. D. Solomon, G. R. Halford, L. R. Kaisand, and B. N. Leis, eds., American Society for Testing and Materials, Philadelphia, Pa., 1988.
67. R. Iannuzzelli, "Predicting Solder Joint Reliability—Model Validation," *Proc. IEEE 43rd ECTC,* 1993, pp. 839–851.
68. V. Sarihan, "Energy-Based Methodology for Damage and Life Prediction of Solder Joints under Thermal Cycling," *Proc. IEEE 43rd ECTC,* 1993.
69. J. H. Lau, "Thermal Fatigue Life Prediction of Flip Chip Solder Joints by Fracture Mechanics Method," *International Journal of Engineering Fracture Mechanics,* 1993, pp. 643–654.
70. *ANSYS Users Manual,* vol. IV Theory, Swanson Analysis Systems, Houston, Pa., 1992.
71. W. Engelmaier, "The Use Environments of Electronic Assemblies and Their Impact on Surface Mount Solder Attachment Reliability," *IEEE Trans. on CHMT,* vol. 13, no. 4, December 1990, pp. 903–908.
72. M. Petrucci, R. Johnson, A. Mawer, T. McQuiggin, B. Nelson, and D. Rosckes, "Feasibility Study of Ball Grid Array Packaging," *Nepcon East,* 1993.
73. T. Ejim, "AT&T BGA Reliability Assessment Program," *SEMI/HDP Users Group Array Packaging Workshop,* Boston, Mass., 10/21/93.
74. B. Nagaraj and M. Mahalingam, "Package-to-Board Attach Reliability—Methodology and Case Study on Ompac Package," *Proc. ASME International Electronic Packaging Conference,* 1993, vol. 1, pp. 537–543.
75. B. Nagaraj, "OMPAC Package C5 Reliability—Parametric Study," *Proc. Motorola 1993 Summer AMT Symposium,* pp. 133–141.

Chapter

14

Area Tape Automated Bonding Ball Grid Array Technology

Chin-Ching Huang and Ahmad Hamzehdoost

14.1 Introduction

As IC devices increase to higher frequency and higher power dissipation, traditional packaging faces drastic challenges. The traditional packaging approaches which use Quad Flat Pack (QFP) and Thin Quad Flat Pack (TQFP) for Surface Mount Technology (SMT) have many physical constraints. The constraints are relatively large body size (see Fig. 14.1),[1] with long leads having high inductance and capacitance, high profiles, and high thermal resistance. These constraints face major challenges in high-end applications and next-generation ASICs which will expand device lead counts beyond 800 and speeds from 100 MHz to 400 MHz.

The Area Tape Automated Bonding (ATAB) Ball Grid Array (BGA), or TBGA, provides high lead counts (Table 14.1), a thin, lightweight, high electrical and thermal performance, and BGA surface mount solution to today's package challenge.[1] This chapter will focus on the structure, assembly of this package, and thermal and electrical management of TBGA.

14.2 Structure of TBGA

MQFPs have been in existence for only a couple of years, and their structure and dimensions have already been standardized within the packaging industry. Figure 14.2*a* shows a cross section of a MQFP with a die-up orientation.

Recently Plastic BGA (PBGA), shown in Fig. 14.2*b*, was introduced by Citizen and Amkor Electronics. This package family has gained a lot of popularity mainly due to its small footprint and high leadcount capability. In PBGA, an IC chip is placed on a substrate made of bis-

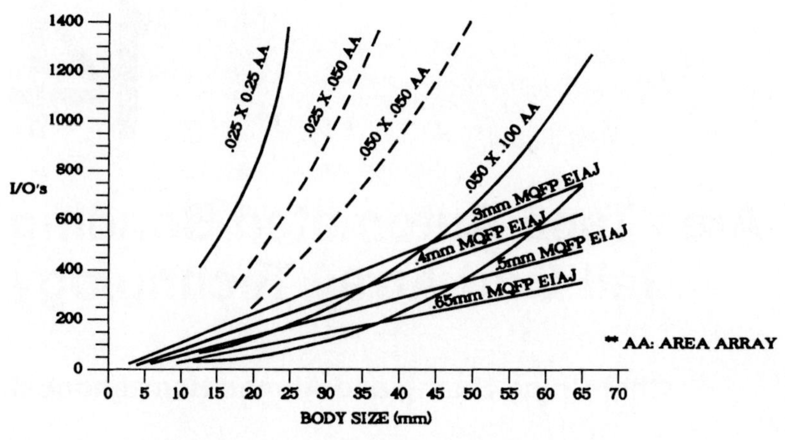

Figure 14.1 I/O Capability of ATAB Package and QFP versus body size (mm) after Ref. 1.

maleimidetriazine (BT) resin. After wire-bonding, the top side of the structure is molded followed by the solder ball attachment. Solder balls in PBGA are commonly 37/63 percent alloy of PB/Sn. Peripheral ATAB BGA and ATAB BGA, shown in Fig. 14.2*c–d,* are a different version of BGA with TAB interconnect used in place of the standard gold-wire bonding process. Here a TAB tape made of copper with a carrier such as upilex film is used as a conducting bridge between silicon die and the substrate. Solder balls made of 10/90 Sn/PB alloy provide the right contact from the copper leads to the printed circuit board. An external heat sink is attached to the back of the die to remove the generated heat within the chip. Size of this external heat sink can be changed to meet the power dissipation requirement of the silicon chip.

TABLE 14.1 ATAB I/Os for 50 Mil Grid OLB Array

Package size (mm)	Array	I/O Max
21	16	256
23	18	324
25	19	361
27	21	441
29	22	484
31	24	576
33	26	676
35	27	729
37.5	29	841
40	31	961

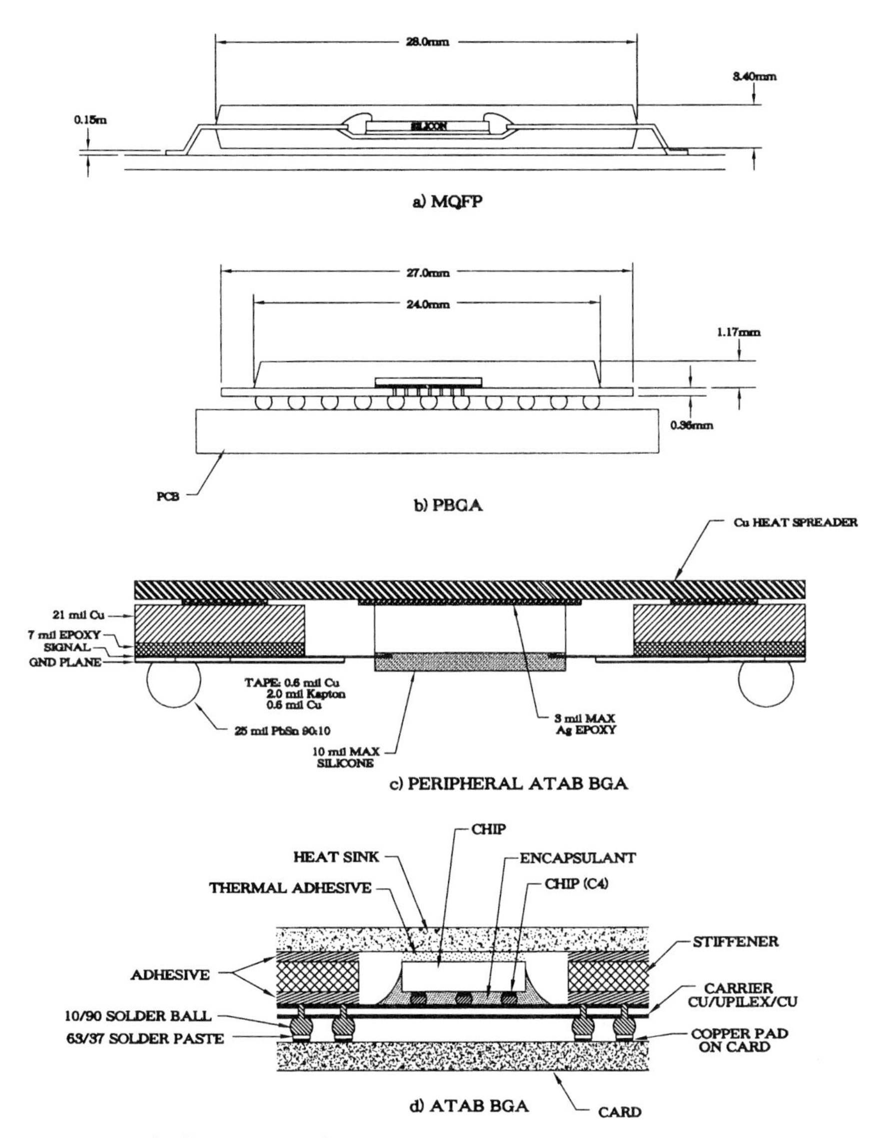

Figure 14.2*a–d* Cross section of MQFP, PBGA, peripheral ATAB BGA, ATAB BGA.

14.3 Assembly of TBGA

The ATAB package consists of a tape with 2-layer copper sandwiched between a 2-mil-thick polyimide called Upilex dielectric to interconnect the die with gold bumps or C4 bumps to PCB or FR4-board through 10/90 (Sn/Pb) solder balls. The schematic structure is shown in Fig. 14.2*d*. The heat sink can be placed on the backside of the die and stiffener by filled thermal adhesive on the back of the die and adhesive on one side of the stiffener. The stiffener is made of materials with the coefficient of thermal expansion (CTE) matching the CTE of the carrier (PCB or substrate) to minimize the stress of the joints during temperature cycling or stress testing.[1] For mounting to FR-4 PCB, the tin-plated copper or FR-4 materials are used for stiffener. The main functions of the stiffener are to provide rigidity and planarity of this package.

The conventional peripheral Inner Lead Bonding (ILB) process (usually the gold bump process) and the so-called solder attach tape technology (SATT)[2] can be used for growing the bumps or balls for the die to connect to the tape. The advantage of using SATT is that the SATT balls (97/3/tin/lead) can be used for joining die to ceramic substrate. An encapsulation process is needed for the protection of die, balls, and lead (tape). Usually, an epoxy encapsulant is coated on top of the die covering balls, bumps and tape leads by dispensing.

The material for encapsulation has to be good enough to pass all the stress test. The ATAB BGA package can be assembled on the PCB board by using the standard Infrared Reflowed (IR) process by melting 63/37/tin/lead solder paste.

14.4 Reliability of TBGA

14.4.1 Reliability concerns in TBGA

Figure 14.2*d* shows a schematic drawing of ATAB BGA structure and assembly. It shows some reliability concerns for electronics packaging. The Inner Lead Bonding (ILB), Outer Lead Bonding (OLB), adhesive, encapsulant sites, and solder-ball-to-PCB junction may be the risk sites of reliability.

The reliability of inner lead bonding with traditional gold bump to tape have been discussed by Evans.[3] The reliability of solder ball to PCB were also discussed by Kromann.[4] It seems that the reliability concerns of these sites in TBGA are already resolved. The following section will discuss the overall accelerated stress tests and some results provided by the vendor (IBM).

14.4.2 Accelerated stress test conditions and test results for TBGA reliability

The purposes of accelerated stress tests for electronics packaging are to estimate the lifetime of products (packages) in field environments and to qualify the products (packages) in simulated conditions of applications.

TBGA packages have been under several accelerated stress tests for reliability. According to IBM-released data, (IBM is the only vendor who provides the package) the JEDEC moisture precondition, accelerated temperature cycle (ATC), thermal age, temperature/humidity/Bias, mechanical vibration plus ATC have been done to quality this package.

The JEDEC moisture preconditions for TBGA is a sequential test to simulate the surface mount process. The conditions are ship shock test (5 cycle, –40 to +50°C), bake (24 hours, 125°C), moisture soak (96 hours, 30°C/60 percent RH) and IR reflow (3 passes). Table 14.2 shows the initial ATC fails data, Table 14.3 shows ATC MIL STD 883 condition B fails.

14.4.3 Summary

The data shown in Sec. 14.4.2 indicates that TBGA will be a reliable package for electronics industry, even though more data in reliability are needed in the near future.

14.5 Thermal Management of TBGA

14.5.1 Introduction

What is the maximum power that can be dissipated in a microelectronics package? How high will the junction temperature reach? Can the package handle high temperatures? What can be done to improve

TABLE 14.2 55 to 125°C ATC (Wet) Fails

Encap	CP	Device size (mm)	T_o	250	500	750	1000
A*	No	8.9	0	0	0	0	0
B*	No	8.9	0	0	0	0	0

* Encapsulant was provided by different vendors.
NOTE: CP = cover plate
T_o = before stress test
Encap = encapsulant

TABLE 14.3 –55 to +125°C ATC MIL STD 883 Condition B Module Fails

Encap	CP	T_o	140	278	383	490	593	672	780	879
B*	Yes	1	0	0	0	0	0	0	0	0
B*	No	0	0	0	0	0	0	0	0	0

* Encapsulant was provided by different vendors.
NOTE: CP = cover plate
T_o = before stress test
Encap = encapsulant
SOURCE: Data provided by IBM

the thermal performance of a given package? These are the typical questions and concerns in today's microelectronics industry where too often limits of IC designs are easily reached and designers are asked to maximize on every aspect of their product.

With current trends toward faster-operating integrated circuits, higher powers need to be dissipated often in ever smaller and less expensive microelectronics packages. The higher the power dissipated by the IC chip, the higher its temperature will reach. The question is if the package will help the IC chip to dissipate the power generated by the circuitry.

In order to understand and provide a suitable thermal management solution for each specific case, one needs to understand the fundamental of heat transfer within a package. Various benchtop tests performed by package and IC vendors can be used to assist the designers to predict and understand further the thermal performance of various packages along with their operating limits. Once these limitations are well understood, then available options can be compared for choosing the best candidate based on factors such as cost, reliability, and cycle time.

14.5.2 Heat transfer in microelectronics packaging

The generated heat within IC is transferred to the outside by various modes of heat transfer known as conduction, convection, and radiation, as shown in Fig. 14.3*a–d*. In general, heat flow through solids is by means of conduction. In packaging this heat transfer mode is through various internal mediums such as leadframe, plating surfaces, vias, solder bumps, and molding material.

At the interface between solid and fluids, heat transfer mechanism is by convection. Airflow or liquid cooling, widely used in the packaging industry, are good examples of convection heat transfer. In radiation mode, thermal energy is transported from one surface to another by means of heat waves. In general, black surfaces are more efficient to emit the heat away from the package compared to light-colored surfaces. In most IC packaging applications, a small portion of the heat is carried away by radiation.

The area right under and above the chip is very critical in distributing the heat. The key for superior thermal performance is utilizing this area to its best. Heat slugs placed in this area can be used in MQFP to improve the thermal performance of the package.[5] These heat slugs basically take the place of the die attach pad. In cases such as TBGA BGA, where the die is accessible from outside, external heat sinks can be attached directly to the back of the die and large amount of the heat is transferred through this path.

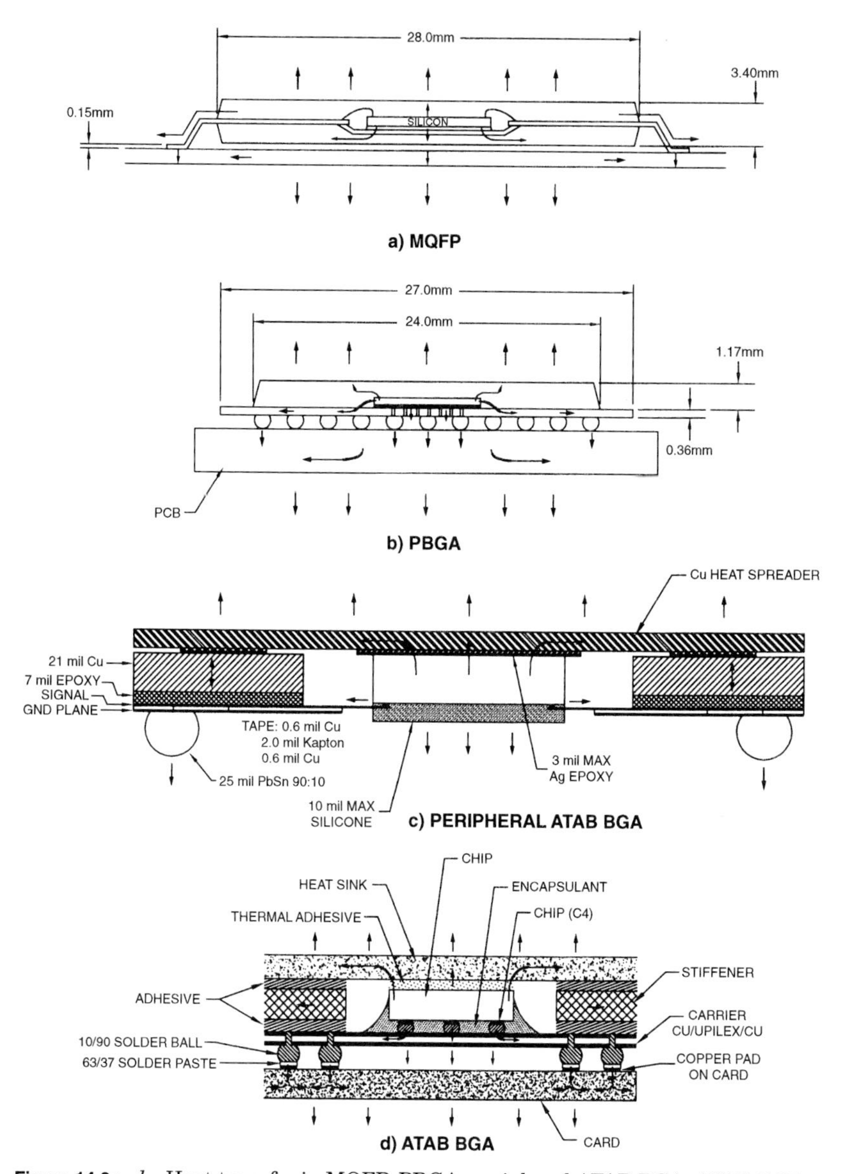

Figure 14.3*a–d* Heat transfer in MQFP, PBGA, peripheral ATAB BGA, ATAB BGA.

14.5.3 Thermal resistance concept

The thermal resistance of a package is determined by adding all the individual thermal resistances seen in the path of the heat flow, both internal as well as external, to the package. This thermal resistance commonly is known as theta jx and is defined as:

$$\theta_{jx} = (T_j - T_x)/P \qquad (14.1)$$

where θ_{jx} = thermal resistance from junction to a reference point x such as ambient air, specific location on the PCB, or top of the heat sink
T_j = junction temperature
T_x = the reference temperature
P = power generated by the IC

Generally, resistances from silicon junction to the package surface (θ_{jc}) and also from junction to the ambient air (θ_{ja}) are more commonly used. Once these parameters are known, maximum junction temperatures can easily be determined for a set of given conditions using Eq. (14.1). The relation between these two parameters is shown below:

$$\theta_{ja} = \theta_{jc} + \theta_{ca} \tag{14.2}$$

where θ_{ca} = thermal resistance from case to ambient air.

The numerical approaches could also be used to determine the thermal resistances. These approaches vary widely from use of simple one-dimensional classical heat transfer equations to more sophisticated computer modeling software tools. Lets first discuss some of the simple one-dimensional approaches.

14.5.4 Conduction

As discussed above, heat transfer within the fully encapsulated package is by conduction. Packages with cavity could have some type of convection and radiation heat transfer depending on material used in the cavity.

Internal thermal resistance of packages can be calculated by adding all the conductive thermal resistance against the flow of heat. This resistance is calculated with the following equation:

$$\theta_{jc} = t/KA = (T_j - T_c)/P \tag{14.3}$$

where t = thickness of the material that heat will travel through
K = thermal conductivity of the material
A = surface area of the material normal to the heat flow

There are various types of microelectronic packages used in the industry. Each one has its own thermal characteristics often quite different from other packages. Table 14.4 shows various materials used in microelectronic packaging along with their thermal conductivities. Typically, metallic materials are used for interconnecting and plating purposes while ceramics and FR4 are used as a substrate. Molding compounds are used as encapsulation and protective compounds for the IC chip.

TABLE 14.4 Common Material used in Microelectronic Packaging

Material	Thermal conductivity (W/m · K)
Alloy 42	16
Aluminum	205
Copper	390
Copper alloy	100–200
Gold	320
Kovar	17
Lead	34
Nickel	90
Molybdenum	138
Platinum	69
Silver	418
Solder (95%PB5%Sn)	36
Solder (37%PB63%Sn)	53
Tin	63
Alumina	17–21
Aluminum nitride	200–230
Beryllium	250–370
Silicon	84
FR4	.2–1
Molding compound	.6
Epoxy	1.6–2
Air	.02

Conduction example. As an example let's calculate the internal thermal resistance of a standard 28-mm MQFP using Eq. (14.3). A cross section of a MQFP is shown in Fig. 14.2*a*. For the sake of simplicity, let's neglect the effect of bond wires, and furthermore, let's assume the lead frame is similar to a solid sheet as wide as the package. Figure 14.4 shows the resistive network for finding the total resistance between the junction and the lower case.

This first order approach indicates most of the internal resistance is primarily due to the mold compound material. If the mold compound is replaced by a more thermally conductive material, such as alumina, with similar thickness, the overall thermal resistance from junction to case would drop to only .28 C/W as follows:

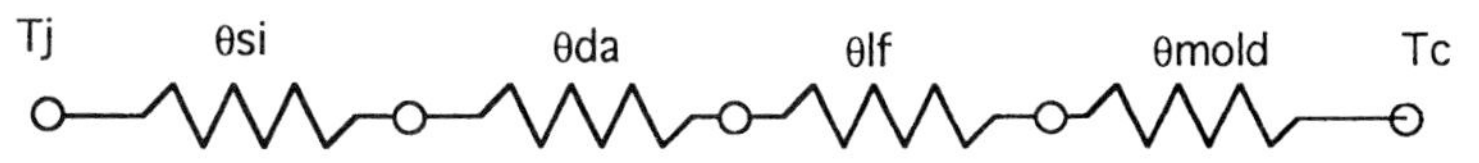

Figure 14.4 Resistive network from junction to case of MQFP, where $\theta_{jc} = t/KA$ = Sum of internal resistance for each layer; θ_{si} = [.000508]/[84(.00914)^2] = .07 C/W; $\theta_{\text{die attach}}$ = [.0000508]/[1.6(.00914)^2] = .38 C/W; $\theta_{\text{leadframe}}$ = [.000152]/[100(.028)^2] = .002 C/W; θ_{mold} = [.0018]/[.6(.028)^2] = 3.8 C/W; $\theta_{jc\text{ total}} = \theta_{si} + \theta_{da} + \theta_{lf} + \theta_{\text{mold}}$ = 4.2 C/W.

$$\theta_{si} = .07 \text{ C/W}$$
$$\theta_{\text{die attach}} = [.0000508]/[6(.00914)^2] = .1 \text{ C/W}$$
$$\theta_{\text{leadframe}} = \text{negligible}$$
$$\theta_{\text{ceramic}} = [.0018]/[20(.028)^2] = .11 \text{ C/W}$$
$$\theta_{jc\text{ total}} = \theta_{si} + \theta_{da} + \theta_{lf} + \theta_{cer} = .28 \text{ C/W}$$

In ATAB BGA or flip-chip structures the heat flow to the substrate is through solder bumps. In these designs there is an air gap between the silicon die and base substrate which is filled with filler materials. The conduction heat flow through the solder bumps and filler material act in a parallel path to transfer the heat to the substrate. This path functions similar to the die-attach layer. The resistive network for this path is shown in Fig. 14.5 with N number of solder bumps. Corresponding thermal resistances are calculated below assuming the solder bump of resistance is 643°C/W[6] and the filler material has thermal conductivity of 1 W/m^2-K with an effective area of equal to half of the die.

As shown before, if the silicon were die-attached to the substrate by epoxy, the thermal resistance from junction to leadframe would have been .17°C/W. On the other hand, if silicon is attached to the substrate by the flip-chip method via solder bumps, the thermal resistance is 1.6°C/W from junction to substrate, which is a fairly good jump. This would have been much worse if the air gap were not filled with any thermally conductive material.

Limitations of one-dimensional approach. All the calculation done in the simple model described above was based on Eq. (14.3). This equation is assuming the heat flow from the source to the outside world is only along the thickness of the package. In reality the generated heat is dispersed through the package in all directions. Other assumptions made for Eq. (14.3) are that uniform heat source exists across the package and each layer of material is isothermal.

Heat source is also assumed to be time-independent. In reality the silicon chip cycles on and off causing discontinuity in the heat pattern. Of course the approach shown above is a worse-case scenario and gives the maximum temperatures.

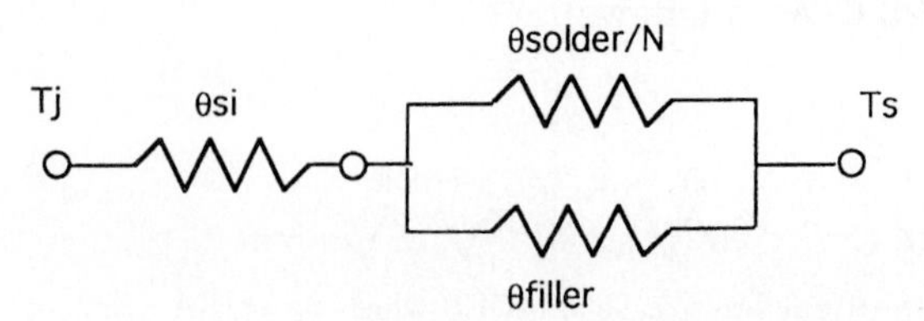

Figure 14.5 Resistive network for junction to substrate of a flip chip structure, where $\theta_{si} = .07$ C/W; $\theta_{\text{solder}} = 643/100 = 6.43$ Deg C/W[3]; $\theta_{\text{filler}} = [85\text{e-}6]/[1 \times .5(.00914)^2] = 2$ Deg C/W; $\theta_{jc\text{ total}} = \theta_{si} + (1/\theta_{\text{solder}} + 1/\theta_{\text{filler}})^{-1} = 1.6$ Deg C/W.

The major shortcome of the above approach is that it does not take the spreading of the heat into consideration. According to the above analysis the same resistance of about 4.2°C/W would have been obtained even without the presence of the copper leadframe. On the other hand, elimination of the copper leadframe should have resulted in a drastic jump in the overall theta *jc*. Here copper leadframe acts as a distributing agent dispersing the heat to the far corner of the package and reducing the temperature gradient of the package. In other words the package is more isothermal with the copper leadframe than without it.

One correction to the above analysis is to correct for the spreading resistance caused by heat flow from a small surface to a much larger surface. Kennedy[6] analyzed this phenomena for circular heat source and introduced a correction factor series which could be used to reach a better answer for the internal thermal resistance. These graphs are shown in Fig. 14.6*a*–*c* for various aspect ratios.

$$\theta_{jc} = H/K\pi a \tag{14.4}$$

where H = correction factor obtained from the Kennedy's graphs[6]
a = equivalent radius of the heat source

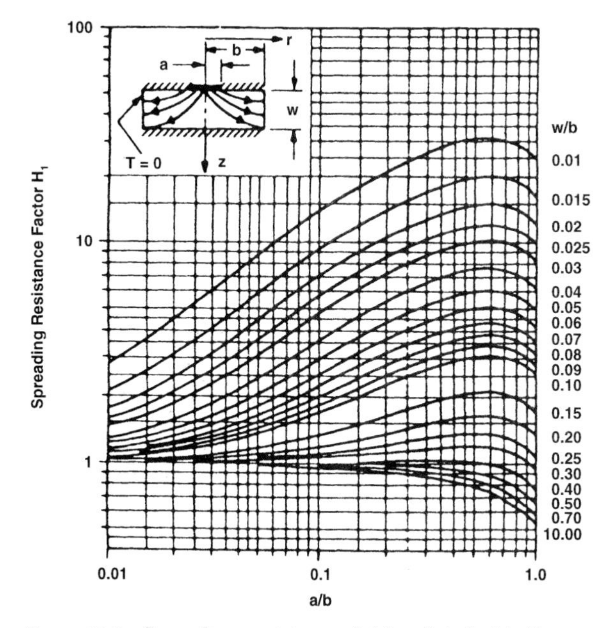

Figure 14.6 Spreading resistance: (*a*) insulated at bottom.

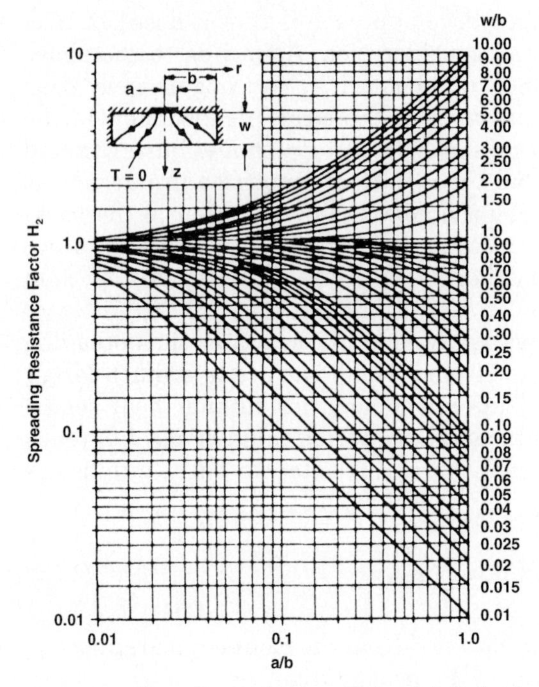

Figure 14.6 Spreading resistance: (*b*) insulated at side.

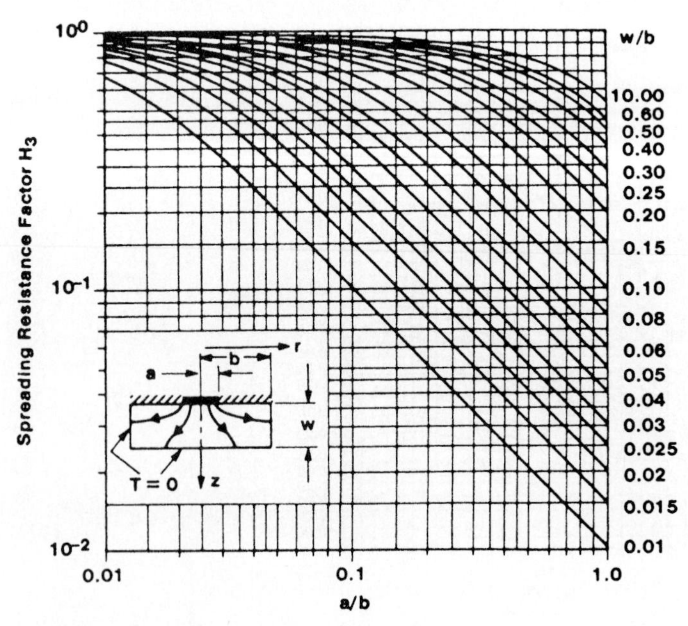

Figure 14.6 Spreading resistance: (*c*) insulated at top.

The corresponding parameters used in Fig. 14.6*b* for MQFP are:

$$a = [(.009)^2/\pi]^{.5} = 5.16 \text{ mm}$$
$$b = [(.028)^2/\pi]^{.5} = 15.8 \text{ mm}$$
$$a/b = .32, \ w/b = .11, \ H = .3$$
$$\theta_{\text{mold}} = .3/.6\pi(.00516) = 30.8 \text{ C/W}$$

This is the internal thermal resistance of the mold layer assuming IC heat source is at the center of the mold. The reason for such a sharp jump from 3.8 C/W to 30.8 C/W is mainly here it was assumed that there is no copper leadframe present between the silicon and molding compound. Adding the copper layer and assuming the effective heat source area is 75 percent of the whole package area the thermal resistance of mold layer can be calculated as:

$$a = .75\text{b} = 11.85 \text{ mm}$$
$$b = [(.028)^2/\pi]^{.5} = 15.8 \text{ mm}$$
$$a/b = .75, \ w/b = .11, \ H = .15$$
$$\theta_{\text{mold}} = .15/.6\pi(.0118) = 6.7 \text{ C/W}$$

This value of 6.7°C/W compared to the one-dimensional value of 3.8°C/W still is a major correction and determines how the one-dimensional approach could lead to erroneous values.

14.5.5 Convection

At the surface of the package, heat is transported to the surrounding area by a convection mechanism. The thermal resistance at the interface often is known as external resistance and is governed by Eq. (14.5).

$$\theta_{ca} = 1/hA \tag{14.5}$$

where h is the convective heat transfer film coefficient and A is the convective surface area in contact with the fluid surrounding the package. This shows the larger surface area will provide lower external resistance.

The heat transfer film coefficient is determined by fluid type and its velocity along with package temperature, dimension, and orientation of the package. Since there is temperature variation along the surface of the package, the heat transfer film coefficient will also vary along the surface. For at least the first iteration, average heat transfer coefficient can be used. Due to its temperature dependency and temperature variation along the surface of the package, an average coefficient is commonly used.

The film coefficient ranges from 2 to 20 W/m^2-K for free air convection to 20 to 200 W/m^2-K for forced airflow. Liquid cooling with forced circulation can easily have film coefficient of over 1000 W/m^2-K leading to very low external thermal resistance and the overall thermal resistance values. Listed below is the calculation for a 28-mm convective surface area and average film coefficient of 14 W/m^2-K for upper and lower surfaces of the package.

For natural convection $\theta_{ca} = [1/14(.028)^2]/2 = 45°C/W$

For forced air 200 fr/min $\theta_{ca} = 1/[30(.028)^2]/2 = 21°C/W$

Determining exactly what coefficient to use is not quite trivial. In the above case it was assumed that both the upper and lower surface of the package see the same airflow. This is not quite an accurate assumption since only the upper surface is exposed to the airflow. The lower surface hardly sees any airflow. In fact, in real applications, airflow is not uniform and often blocked by various other packages and units on the PCB board. This heat transfer film coefficient can be estimated from classical heat transfer[4] for horizontal or vertical flat plates as follows:

Forced airflow over flat surface:

$$h = .664(k/L)(P_r \rho \; VL/\rho)^{1/2} \tag{14.6}$$

where P_r = constant depending on the fluid type
ρ = fluid density
ρ = fluid dynamic viscosity

14.5.6 Thermal measurement

Introduction. The main difficulty in accurately determining the thermal resistances is measuring the junction temperature itself. Sticking a thermocouple into the package is simply not possible since the chip is encapsulated. In order to measure the temperature by a thermocouple, the silicon chip needs to be accessed from outside by deliding or partially removing the encapsulation material. However, both of these methods alter and unbalance the heat transfer mechanism.

Another method to measure the junction temperature is by means of monitoring the voltage drop of a diode within the IC chip. The forward voltage of most diodes is known to decrease by about 2 mV/1°C rise in the temperature. This factor is process-dependent and can be measured by calibrating the voltage of a diode at various environment temperatures. Following the calibration procedure, fixed power is applied to the silicon chip and consequently its voltage change is monitored. Knowing the final voltage, the calibration curve is used to determine the junction temperature for the specific power applied to the silicon chip. Ambient and case temperature are simply monitored using thermocouples.

Thermal Resistance of MQFP, PBGA, and ATAB BGA. Typical theta *ja* values versus airflow are shown in Fig. 14.7 comparing standard MQFP, PBGA, and peripheral ATAB BGAs. The high thermal resistance of MQFP is mainly due to the molding compound and too little copper leadframe used in these packages. The performance of these MQFP package can be easily improved by attaching internal heat spreaders. While PBGA's performance is similar to MQFPs, peripheral TBGA's performance is superior and is a function of the heat sink attached directly to the chip. The larger the heat sink the better ATAB BGA's thermal performance will be. In fact obtaining 3–5°C/W theta ja values can be easily accomplished for TBGA with moderate airflow and heat sinks attached directly to the back of the die. Theta jc values are less than 1°C/W. These types of TBGA packages can handle high-power applications of 20 W for most typical office computer applications.

14.6 Electrical Management of TBGA

14.6.1 Introduction

The electrical parasitics (resistance, inductance, and capacitance) of TBGA packages are critical to electrical management for these packages. Basically, the tape of TAB design is electrically dominant to the

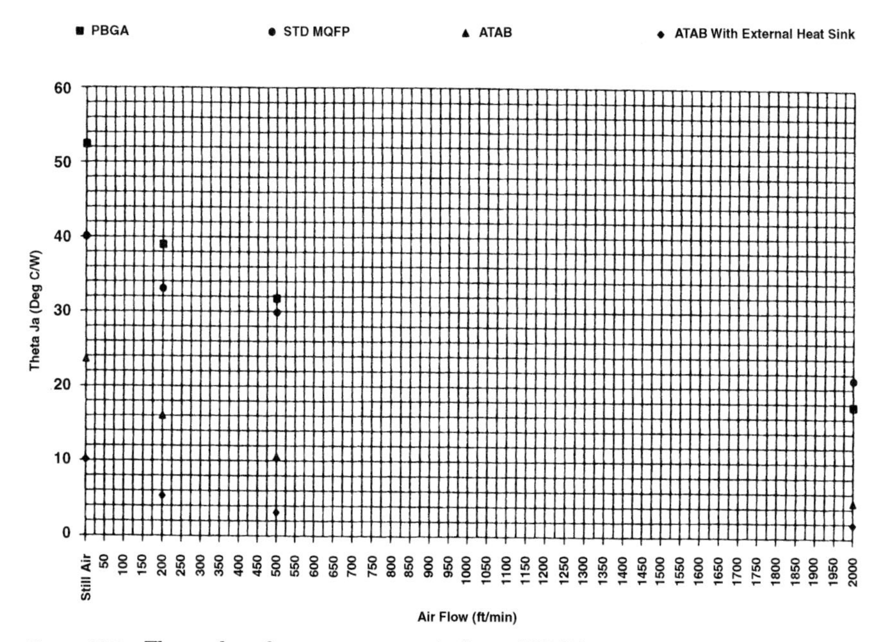

Figure 14.7 Thermal performance versus air flow of TBGA.

electrical performance of these packages. The simple calculation formulas of resistance, inductance, and capacitance for low frequency as well as transmission line properties for high frequency will be discussed in the following.

14.6.2 Basic analytical (closed-form) formulas

Resistance. The dc resistance is calculated using Eq. (14.7)

$$R = l/\sigma A \tag{14.7}$$

where l = length of the metal trace
σ = electrical conductivity of the metal trace
A = cross-sectional area of the metal trace

Inductance—Isolated metal trace. If the skin effects are neglected and, assuming the current in the metal trace (lead) is uniform, Eq. (14.8)[8] can be used to estimate the inductance.

$$L = 0.2l\,[ln\,((2l/(t + w)) + 1/2 - lnk](\mu H) \tag{14.8}$$

where L = self-inductance of the metal trace in μH
l = length of the metal in meters
t = thickness of the metal trace
w = width of the metal trace

lnk is the function of t and w and can be obtained from the table of Ref. 8. For the parallel metal conductors, the mutual inductance between the adjacent traces is given by Eq. (14.9)[8]

$$M = 0.2l\,[ln\,(2l/d) - 1 + d/l - 1/4\,(d^2/l^2) - lnk]\;\mu H \tag{14.9}$$

where l = length of the metal in meters
d = separation distance between the metal trace
t = thickness of the metal trace
w = width of the metal trace

lnk is the function of t, w, and d and can also be obtained from the table of Ref. 8.

Inductance—Metal trace with ground plane. The inductance of microstrip structure with ground plane is given by Eq. (14.10)[9] and (14.11).[10]

For $w/h \leq .8$

$$L = 0.2l\,[ln\,(6h/(0.8w + t))]\;\mu H \tag{14.10}$$[9]

For $w/h \leq 1$

$$L = 0.2l\,[ln\,(8h/(w + t)) + (w + t)/4h]\ \mu H \qquad (14.11)^{10}$$

where l = length of trace of microstrip
w = width of the metal trace
t = thickness of the metal trace
h = distance between the metal trace and ground plane

These two equations are based on TEM approximation and are only good for the conditions provided.

Capacitance. The capacitance of the microstrip can be calculated by Eq. (14.12)[9,11] and (14.13).[10,11] Eq. (14.6) is only valid as $w/h < 0.8$.

$$C = 1.43\ \varepsilon_{eff}\,[1/(ln\,(6h/w_{off}))] \qquad (14.12)$$

where $\varepsilon_{eff} = 0.475\varepsilon + 0.67$
$w_{eff} = 0.8w + t$
$C = p_{F/inch}$
ε = dielectric constant of the material.

for $w/h \leq 1$

$$C = 1.43\ \varepsilon_{eff}\,[1/ln\{(8h/(w + t)) + ((w + t)/4h)\}] \qquad (14.13a)$$

for $w/h \geq 1$

$$C = 0.224\ \varepsilon_{eff}\,[w/h + 2.42 - 0.44\,(h/w) + (1 - h/w)^6]$$

$$\varepsilon_{eff} = (\varepsilon + 1)/2 + (\varepsilon - 1)/2\,(1 + 10h/w)^{-1/2} \qquad (14.13b)$$

where ε_{eff} = effective dielectric constant
h = the distance between the metal trace and ground plane
$C = p_{F/inch}$

Impedance. Equations (14.14) and (14.15) can be used for characteristic impedance calculation of microstrip structure.

for $w/h \leq 0.8$

$$Z_0 = [60/\sqrt{0.475\varepsilon r + 0.67}]ln[6h/(0.8w + t)]\ \text{ohm} \qquad (14.14)^9$$

for $w/h \leq 1$

$$Z_0 = 60\ ln\,[8h/(w + t)) + ((w + t)/4h)]\ \text{ohm} \qquad (14.15a)^{10}$$

for $w/h \geq 1$

$$Z_0 = \frac{120\pi}{w/h + 2.42 - 0.44\, h/w + (1 - h/w)^6} \text{ ohm} \qquad (14.15b)^{10}$$

Equation (14.15) is for air dielectric only. If there is a dielectric material other than air, Eq. (14.16) is needed for characteristic impedance calculation.

$$Z = Z_0/(\varepsilon_{\text{eff}})^{1/2} \qquad (14.16)$$

where $\varepsilon_{\text{eff}} = (\varepsilon r + 1)/2 + [(\varepsilon r - 1)/2](1 + 10h/w)^{-1/2}$
εr = dielectric constant of material

The closed-form equations are useful to calculate the inductance, capacitance, and impedance of metal trice in the TBGA package. However, to use these equations, certain conditions are critical for accuracy of estimation. For example, w/h ratio has to be less than 0.8 and TEM mode approximation is needed to use Eq. (14.14). For more precise calculation of parameters, 3-dimensional finite-element softwares are ultimate tools.

14.6.3 Inductance, capacitance, resistance, and impedance of TBGA

One TBGA has been designed for CMOS ASIC applications. It has 432 leads with 31-mm body size and peripheral pad for die. The cross section is shown in Fig. 14.2*c*. It has two layers of metal and 2 mil kapton. The ground plane is shown in Fig. 14.8 and signal plane is shown in Fig. 14.9.

Table 14.5 shows the measured inductance, capacitance, and resistance of this package. To verify the measured data, 3-D finite-element software was used to simulate the parameters. Table 14.6 shows the results. It is shown that the measured data are consistent with the simulated values. However, the calculated values are much different from the measured data. The w/h ratio is exactly equal to 1, which violates the condition of w/h < 0.8[9] called *end-line effects.*

To resolve this inconsistency, the 3-D finite-element software tool is the best tool. For comparison, Table 14.7 is made from Ref. 1 and Table 14.5.

The discrepancy is due to the different design which will make the different electrical performance of the package.

14.7 Summary

The innovative TBGA is made by a combination of C4, TAB, and solder ball surface mount technology. The reliability of the package will not be

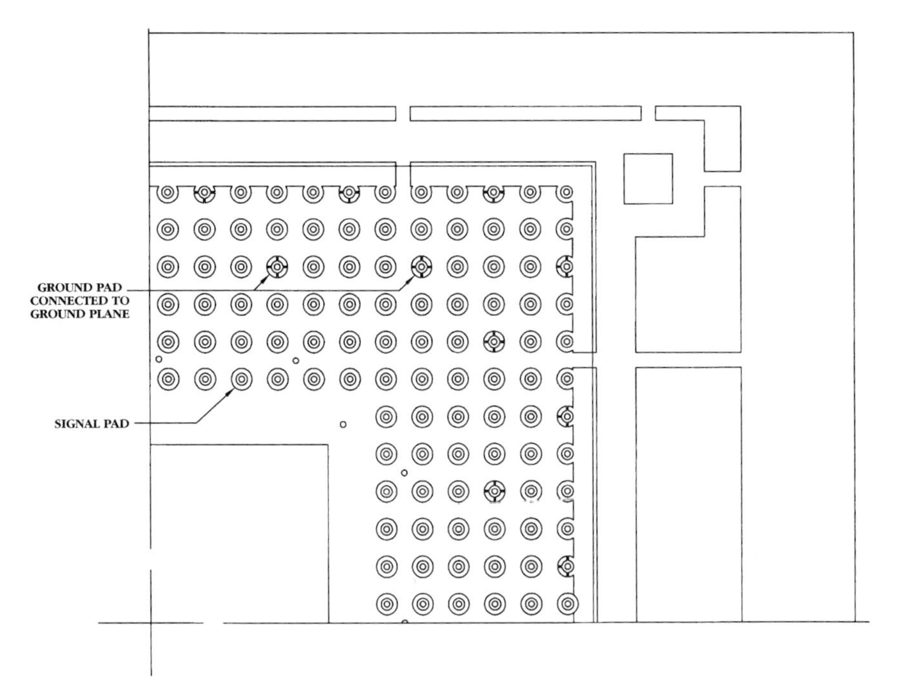

Figure 14.8 Ground plane of 432 lead ATAB ball grid array package.

an issue in industry if a cost-effective, manufacturable process can be developed to take into consideration the thermal and electrical advantages. If so, the TBGA will be package of the future.

For thermal aspects, first-order approaches can be used to approximate the thermal performance of the package. More accurate results can be obtained either by measurement or via computer simulation tools.

The TBGA have superior thermal performance over standard packages such as MQFP, and there are various methods to build up on their potential to achieve very low thermal resistance:

1. Use highly thermally conductive materials for adhesive, leadframe, coating, and substrate. Avoid structures designed with cavity and air gaps between various layers.
2. Transfer the heat directly from the back of the die by attaching external heat sinks to the die in conjunction with airflow.
3. Utilize the PCB board itself as a heat sink by connecting the BGA solder balls to various ground, power, or controlled temperature planes.

For electrical aspects, there are two advantages for electrical management of ATAB ball grid array. The first one is that the capability of

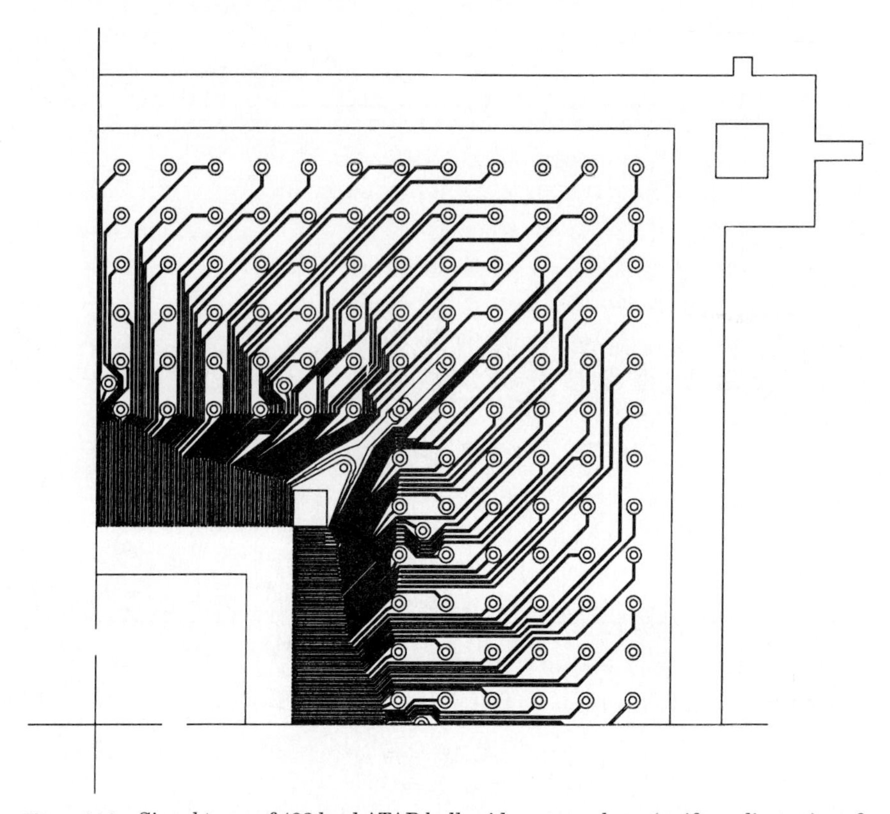

Figure 14.9 Signal trace of 432 lead ATAB ball grid array package (uniform dimension of trace-to-control impedance).

increasing I/Os in the certain body size[1] (see Table 14.1) is much easier to manage for high-speed, high-lead count devices. The second is that lead length of trace in the package can be easily cut short by using TAB for die and solder ball for PCB as interconnect.

However, the following steps still need to be followed to optimize the electrical performance of the package.

TABLE 14.5 Electrical Parameters (Measured) of 432ld ATAB, Ball Grid Array Package

Impedance (ohm)	Self-inductance (nHo)		Capacitance (pF)		Resistance (mohms)	
	short	long	short	long	short	long
65*	2.3	9.5	0.4	0.6	87	240

* The data is simulated. The inductance of Vss (ground pin) is 1.3nh.

TABLE 14.6 Measured, Simulated, and Calculated Inductance of 432ld ATAB Ball Grid Array

Trace	Simulated (nH)	Measured (nh)	Calculated* (nh)
Long	8.9	9.5	4.5
Short	1.9	2.3	1.1

* Equation (14.10) is used for this calculation.

TABLE 14.7 Comparison for Electrical Parameters of ATAB Ball Grid Array

Body Size (mm)	Self-inductance (nh)		Capacitance (pF)		Resistance (mohms)	
	short	long	short	long	short	long
25	1.1	2.6	0.3	1.3	85	150
36	1.3	5.5	0.4	2.4	102	360
31	2.3	9.5	0.4	0.6	87	240

1. Utilize the basic closed-form formulas to estimate the electrical parameters of the package.
2. Place ground pins for uniform current distribution in the plane using criteria of the effects of the holes, and mesh patterns on the ground plane[9] and simultaneous switching output (SSO) guideline provided by the device designer.
3. Simulate the electrical parameters (inductance, capacitance, resistance, impedance, and propagation time) using 3-D finite element software and simulate the signal integrity of devices, packages, and PCB to verify the ATAB ball grid array package design.

14.8 References

1. Andros, F. E., and Hammer, R. B., "Area Array TAB Package Technology," *Proc. ITAP,* 1993.
2. Anderson, S. W. "Solder Attach Tape Technology (SATT) Inner Lead Bond Process Development," *Proc. 4th ITAB Symp.,* Feb. 1992, pp. 158–172.
3. Evans, C. H., O'Hara, J., and Viswanadham P., "Reliability Aspects of Tape Automated Bonding," *Handbook of Tape Automated Bonding VNR,* New York, 1992.
4. Kromann, G., Gerke, D., and Huang, W., "A Hi-Density C4/CBHA Interconnect Technology for a CMOS Miscroprocessor," *Proc. 44th ECTC,* 1994, pp. 22–28.
5. Tanaka, M. et al., "Thermal Analysis of Plastic QFP with High Thermal Dissipation," *ECTC 1992,* pp. 332–339.
6. Tummala, R. *Microelectronics Packaging Handbook,* pp. 853–921.
7. Chapman, A. J. "Heat Transfer" McMillan Publication Co., New York, 1976, pp. 332–388.

8. Grover, F., *Inductance Calculations,* Dover Publication, Inc., New York, 1980, pp. 34–35.
9. Kaupp, H. R., "Characteristics of Microstrip Transmission Lines," *IEEE Transaction on Electronic Computer,"* vol. EC-16, no. 2, April 1967, pp. 185–193.
10. Schneider, M. V., "Microstrip Lines for Microwave Integrated Circuits," *Bell Syst. Tech. J.,* vol. 48, no. 5, p. 1421, May 1969.
11. Bogatin, E., "Design Rules for Microstrip Capacitance," *IEEE Transactions on CHMT,* vol. 11, no. 3, Sept. 1988, p. 253.
12. Omer, A., Swaminathan, M., Iqbal, A., and Nealon, M., "Effect of Mesh Planes on Striplines in High Speed Packaging Applications," *Proc. IEPS,* 1993, pp. 390–399.

Chapter

15

Inspection of Ball Grid Array Assembly

John A. Adams

15.1 Introduction

Lord Kelvin, a famous British physicist and electrical engineer (1824–1901), has been quoted:

> I often say that when you can measure what you are talking about and express it in numbers, you know something about it; but when you cannot measure it, when you cannot express it in numbers, your knowledge is of a meagre and unsatisfactory kind.[1]

The need for inspection of goods is a very human trait. Some form of inspection has been used since prehistoric times, especially during trading of goods. The trade item usually underwent a visual inspection that also included some form of quantitative measurement, such as the item's weight or length.

This chapter is about the inspection of Ball Grid Array (BGA) assemblies after solder reflow.

Early electronics featured large connections that relied upon the mechanical wrapping of the lead around a connection post and the encapsulation of the joint using solder. Inspection was easily accomplished using human visual inspectors.

In the late 1950s, the advent of printed wiring boards—along with component pins placed in holes through the board—brought new levels of interconnection densities and required new criteria for inspection. Later, the use of machine wave flow soldering allowed printed circuit card assemblies to be manufactured at high production rates.

Visual inspection by humans became quite tedious and results were prone to have errors. A report was published in 1988 of a study conducted by AT&T Bell Laboratories for the AT&T Federal Systems Division, Burlington plant, for the purpose of process improvement of the wave flow soldering machine used to assemble military electronics. During this study, it became apparent that the visual inspectors were inconsistent in their assessment of the defects present on the assemblies. "If you cannot measure the results, you cannot make improvements."[3]

15.2 Short History of Inspection

A 1791 report entitled "Report on Manufacturers" to the U.S. House of Representatives by Alexander Hamilton, then Secretary of the Treasury, calls for the "Judicious Regulation for the Inspections of Manufactured Commodities" to "prevent fraud . . . improve the quality . . . preserve the character of the national manufacturers . . ." and calls for a ". . . judicious and uniform system of Inspection . . ."[2]

Early inspection methodology was developed by the American Bell Telephone Company and Western Electric Company before the turn of the century. In the mid-1920s, quality assurance programs were developed through the work of Bell Telephone Laboratories. The growth in quality control and quality assurance was dramatic up through the war years. Dr. W. Edwards Deming worked throughout the war years to help U.S. manufacturers achieve increased throughput while maintaining quality.

Early electronics used in radios and telephones were routinely inspected. The simple sets, with their Flemming valves and the numerous interconnections, were inspected for quality, reliability, and cosmetic looks.

15.3 Visual Inspection Rated

The AT&T study was a double-blind experiment, with three of the four inspectors inspecting the four printed wiring card assemblies twice, not knowing that the assembly was the same card and not knowing when the card would be under inspection. The boards inspected averaged 4000 connections. The AT&T study found that on average, the three best inspectors were consistent with themselves only 44 percent of the time. Of the four evaluators, the consistency between any two evaluators was only 27.8 percent. The consistency among all four evaluators was only 5.6 percent. See Fig. 15.1.

In the concluding remarks, the AT&T study said it was easy to see that any one inspector could find as few as 17 defects or as many as 302 defects. They concluded that they were using an inspection tool with a variation that was much larger than the process they were trying to control. They quoted their process at 0.1 percent solder defects or 1000 parts per million (ppm). They also concluded that it was impossible to

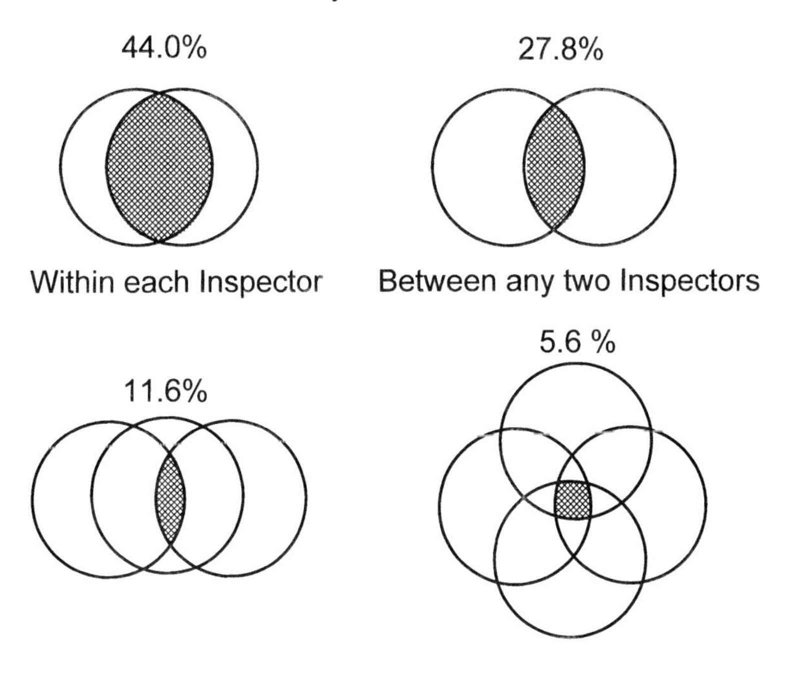

Figure 15.1 The consistency of using human inspectors to inspect printed wiring assemblies in the AT&T study.

effectively and efficiently interpret the visual solder criteria, and recommended work on the standards to help reduce the visual inspection problems.

15.4 New Packaging Technologies Cause Difficulties

In the last ten years, surface mount assembly of components on printed wiring boards has grown dramatically. The numbers of soldered connections on a single board have grown from the typical 250–4000 for a through-hole board assembly to typically 1000 to 10,000 connections on a surface mount assembly. Some companies have achieved upwards of 200,000 solder connections on a square foot using small pitch solder bump array technology pioneered by IBM.[4]

With the increasing numbers of connections on the circuit board, it has become critical that manufacturers improve their process to further reduce the defect levels so that the process gives a high functional yield of working boards. The effect of increasing numbers of joints on board yield can be seen in Fig. 15.2. As manufacturers move to higher connection densities, they have encountered even more difficulties in the process. It seems that every new board design initially results in even higher defect levels than past boards. This makes the board yield even more critical and it increases the need for good quantitative measurements of the assembly process. Reducing the time that it takes a new product to be production-ready is extremely important in the aggressive world of electronics manufacturing.

The role of the human visual inspector for soldered assemblies is rapidly diminishing. In addition to the densities of connections growing, many of the new solder joint mounting methods have hidden or partially hidden solder connections. Solder joints using the J-lead, ball grid array, flip chip area bump, and joints under connectors are mostly invisible not only to human inspectors, but also to many machine vision systems. Some of the newer fine-pitch TAB connections have solder connections obscured by the lead heel. For these new mounting technologies, automated x-ray has come to the rescue as a tool for inspection.[5,6]

15.5 Inspection Methods

The definition of a good solder joint has changed from concern purely about the electrical connection to interest in both the mechanical and thermal integrity. The role of inspection has also changed, from locating joints that needed to be repaired to one of being used as a metrology tool to gauge the actual process itself. Properly applied, inspection can actually help shift the emphasis from "ship working product" to

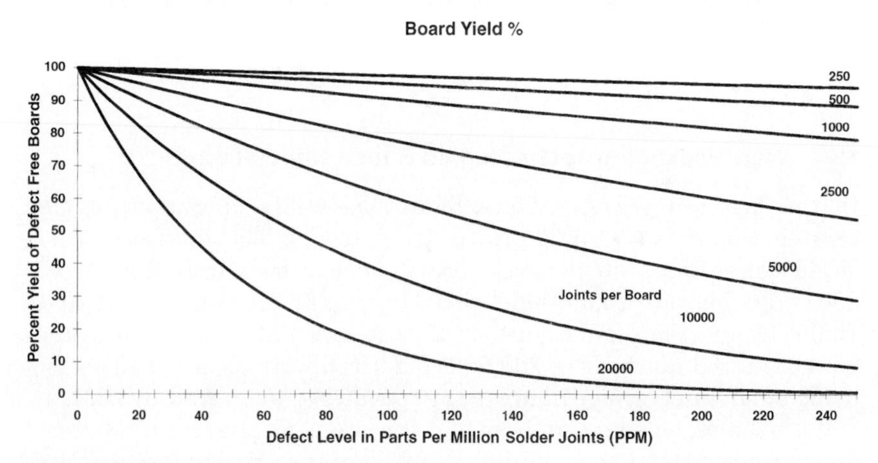

Figure 15.2 This chart shows how the number of joints on a board and the defect level affect the total board yield.

manufacturing with process control for production of a lower-cost and higher-quality product.

Classically, the definition of a good solder connection has been solely based on the external shape and look of the solder. Many inspection criteria are based upon gross defects and provide little information about the process that manufactured the solder connection. Solder inspection criteria are slowly being shifted to measurements that correspond to lifetime and reliability. Visual inspection should be used only in an audit role, to verify that the automated inspection system is performing well and to sample for any strange or undetected defects that may manifest themselves.

A variety of nondestructive test methods have been used to inspect solder joint connections with little to great success. Equipment to evaluate solder joint quality can be roughly divided into two types of systems, those that use surface illumination and those that use penetrating radiation. The systems can also be described as one- to three-dimensional systems in the way they obtain their data. A system that uses a single signal is generally referred to as one-dimensional (1-D). A system that uses a camera and takes gray scale information as a function of *XY* position is called a 2-D system, and one that uses *X, Y,* and *Z* height information is a 3-D system. Table 15.1 lists some of the various inspection systems.

15.5.1 Surface illumination techniques

The eye. The most basic surface illumination inspection is that of the human eye aided by a low-power lens or stereo microscope. The advantages of this type of inspection is that it appears to be low-cost and high-speed. With the BGA connection, however, special mirrors, fiberoptic lenses, or probing is required to try to see under the devices. The previously referenced AT&T study highlights the major problems with this approach.

TABLE 15.1 Process Inspection or Test Tools for Circuit Board Assembly Operations

Surface illumination	Penetrating radiation*
Aided human visual inspection	1-D single-pulsed infrared
2-D automated optical inspection	2-D infrared imaging
3-D AOI using structured light	2-D ultrasound SLAM
3-D AOI using laser line scanning	3-D ultrasound SLAM
	2-D x-ray
	2-D automated digital x-ray
	3-D X-ray tomosynthesis
	3-D x-ray laminography

* The items are listed in order of information provided and depth of penetration.

Automatic Optical Inspection (AOI). Two-dimensional (2-D) systems using a camera to capture and digitize the optical image into a gray scale image are an improvement over aided human vision. Typically, each pixel in an array represents a brightness level of 0 to 255. Various techniques can evaluate the inspected solder joint in comparison to a good joint or to a set of rules that define a good or bad solder connection. The use of the computer eliminates the human subjectivity, but the lighting setup and inspection rules can be very difficult to program. The reliability of the inspection, however, is usually also a function of unpredicted variations in the product under test.

Three-dimensional surface illumination systems generally use either structured light or laser scanning to generate height information along with a gray-scale image. These systems tend to be less sensitive to lighting conditions and tend to provide more surface data for analysis. These systems tend to be slower than 2-D systems using light. Some success has been reported using a 3-D laser inspection to measure the solder paste deposition on the board before the BGA device is placed. The combination of solder paste inspection using 3-D laser with solder joint inspection using a cross-sectional x-ray technique has allowed IBM in Toronto to significantly improve their BGA process as reported.[7]

15.5.2 Penetrating radiation techniques

Infrared. Irradiation of the solder connections with an infrared source and then monitoring the time-rate of decay has been used with some success to gauge solder joint integrity. Either one joint at a time is monitored with a single laser beam (a 1-D system), or the entire board is thermally shocked and the thermal energy die-away is monitored by an infrared camera for a 2-D map. This technique has been used with success where uniformity of the solder connection type and the design of boards have been tightly controlled. Additionally, extreme cleaning of the boards has been necessary to produce reproducible results. With the advent of no-clean soldering, this technique may no longer work.

Ultrasound. Ultrasound using the SLAM technique (surface laser acoustic microscope) has been successful in the test of die bond attach, some inner lead TAB, and some outer lead TAB devices. The basic limitations of this imaging technique are its ability to penetrate complicated structures. If the item under test has at most one or two layers of different materials, this technique can do quite well. The drawbacks are that it is still relatively slow for high-resolution imaging, and that the test sample has to be immersed in a coupling fluid.

The Ultrasound CSAM (cross-sectional acoustic microscope) has the ability to slice the signal to isolate a layer in the sample. Hence, the die bond layer may be separated from the chip circuitry. This technique has the same penetration limitations as SLAM, and is a lot slower. As a tool for process development, it has proven to be of great value. This technique has not been automated as fully as the optical or x-ray techniques, however.

X-rays. The contrast between the nineteenth-century physicists and their successors in the twentieth century is very dramatic. One scientist in 1893 said it was likely that all great discoveries in the field of physics had been made. The physicists in the future would have nothing to do but repeat the experiments of the past.

Two years later, on December 28, 1885, Professor Wilhelm Konrad Roentgen announced the discovery of x-rays. He published photographs of the bones in his hand, and of keys and coins photographed through a leather pocketbook. There was nothing in the nineteenth-century physics to explain this phenomenon.

It wasn't long afterward that the French demonstrated a real-time baggage inspection system for use in the railroad stations in 1897.[8] Figure 15.3 shows a high-voltage generator, with the two-wire leading to the x-ray tube. The tube is mounted on a stand in the center of the picture. While one inspector is holding the suitcase in front of the bare x-ray tube, the inspector to the right is looking through some optics to view a fluorescent image screen. It wasn't long after that people realized this was not the best way to use x-rays for inspection.

For many years, the basic cabinet radiography system—consisting of a high-voltage power supply, controls, x-ray tube, a shelf for mounting a circuit board, an x-ray film holder, and a lead-lined cabinet—has allowed inspectors to see hidden connections.

With the advent of the microprocessor, improvements in image intensifiers, and the dramatic lowering of the cost of computer calculational power, digital real-time x-ray became possible. The application of vision image processing techniques and image analysis with a digital computer allows for fully automated inspection without the need for human interpretation.

15.6 Automated X-ray Inspection

One of the first published records of a fully automated digital x-ray imaging inspection system was the Navy ARIES fuse inspection system. This system was a recipient of R&D magazine's IR-100 award for 1984. The ARIES system fully inspected a military mechanical fuse for several critical parameters at a rate of 50 seconds per fuse. Humans were not required for the accept/reject decision.[9]

Figure 15.3 Railway baggage inspection in France in 1897.

During the 1980s, many companies began developing automated visual inspection systems using light, structured light, strobe lighting, laser scanning, and infrared measurements as previously discussed. Many of these systems were never really successful on the actual assembly line for printed circuit card assemblies. However, some of the better visual and x-ray technologies are being used on printed circuit card assemblies today. Visual inspection is being used in the inspection of solder paste, placement of components, and after-reflow solder inspection. For devices with hidden connections, such as the J-lead, surface mount pin grid array, very fine pitch TAB, flip chip using the C4 process, and the various types of area array connections such as Ball Grid Array (BGA), Ceramic Ball Grid Array (CBGA), Plastic Ball Grid Array (PBGA), Solder Column Connect (SCC), and pad area array, x-ray or cross-sectional x-ray inspection has worked very well.

Delco Electronics of Kokomo, Indiana, reported success using fully automated x-ray inspection of surface-mounted components assembled onto a single side of a printed wiring board.[10] They reported that the x-ray inspection found more of the real faults on the board than did electrical in-circuit test. They also reported a significant savings in inspection labor.

An automated x-ray inspection system that uses cross-sectional images to analyze both single-sided and double-sided surface mount assembly circuit cards was presented in early 1989 at Nepcon West.[11]

IBM reported success in the implementation of an automated x-ray inspection machine for the inspection of surface mounted assemblies with components on both sides. To confirm the effectiveness of the cross-sectional x-ray machine at screening solder-related defects, one year of card inspection results were tallied. The sample was divided into two groups for analysis, the period before and the period after the machine's integration into the inspection process. The data was further divided into the number of defects found after reflow and the number found at final inspection. The data was analyzed. The results showed that the number of defects after reflow decreased by 8 percent with the use of the machine. This indicates an improvement in the process or that the machine inspection process is less likely to generate false calls. The number of defects making it to final inspection decreased by a factor of 10 with the use of the machine. This result shows that the machine is much better at screening defects than human inspectors . . . a factor of 10 better.[12]

The use of a cross-sectional x-ray system to provide process control data was discussed in a paper given at the Institute for Interconnecting and Packaging Electronic Circuits (IPC) Fall meeting in 1989. This paper described how the solder process can be monitored and controlled by the analysis of measurement data provided by an automatic cross-sectional x-ray system.[13]

Recent developments have resulted in a growing list of suppliers for fully automated equipment for inspection of soldered components on printed wiring boards using all forms of inspection technology from vision through cross-sectional x-ray. Since the Ball Grid Array (BGA) is a hidden connection, we will concentrate on the use of x-ray and cross-sectional x-ray for inspection.

15.7 To Inspect BGA or Not?

The BGA connection has been reported in industry as having an intrinsically low defect rate.[14] In this reference the author refers to IBM's CBGA and Column Grid Array (CGA), and reports that the assembly solder joint defect level with the CBGA connection is under 5 ppm. He also references IBM's use of x-ray inspection to maintain high quality.

There are many in industry who say the BGA connection does not require 100 percent inspection or 100 percent test. There are others that routinely use x-ray to perform 100 percent inspection or test of the BGA connection. Who is right? Both are.

The use of a single BGA product on a circuit board that has an overall solder joint defect level of under 5 ppm does not add significantly to the overall board yield reduction. However, in order to manufacture a surface mount board that is under 5 ppm after reflow, very strict process control must be practiced, strict design rules must be used, and a means to gauge progress in process improvement must be provided.

In the startup of a new manufacturing process, many manufacturers can receive benefits from having an x-ray system to provide instant feedback on the immature process. As the process comes under control, then a sampling of product to ensure continued low defect rates is probably all that is necessary. These manufacturers allow the functional test or the customer to uncover the few defective boards that a 1-5 ppm process means.

For a very high quality manufacturer, or for circuit boards that are very complex and expensive, no level of defects in the final product is acceptable. These manufacturers generally would test the BGA 100 percent with x-ray to ensure that the BGA is free from shorts, opens, and mechanically marginal joints. This test method reduces to practically zero the field returns or latent field failures due to weak soldering of the joints.

Considering how applicable Lord Kelvin's statement is here, we must be able to measure the process if we are to have any confidence in it. One of the very powerful tools for measuring the soldering process is the automatic x-ray inspection system.

15.8 What Does Cross-sectional X-ray Measure on a BGA?

All automated systems for inspection of assembled print wiring boards have a lot in common. There is a measured signal that has the desired information contained within, mixed with noise. A detector captures both the noise and the signal information. A computer system analyzes the captured signal and extracts useful information from the noise. The useful information can be critical measurements or the presence or absence of a particular feature. A decision is then made as to whether the connection is good or bad, and the results of the decision are identified with the joint.

X-rays are unique in that they penetrate through the connection and give density information throughout the volume of the connection. Using a cross-sectional x-ray technique, the measurements can be isolated to a horizontal layer of the solder connection, thus separating components on opposite sides of a surface mount assembly.

Figure 15.4 illustrates some of the typical CBGA-type connections. The CBGA connection uses a high-lead content solder ball that is connected to the package and to the board with eutectic solder. TBGA uses a slightly smaller high-lead ball that is fused directly into TAB carrier vias. The TAB package is stiffened with constraining material and applied to the board using eutectic solder. The SCC uses a high-lead-content solder column that is connected to the package and the board with eutectic solder. The PBGA-type connection is just a eutectic solder ball reflowed to the plastic package and then attached to the board with eutectic solder.

15.9 CBGA Typical Defects

The CBGA connection is rather robust; however, certain process conditions can cause the connection to be unreliable. Figures 15.5 to 15.11 illustrate typical defects. Defects are similar for the PBGA connections.

Bridging. If there is too much solder present, then bridging to the adjacent ball is possible. See Fig. 15.5.

Insufficient strength. Too little solder provides insufficient strength in the ball/board connection. See Fig. 15.6.

Voiding. Contamination or paste problems can cause voiding to be present. See Fig. 15.7.

Opens. Excessive board warp can cause opens, since there is not enough solder to bridge the open gap. The BGA allows approximately 5 mils of local board warp, center of package to edge, before the open condition could occur with normal amounts of solder (this corresponds to a warp of about 1 percent). See Fig. 15.8.

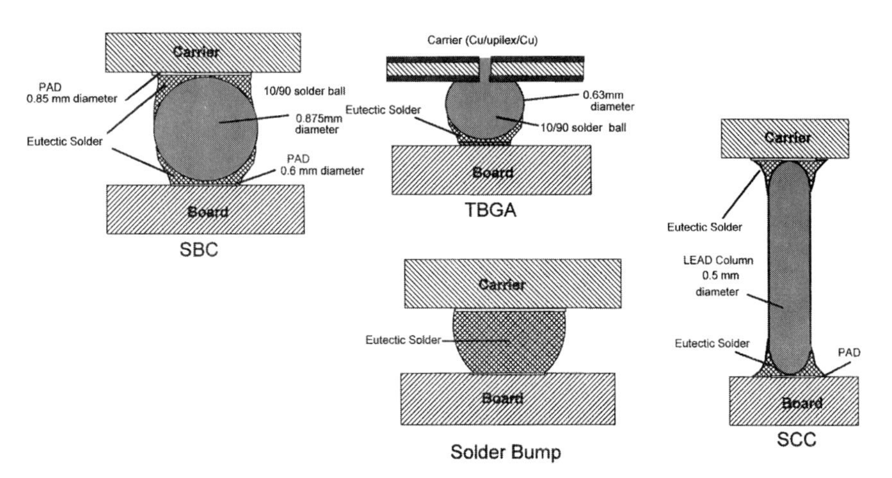

Figure 15.4 Schematic of typical good BGA connections.

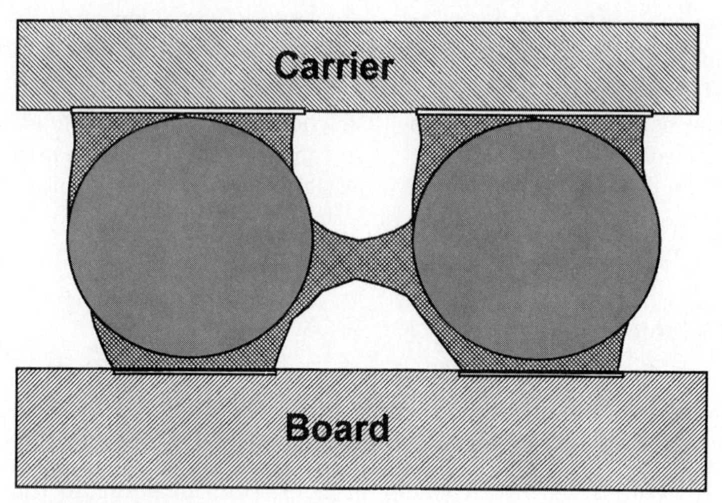

Figure 15.5 Bridging.

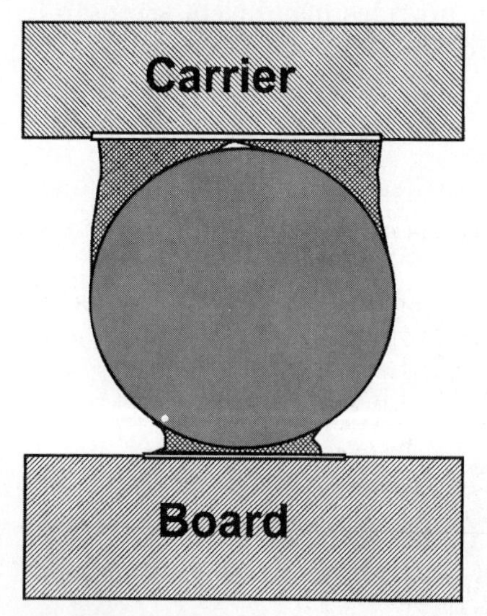

Figure 15.6 Insufficient.

Poor wetting. Poor pad wetting causes the solder to flow up the side of the ball. Poor ball wetting prevents the solder from flowing up the ball. See Fig. 15.9.

Solder balls. Small solder balls are caused by material ejected from the solder during reflow. See Fig. 15.10.

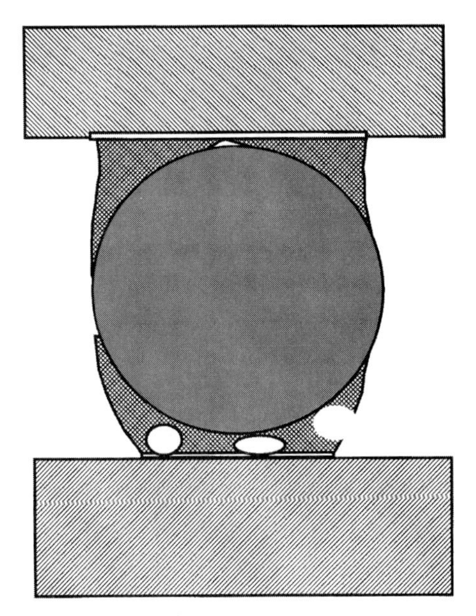

Figure 15.7 Voiding.

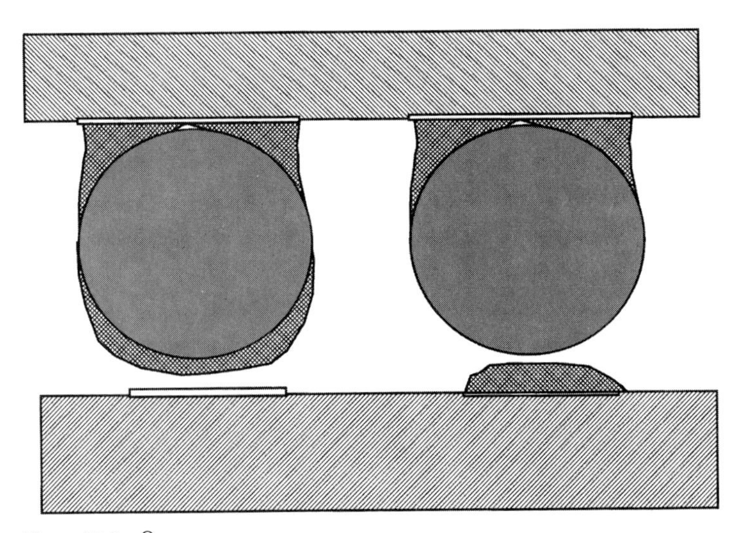

Figure 15.8 Opens.

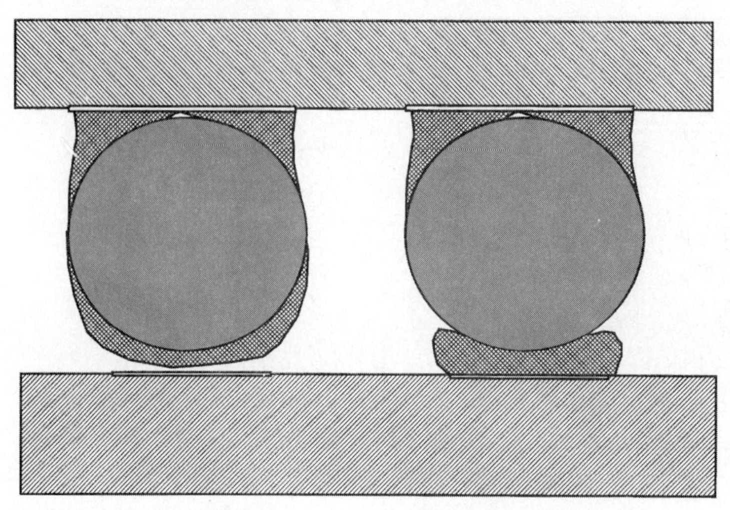

Figure 15.9 Poor wetting.

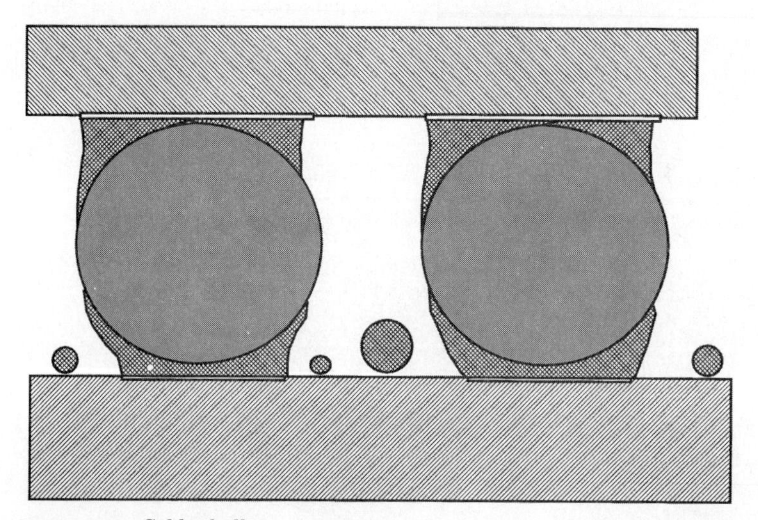

Figure 15.10 Solder balls.

Misregistration. The ball is not centered on the pad. The BGA has tremendous self-alignment capabilities; however, it is possible that there can exist a registration error between the ball and the pad. See Fig. 15.11.

15.10 SPC Measurements of the CBGA Device

The CBGA connections are especially hard to image with x-rays, due to the thickness of the solder ball and the difficulty of even 160,000-volt

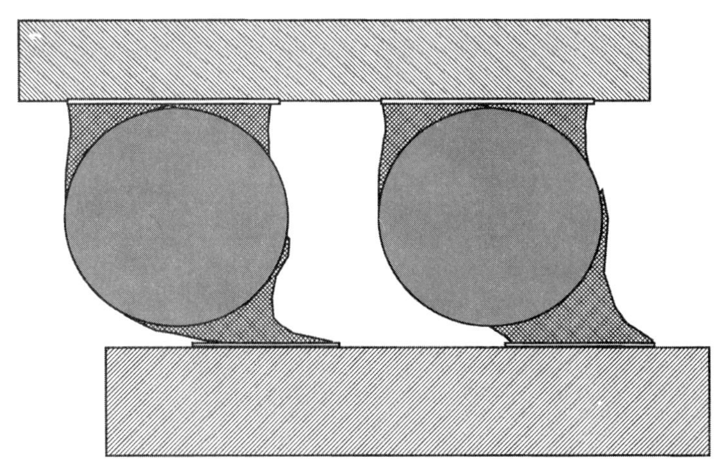

Figure 15.11 Misregistration.

x-ray systems to penetrate the balls. In a conventional x-ray system, with the beam perpendicular to the plane of the balls, very little information about the soldered connection can be obtained. A cross-sectional x-ray technique allows the testing of the CBGA-type connections.

15.10.1 SPC ball measurements

The basic measurements on CBGA and PBGA solder connections begin with the Statistical Process Control (SPC) measurements. Figure 15.12 identifies two basic levels on the assembly where cross-sectional slicing provides images to analyze.

In addition to the basic measurements from the slice levels just identified, two additional slices allow for a higher level of defect detection and provide some additional process information. Figure 15.13 shows

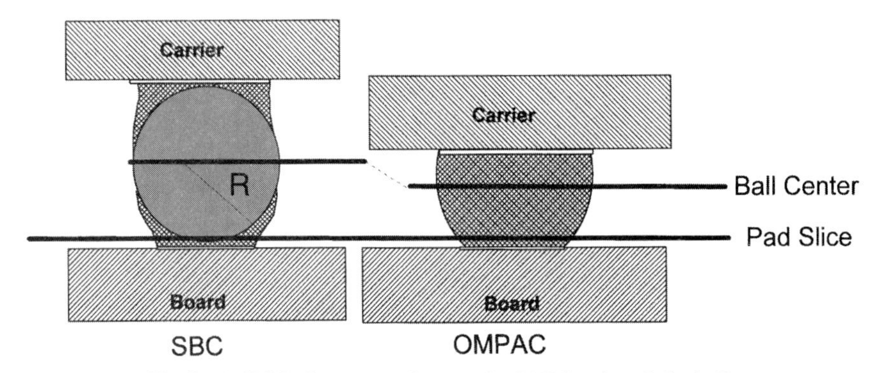

Figure 15.12 The basic BGA slices are taken at the PAD level and the ball center.

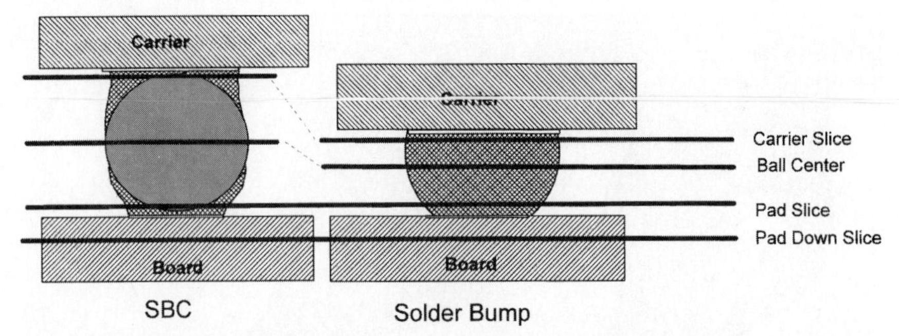

Figure 15.13 The slices for complete analysis of the CBGA and PBGA. The carrier slice is at the ball/carrier interface, the center slice between the carrier and the board, and the pad slice at the ball/board interface. The pad-down slice is approximately 0.3*radius of ball into the board from the pad slice.

additional slice levels for the BGA to obtain all of the process information. The addition of the carrier slice allows the test for voiding, alignment, insufficient, excess, opens, and shorts for the carrier-to-ball connection. The pad-down slice allows for additional accuracy in the decisions in the algorithms for insufficient and open. Once an assembly process is stable, the board could be tested with only one slice for an increase in throughput.

Figure 15.14 shows a composite image consisting of a cross-sectional x-ray image of three connections on a CBGA device. The device was turned on end and the side-view slice was generated. The device was then placed in the system and three more slices were taken, one at the level of the carrier/ball interface, one through the center of the ball, and one at the ball/pad interface. The darker areas in the image correspond to solder, while the light gray sections correspond to low-level smear background from the out-of-plane features of the connection.

Since a CBGA or PBGA probably will be removed from the board whenever a serious reject condition is determined on any ball on the device, it is extremely important to have a high signal-to-noise ratio for any reject condition. It is very important that all opens and shorts are detected. All other defects types seem to be less critical.

15.11 CBGA Inspection Results

Test CBGA devices were prepared by two manufacturers to characterize the CBGA mounting process using the measurements and test results from cross-sectional x-ray. One manufacturer produced 20 sample boards with multi-up CBGA and SCC devices on the board. Another manufacturer produced small test boards with a CBGA on each board.

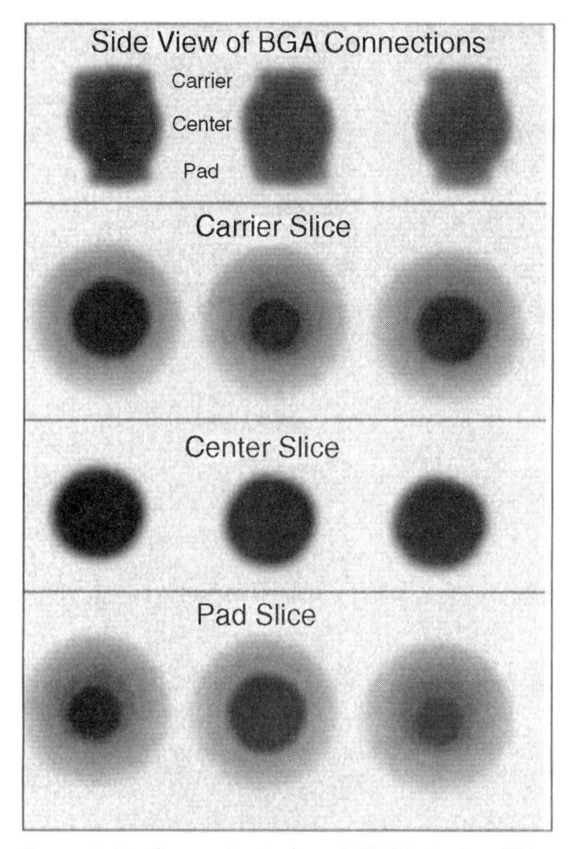

Figure 15.14 Composite image of CBGA device. Side view and three slices through various Z levels.

Several CBGAs were mounted to intentionally create faults. Measurements were taken on the test boards and some of the results are reported here. The measurements consist of determination of the radius of the ball center, pad, and pad-down slices, along with analysis of centroids for calculation of off-position. The comparison of the shape of the ball or pad image to the standard circle was made to measure noncircularity. Other pad designs require analysis to their own particular shapes, such as square or diamond pads. To gauge the repeatability of the radius measurements, one CBGA ball was repeatedly tested. The radius of the solder connection at the pad was quite repeatable, with the value of 0.0142 ± 0.00011 inches (0.36 ± 0.0028 mm) for a repeatability of 0.8 percent. The other measurements associated with the CBGA and PBGA balls have also shown high repeatability.

The CBGA shown in Fig. 15.15 has two joints, shown in the lower right corner of the image, that have missing balls, forming one type of

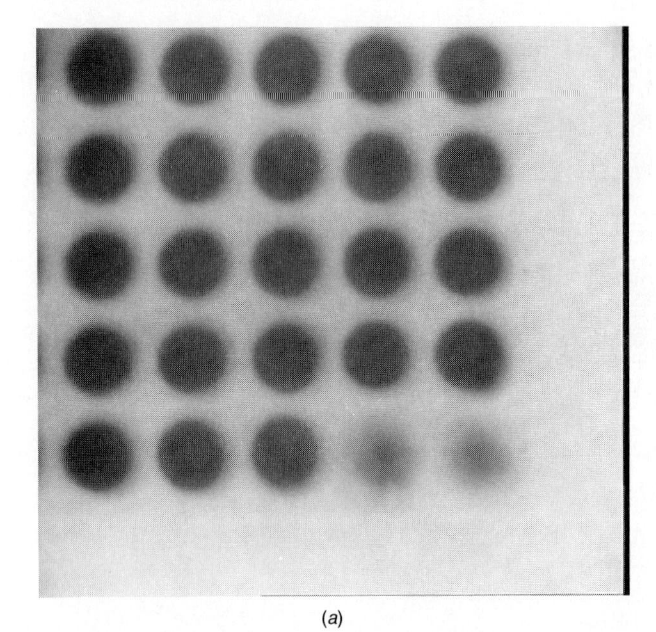

(a)

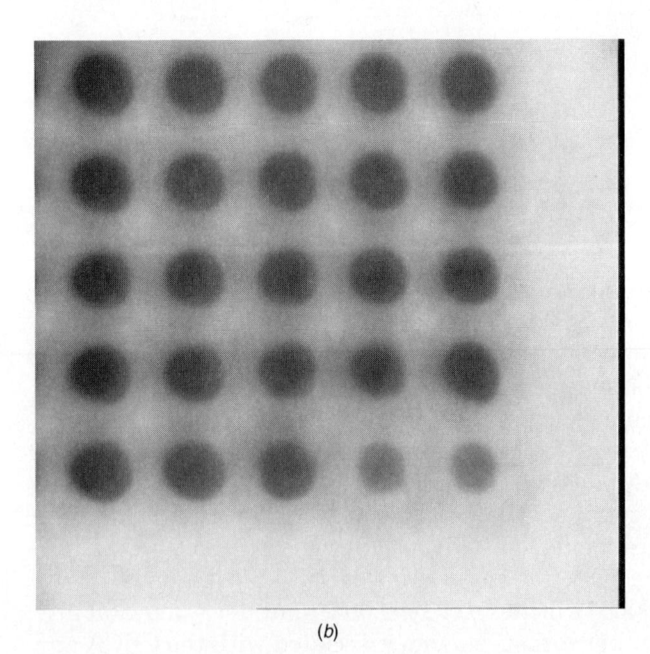

(b)

Figure 15.15 Lower right corner of CBGA showing two missing balls. Slices shown are for the ball center (*a*), pad (*b*), and pad-down (*c*). Close examination reveals some slight off-positioning of the balls to the pads.

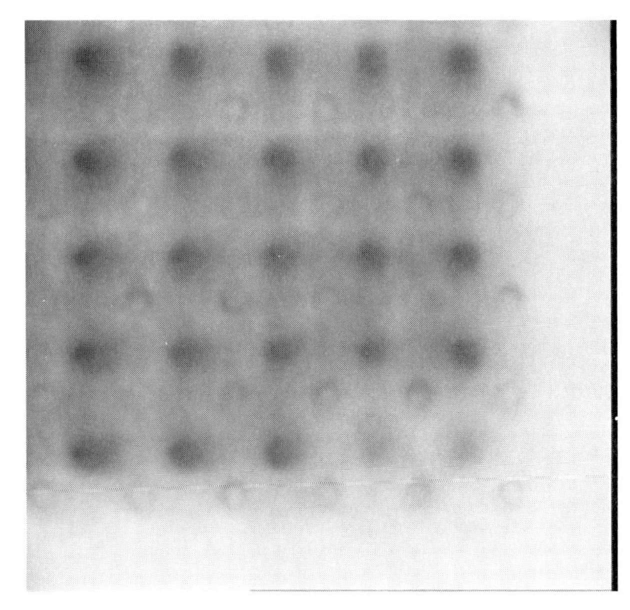

(*c*)

Figure 15.15 (*Continued*)

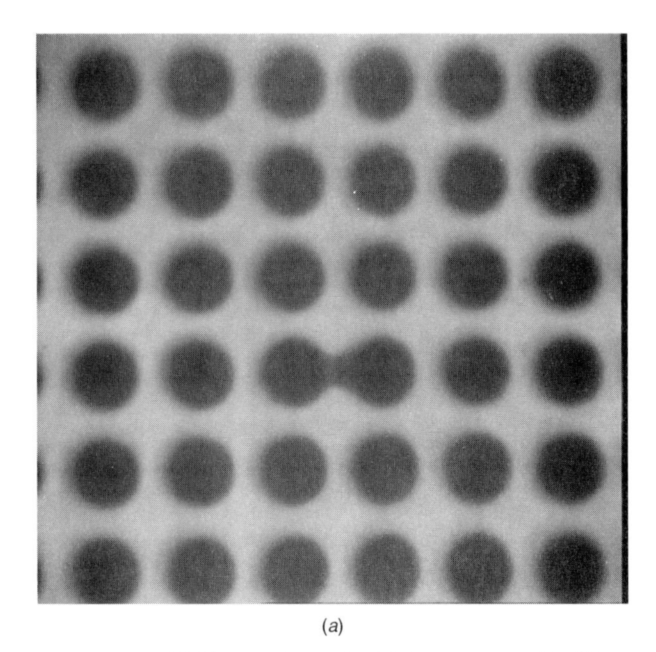

(*a*)

Figure 15.16 CBGA showing a short between two balls. Slices shown are for the ball center (*a*), pad (*b*), and pad-down (*c*).

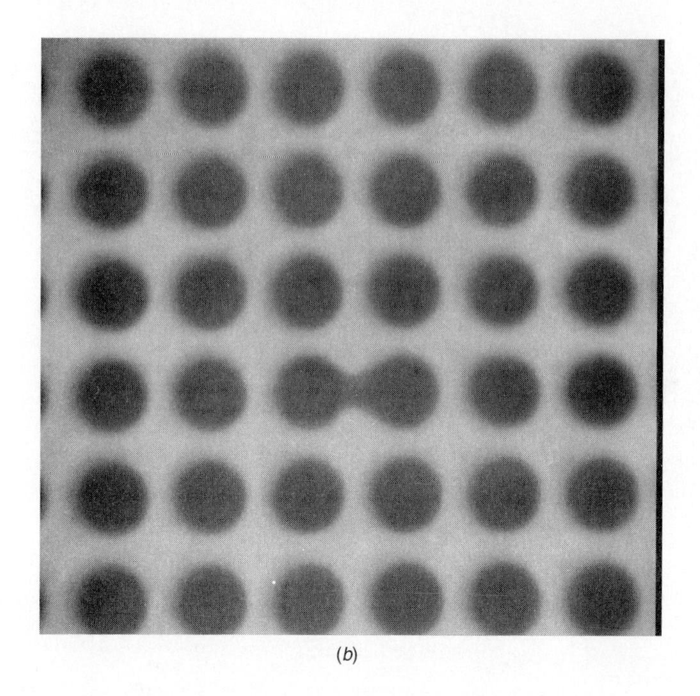

(b)

(c)

Figure 15.16 *(Continued)*

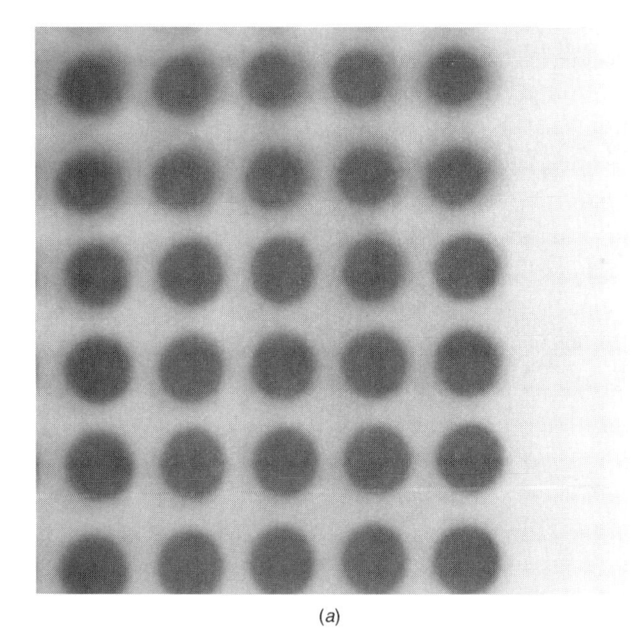

(*a*)

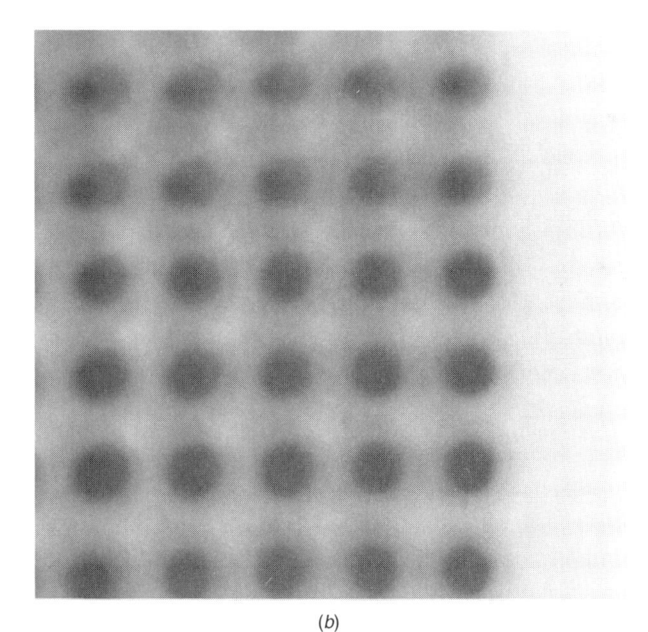

(*b*)

Figure 15.17 CBGA showing the effects of board warpage on the quality of the connection. As one views the image from top to bottom, the quality of the ball-to-pad connection decreases, due to increasing distance from the ball bottom to the pad. The top two rows are open connections.

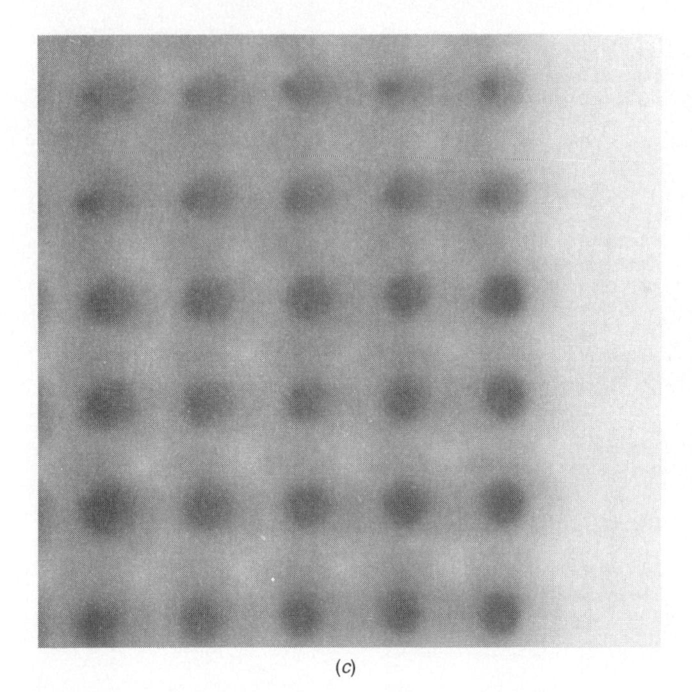

(c)

Figure 15.17 *(Continued)*

open connection. The solder paste deposited was within normal ranges. Figure 15.15 shows the slices taken at the ball center, pad, and pad-down locations. Note how easily missing balls are detected.

The CBGA shown in Fig. 15.16 has a bridge between two balls. The bridge is between the balls at the ball center slice. The bridge is also evident at the pad slice. Figure 16 shows the second CBGA sample.

The CBGA shown in Fig. 15.17 is mounted to show warpage effects in creating the open condition. There is a tilt of the circuit board away from the carrier such that as one views the image from bottom to top, the connections become less and less reliable. In the last two top rows, there is absolutely no connection between the ball and the solder on the pad. Figure 15.17 shows the three slices for this CBGA.

Figure 15.18 shows an image of a PBGA device. When testing this type of connection, only one "slice" is usually required. The image shows a lot of defects and variation, with a short between two connections and several insufficient connections (smaller diameters and lighter). The image also shows several connections that have been damaged by gas formation during the reflow process, causing voiding and resulting in poor connections.

Figure 15.19 shows the effects of contouring the image in Fig. 15.18. Here it is very easy to see how a computer can easily identify process conditions.

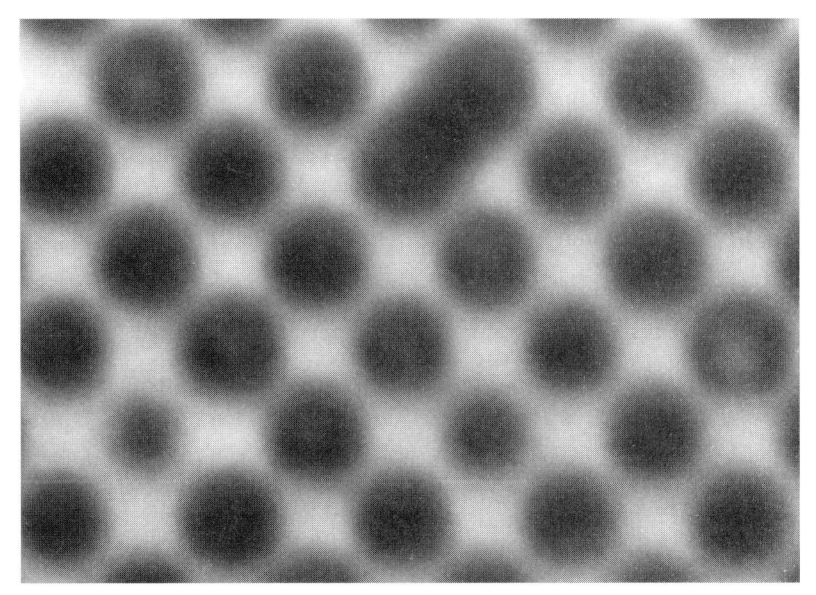

Figure 15.18 An image of a defective solder-bump-type PBGA connection. Here several defects are present.

In another study, five PBGA 360 I/O devices were manufactured under various process conditions. The radius of the solder connection at the pad was charted to determine if the process parameter changes had any measurable effects. Figure 15.20 shows the histograms of the solder radius at the pad level for the five PBGAs. Table 15.2 gives the measured average solder radius on the pad and the standard deviation of the measurement. It can be seen that the BGA connections have measurable differences caused by the process conditions.

15.12 Summary

Testing of assembled CBGA and PBGA devices has been routinely performed for over a year at several sites. Additionally, a manufacturer is using a combination of solder paste inspection before placement with

TABLE 15.2 The Average Solder Radius (in mils, 0.001 inch) at the Pad Slice for the 5 PBGA Devices under Test

PBGA	Radius mils	Std. dev. mils
1	9.47	0.58
2	10.19	0.65
3	10.33	0.56
4	10.51	0.43
5	10.38	0.56

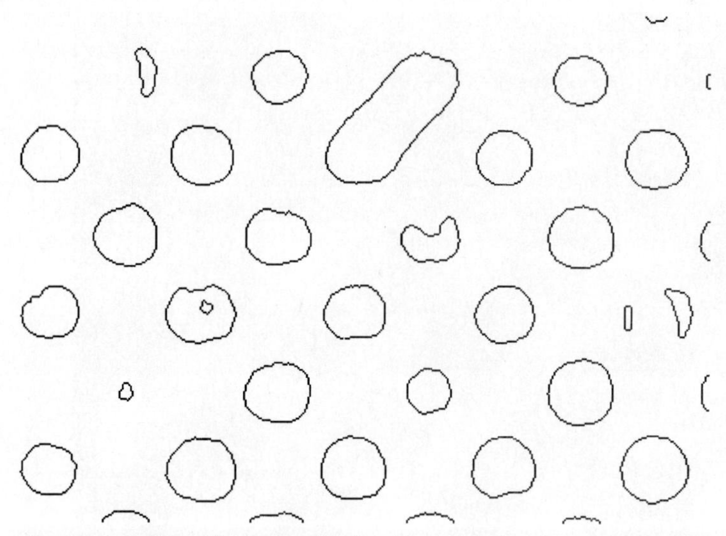

Figure 15.19 Shows the image in Fig. 15.18 that has been contoured to illustrate that ease of identification of defects. A well-formed bump should form a perfect circle.

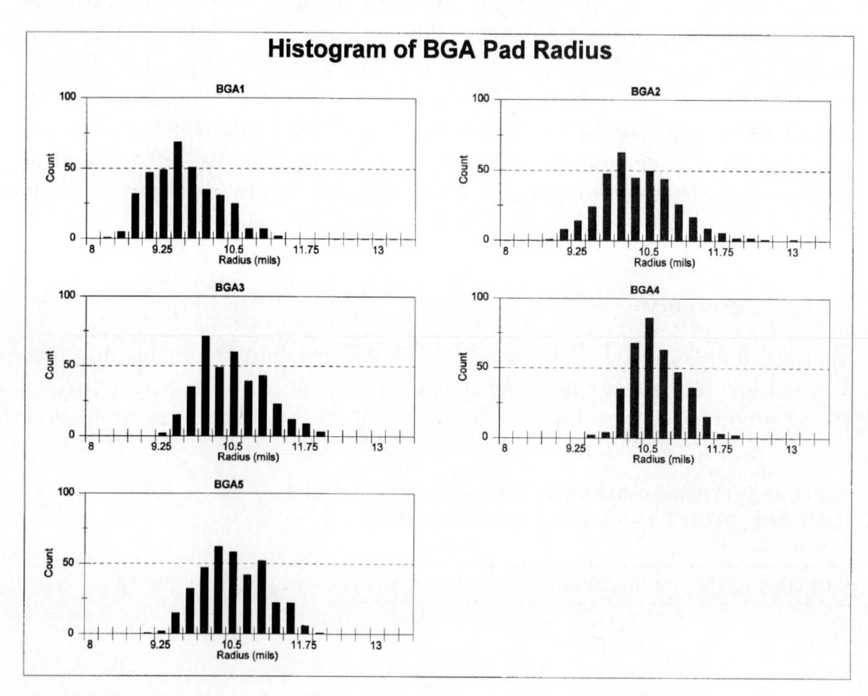

Figure 15.20 Histogram of solder radius on pad for five different PBGAs.

solder joint inspection after placement and reflow. As new mounting devices go through prototype production and process development, many startup problems are being quickly resolved with the use of cross-sectional x-ray inspection. It is expected that some form of automated process test will be used on BGA-type connections to keep the process in control. These new BGA mounting technologies have an excellent chance to drive the solder joint defect rates to the low parts per million (ppm) level, when they are made with a well-understood and controlled process.

15.13 References

1. Wadsworth, Harrison M., et al., "Modern Methods for Quality Control and Improvement," John Wiley & Sons, TS156.W25 1985 658.5′62 85-20285 ISBN 0-471-87695-X p. 25.
2. Wadsworth, Harrison M., et al., "Modern Methods for Quality Control and Improvement," John Wiley & Sons, TS156.W25 1985 658.5′62 85-20285 ISBN 0-471-87695-X p. 4.
3. Donnell, A. J., Fanelli, C. P., and Thomas, J. W., "Technical Memorandum, Visual Soldering Inspection Inconsistencies," AT&T Bell Labs, May 6, 1988.
4. Blodgett, Albert, J., Jr., "Microelectronics Packaging," *Scientific American,* July 1983, pp. 86–96.
5. Adams, John A., "Using Cross-sectional X-ray Techniques for Testing Ball Grid Array Connections and Improving Process Quality," *Proceedings Nepcon West,* 1994.
6. Banks, Don, "Letter to Editor," *Circuits Assembly,* August 1993 p. 8. (In this reference the author refers to IBM's BGA and CGA, with BGA at under 5 ppm, and reference to x-ray inspection.)
7. Clark, David, "3-D Laser Inspection Cuts BGA Rework," *Evaluation Engineering,* February, 1994, pp. 72–73.
8. Bryant, Lawrence E., "Radiography and Radiation Testing," *Nondestructive Testing Handbook; V3,* TA417.25.R237 1985 620.1′272 84-24438, ISBN 0-931403-00-6.
9. Adams, J. A., Ross, E. R., and Trippe, A. P., "Automated Filmless Artillery Fuse Inspection," Technical Paper, *Autofact Europe Conference,* September 24–27, 1984, Basel, Switzerland. SME publication MS84-621.
10. Goodwin, Charles D. and Reeder, Galen J., "Real-Time Process Control for Solder Joint Integrity," *Proceedings of the Technical Program, Nepcon East,* June 9–11, 1987, Boston, Mass. pp. 357–366.
11. Adams, J. A., "Scanned Beam Laminography Breaks through the 3D Barrier," *Proceedings of the Technical Program, Nepcon West,* March 6–9, 1989, pp. 112–116.
12. Sack, Thilo, "Implementation Strategy for an Automated X-ray Inspection Machine," *Proceedings of the Technical Program, Nepcon East,* June 10–13, 1991, pp. 65–73.
13. Adams, J. A., and Malloy, Dennis, "Discussion of 3-D X-ray Laminography as a Provider of Process Control Data for Real-Time Quality Monitoring," Technical Paper IPC-TP-851, presented at IPC Fall Meeting, September 10–15, 1989, New Orleans, La.
14. Banks, Don, "Letter to Editor," *Circuits Assembly,* August 1993, p. 8.

Chapter

16

Rework of Ball Grid Array Assemblies

Tom C. Chung and Paul A. Mescher

16.1 Introduction

Ball grid array (BGA) assemblies are usually expected to have much higher assembly yield than conventional surface mount assemblies, e.g., plastic quad flat pack (PQFP) or plastic leaded chip carrier (PLCC), due to BGA's unique attributes such as larger interconnect pitch, e.g., 1.0, 1.27, or 1.5 mm pitch for BGAs versus 0.65, 0.5, or 0.4 mm pitch associated with typical fine pitch surface mount components, easy self-centering alignment between solder balls on a BGA component, and corresponding pads on board to which a BGA component is attached.[1–3] Therefore, it seems that rework of BGA assemblies should be rare, or at least not an issue. However, in the real world, rework of BGA assemblies is not only needed but also required because of a variety of reasons such as engineering change requests, device or system upgrade, defective solder joints (see Ref. 4 of this chapter and Chap. 15 in this book for further details), misregistration, wrong component orientation, etc. Moreover, rework of BGA assemblies is not a simple issue at all. This is because BGA's solder joints are hidden underneath the package, which are quite different from those of a peripherally leaded surface mount component as shown in Fig. 16.1*a* and *b*. Therefore, not only inspection but also rework is very challenging for a BGA assembly. In addition, the proliferation of types of BGA assemblies meaning different solder geometry (diameter, height, width) and composition (e.g., eutectic solder material for plastic BGAs versus composite solder

(a)

(b)

Figure 16.1 BGA versus conventional surface mount assemblies: (*a*) a typical BGA assembly with solder joints hidden underneath the package; (*b*) a typical PQFP assembly with peripherally soldered leads.

material for ceramic and tape BGAs) has further increased complexity of the BGA rework materials and processes, not to mention availability of reliable and repeatable rework equipment.

In this chapter, the definition of BGA rework is first discussed along with overview of BGA rework process, equipment, and other key considerations, followed by discussion on rework of plastic BGA (PBGA) assemblies, including a typical PBGA rework process, and issues and approaches for rework of large-size PBGA assemblies. Then, detailed discussion of both rework of ceramic BGA (CBGA) and tape BGA (TBGA) assemblies are reviewed and discussed, including package descriptions, defect mechanisms, component removal, site preparation, solder replenishment, component replacement and reflow, and cleaning and inspection.

16.2 Overview of BGA Rework

Before detailed discussion of rework of BGA assemblies, it is important to understand the meaning of *rework*. In general, rework includes repair and replacement. *Repair* refers to performing correction of local defects (e.g., solder joint opens or shorts) detected on a BGA assembly; *replacement* refers to removal and replacement of an existing BGA assembly which either has unrepairable defect(s) or simply needs be replaced. For conventional surface mount components, rework (including both repair and replacement) has been practiced for many years.[5–7] Repair is a feasible process for conventional surface mount assemblies because their peripheral solder joints are physically inspectable and accessible. For a BGA assembly, repair is neither feasible nor practical because all solder joints except for the outermost rows of a BGA assembly are hidden underneath the BGA package. Therefore, in this chapter, for rework of BGA assemblies, we focus on removal and replacement of BGA assemblies.

A typical BGA rework process is shown in Fig. 16.2. In general, rework of solder reflowed assemblies requires, as a minimum, a heat source and a component pickup tool. The heat source can be hot gas or air, focused IR, or a conductive heating device. The pickup tool can be a vacuum cup, tweezer, or a mechanical pickup tool, etc. It has been found that a hot gas system along with a vacuum pickup tool is the most common approach for rework on BGA assemblies. The BGA rework process shown in Fig. 16.2 can be broken into three major areas, i.e., removal of the existing component, preparation of the board for component replacement, and the replacement with a new "known good" component. The complexity of each step shown in Fig. 16.2 varies depending upon type of BGA, solder metallurgy, geometry, etc. It is important to closely monitor/control all the process steps/parameters to ensure that the board and neighboring components are not damaged during the entire rework process. A repeatable and reliable BGA rework process can be accomplished only by using rework equipment which provides not only component pickup and reflow capabilities but also vision-assisted alignment capability for component placement. To date, almost all the BGA rework machines evolve from their surface mount rework machines. The typical rework equipment vendors and their capabilities for surface mount components are listed in Table 16.1.[7,8] Companies such as Austin American Technology (AAT), Air-

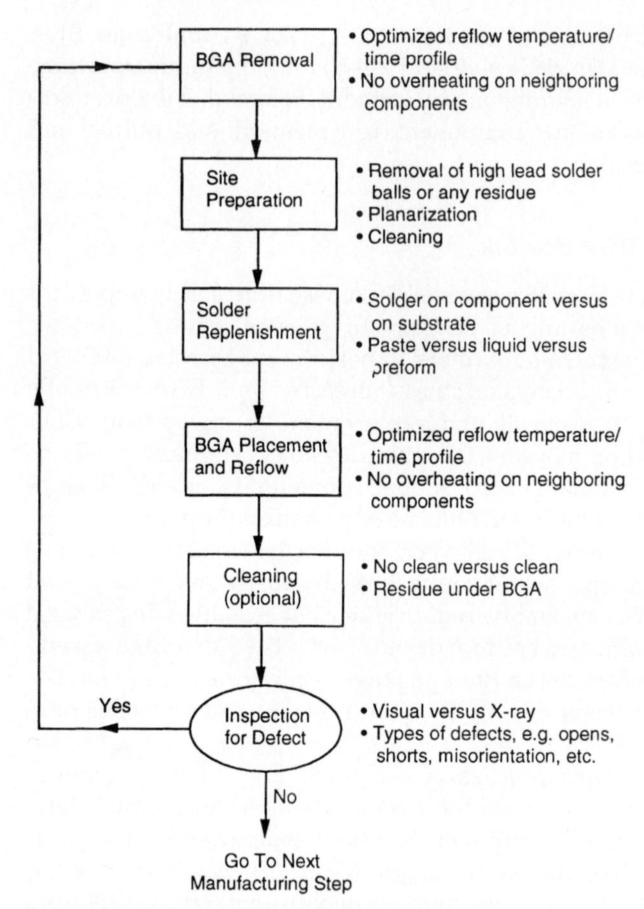

Figure 16.2 A typical BGA rework process.

Vac, Conceptronic, Manix Manufacturing, PACE, Semiconductor Equipment Corp. (SEC), and Sierra Research and Technology (SRT) utilize hot gas to reflow solder joints, while removing the component under rework with either vacuum or a manual pickup tool. In addition, AAT, Conceptronic, PACE, and SRT offer focused IR (infrared radiation) and/or thermode (or hot bar) as the heat source to reflow solder joints. AAT also offers a special component pickup method using thermoplastic adhesive to provide enough removal force, which is needed for rework on tape automated bonded components with die attach material. In general, the main difference between rework on conventional surface mount and BGA assemblies is most BGA solder joints are hidden underneath the BGA package, which makes not only rework on BGA much more challenging but also convective hot gas a much more effective heat source for solder joint reflow than radiative-focused IR or directly conductive thermode. In addition, the BGA rework equipment requires innovative improvements in component handling, alignment, reflow, and process control in order to perform repeatable and reliable rework on BGA assemblies. While most rework equipment vendors shown in Table 16.1 claim the availability of BGA rework capability, only a few have physically demonstrated feasible rework capabilities on all types of BGA assemblies, i.e., PBGA, CBGA, and TBGA. One example is the BGA rework equipment shown in Fig. 16.3, made by Air-Vac Engineering Co., Inc., located in Milford, Connecticut, which offers an innovative nozzle design, i.e., the horizontal flow control (HFC) nozzle[9] as shown in Fig. 16.4. Sealing its bottom surface against the board surface surrounding the BGA component under rework, the HFC nozzle offering the capability to distribute heated (inert) gas flow underneath the BGA package directly to the solder joints provides significant process advantages versus heating directly through the component from the top side. In addition, the heated gas exits the nozzle through side exhaust ports which direct the gas up and away from neighboring components. This can be very important because the neighboring components may be damaged (e.g., popcorn effect, discoloration, etc.) if they are exposed to excessive heat.[3,10] The alignment function of the BGA rework equipment is also very important because the BGA rework equipment is also used for placement of the new "known good" BGA component. It has to provide alignment between solder balls underneath the component and corresponding pads on board using a high-resolution vision system equipped with a high-quality stereo microscope or camera. One example is the beam-splitter vision system used in Air-Vac's BGA rework system, as shown in Figs. 16.5 and 16.6, which illustrate an overall configuration and a close-up view of the system, respectively. With the BGA component held firmly in the nozzle with vacuum, the image of

TABLE 16.1 Examples of Typical Surface Mount Rework Equipment and Their Capabilities

Surface mount rework equipment vendor	Solder reflow heat source	Component removal method/tool
AAT	Hot gas or thermode	Vacuum or adhesive
Air-Vac	Hot gas	Vacuum
Conceptronic	Hot gas or focused IR	Vacuum
Manix Manufacturing	Hot gas	Vacuum
PACE	Hot gas or thermode	Vacuum or pickup tool
SEC	Hot gas	Vacuum
SRT	Hot gas, focused IR, or thermode	Vacuum

Figure 16.3 This figure shows the Air-Vac DRS26 hot-gas BGA rework system. (Note the component nozzle is located directly above the board, the alignment/vision monitor on the left, and the process control monitor on the right.)

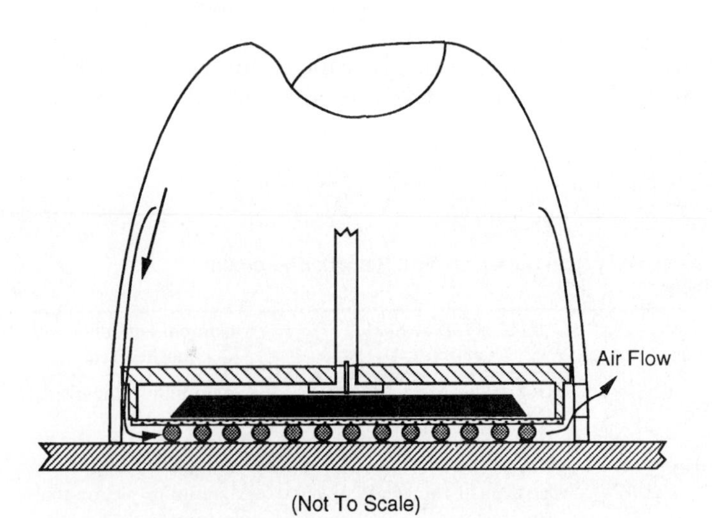

Figure 16.4 This figure shows the concept of the the horizontal flow control nozzle patented by Air-Vac Engineering. (Note the horizontal air flow underneath the BGA component.)

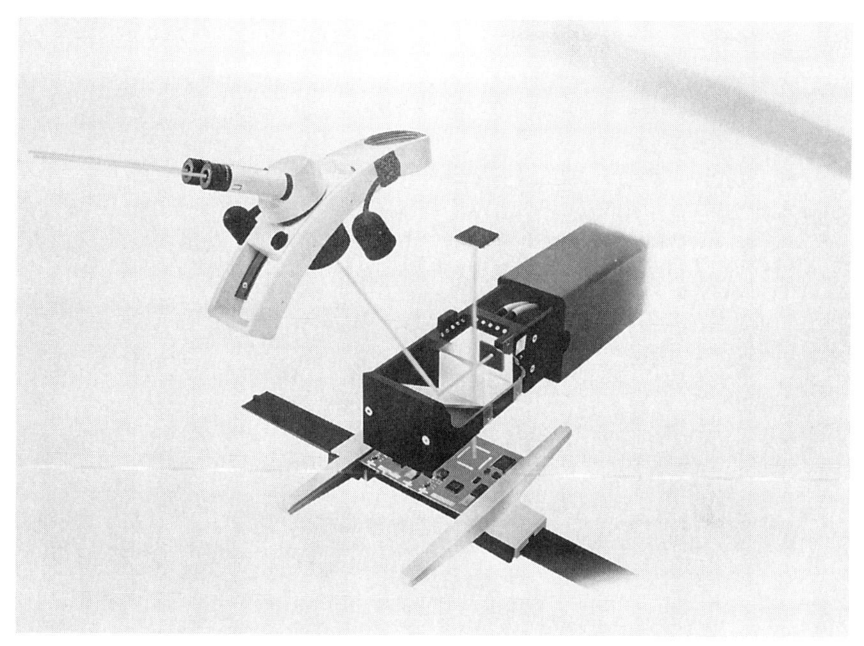

Figure 16.5 This figure shows configuration of the beam-splitter vision system built in the rework system shown in Fig. 16.3. (*Courtesy of Air-Vac Engineering.*)

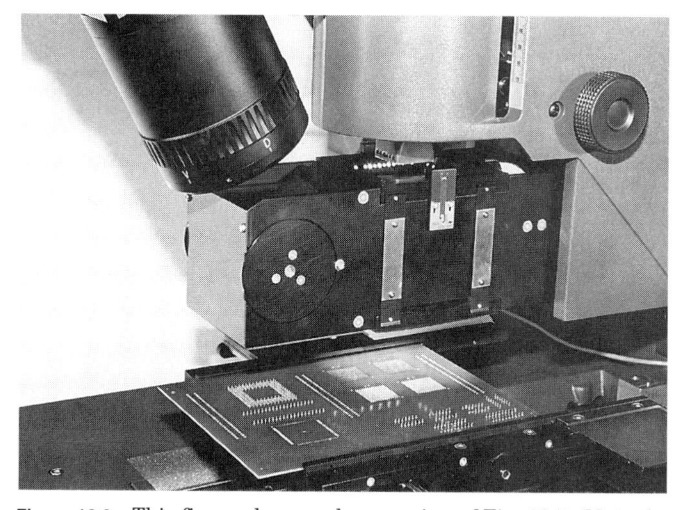

Figure 16.6 This figure shows a close-up view of Fig. 16.5. (Note the beam-splitter shuttle assembly which is positioned between the component nozzle and board.)

(a)

(b)

Figure 16.7 Examples of BGA alignment: (*a*) before, and (*b*) after proper alignment of the projected images of both solder balls underneath a BGA package and corresponding pads on board.

the board pads is projected up through the beam-splitter to the stereo microscope or camera. Simultaneously, the image of the BGA solder balls is projected directly over the pads. The superimposed image provides the capability to see and align every solder ball and pad as shown in Fig. 16.7*a* and *b*.

In addition to the rework process and equipment previously discussed, there are other important rework considerations:

- The physical/thermal disturbance or damage to nearby components, leads, or board surface during rework process is not acceptable, and must be controlled.
- Proper electronic discharge prevention procedure must be followed in order to avoid electrical damage to the reworked component.
- Acceptable maximum number of reworks per site must be characterized and specified not only without degradation in performance and reliability but also with no significant cost penalty.[11]

16.3 Rework of Plastic BGA (PBGA) Assemblies

The PBGA package and assembly are discussed in detail in several of the chapters in this book. In addition, rework of typical PBGA assemblies is also discussed in Chap. 10. Therefore, in addition to a brief review of the typical PBGA rework process, issues particularly associated with rework of large-size, high-pin-count PBGA assemblies and suggested approaches to address these issues are discussed as follows.

16.3.1 Typical PBGA rework process

In general, rework of PBGA assemblies is relatively straightforward, and is carried out as follows[1]:

- Dispense liquid flux under the component which is to be reworked.
- Use a hot gas rework machine to deliver heat primarily from the top side of the component.
- Position the nozzle (with the same size as the component or smaller) 0.125 to 0.25 inches above the component, or use the HFC nozzle as shown in Fig. 16.4.
- Ramp the heater for 45 to 60 seconds with a maximum temperature between 210 and 220°C.
- Remove the component using a vacuum tip once the solder has liquified.
- Use a solder wick and iron to remove excess solder on pads.
- Clean (optional), dry, and inspect the reworked site.
- Install a new PBGA component.
- Clean (optional) and inspect the reworked assembly for any defect, including the BGA component site, board surface, and neighboring components, which completes the PBGA rework process.

16.3.2 Issues for rework of large-size PBGA assemblies

Based on the typical PBGA rework process described previously, rework of PBGA assemblies is usually considered to be less challenging than rework on CBGA or TBGA assemblies, which are to be discussed in detail in the latter part of this chapter. This is because a PBGA assembly contains only one solder system, e.g., 62wt%Sn/36wt%Pb/2wt%Ag, which makes the rework process of PBGA assemblies relatively less complicated than that of CBGAs or TBGAs. However, it can become more challenging for rework on PBGA assemblies as the PBGA package size grows larger and larger, e.g., more than 27 mm square. The typical issues associated with rework of large-size PBGA assemblies are component handling and alignment, solder reflow process control, and solder joint integrity and reliability.

16.3.3 Approaches for rework of large-size PBGA assemblies

The issues of rework on large-size PBGA assemblies may be resolved through improvements in board design, rework process, equipment, or tooling design, depending on application requirements. While component handling and alignment is definitely related to rework equipment/tooling, good solder reflow process may be accomplished through improvement in equipment and/or process characterization and optimization. For example, it is not only important but also desirable to incorporate an anticrushing feature in the component pickup nozzle/tip design, which may prevent excessive force from being applied to the large PBGA package. This is also required in order to prevent "pancaking" the PBGA solder joints due to weight of the large-size PBGA package and/or incorrect force application algorithm. In addition, it may provide room for any expansion of the board or component during the rework process. For alignment, regardless, it is a vision system with complete view or split image of a component; it is a must to be able to precisely view and align the superimposed images of both solder balls on the package and pads on the board as shown in Fig. 16.7 in order to perform a defect-free PBGA assembly. Furthermore, it is important to check and understand the field-of-view capability of a vision system so that the size limitation of a rework equipment can be defined.

For the issue of solder reflow process control, it is very important to provide repeatable, reliable, and uniform reflow heat source across the entire array of solder balls without overheating package and adjacent components. This is usually accomplished through removal/reflow process characterization or thermal profiling. The concept is to use a thermal profile (a temperature versus time curve) which is characterized and optimized by correlation of a fixed-position temperature (usually measured by a thermocouple) with temperatures of various points underneath the BGA component for rework. This is very similar to thermal profiling of a conventional surface mount component, except for inspectability of the reflowed solder joints. As stated before, the BGA solder joints are hidden underneath the package, definitely increasing the difficulty of thermal profiling. For a large-size PBGA, it may be necessary to provide a real-time process control of temperature, time, vacuum activation, and flow rate parameters in order to achieve an optimized reflow process. One example is the method of real-time process characterization which generates the correlation curves as shown in Fig. 16.8. A thermocouple is mounted to the nozzle exhaust to measure the nozzle exhaust gas temperature. If the nozzle exhaust temperature has a direct correlation to the gas temperature flowing underneath the component, the exhaust thermocouple can provide very accurate information as to when reflow of the solder joints is occurring. The other potential approach for achieving good solder reflow process is the BGA rework utilizing condensation inert heating (CIH), which is currently under development at Centech (located in Minneapolis, Minnesota). CIH is essentially a method of vapor phase reflow in which a perfluorinated fluid is heated to its boiling point (usually 215°C) and a saturated vapor layer is maintained above the boiling fluid. The BGA assembly is then placed in the saturated vapor where the latent heat of vaporization "uniformly" heats the assembly as the vapor condenses on its surface. Typically, the assembly is pre-

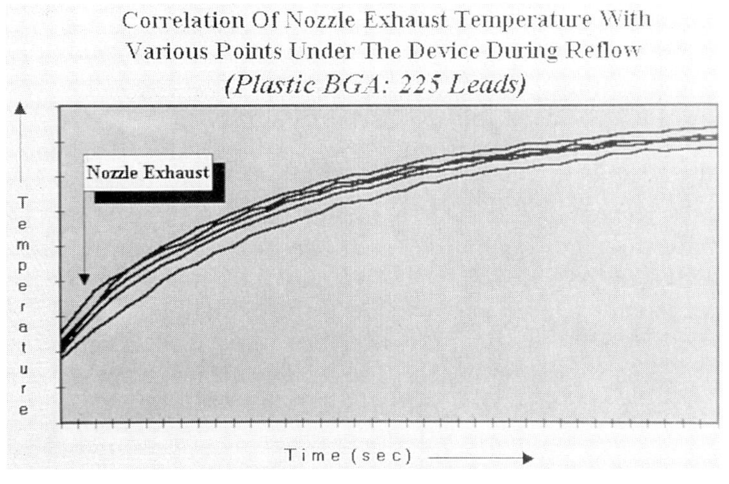

Figure 16.8 This figure shows correlation of a component pickup nozzle exhaust temperature with temperatures of various points underneath a BGA component.

heated using IR heating; then saturated vapor at 215°C is brought up to the board level so that the board is heated uniformly in the inert vapor that is oxygen-free. The solder joints of the BGA assembly reflowed at the same time and the component can be removed from the board by an electromechanical pickup tool without stress. The replacement part can then be placed on the board and reflowed into place. The process is quick and simple, providing a relatively easy rework process with minimum stress to the board. However, the drawback of this approach is that all other components on the reworked board are also reflowed.

The issue of solder joint integrity and reliability, which is a concern for all types of BGA rework, is particularly challenging for a large-size PBGA due to its large DNP (distance to neutral point). For a fully symmetrically designed BGA package, all four corner solder balls of the package have the same and largest DNP. Generally speaking, the larger the DNP a solder joint has, the higher the stress/strain it has. The stress/strain is generally caused by CTE (coefficient of thermal expansion) mismatch between two or more materials and a temperature change. The good solder joint integrity and reliability may be achieved in two ways. One is by designing noncircular pads instead of standard circular pads on the board to which a PBGA package is attached. There are two potential advantages using this approach. One is that the solder joint integrity of this type of pad design will be easier for inspection by conventional x ray. When solder balls are connected to these pads, the wetting and formation of fillet create an x-ray image in the shape of the pad, thus indicating that the solder joint has been formed.[1,10] The other "potential" advantage for this approach is that the shape of the solder joint created by this type of pad design may improve fatigue strength of the solder joint and, ultimately, reliability of the joint.[12] The tradeoff for this approach is the routing space on the board and potential difficulty for solder paste application. The other approach which may improve the solder joint integrity and reliability of the large-size PBGA rework is to use the solder joints with the largest DNP, e.g., four-corner solder joints of a fully symmetrically designed BGA package, for mechanical joints only. In other words, use no electrical circuitry designed for these solder joints, which are mainly used to mechanically fasten the package to the board. Since the highest stress/strain in a BGA assembly is always on the solder joints with the largest DNP, this approach can result in a significant improvement in reliability of the BGA assemblies, especially for a large-size BGA. The drawback of this approach is that a small percentage of the BGA solder joints cannot be used for electrical connection.

16.4 Rework of CBGA and TBGA Assemblies

16.4.1 CBGA and TBGA assemblies

The CBGA package (sometimes also called *module* or *component*) shown in Fig. 16.9 has been discussed in great detail in the previous chapters of this book and in several papers.[13,14] The ceramic-based BGA package is attached to the board via a 0.050-in grid of 0.035-in diameter, 10wt%Sn/90wt%Pb solder balls in a eutectic solder joint. The solder balls are also connected to the ceramic package using eutectic solder. The dimensions (e.g., solder volume, pad diameter, DNP) of the ball-to-board solder joint are the critical limiter of the overall thermal cycle reliability of CBGA package.[15] Thus, the details of the rework process are important not only from a yield perspective, but as a critical step in producing a product with predictable reliability.

The TBGA package shown in Fig. 16.10, as discussed in previous chapters, is a hybrid of TAB (tape automated bonding) and BGA.[16] The resulting assembly structure, with 10wt%Sn/90wt%Pb balls connecting the package to the board by eutectic solder joints, is similar to that of the CBGA. However, the TBGA solder balls are connected to the package body by partial reflow of the 10wt%Sn/90wt%Pb solder balls, which is different from the eutectic connection used on the CBGA package. In addition, the package cross section is markedly different, and the requirements of rework reflect this. First, the TBGA has quite different behavior due to the multiple layers of the package. The polyimide tape/substrate, derived from TAB technology, is attached to a copper stiffener using a pressure-sensitive adhesive. The adhesive and polyimide layers act as thermal barriers to direct conduction. Additionally, the solder balls are 0.025 inch in diameter, as opposed to 0.035 inch for the CBGA. Finally, and most significantly, the TBGA is not reliability-sensitive to solder volume. This is because the TBGA package is dominated by the copper stiffener, with a CTE that is closely matched to standard PCB (printed circuit board) materials to which it would be attached. The small level of strain that is generated by either subtle CTE mismatch or temperature differences from package to board is absorbed in the adhesive layer, and results in negligible stress/strain in the solder joints.[16,17] Thus, while process control is still necessary to guarantee high yields, there is more room for flexibility in the determination of most TBGA assembly process windows.

16.4.2 Why rework is needed for CBGA and TBGA assemblies

The primary causes for rework in CBGA or TBGA are not assembly defects but, instead, engineering change requests or package- (or more

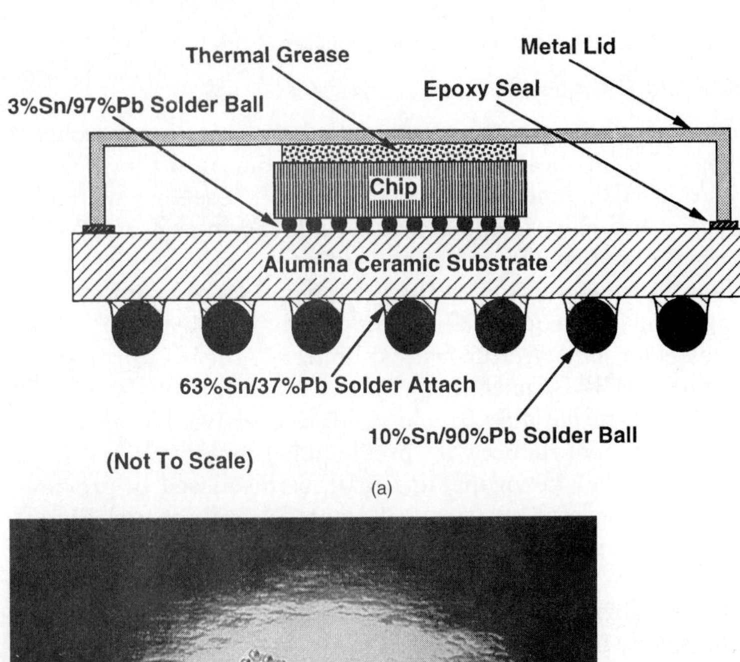

(a)

(b)

Figure 16.9 Examples of a CBGA package: (*a*) a CBGA schematic cross section; (*b*) a CBGA package with 361 balls and a 25-mm body size.

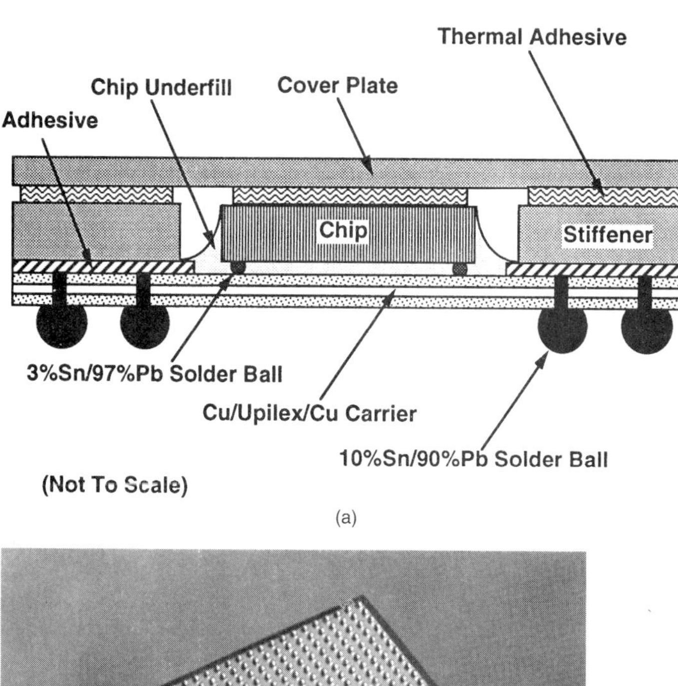

(a)

(b)

Figure 16.10 Examples of a TBGA package: (*a*) a TBGA schematic cross section; (*b*) a TBGA package with 736 balls and a 40-mm body size.

pointedly, semiconductor-) level defects. However, assembly processing defects can and do occur. The most predominant CBGA assembly-related failure is shorting between solder balls. Shorting generally is indicative of process problems in the initial attach process, but (like most things) the problem is often overlooked until it begins to produce defects. Solder shorts can be detected using top-down x ray. Prior to rework, electrical shorts should be verified by x ray, as shorts can be generated by conditions other than solder bridging.[14] The most common defect in TBGA assemblies is package tilt relative to the board, resulting in opens—usually as a result of improper placement during initial attach. In addition, package misorientation, either by misregistration or theta twist, can require removal and replacement of a CBGA or TBGA. A nonwetted joint (shown in Fig. 16.11), due to inadequate flux or ball contamination, is particularly insidious, as it still conducts electricity and often does not show up on initial testing, but results in electrical failure during early product life. The bottom line is that any manufacturing or assembly-related defects such as low solder volume, contamination, missing ball, or misorientation that may cause a component malfunction in the short or long term will require component removal and replacement.

16.4.3 Rework processes for CBGA and TBGA assemblies

Because of the presence of the 10/90 ball, the rework process can have a somewhat different overall flow from a pure eutectic ball system, and has several items that must be dealt with that are not present in a eutectic solder ball system. The general process flow for both CBGA and TBGA is the same, and involves four basic steps: (1) component removal, (2) site preparation, (3) solder replenishment, and (4) component replacement and reflow, cleaning, and inspection.

Figure 16.11 A cross-sectional view of a nonwetted solder joint in a TBGA assembly, which is caused by either inadequate flux or solder ball contamination.

Component removal. The first important step in the BGA rework process is component removal. Like all BGA components, the removal of a CBGA or TBGA involves getting adequate temperatures to all the solder joints to generate reflow. At the same time, however, it is desirable to keep the maximum joint temperatures as low as possible, to limit dissolution of the high-lead ball into the eutectic joint. If too much lead dissolves into the liquid, it can result in a higher-melting solid, which is more difficult to work with during subsequent processing. The control of temperature is also important in minimizing thermal shock and warpage to the board itself.

The primary challenge with CBGA is the thermal mass of the component. The ceramic body acts as a tremendous heatsink when trying to heat the solder joints. Thus it is necessary to heat the component body along with the joints. With TBGA, it is not the thermal mass, but the thermal gradient generated by the multiple materials in the package cross section, some of which are not particularly good thermal conductors. Generally, to accomplish the desired reflow requires forced convective hot gas to raise the temperature. Custom-gas-nozzle designs for specific body sizes are generally required, and additional customization for specific board designs to help shield neighboring components from overheating can also be helpful. Focused IR is not a preferred method of heating, as it forces high thermal gradients from top to bottom, and often cannot achieve adequate temperatures across the entire part.

The process settings used on a given tool will vary depending on design variables such as component type, body size, part thickness, board thickness, board cross section, and tool-design variables. The settings should be determined for every significantly different design by careful thermal profiling. This usually involves burying thermocouples at locations across the part and checking joint temperatures against the profile used (as discussed in Sec. 16.3). One danger in profiling is failure to include power/ground connected joints. The board can act as a significant heatsink for joints connected by plated through-holes to large copper planes.[18] These locations must be taken into account when determining which solder joints to use for representative temperature monitoring.

In the CBGA, the presence of eutectic solder on both the ball-to-board joint and ball-to-package joint results in a certain percentage of the solder balls being left on the board when the component is removed. The specific percentage left behind will be determined by the details of the process used to remove the component. An example surface is shown in Fig. 16.12. Because balls will be missing from the component once it is removed, the component is not immediately reusable. The component can have balls reattached by returning the entire com-

Figure 16.12 This figure shows a CBGA component site after removal. (Note solder balls remaining.)

ponent to the manufacturer. The TBGA solder balls are attached using a partial reflow of the 10/90 solder, and thus are all removed with the component.

One process variation that should be avoided at component removal is maintaining the component under tension while heating. The motivation for doing this would be to remove the component as soon as all the joints are liquid, thus minimizing thermal problems with the board. The downside of tension has two manifestations. First is in number of solder balls remaining on the board for CBGA. If the part is being heated from the top (as is usual with hot gas), then the joints in contact with the ceramic will reflow before the joints on the board. The result can be an abnormally high percentage of solder balls left on the board rather than being removed with the component.

The second danger of tension is increasing the probability of lifting pads. As the component is heated and joints begin to reflow, the number of solid joints begins to decrease. Eventually the number will be low enough that the tension can result in a fairly significant stress on the remaining pads. Depending on the copper-to-board adhesion qualities, this can cause the copper to lift off the surface and, in the worst case, come completely free of the board. This destroys the component site and therefore the board.

Site preparation. The next step in the rework of CBGA or TBGA is the board site preparation, that is, having the site ready to accept a new component in a local replacement process. This step has several important functions: removal of the remaining 10/90 solder balls (CBGA only), removal of the high-lead residue from the mounting pads, and overall planarization of the component site.

First and most obvious, in the case of CBGA, the remaining 10/90 balls must be removed. This can be accomplished in several ways, the most simple being a soldering iron and tweezers. Such a process is tedious but effective for very low volumes of parts. In general, it is most efficient to combine the removal of the balls with the removal of the high-lead residue. The residue is the result of dissolution of the high-lead ball into the eutectic solder. When the solder ball is removed from the solder joint during initial component removal, the remaining solder is no longer of pure eutectic composition. Instead, large areas of proeutectic lead now exist in the eutectic matrix. This has the effect of raising the melting point of the solder pad, thus making it more difficult to reattach a new component. Also, if multiple reworks are done on the same site, the amount of dissolved lead will increase with each rework, eventually raising the melting point to where even component removal is questionable. The specific process used to remove the high-lead residues can involve wicking, vacuum, solder immersion, or combinations of these processes. In general, the intent is to leave behind a surface composed of some volume of pure eutectic solder on the pad. An example of a redressed site surface is shown in Fig. 16.13. Care must be taken not to remove too much liquid as, if the Cu-Sn intermetallic layer is exposed, the pad is likely to be unsolderable for reattach.

The other site preparation step that may be necessary is planarization. As with initial attach, BGA rework is a Z-tolerance stack. If the board is warped significantly in the component site, the component will not seat, and opens and/or defective solder joints will result. Some pro-

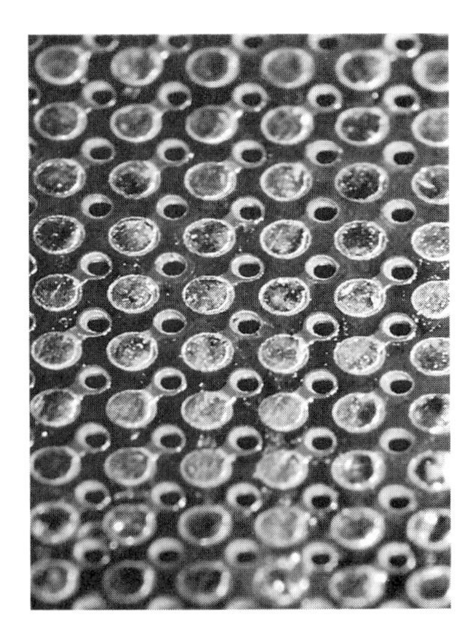

Figure 16.13 Example of a redressed board site surface.

cesses used for component removal or solder ball removal can add to the warpage of the component site. Also, the variation of solder heights on the remaining eutectic solder will add to the tolerance stack. The specific limits used will depend on user discretion, but if sites are detected outside some preset limit, additional processing to planarize will need to be performed. The process could involve local touch-up, global flattening, local grinding, or any number of other possible planarization techniques. The desired final result is a eutectic solderable surface that is as near to flat as possible, within reasonable limits.

Solder replenishment. As previously stated, the volume of solder used to connect the CBGA solder ball to the board is the determining factor in the thermal cycle reliability of the assembly. In the case of TBGA, the solder volume is not reliability-critical. However, the existence of a consistent solder joint is the most obvious method to ensure high rework yields. Most of the eutectic solder that attached the original component to the board will have been removed during the component removal and/or site preparation operations. Therefore, a *controlled* reapplication of solder is necessary to reattach the component to the board and, in the case of CBGA, to provide the component with equivalent reliability to a nonreworked component.

Fundamentally, there are three ways that solder can be added to the system: in paste, as liquid, or as a solid. Solder paste printing is the most obvious choice for board assembly, as the general know-how is practiced daily as part of the manufacturing process. However, there are new challenges involved in this approach. Geometry of the board can make location of a stencil in the component area difficult to impossible. Then there is the danger of accidental paste contamination of surrounding components from "sloppy" errors. Finally, there is the question of the volume control necessary to guarantee reliability. The laser tools that are used for initial paste inspection may be blocked by adjacent components, and thus some alternate process may have to be defined. A variation to paste printing on the board is paste printing on the package itself. By applying the paste to the package, controls can be somewhat more flexible. But in this case, the fixturing, printing process control parameters, and general practices all have to be defined as unique processes. In either case it is important to remember that a single low-solder-volume deposit can jeopardize the reliability of the final product, and thus diligence in process control is critical.

Liquid solder deposition is the easiest to control, but the most difficult and costly to practice. The process involves depositing a drop of liquid solder onto each mounting pad. To perform this type of operation, a specialized tool is required. Following the deposition onto all sites, the site and package would both be fluxed prior to placement and reflow.

The choice of flux is important, as it needs to serve in this case as a tack vehicle as well as fluxing agent.[19] Floating of components on flux can become a problem as well, with the component riding on the liquid flux and moving off location prior to reflow. Finally, the question of the Z-tolerance stack must again be considered. Obtaining contact between the solder ball and the eutectic solder deposit is required. If variations in the package or the solder deposits are not understood and appropriately controlled to guarantee contact, opens will result.

Solid solder introduction is most commonly referred to as *preforms*. Preforms are fixed solder shapes that can be used to add solder to a specific location. The most common application is solder rings used to supplement solder on pinned components. For CBGA or TBGA rework, the concept is somewhat different. The idea is to locate a fixed solid solder volume over each mounting pad. Rather than using individual pieces, a single preform can be created to cover the entire component site. A thin sheet of solder can be etched chemically into a web, as shown in Fig. 16.14. The web is placed on flux which has been coated onto the board site. Depending on type of flux, additional application may be necessary between the web and the component as well.

The web between pads needs to pull back during reflow to form discrete joints, and therein lie the major challenges of using this technique. If the web does not separate, but instead draws solder to it, then a short between adjacent balls will be created. The extent of this phe-

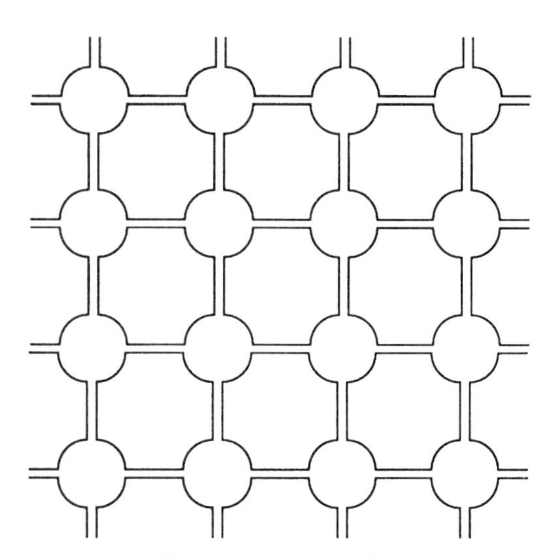

Figure 16.14 This figure shows a design configuration of a solid solder web which can be made for solder replenishment by chemically etching of a thin sheet of solid solder.

nomenon will depend on preform design, board design, flux type and control, and solder composition. Having considered the possibility of shorting, the process needs to be additionally controlled to guarantee adequate solder volume. The percentage of the web that pulls back into each pad will help determine the solder volume in the joint. By nature, this will create some statistical variance in the solder joint. Thus, development of a successful preform process will also take into account this variance and compensate for it in the design of the preform. Finally, as in all cases, opens can result if attention is not given to taking care of the tolerance stack to guarantee solder contact with both the board and the component. Opens can result from overapplication of flux as well as underapplication as too much flux will allow the solder to preferentially wet the board or the ball, and may prevent contact with the opposing surface.

As stated, solder reapplication is the most reliability-critical of the rework steps. It is in this area that the most time should be devoted to developing, understanding, and exercising the proper process controls to guarantee high yields and good reliability of the finished product. If these controls are not put in place, the costs will be measured in the short term as low rework yields that drive costly additional processing, and in the long term as increased field-failure rates that are costly both in terms of repair and reputation.

Component replacement and reflow. Having already defined a basic reflow process to remove a component, the additional intricacies of replacement and reflow are generally variations on a theme. The additional challenges lie in dealing with component-to-board alignment, component placement process flow and control, and paste/flux/solder reflow considerations.

Component alignment challenges for all BGA components are the same. Some form of alignment of ball to pad for the array is necessary to locate the component to the site. Specific placement tolerances vary depending on the specifics of the board design and solder reapplication techniques used. But, in general, hand and eyeball placement will not be adequate. Alignment of an array of solder balls can be either mechanical or optical. Mechanical registration is usually used on semiautomated tools that are programmed to precisely locate the board site to be replaced. A locating fixture, or "nest," is used to precisely position the component relative to some global reference. The tool then locates the component site via vision software relative to the same global reference and does a direct pick-and-place of the component. Optical registration most commonly takes the form of split-optics tooling, where images of the board site and component are superimposed and aligned manually to provide accurate placement. Either method will provide an adequate solution for a wide range of BGA and non-BGA component placements.

Reflow of the component has similar constraints to removal, but with the added limitations caused by the use of either solder paste or flux. As with component removal, tool settings will have to be optimized for each design (or design family) due to the variable thermal behavior between designs. When solder paste is used, the settings can be optimized to the same temperature profile specification limits as are used with standard reflow, either as recommended by paste manufacturer or specially developed by the assembler. If flux and solid solder are used, reflow profile limitations usually have to be developed for the specific process and materials set by the assembler prior to optimization of tool settings. The reason for this is that profiles can have a dramatic effect on the activity, performance, and physical properties of flux, and therefore will play an important role in eliminating the problems mentioned previously relative to these systems (shorts, opens, component float). The profile limits recommended by the flux manufacturer do not necessarily take this into account, and thus may be too lax for use in CBGA or TBGA rework.

Once the limits of the system are determined, the tool optimization to the limits is the same as with optimization for component removal. The tool settings used for replacement should be usable for component removal as well. However, the constraints of the replacement profile may drive a longer cycle to do replacement than is necessary for removal. Thus, if overall cycle time is important, then different settings on the tool can be used to shorten removal relative to replacement.

Cleaning and inspection. Cleaning after rework will be required if water-soluble or rosin-based materials are used. Cleaning parameters post-rework should be the same as used for initial attach processes. For water cleaning, this means guaranteeing the dryer profiles are adequate to remove large pockets of trapped water from under the component. This can be easily accomplished by blowing compressed air across and under the component upon removal from the cleaner. Water splatter indicates inadequate drying. In the case of rosin materials, the adhesive used in producing the TBGA may be soluble in some solvents. A check on specific material compatibility is suggested prior to using organic solvents on TBGA. The CBGA has no solvent limitations.

As with all BGA components, the interior joints are considered uninspectable in the conventional sense. However, some level of inspection after rework is considered wise to ensure that no gross defects have been missed. The two inspection methods that provide the best level of data are x ray and visual.

A top-down x ray will show up shorts very clearly, and is fairly simple to perform.[4,14] However, x-ray equipment is costly and requires an investment in training that may not justify the number of defects found. Such inspection is probably best considered only if the tooling already exists in the assembly plant for alternate reasons, or as a vended service. In addition, there are x-ray laminography systems (such as the one made by 4-Pi[4]) that can provide a 3-D map of the solder ball and joint. While the resolution makes it difficult to use these tools to determine a too-small joint from a good one, they can detect opens and shorts both with a very high degree of accuracy. The downside to these tools is both the cost of the tool and the run time for inspection. Again, if the tooling is already in use or can be hired on a per part basis, it can provide valuable insight, especially at the development stages.

Visual inspection should include alignment and outside row joint quality. While alignment may have been checked during component placement, reflow floating problems (depending on assembly/rework process used) can cause the reflowed component to shift rows or twist relative to the board site. A recheck takes seconds, and can detect the occasional defect without escape into test. Checking outside row joint quality is more time-consuming, but can help reassure the process control skeptic that the process is working correctly. Particular attention should be paid to the solder joints at or near the corners, as these are the highest stress points and therefore most sensitive to defects in terms of reliability.

16.5 Summary

Rework of BGA assemblies can and should be performed in a controlled fashion with a constant eye toward yield and reliability. It is important to remember that a "quick and dirty" rework process will likely result in the need to rework the part again. For large PBGA, special attention needs be paid to board designs, rework equipment capability, and process control. For PBGA and CBGA, thermal cycle or similar testing must be used to determine the validity of any rework process or changes to existing processes. For either CBGA or TBGA assembly, the solder reapplication process is *the* critical step to creating a solder joint equivalent to an unreworked assembly. It is on this step that primary attention and control should be focused. In addition, the thermal management of the removal and replacement processes is also very important. It can be accomplished using clever hot gas tool design as stated in the early part of this chapter, and tailoring equipment/tool settings around component and board design variables. The successful rework process will take all these things into consideration, and should result in a high-yield process that provides excellent reliability performance results. In summary, only a well-characterized rework process carried out by a reliable rework machine equipped with correct capabilities will result in repeatable and reliable rework of BGA assemblies.

16.6 Acknowledgments

The authors would like to express appreciation to Mr. Brian Czaplicki of Air-Vac Engineering, Inc., who provided valuable discussion and several pictures for this chapter, and Mr. Nile Plapp of Centech for discussion of their developmental work on BGA rework. In addition, we would like to thank and recognize the team of people at IBM who have been part of BGA rework development: Russ Lewis, Chuck Kephart, Craig Hiem, Keith Vanderlee, Leo Anderson, and Karl Putlitz. Special thanks go to Mr. Aurangzeb Khan at Tandem Computers for providing management support, and Dr. John Lau for organizing this book and thus providing us the opportunity of writing this chapter.

16.7 References

1. Houghten, J., "Capturing Design Advantages of BGAs," *Surface Mount Technology,* March, 1994, pp. 36–43.
2. Ries, M. D., et al., "Attachment of solder ball connect (SBC) packages to circuit cards," *IBM J. Res. Develop.,* vol. 37, no. 5, September 1993, pp. 597–608.
3. Lau, J. H., et al., "No Clean Mass Reflow of Large Over Molded Plastic Pad Array Carriers," *International Electronics Manufacturing Technology (IEMT) Symposium,* October, 1993, pp. 63–75.
4. Adam, J., "Using cross-sectional x-ray techniques for testing ball grid array connections and improving process quality," *NEPCON-West,* March 1994, pp. 1257–1265.
5. Abbagnaro, L., "Process development in SMT rework," *Circuits Manufacturing,* December 1988.
6. Chang, J., et al., "Rework of Multichip Modules—Die Removal," *NEPCON-West,* February, 1992, pp. 512–522.
7. Lau, J. H., *Handbook of Fine Pitch Surface Mount Technology,* Van Nostrand Reinhold, 1994, pp. 505–519.
8. Economou, M., "Rework System Selection," *NEPCON-West,* February, 1993, pp. 1101–1118.
9. Patent pending by Air-Vac Engineering Company, Inc., Milford, Conn.
10. Tuck, J., "BGAs: A Trend in the Making," *Circuits Assembly,* December, 1993, pp. 20–21.
11. Mavroides, J., "Cost Saving Opportunities with Multichip Modules," *IEMT Symposium,* September, 1991, pp. 413–416.
12. Hedemalm, P., and Salonen, J., "Design-for-Reliability of Flip-Chip Solder Joints," *ITAP & Flip-Chip Proceedings,* February, 1994, pp. 111–116.
13. Bartley, J., et al., "CBGA: A Packaging Advantage," *Surface Mount Technology,* November, 1993, pp. 35–40.
14. Iversen, W., "Ball Grid Arrays Emerge," *Assembly,* November–December 1993, pp. 20–23.

15. Phelan, G., "Solder Ball Connection Reliability Model and Critical Parameter Optimization," *43rd Electronic Components & Technology Conference,* June, 1993, pp. 858–862.
16. Andros, F., and Hammer, R., "Area Array TAB Package Technology," *ITAP & Flip-Chip Proceedings,* February, 1993.
17. Mescher, P., "Card assembly implications in using the TBGA module," *MEPPE Focus Proceedings,* May, 1993.
18. Groover, R. L., et al., "BGA—Is it Really the Answer?" *ITAP & Flip-Chip Proceedings,* February, 1994, pp. 57–64.
19. Lau, J. H., *Chip on Board Technologies for Multichip Modules,* Van Nostrand Reinhold, 1994, pp. 364–365.

Chapter

17

Burn-In Sockets for Ball Grid Arrays (BGAs)

Christopher A. Schmolze

17.1 Introduction

Ball Grid Arrays (BGAs) are a new and exciting form of semiconductor packaging. BGA packages are attractive to semiconductor packaging for improving the electrical performance of their design and reducing the area which the package currently occupies on the printed circuit board. Manufacturing engineers are finding BGA packages an option for replacing fine-pitch quad flat packages (QFPs) and increasing manufacturing yields or for converting through hole pin grid array (PGA) packages to a surface mount package. Competitive BGA package alternatives are currently proliferating at a fast rate. Currently available packaging options include: plastic overmolded packages, plastic laminate packages, tape automated bonding (TAB) based packages, and ceramic material packages. These packaging technologies remain relatively new, and with new technology will come new questions regarding the reliability of the technology. Burn-in is one method that will be used to answer many of these reliability questions.

17.1.1 What is burn-in?

As semiconductors and semiconductor packaging continues to grow in complexity, increased reliability of these products is needed.

> The reliability goals of the semiconductor industry are typically discussed with reference to the traditional reliability curve of component life. This reliability curve describes the failure rate distribution of all integrated circuits (ICs) and is known as the 'bathtub' curve, due to its characteristic shape when charted in a line graph. Manufacturing defects are generally

expected in a small percentage of ICs and because of these inherent manufacturing defects, the affected devices have shorter lifetimes than the remaining population. Known as *infant mortalities,* these devices constitute a small fraction of the total population, but are the largest contributor to early-life failure rates. Once infant mortalities have been removed, the remaining devices have a very low and stable field-failure rate. The relatively flat, bottom portion of the bathtub curve represents stable field-failure rates after the infant mortalities have been removed and before device wear-out occurs. It is referred to as the *random failure* portion of the curve. Eventually, as wear-out occurs, the failure rate of the devices begins to increase rapidly. The average lifetime of an IC is not clearly understood, since most lab tests simulate only a few years of normal device operation.[1]

Burn-in refers to a process in which elevated temperature is used to provide a stress to the IC for the purpose of accelerating the rate at which the infant mortalities occur. Additionally, a biased voltage across the signal input and signal output pins is included at the elevated burn-in temperature. The weeding out of these infant mortalities, allows for product to be shipped that is much more certain to have reliable performance.

17.1.2 Types of burn-in and burn-in conditions

Some standard temperature and bias conditions associated with the burn-in process are in Military Standard 883, "Test Method and Procedures for Microelectronics."

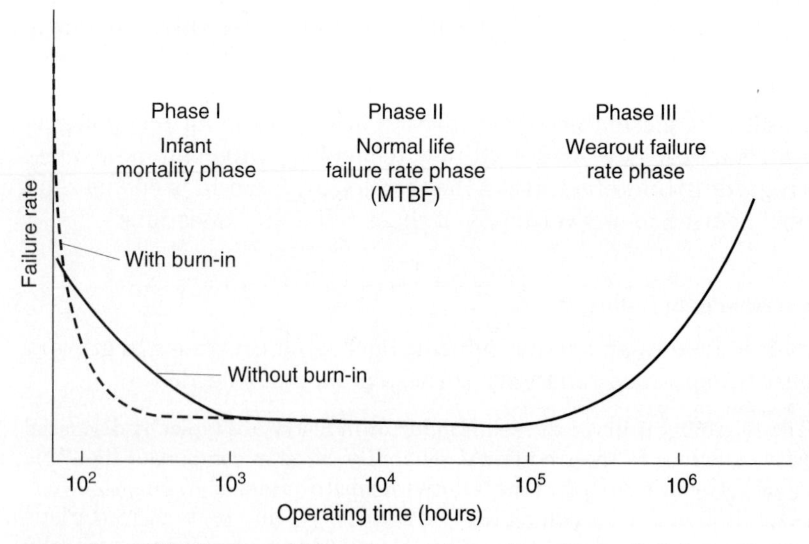

Figure 17.1 Bathtub curve. (*Micron Semiconductor Inc.,* Quality/Reliability Handbook, *1993, pp. 8–10.*)

> This standard establishes uniform methods, controls, and procedures for designing, testing, identifying and certifying microelectronic devices suitable for use within Military and Aerospace electronic systems.[2]

These are three types of burn-in procedures. *Static* or steady-state burn-in refers to an elevated temperature condition where some of the device pins receive a bias voltage and the device is not excercised (data written) during the time that the IC is at elevated temperature. In *dynamic* burn-in, pins are biased at elevated temperature and data is written to and read (monitored) or not read (unmonitored) during the time that the IC is at elevated temperature. *Test during burn-in* (TDBI) pins are biased at elevated temperature; and data is written to the device to perform functional and programmable logic tests prior to the data being read from the device, all occuring while it is at elevated temperature.

> Testing at temperature can off load an expensive tester, and a built-in self-test circuit allows the ability to identify the failure and when the failure occurred, thereby allowing infant mortality rates to be computed as a function of burn-in time. As a result, an optimal burn-in time for each product family can be established. This data also allows the ability to correlate the burn-in failure rates with life test data, typically obtained by IC manufacturers to determine the field failure rates of their products.[1]

To further accelerate the rate at which infant mortalities occur, additional stresses to the IC can be introduced. These stresses may include: a

A burn-in time and temperature regression table is given in this document. Common temperatures for burn-in are 125, 150, and 175°C. To a large extent the commercial IC industry has adapted the principles of Military Standard 883, which was originally written with an emphasis on hermetic ceramic material packages.

Today over 90 percent of all integrated circuits produced today are packaged in plastic encapsulant materials.[3]

The continual improvement in the reliability of these packages has lead the military to create Appendix B for plastic packages to Military Standard 1835, "Microcircuit Case Outlines."

TABLE 17.1 Types of Burn-In

Common names	Temperature	Pins biased	Data transfer
Static, steady-state	Elevated	Some	No
Unmonitored dynamic	Elevated	Yes	Written not read
Monitored dynamic (Intelligent, institute, output monitoring)	Elevated	Yes	Written and read
Environmental stress screening	Cycled	Yes	Varies
Test during burn-in	Elevated	Yes	Written, changed and read

SOURCE: Nicolaou, C. A., "Plastic Package Failure Modes," *Surface Mount Technology,* Nov. 1993, pp. 45–50.

TABLE 17.2 Types of Environment Stress Screens

Test	Description	Failure mechanism
Thermal cycle	Cycle between several temperature extremes, ex. –40°C to 125°C 10 min dwells 1000 cycles	Thermal mechanical stress of wire. Die bonding stress.
Autoclave/pressure pot	Unbiased storage in saturated steam, ex. 121°C, 100% RH, 15 psig, 48 hrs.	Galvanic corrosion of bond pads.
Temperature/humidity/bias	Biased or reverse-biased operations in high-humidity, high-temperature conditions, ex. 85°C ambient, 85% RH, 1000 hrs.	Aluminum corrosion of bond pads. Aluminum metallization corrosion. Ionic impurity of molding compound.
High-temperature life	Biased operation high-temperature environment, ex. 1000 hrs, 125°C	High temperature failure mechanism. Aluminum/gold intermetallic metal/silicon diffusion

bias voltage across the signal input and signal output pins greater than or in the opposite direction (forward) than the normal (reverse) bias, thermally cycling, power cycling, raised humidity, vibration, and raised pressure. The introduction of one or more of these stresses to the burn-in process is often referred to as *environmental stress screening* (ESS).

17.1.3 Types of failures burn-in identifies

Infant mortalities can be classified as two types: those which are *intrinsic* to the die (inherent manufacturing defects) and those which are *extrinsic* (related to the packaging assembly process). Some of the causes for infant mortality failures are catagorized below.

IC Failure Mechanisms

Intrinsic failures	Extrinsic failures
Ionic contamination	Metallization corrosion
Oxide defects	Bond defects
pin holes	wire bond
uneven layer growth	tape automated bond (TAB)
diffusion	flip chip bond
Silicon defects	Die attach
Dielectric breakdown	Package integrity
Metallization	cracking
Debri contamination	Intermetallic formation
	Ionic migration
	Electromigration
	Cracked die

As long as the maximum die temperature is kept below the glass transition temperature, failure due to thermomechanical stresses will be very rare. Should the junction temperature of a plastic package reach or exceed the glass transition temperature (for epoxies, Tg = 150°C), three phenomena may occur:

- The coefficient of thermal expansion (CTE) makes a sudden two- to fourfold increase, a significant thermal stress.
- The package becomes less resistant to mechanical damage or deformation from shock and impact.
- The mobility of ionic contaminants within the molding material increases markedly.

The electromigration of metallization traces and silicon also is a wear-out mechanism. Primarily a function of current density and temperature, the silicon and metallization actually migrate to form voids, shorting across metallization traces and degrading electrical parameters. However electromigration is usually only seen in those devices operated at very high temperatures.[3]

17.1.4 Why ball grid arrays will be burned in

Along with the lack of availability, cost, and ball inspection questions, the lack of readily available reliability data would have to appear on a

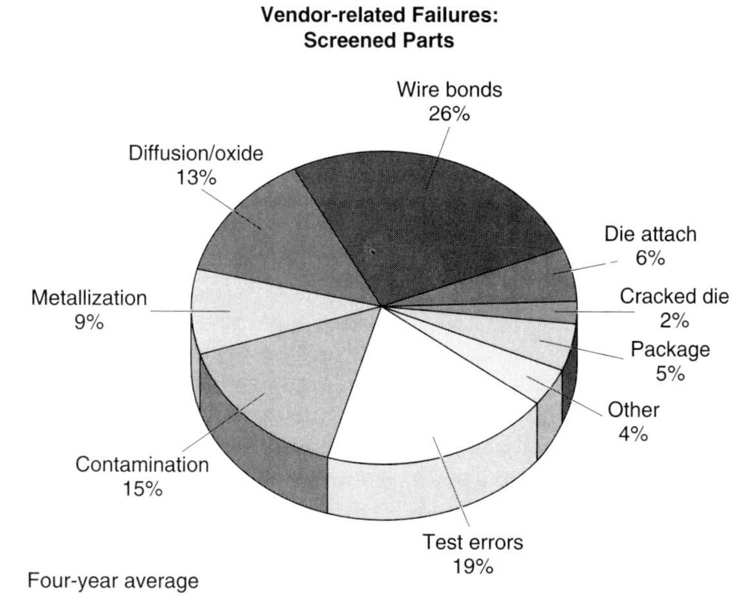

Figure 17.2 Average of IC packaging failures shows wire bonding to be the largest single cause of failure. (*Trigwell, S., "Failure Mechanisms of Wire and Die Bonding,"* Solid State Technology, *May 1993, p. 45.*)

current list of concerns relative to BGA packages. Though many of the new materials used today in the electronics industry will fail only when subjected to stresses that would have caused high failure rates in many of the materials that were used 10 years ago, materials improvement alone does not guarantee that BGA packages will not fail. By definition, the intrinsic failures are not related to the package. The intrinsic failure data for a die that has a history of use will not change simply because it is being embodied in a new package such as the BGA; therefore the reliability data which will be new and need to be understood will be regarding extrinsic failures.

> Reliability concerns vary among the different BGA technologies. The primary concerns are: (1) for ceramic BGAs (CBGA), the thermal-coefficient-of-expansion (TCE) mismatch between the ceramic substrate and the card and (2) for tape BGAs (TBGA), the moisture sensitivity and the TCE mismatch between the silicon chip and the printed circuit board (PCB) material. TBGA and plastic BGAs (PBGA) packages are moisture-sensitive, while CBGA packages are not. TBGA packages are sensitive to moisture absorbed in the epoxy between the flip chip and the tape substrate or the epoxy over the tape automated bond (TAB) inner lead bond interconnections. PBGA packages are sensitive to moisture absorbed by the mold compound, die attach, and PCB material. As the packages are heated to card-assembly reflow temperatures, the expansion of the absorbed moisture as it escapes causes loss of adhesion and/or cracking. PBGA packages, particularly with large chips, are susceptible to solder-joint fatigue failures caused by the TCE mismatch between the silicon chip and the PCB materials. The silicon constrains the expansion of the PCB.[8]

Though optimum material selection and good design practices can minimize many of these extrinsic failure mechanisms, as material and design changes are implemented burn-in will be used to quantify the improvement and provide feedback to the package designer, as material and design changes are implemented.

17.2 Burn-In Socket Attributes

A *socket* is a group of contacts designed to receive an IC package and interface the package to a printed circuit board. In the case of a BGA burn-in socket, the IC specifically takes on the attributes of a BGA package, and the socket's group of contacts is arranged correspondingly to mate to a high-temperature printed circuit board known as a *burn-in board*. In addition to their ability to withstand high temperatures, burn-in sockets have many other attributes to which the burn-in engineer must give consideration prior to selecting the optimal burn-in socket for the burn-in application. These attributes include: mechanical

TABLE 17.3 Burn-In Socket Attributes

Issues	Attribute
Thermal	Venting, heat dissipation, maximum temporary exposure temperature, maximum continuous operating temperature
Mechanical	Durable, actuation life, forces transferred to package
Service & supplier	Field repairable, deliverable, trackable, cost packaging, credibility, location
Ergonomic	Maximum actuation force, obvious to use, orientation features, board assembly, ancillary tool requirements, probable
Electrical	Low and stable contact resistance, adequate bandwidth, minimal crosstalk, target impedence, minimal lead inductance
Material	Ionic impurities, hazardous outgassing, solderable, chemical resistance, plating requirements
Contact point	Package lead damage, wipe, location, normal force.
Product	Low insertion vs. zero insertion, open top or lidded, breadth of pin counts, probable, size (L × W × H), footprint, standoffs, alignment means, receptacles.

reliability, thermal requirements, ergonomics, contact to IC package lead interface, electrical requirements, material selections, product variations, service issues, cost issues, and basic product requirements.

17.2.1 Ball grid array burn-in socket issues

There are many attributes to consider when selecting a burn-in socket. The attributes of primary concern are often weighted differently for specific IC package types. When selecting a burn-in socket for a BGA package which offers increased board densities and improved electrical performance over peripheral leaded packages and opportunities for improving manufacturing yields, specific attention should be taken to ensure that these package attributes are not compromised by the burn-in socket. Therefore, the burn-in socket for a BGA package should:

1. Not compromise the electrical performance of the BGA package during the burn-in process, i.e. adequate bandwidth, minimum crosstalk and low contact resistance.
2. Not effect the BGA package yields during the original equipment manufacturer's (OEM) assembly process, such as damage to the substrate or deformation to the solder balls.

17.2.2 Ball grid array burn-in socket materials

Burn-in socket construction consists of three general components:

1. Insulator material
2. Conducting material and its protective plating
3. Assembly hardware

The insulator material is commonly a high-temperature thermal plastic, selected for its ability to withstand stress at elevated temperature and the ease at which it can be molded or formed into the necessary geometry required by the design. Thermal plastics which fall into this group include: polysulfone, polyethersulfone, polyetherimide, polyamide-imide, and polyphenylene-sulfide.

Polyethersulfone material is being used by three BGA burn-in socket manufacturers. Glass-filled grades of polyethersulfone offer higher stiffness and dimensional stability, which benefits creep resistance and lowers thermal expansion.

The conductive material used for the contact of a burn-in socket should be a good electrical conductor, have a high yield strength, and have properties which allow it to be stamped or formed into a spring design and resist stress relaxation at the elevated burn-in temperature.

> *Stress relaxation* is defined as the time-dependent decrease of stress in a solid under constant constraint at constant temperature. If excessive relaxation occurs, the contact may malfunction leading to a failure in the circuit.[4]

TABLE 17.4 Typical Plastic Properties

Plastic material						
Physical	Units	Polysulfone	Polyether-sulfone	Polyphenylene-sulfide	Polyetherimide	Polyamide-imide
Specific gravity	—	1.45	1.60	1.65	1.51	1.60
Water absorption						
24 Hours 73°F	%	0.20	.34	.02	.18	.20
MECHANICAL						
Tensile strength	psi	18,000	18,000	23,000	24,500	33,000
Elongation	%	3.0	1–3	3–4	3	4.7
Flexural strength	PSI	24,000	27,000	32,000	37,000	46,000
Flexural modulus	PSI	1,200,000	1,220,000	1,800,000	1,230,000	1,400,000
Compressive strength	PSI	24,000	—	24,700	23,500	—
Izod impact notched	ft-lbs/in	1.8	1.4	1.5	1.7	1.5
Rockwell hardness	—	M-92	—	R-123	M-125	C-74
ELECTRICAL						
Dielectric strength S/T	volts/mil	480	—	375	630	560
Dielectric constant 10^6 Hz	—	3.49	—	3.80	3.5	—
Dissipation factor 10^6 Hz	—	0.0049	—	0.0013	.0015	—
Arc resistance	secs	115	—	50	85	—
Volume resistivity	ohms-cm	10^{17}	10^{15}–10^{16}	10^{16}	3×10^{16}	10^{17}
THERMAL						
Heat deflection temp.						
264 PSI	°F	365	420	505	410	530
Thermal conductivity	Btu-in hr.ft.^{2}f	2.2	—	3.1	1.5	—
Coeff. of linear exp.	in/in/f	1.4×10^{-5}	—	1.5×10^{-5}	1.1×10^{-5}	1.7×10^{-5}
Flammability	—	94V-0	94V-0	94V-0	94V-0	94V-0

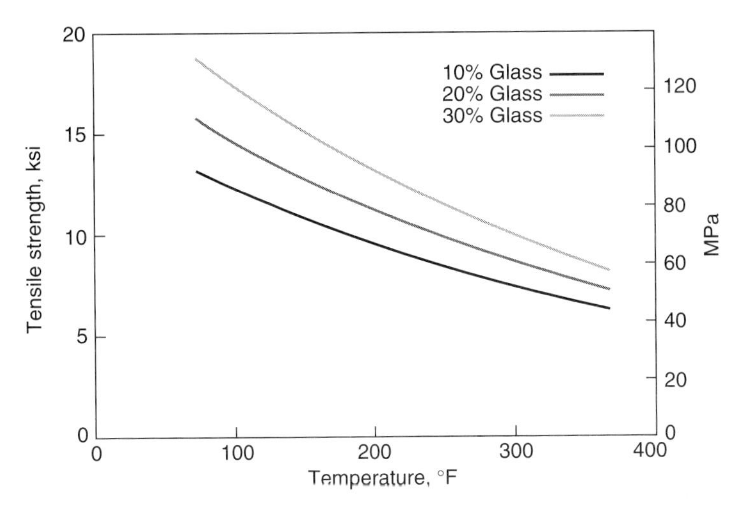

Figure 17.3 Tensile strength versus temperature for glass-filled polyethersulfone resins. (*Amoco Performance Products, Radel™. Radel is a Trademark of Amoco Performance Products Inc. Engineering Data, Technical Literature #R-F-50029.*)

Three materials which exhibit resistance to stress relaxation and which are used extensively by designers in burn-in sockets are beryllium copper, beryllium nickel, and spinodal copper alloy.

Beryllium copper is being used by three BGA burn-in socket manufacturers. Tempered grades of beryllium copper offer higher yield strengths and resistance to stress relaxation.

Precious metal platings are used on burn-in socket contacts to eliminate insulating films which increase contact resistance. Electrodeposited gold is the most widely used finish for burn-in socket contacts. It is generally plated .76 μm (.030 μin) thick and in this range is usually porous. Nickel is commonly used as an underplate to the gold surface plate in order to prevent base metal corrosion through these pores. This nickel underplate or "barrier" layer is generally plated from 1.27 to 3.81 μm (.050 to .150 μin) thick.

The assembly hardware (screws, roll pins, and springs) should be resistant to both corrosion and stress relaxation at the elevated temperature and atmosphere of burn-in.

> Certain alloys of iron and chromium are highly resistant to corrosion and oxidation at high temperatures and maintain considerable strength at these temperatures. These alloys sometimes contain nickel and small percentages of silicon, molybdenum, tungsten, copper, and other elements. This large and complex group of alloys is known as *stainless steel.* For small springs, steel is often supplied to spring manufactures in a form that requires no heat treatment. It is especially important for springs that the surface of the steel be free from all defects.[5]

TABLE 17.5 Physical Contact Properties

	Pfinodal®	Mill hardened BeCu	Brushform 290	BeNi[1]	301SS[2]
Melting point (°F)	1742–2039	1590–1800	1590–1800	2100–2300	2500–2590
Density (lb/in^3)	.323	.302	.302	.302	.29
Coefficient of thermal expansion /·°F(68°F–392°F)	9.1×10^{-6}	9.7×10^{-6}	9.7×10^{-6}	8×10^{-6}[3]	9.4×10^{-6}
Thermal conductivity (Btu/ft · hr · °F)	17	62	60	28	9.4
Thermal capacity (Btu/lb. · °F at 68°F)	0.09	0.10	0.10	—	0.12
Electrical conductivity (%IACS at 68°F)	8	20[4]	17[4]	7	2.5
Temperature coefficient of electrical conductivity (%IACS/°F, 68°F–392°F)	1×10^{-3}	—	—	—	—
Electrical resistivity (ohm circ. mil/ft)	129	52	61	148	433
Elastic modulus (psi)	18.5×10^6	18.5×10^6	19×10^6	29×10^4	28×10^6
Rigidity modulus (psi)	7.5×10^6	7.5×10^6	7.3×10^6	11.2×10^6	10.8×10^6
Fatigue strength (ksi) 10^8 cycles in reverse bending	28	40	—	87[5]	—

(1) Cabot Berylco Bulletin 306 2-PD1
(2) Materials Engineering 1983 Materials Selector
(3) 68–1022°F
(4) Given as minimums by BeCu suppliers. Samples measured by Pfizer are at or slightly below these minimums.
(5) 10^7 cycles only

SOURCE: Pfizer, Minerals, Pigments & Metals Division, Technical note #2M0883.

17.2.3 How ball grid array burn-in sockets operate

Most BGA burn-in socket designs include a lid which is hinged to the burn-in socket base along one edge, which is intended to clamp the BGA IC package down onto the contact pins of the burn-in socket as the lid is closed.

During closure of certain hinged lids it has been found that the hinged lid results in dynamic components of force being exerted in a nonnormal direction on the IC package. The normal direction is defined by a vector perpendicular to the plane of the BGA package. This force vector is often referred to as the contact's *normal force.*

Any applied force in other than the normal direction may result in movement of the BGA package relative to the contact pins of the burn-in socket. This could damage the contact pins of the socket, the balls of the BGA package, or cause cracking or breakage of the ceramic or plastic BGA package body material or lids or lid seal rings.

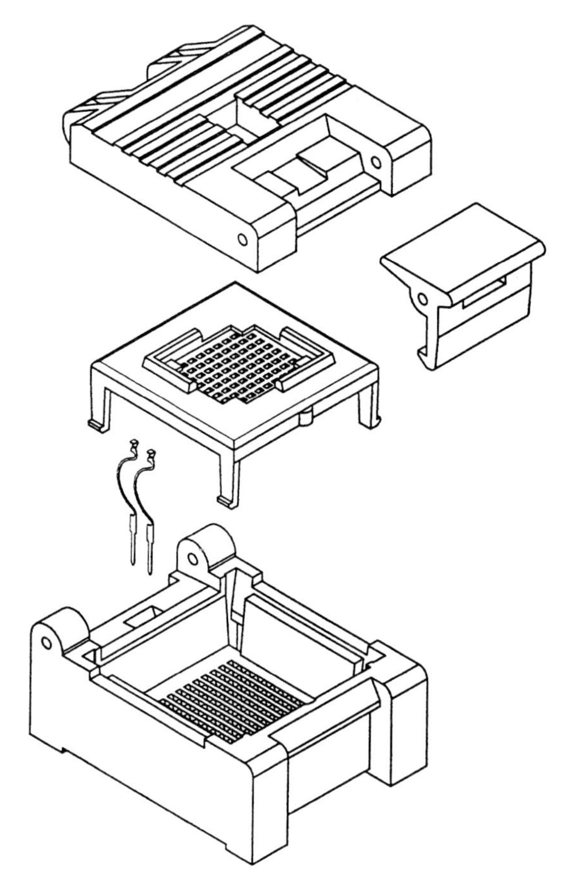

Figure 17.4a BGA socket components. (*Rios, J. P., 3M Company, Austin, Texas, 1993.*)

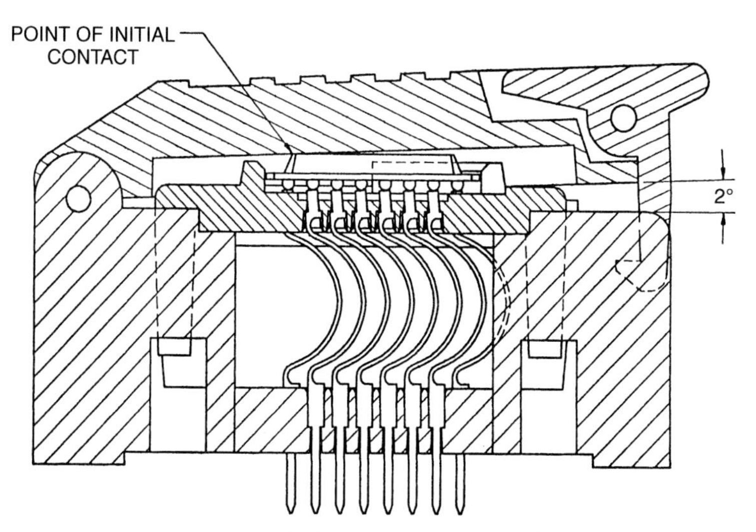

Figure 17.4b Section view of BGA socket with BGA package inside. (*Rios, J. P., 3M Company, Austin, Texas, 1993.*)

BGA Test & Burn-in Socket
CONTACT CONDITION & CONTACT FORCE

SAMPLE

SOCKET: ***3M*** *BGA Test & Burn-in Socket 225 pin*
2-0225-08172-000-019-002
PACKAGE: ***JEDEC*** *BGA 225 pin*
Pkg. Height 2.13mm nom.

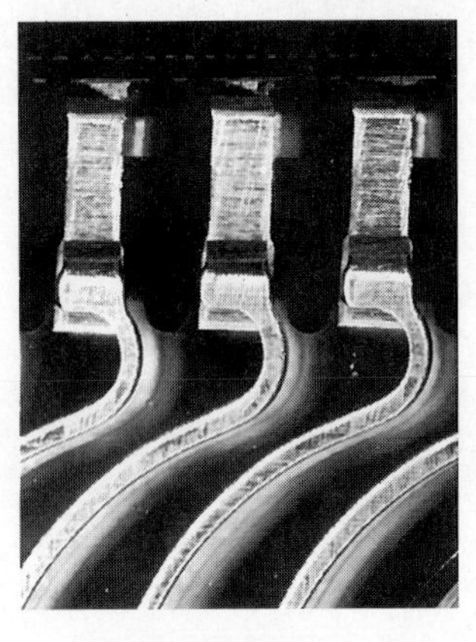

UNLOADED CONDITION

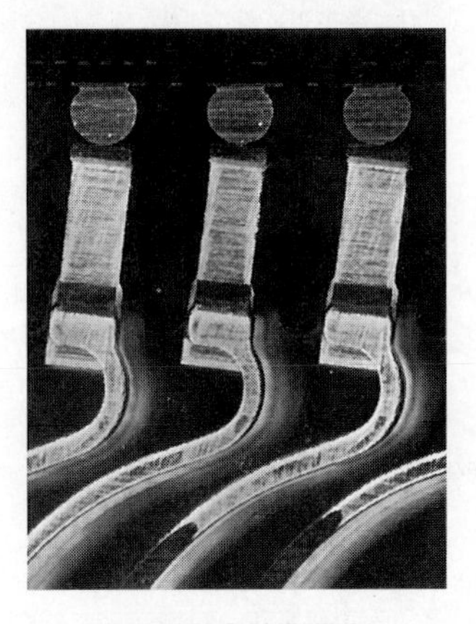

LOADED CONDITION

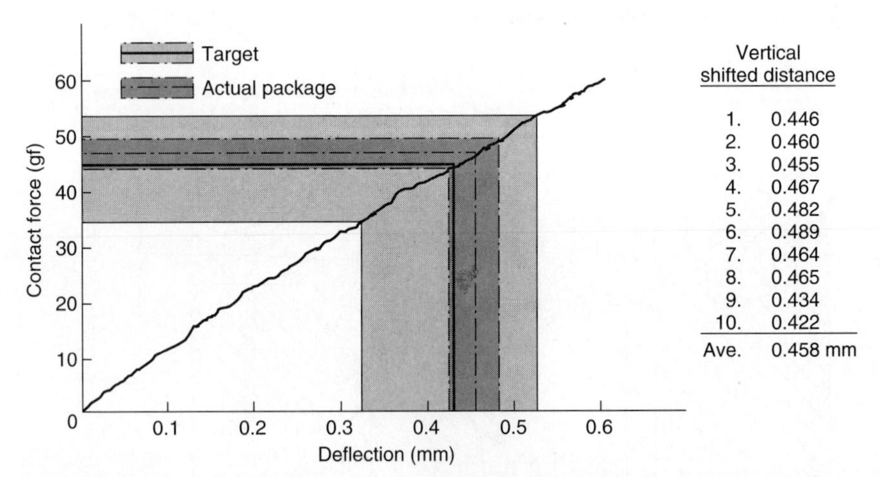

Figure 17.5 BGA socket contact and normal force. (*Nakamura, S., Sumitomo 3M Ltd., Japan, 1994.*)

Current BGA socket designs compensate for nonnormal forces caused by the closure of the burn-in socket's lid by providing a burn-in socket which includes a platform for supporting the BGA package. The platform is generally square in shape and includes two raised coarse alignment ridges which define a nest for the BGA package. Interior to these raised ridges, the nest top surface may contain additional ridges or walls to accomplish finer alignment of the BGA package balls to the burn-in socket contacts. An alternative method of fine alignment is accomplished by means of a hole which is punched through the head or top surface of the socket contact. This hole then provides a cup in which the BGA package ball may rest.

Extending from the flat portion of the nest are four legs which terminate in outwardly projecting ends which engage recesses in the base walls. This engagement between the nest leg ends and the base recess acts to prevent the nest from moving in a direction away from the lower surface of the base. The nest is supported above the lower surface of the base by the contact pins. These contacts are formed with a bowed central portion to provide the contact pins with a spring action. The contact pins extend through slots or holes in the nest to contact the balls on the BGA package. The ends of the contact pins opposite the balls on the BGA package are formed as blades which extend through the lower surface of the base for electrical connection to holes in the burn-in board to which the burn-in socket is solder-attached.

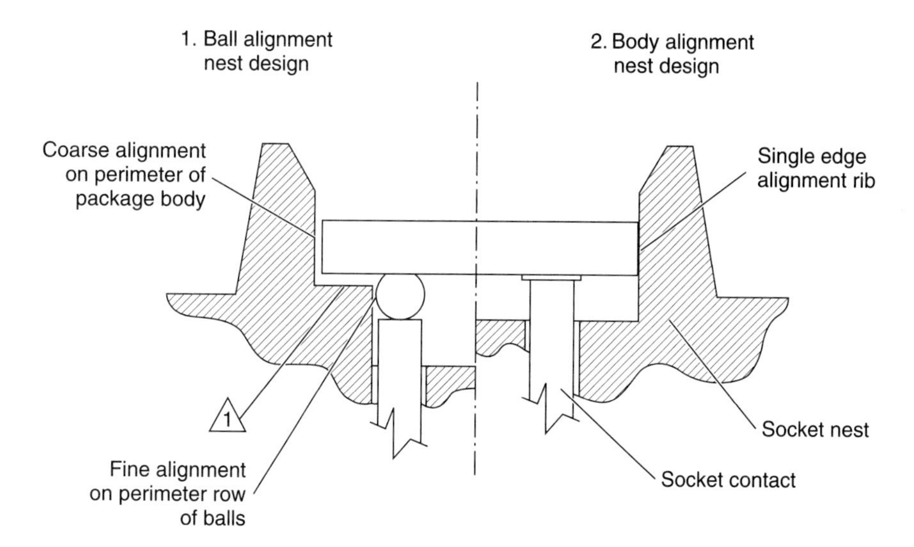

Figure 17.6 Cross-sectional view of socket-to-package-alignment method. (*Rios, J. P., 3M Company, Austin, Texas, 1993.*)

Figure 17.7*a* 3M BGA socket contact head. (*Schueller, R. D., "BGA Burn-In Socket Contact Comparison Study," Technical Paper, Aug. 4, 1993, pp. 6 & 9.*)

Figure 17.7*b* Alternative BGA socket contact head.

Since the lid is attached to the base at a hinge disposed at one end of the lid, contact between the underside of the lid and the top surface of the BGA package is not simultaneous across the surface of the BGA package but, rather, concentrated along the edge of the BGA package closest to the hinge pin. To eliminate the problem caused by concentration of forces between the socket's lid and the BGA package, the socket nest is free to "float" relative to the base (that is, any portion of the nest may be depressed relative to the base independently of any other portion of the nest). Thus the nest is free to tilt in any direction in response to forces which are unevenly applied to the BGA package, as in the case where an initial force is applied to the BGA package which is nearest the BGA socket hinge during the latching operation. The legs of the nest adjacent the hinge will move downward toward the lower surface of the base, while the legs farther from the hinge will not have moved relative to the base. Thus the nest has tilted in response to the uneven forces applied by the cover to the BGA package and, by doing so, will prevent any damage to the BGA package. Further movement of the lid toward the base will cause the nest legs distant from the hinge to likewise move toward the lower surface of the base to a position where the lower surface of the base, the nest, and the lid are all parallel. At this final position, forces applied to the BGA package by the lid are evenly distributed across the surface of the BGA package and will result in good contact between the socket contacts and the solder balls on the BGA package. The operation of this free-floating nest design is analogous to the operation of multipieces of pivoting plastic which have been common design features on the underside of many quad flat pack (QFP) socket lids.[6]

17.2.4 Ball grid array burn-in socket contact technologies

The contact is the conducting member of the burn-in socket. To not affect the electrical signal it should have a low and stable electrical resistance. The contact is a type of spring, with a "head" end, which contacts the solder ball of the BGA IC package and a "tail" end which terminates to the burn-in printed circuit board (PCB).

> The contact base metal provides a smooth surface for the contact coating which is metallurgically bonded to the base metal by plating. At the head end, the base alloy spring supplies a predetermined force to the ball of the BGA IC package. This force must be high enough to break through thin surface oxides on both the plated contact and solder ball surfaces, yet low enough to minimize damage and wear. At the tail end the contact must protrude through the burn-in PCB and be capable of attachment to the PCB by various soldering methods.[7]

BGA Test & Burn-in Socket
SOLDER BALL DEFORMATION

SAMPLE

SOCKET: ***3M*** *BGA Test & Burn-in Socket 225 pin*
2-0225-08172-000-019-002

PACKAGE: ***JEDEC*** *BGA 225 pin*
Pkg. Height 2.13mm nom.
Solder Ball Composition is 62/36/2 SnPbAg.

TEST CONDITION

125˚C

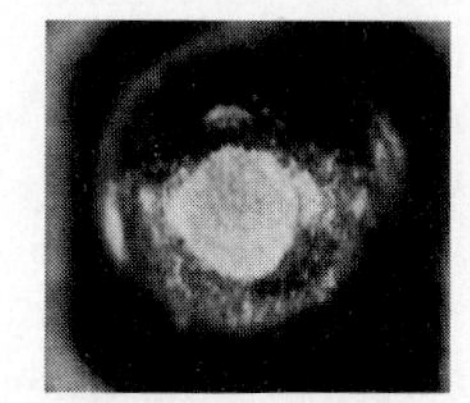

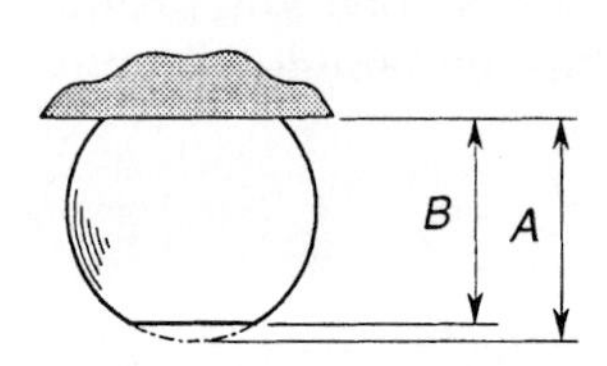

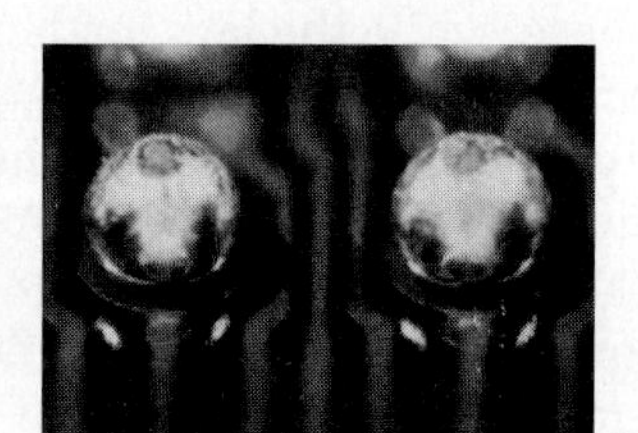

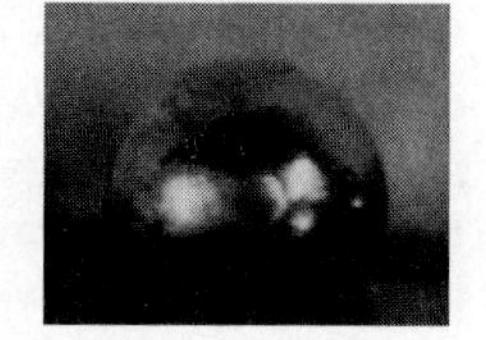

INITIAL *AFTER 72 HOURS*

SOLDER BALL HEIGHT

	A	B (After Test)			
	Initial	12 hrs.	24 hrs.	48 hrs.	72 hrs.
1.	0.594	0.582	0.581	0.579	0.577
2.	0.590	0.582	0.584	0.577	0.577
3.	0.570	0.563	0.561	0.562	0.563
4.	0.588	0.570	0.570	0.569	0.569

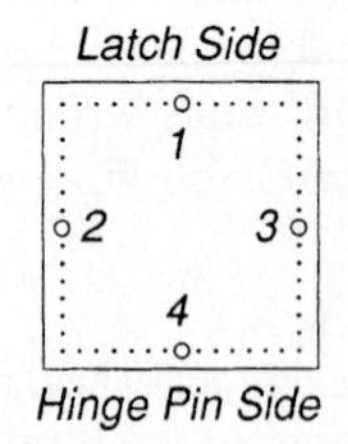

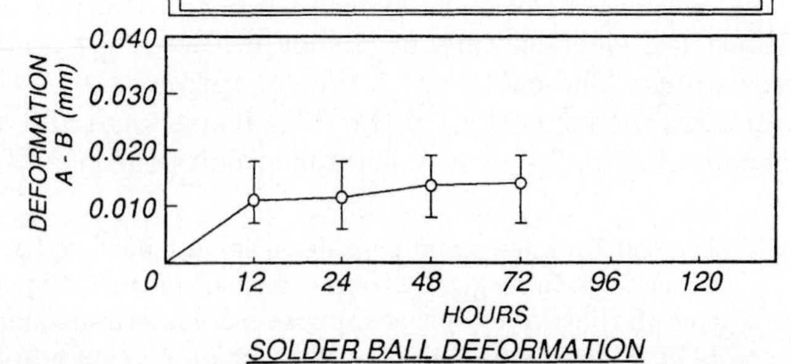

Figure 17.8*a* Solderball deformation photos and graph. (*Nakamura, S., Sumitomo 3M Ltd., Japan, 1994.*)

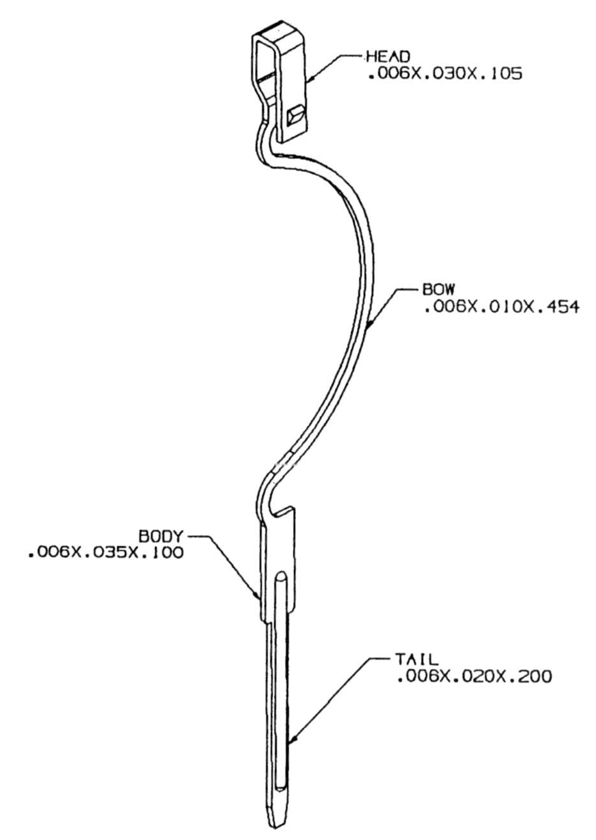

Figure 17.8*b* Isometric contact drawing. (*Rios, J. P., 3M Company, Austin, Texas, 1993.*)

The quality of the contact connection is defined by the nature of the contributing electrical and mechanical elements of the design.

> This design of electrical contacts and the selection of contact materials are based on a body of interrelated physical, metallurgical, and chemical principles. The surfaces of solids are irregular on a microscopic scale. Even nominally plane, smooth surfaces have a large-scale waviness on which is superimposed a roughness with peak-to-valley distances of several micrometers. When two metallic bodies are placed in contact at a light load, they touch at only a few small spots, or asperities (a spots). As the load is increased, more and more asperities come into contact and the surfaces move together. The true area of contact depends, therefore, on normal force surface roughness and the hardness of the metal. The real area of contact is only a fraction of the apparent area in most cases, except at very high loads. If metallic surfaces are covered by a nonconducting layer, such as an oxide film, the area of contact will be zero provided that

the film is unbroken. If the nonconductive layer on the surface is discontinuous or is punctured, the mechanical load is borne by both the film and the metal. Current then flows through the asperities.[9,10]

The lines of electric flow converge at these spots. Constriction resistance, the increase of resistance beyond that of a continuous solid, not having an interface,[10] originates at this convergence. Since the length of metal associated with the connector contact is ordinarily in the path between the contact head end and the PCB to which the contact tail end terminates, its resistance (bulk resistance) must be added to the contact resistance when considering the connector as a circuit element.[9] This resistance can be expressed as

$$R = R_c + R_f + R_b$$

where R = connector resistance, R_c = constriction resistance, R_f = film resistance, and R_b = bulk resistance.

The contact resistance

$$R' = R_c + R_f$$

typically ranges from tens of milliohms to tenths of ohms for BGA burn-in sockets. R' should remain stable, i.e., not change during the burn-in

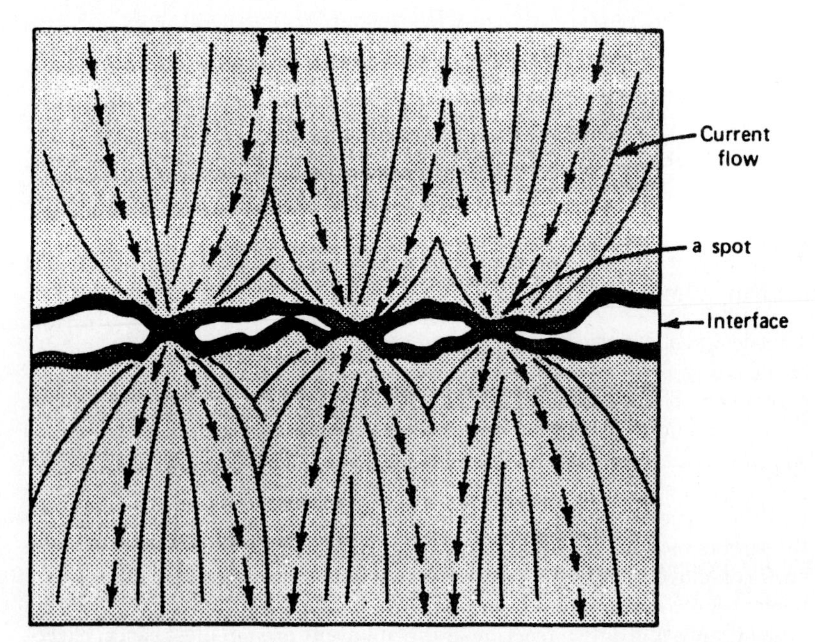

Figure 17.9 Microscopic view of contact interface (schematic); constriction resistance of the constriction of current flow through the touching metallic junctions of the mating surface. (*From* Encyclopedia of Chemical Technology; *see Ref. 9.*)

cycle. Changes greater than 50.0 milliohms[11] may affect the electrical signal which is input to the IC during dynamic burn-in. An affected electrical signal in which a data bit "1" has erroneously changed to a data bit "0" or which a data bit "0" has erroneously changed to a data bit "1," is called a *soft error*. If undetected, this could lead to false negatives and lower burn-in yields.

The contact material, contact area, and its heat-dissipating ability, as well as the heat-dissipating ability of the structure to which it is attached, limit the amount of current which a contact can transport. Excessive current will heat and soften the metal contact, and this softening will result in an increase in the surface area of the contact and a corresponding reduction in contact resistance. At higher currents, the mating junction will melt. Both softening and melting occur at characteristic voltages. A typical value for tin (Sn) softening is 100°C, and 232°C for melting, or 0.07 and 0.13 V, respectively.[10] If a significant voltage can be passed by a circuit across a film-covered contact, the film, depending on its thickness and composition, may break down electrically. Puncturing is in the order of 0.1 V/nm of film.[9] Burn-in bias voltages range from 100 to 150% of the IC package's rated voltage. In addition to the use of voltage, film resistance can also be reduced by the use of noble metals which do not corrode in the presence of humid, polluted air or by designing the contact such that a mechanical scraping or wiping action takes place at the time of engagement. In this latter case, care must be taken that this wiping action does not create excessive plating wear, which may reduce the life of the burn-in sockets contact. How quickly wear occurs is a function of a number of interrelated variables such as roughness and hardness of the materials involved, contact geometry, surface cleanliness, bearing load, and contact normal force.[12]

Fatigue is a phenomenon that leads to fracture when a material is under repeated or fluctuating stress. A contact's *fatigue strength* is defined as the highest attainable level of stress before fracture, after a given number of cycles.[7] For a burn-in socket the fatigue life of contact should greater than 10,000 cycles.

When a burn-in socket ages, whether operated normally or under accelerated conditions, there is a gradual reduction in the area of metallic contact. When this happens, the constriction resistance rises. If, as usual, current is passing through the connector, the temperature in the contact region also rises. It is generally assumed that a self-accelerating cycle is then initiated: the higher temperature encourages more rapid degeneration at the interface, which in turn leads to still higher temperatures. Eventually the burn-in socket fails, either because all metallic contact is lost so that no current can flow or because the temperature has become dangerously high.[13] This increase in the temperature of the contact can accelerate additional failures of the mechanical contact. Contact materials subjected to mechanical stresses or elevated temperatures in normal operation can relax or suffer permanent deformation and loss in spring force, even at loads below the material's yield strength. *Stress relaxation* is the time-dependent decay of stress (or load) in a material under a constant strain constraint.[7] In burn-in sockets, stress relaxation refers to the decrease in stress in the fixed geometry spring contact. The rate of stress relaxation is a function of the initial applied stress level and the temperature of the operating environment.

Beryllium copper alloy C17200, is commonly used as a contact material in BGA burn-in sockets because of its high yield strength, good electrical conductivity, a high modulus of elasticity and resistance to stress relaxation. Other materials are used for burn-in socket contacts, these include but are not limited to: beryllium nickel and spinodal alloys.

17.2.5 Ball grid array burn-in socket electrical performance

Historically, the performance of a burn-in socket has focused around the mechanical and thermal properties, such as size, actuation force, and the socket's life at the elevated burn-in temperature. Today, dynamic burn-in is common. This, coupled with the continually increasing operating speeds of ICs, has placed an added emphasis on the electrical performance of a burn-in socket. Burn-in socket designers have been forced to develop high-speed burn-in sockets for BGA packages. Whether the BGA package is ceramic, plastic, or TAB-based, burn-in sockets with low crosstalk and broad bandwidth are available. Signal integrity is a must with high-speed digital circuits. During dynamic burn-in the vector pattern or data stream should not be changed by the burn-in socket. The burn-in socket should allow the burn-in system's output driver to generate the fastest edge rate of which it is capable without causing signal degradation or switching limitations.

Time domain reflectometers and network analyzers can be used along with high-frequency fine-pitch probes and appropriate fixture to analyze the electrical performance of a burn-in socket. Inductance, capacitance, and resistance (*LCR*) data can be gathered at representative locations within the burn-in sockets array of contacts.

An equivalent circuit model of the burn-in socket can be created. This model can be combined with the model for the burn-in board and the model for the BGA package. Software is available which can be used to simulate the operating performance of the BGA package during burn-in.

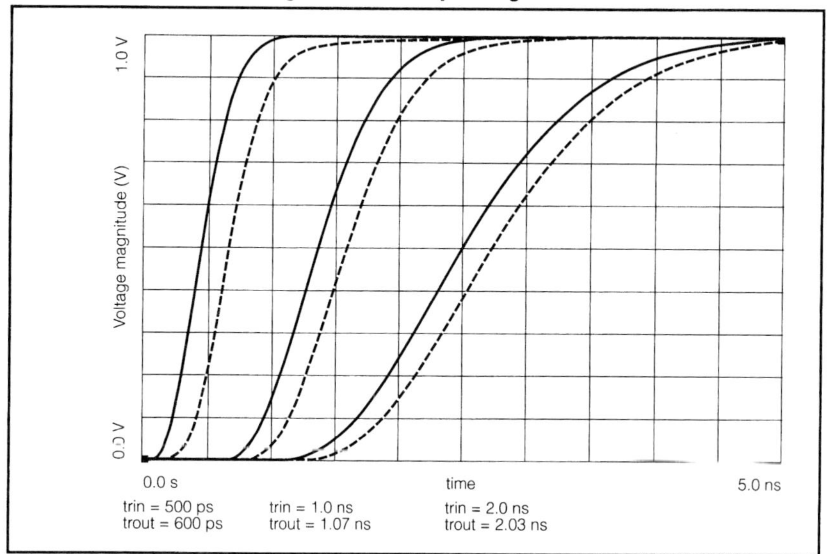

Figure 17.10*a* Risetime degradation. (*3M Tech Brief, "1.5mm BGA Socket Electrical Characteristics,"* #80-6106-0834-3, 1993, pp. 1–4.)

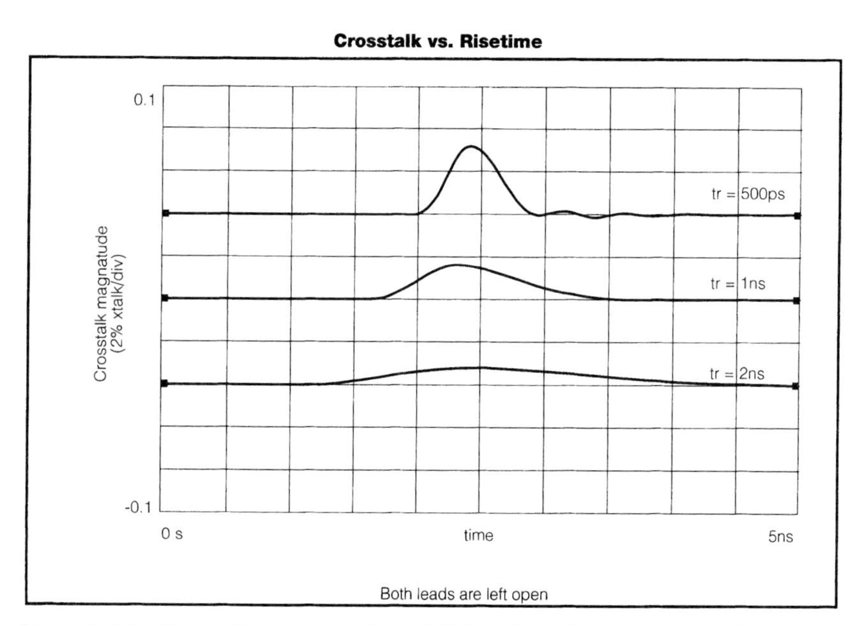

Figure 17.10*b* Crosstalk versus risetime. (*3M Tech Brief, "1.5mm BGA Socket Electrical Characteristics,"* #80-6106-0834-3, 1993, pp. 1–4.)

TABLE 17.6 LCRZ Data

Element values:

Inductance and resistance:

pin	Ls(nH)	M(nH)	Rs(Ω)@50 MHz
center	10.20	1.80	0.30
edge	11.20	2.60	0.30
corner	13.50	3.50	0.30
nonadj.	10.00	0.00	0.30

Capacitance:

pin	Cma(pF)	Cmb(pF)	Cs1(pF)	Cs2(pF)
center	0.27	0.50	0.050	0.100
edge	0.20	0.42	0.062	0.130
corner	0.13	0.32	0.060	0.140
nonadj.	0.27	0.60	0.000	0.000

Characteristic impedance:

pin	$Z_0(\Omega)$
center	115
edge	134
corner	173
nonadj.	107

The s-parameters of the socket lead pins
were measured at four locations:

Center	nonadj.	corner	edge
ooo	ooo	oo	oo
oxo	oxo	ox	ox
oxo	ooo	ox	ox
ooo	oxo		oo
	ooo		

x = measured
o = grounded

SOURCE: 3M Tech Brief, "1.5mm-BGA Socket Electrical Characteristics," #80-6106-0834-3, 1993, pp. 1–4.

17.3 Conclusions and Future Technical Challenges for Ball Grid Array Burn-In Sockets

In response to the proliferation of the new and exciting form of package known as the BGA, a number of burn-in sockets have emerged. The majority of these sockets are lidded and can accommodate CBGAs, PBGAs, TBGAs, and even some "land grid arrays" (CBGAs without solder balls). These sockets are lidded, modularly constructed, and available in a range of sizes to accommodate BGA packages with pitches of

Equivalent circuit model: The s-parameter data and the physical characteristics of the socket were used to arrive at a topology for the equivalent-circuit model. Because of the large physical dimensions, capacitors Cs & Cm were distributed in two places, shown below:

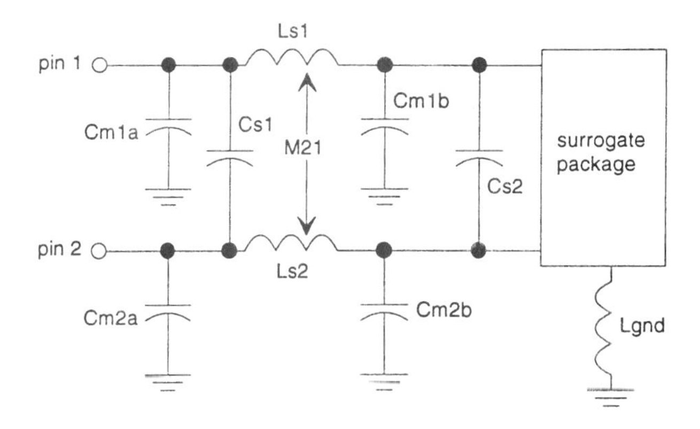

Notes:

1. Inductors Ls1 & Ls2 contain resistances Rs1 & Rs2
2. Fixture parasitics are not shown in this diagram

Element definitions:

Ls: self-inductance of one lead
Cs: lead-to-lead capacitance, divided into two capacitors Cs1 & Cs2
Rs: series resistance of one lead (not shown)
Cm: loading capacitance, divided into two capacitors Cm#a & Cm#b
Lgnd: ground-return inductance (aka common-lead inductance)
M: mutual-inductance between two leads

Figure 17.11 Equivalent Circuit. (*3M Tech Brief, "1.5mm BGA Socket Electrical Characteristics,"* #80-6106-0834-3, 1993, pp. 1–4.)

1.27 mm, 1.5 mm, and .060 in, and over a wide range of BGA package body sizes. The modular design of the BGA socket allows the flexibility needed for this emerging package family. As BGA standards proliferate, this modular design concept will help reduce the cost and time required for burn-in socket modifications. The classic clamshell construction has had a proven field reliability since late 1991, and the contact design is such that the electrical signal integrity during dynamic burn-in is maintained. "Flip-up" lids make these sockets adaptable for both automated and hand operation. The coarse and fine package alignment features of these sockets eliminate the need for special care and handling of the BGA package during socket loading and unloading. The C-shape contact spring with its flat top minimizes solder-ball damage which can create solder voids during the card assembly attachment process.

As the popularity of the BGA package increases, there will be a desire to automate the burn-in process as much as possible to eliminate costs. This will create a need for open top or lidless sockets. Burn-in socket designers will be challenged to design sockets which do not place an excessive bending moment on the BGA package which may lead to substrate materials cracking or delaminating, which can lead to package failure. As lead counts increase the forces required to actuate these sockets and to close the lids on the current lidded socket designs will also increase. New, low-actuation-force lid designs will be needed to maintain these forces within a reasonable limit for actuation by a human hand. These low-actuation-force lids will also need to take care as to where on the top surface of the package that the force opposing the contacts is applied. If care is not taken, product logos and markings may be smeared or cavity-up lids and lid seal rings may be damaged. Each of these could lead to product rejection or failure. There is currently a trend in the burn-in industry toward higher burn-in temperatures. This allows for the burn-in time to be shortened and therefore the throughput to be increased and manufacturing costs to be decreased. These higher burn-in temperatures also allow for additional infant mortalities to occur. These higher test temperatures will place greater demands upon the burn-in socket insulator and contact materials. In addition to accelerating stress relaxation which may lead to contact failure, the rate at which the solder balls soften and deform will be increased. This increased solder-ball deformation will require additional attention to be given to the solder paste screening process which could affect manufacturing yields and package costs. To reap the benefits of higher burn-in temperatures and to minimize the solder-ball deformation, burn-in can be performed prior to the solder ball being attached to the substrate material.

17.4 Acknowledgments

The author would like to thank Denny Aeschliman, and Dr. Carol Jensen, for their support and review of this paper; Dr. Randy Schueller and Dr. Howard Evans for their stimulating discussions relative to BGA package reliability; Sinichiro Nakamura san for his normal force measurements and solder-ball deformation work; Juan Rios for providing CAD drawings and knowledge.

17.5 References

1. Micron Semiconductor Inc., *Quality/Reliability Handbook,* 1993, pp. 8–10.
2. Military Standard 883D, *Test Methods and Procedures for Microelectronics,* Nov. 1991, p. 1.

3. Nicolaou, C. A., "Plastic Package Failure Modes," *Surface Mount Technology,* Nov. 1993, pp. 45–50.
4. Filer, E. W., McClelland, H. T., "Stress Relaxation of Copper-Beryllium and Nickel-Beryllium Alloys," Cabot Berylco, 1983, p. 1.
5. Baumeister, Avallone, Baumeister, Bean, H. S., "General Properties of Materials," *Marks Standard Handbook for Mechanical Engineers,* eighth edition, pp. 6–35 and 6–36.
6. Rios, J. P., Integrated Circuit Test Socket, United States Patent 5,247,250, Sept. 21, 1993, pp. 1–6.
7. IICIT, "Contact Base Metals," *Connectors and Interconnections Handbook,* vol. 1, 1990, pp. 200–211.
8. Cole, M. S., Caulfield, T., "BGA's Are Extending Their Connections," *Electronic Engineering Times,* Feb. 28, 1994, pp. 48 and 63.
9. Kirk-Othmer, "Electrical Connectors," *Encyclopedia of Chemical Technology,* vol. 8., third edition, 1979, pp. 641–659.
10. Holm, R., *Electric Contacts,* 4th edition, Springer-Verlag, New York, 1967, pp. 8–26, 87–92, 124–125, 436–438.
11. Peel, M., "Connector Specifications—Part 1," *Connection Technology,* Sept., 1985, pp. 55–58.
12. Peel, M., "Connector Specifications—Part 2," *Connection Technology,* Nov., 1985, pp. 59–61.
13. Williamson, J. B. P. "Deterioration Processes in Electrical Connectors," *Proc. 4th International Conference on Electrical Contact Phenomena,* Swansea, 1968, pp. 30–40.

Chapter

18

BGA Infrastructure

E. Jan Vardaman

18.1 Introduction

This chapter describes the industry infrastructure for ball grid array (BGA) packages. Industry infrastructure is defined as the relationship and interaction between users and suppliers of the package, material providers, and assembly, test, and rework equipment vendors that allow the production and use of the package.

18.2 Ball Grid Array Class of Packages

Based on package construction, BGAs can be divided first into two distinct categories—organic (including T-BGAs and plastic BGA packages) and inorganic (ceramic). Organic packages include plastic BGAs and tape BGAs (T-BGAs), while inorganic BGA packages can be divided into hermetic and nonhermetic packages (see Fig. 18.1).

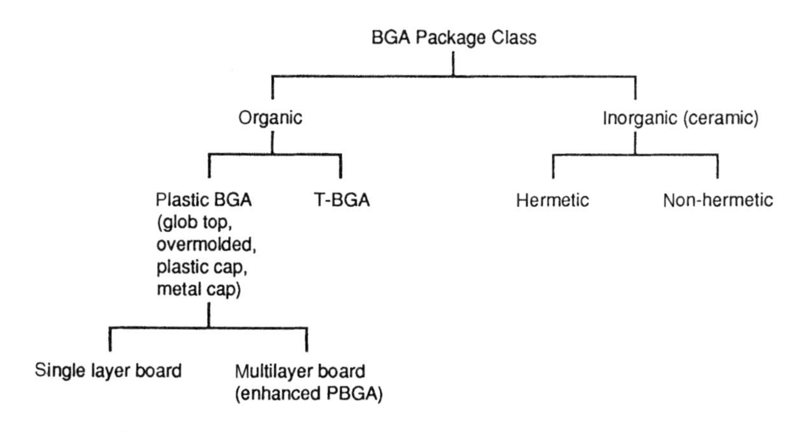

Source: TechSearch International, Inc.

Figure 18.1 BGA package class.

18.2.1 Organic BGAs

The category of organic BGA can further be subdivided into four varieties: overmolded, glob top, plastic or metal cap, and tape-based. The first three categories are often referred to as plastic BGAs. Many of the plastic BGAs used today are low performance packages that dissipate less than 2 watts of power. Several companies are developing multilayer, enhanced multilayer packages for higher electrical and thermal performance applications. The tape-based BGAs have been developed for higher performance applications. Various BGA configurations are shown in Fig. 18.2.

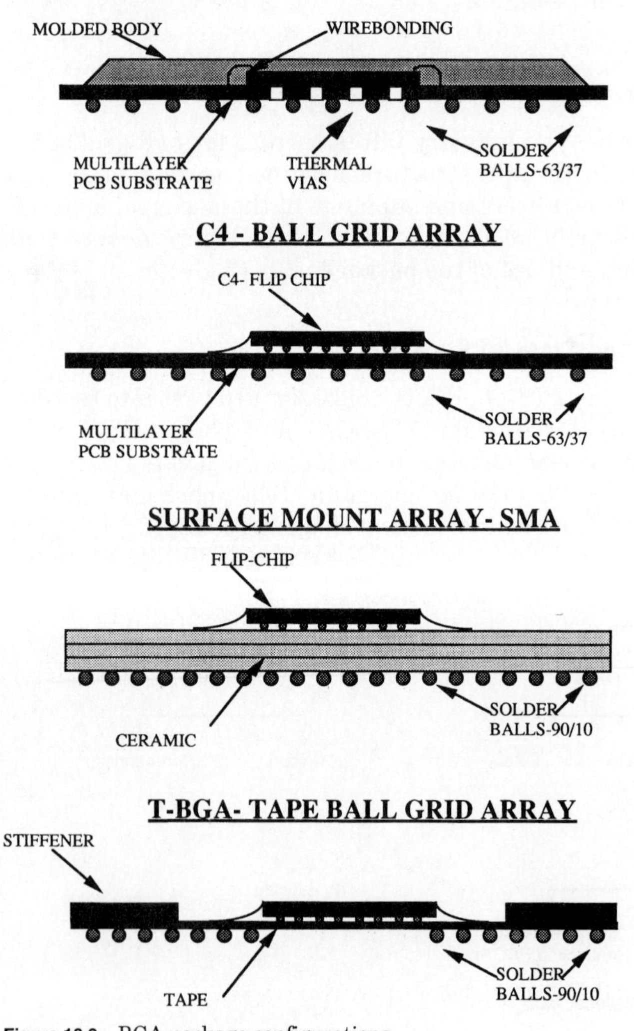

Figure 18.2 BGA package configurations.

1. Overmolded. This version includes the first plastic BGA developed by Motorola and produced by Citizen Watch called OMPAC. It also includes overmolded versions produced by Amkor/Anam, ASAT, Hestia, and Shinko Electric. Additional companies are expected to enter this production category, including Hyundai, Siliconware, and Swire.

2. Glob top. This version includes the package produced by Valtronic containing multiple die. Glob top versions may also be supplied by ASAT, Amkor/Anam, Citizen Watch, and Motorola in the future.

3. Plastic or metal cap. Bosh Blaupunkt has developed a version of a plastic BGA package that uses a plastic cap rather than a glob top or overmolding to cover the chips. In Bosh Blaupunkt's version, the chips are wire bonded, then covered with a silicone gel before the plastic cap is attached. Other companies, including Olin Interconnect Technologies, have developed metal cap BGAs. The metal cap is used as a mechanical protection as well as a thermal enhancement (see Fig. 18.3).

4. Tape-based. The first group to introduce an organic BGA version using TAB tape was IBM's Endicott, New York organization. Several companies in Japan are also examining the use of TAB tape to interconnect the chip to the laminate substrate for plastic BGAs.

18.2.2 A confusing list of names for BGAs for various configurations

There are many different package configurations. Many names have been used for these differing configurations. During the Joint Electron Device Engineering Council (JEDEC) standards committee meetings in 1993, LSI Logic coined the name *ball grid array,* and the electronics industry began referring to both ceramic and plastic packages as BGAs. Previously, several names had been used for plastic BGA packages including Overmolded Plastic Pad Array Carriers (OMPAC) by Motorola and Solder Grid Array (SGA) by Hestia. A multichip glob-top version is called a MultiChip Carrier Module (MCCM) by Valtronics. A

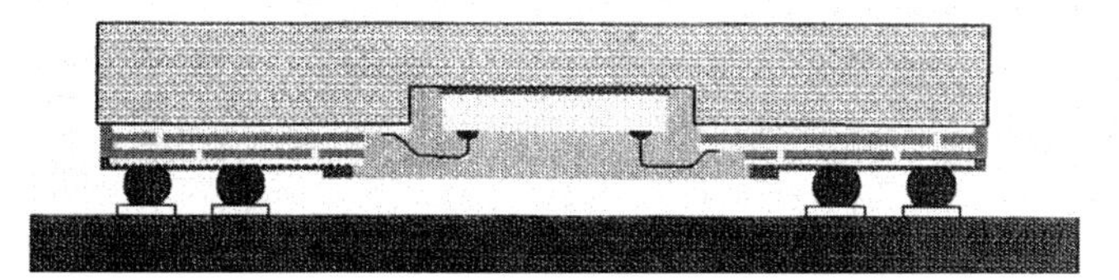

Figure 18.3 Olin's metal ball grid array.

BGA version using TAB tape is called a Tape Ball Grid Array (T-BGA) package by IBM Endicott.

Ceramic BGA packages have been used for many years, and have occasionally been included in the category of Land Grid Array (LGA) packages. The term LGA is also used to refer to packages that are surface-mounted using pads, thus causing some confusion in terminology. Ceramic BGA packages have also been called Pad Array Carriers (PAC) by Motorola. Motorola has developed a leadless solder bumped ceramic package called RISC PAC. It is referred to as an LGA. MicroLithics Corporation in Golden, Colorado has developed a pad array carrier called a PAC that utilizes a ceramic substrate. Both NEC and Hitachi use ceramic chip carriers that are solder-ball-attached to a ceramic substrate with thin film deposited on the top to form a multichip module. NEC calls its package a Flip TAB Carrier (FTC) because TAB is used inside. Hitachi calls its package a Micro Chip Carrier (MCC). IBM calls its ceramic BGAs a Surface Mount Array (SMA) package. IBM has two versions of its SMA package—one with solder balls called an SBC and one with solder columns called an SCC.

18.3 Contract Assembly Package Suppliers

There are several contract assembly package suppliers that provide both organic and inorganic packages. Typically, these companies do not have internal silicon fabrication lines, but package bare die for other companies. A few companies, such as IBM, fabricate semiconductor devices and offer contract package services. Five companies have made an agreement with Motorola to offer packages that might be covered under a broad number of Motorola patents.

18.4 Existing Industry Infrastructure

The plastic BGA industry infrastructure is in the development process. Fortunately, there is some similarity between the industry infrastructure for plastic pin grid arrays (PPGAs) in the materials used, some assembly operations, and the players that may expedite the development of an industry infrastructure for BGAs. A similar situation exists for ceramic packages, with the existing ceramic pin grid array (CPGA) industry. The major difference between the existing pin-based package infrastructure and ball grid array industry infrastructure is the areas of testing, inspection, and rework.

18.4.1 Ceramic BGAs

The infrastructure for ceramic BGAs is similar to that of ceramic pin grid arrays. The primary difference in offering CPGAs and ceramic

Company	Production status	Package type
Amkor Anam*	Volume production	Organic
ASAT*	Future volume production	Organic
Bosch Blaupunkt	Production	Organic
Citizen Watch*	Volume production	Organic
Coors Microelectronics	Production	Ceramic
Hestia	Prototype production	Organic
IBM	Volume production	Organic, Ceramic
Kyocera	Volume production	Ceramic
NTK	Production	Ceramic
Shinko*	Production	Organic, Ceramic
Swire*	Future volume production	Organic
Valtronic	Production	Organic

* = licensed Motorola patents.
SOURCE: TechSearch International, Inc.

BGAs is that balls instead of pins are used in the ceramic BGA. Design, routing, and manufacture of the package up to the point of inserting pins are balls for the ceramic BGA substrate is the same as for the ceramic PGA. (see Fig. 18.4).

The materials, equipment, and facilities for the production of ceramic pin grid arrays are well established. Relying on the same infrastructure for the production of ceramic BGAs allows manufacturers to leverage off existing plant and equipment. Many of the major producers of CPGAs also produce ceramic BGAs, such as Kyocera and IBM. Companies including Alcoa, Coors Microelectronics, and NTK Technical Ceramics have ceramic PGA production lines that can be used for the production of BGAs.

18.4.2 Plastic BGAs

While the plastic BGA is a new package, it can be argued that some of the equipment and materials used to fabricate the package are not new because many are also used in the fabrication of plastic PGAs. There are several differences between the plastic pin grid arrays and the plastic ball grid arrays. Plastic BGAs are typically manufactured in strip form similar to quad flat packages (QFPs). The major difference is the use of balls instead pins or leads.

The first version of the plastic BGA package (the overmolded variety) was jointly developed by Motorola with Citizen Watch. Figure 18.5 shows the production flow chart for Citizen Watch's plastic BGA. This package type uses conventional die bonders, wire bonders, and transfer molding equipment.

Other versions of the plastic BGA use TAB bonding equipment, encapsulation equipment, or lid sealing equipment for plastic caps.[1]

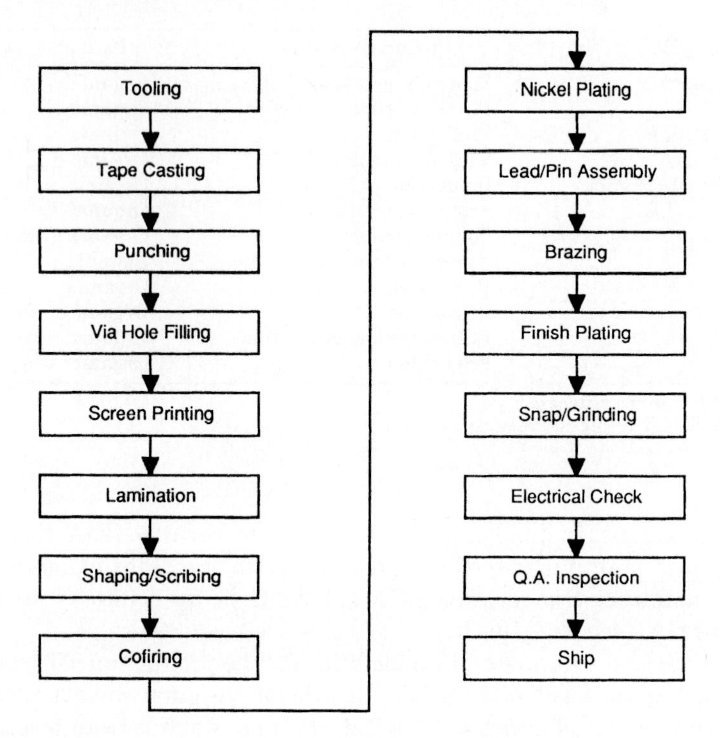

Source: Kyocera

Figure 18.4 Ceramic PGA manufacturing process.

18.5 Production Equipment

While much of the equipment for the production of plastic and ceramic BGAs is borrowed from the existing packaging industry, some equipment is new to the BGA industry, such as the robotic cells and equipment to place the balls on the package. Equipment for BGA package fabrication includes die bonders (if wire bonding or TAB is used); wire, flip chip, or TAB bonders (depending on the method used to interconnect the chip to the substrate of the BGA package); transfer molding equipment; x-ray inspection equipment; electrical test equipment (open/short test); solder ball attachment equipment, and cleaning systems. Chip-to-substrate interconnect equipment, transfer molding, and encapsulation equipment are available from numerous companies and not unique to the BGA industry. X-ray equipment and inspection equipment are rarely used in production, but are often used for process development. Solder inspection equipment is used to determine solder paste volume, especially for ceramic BGAs.

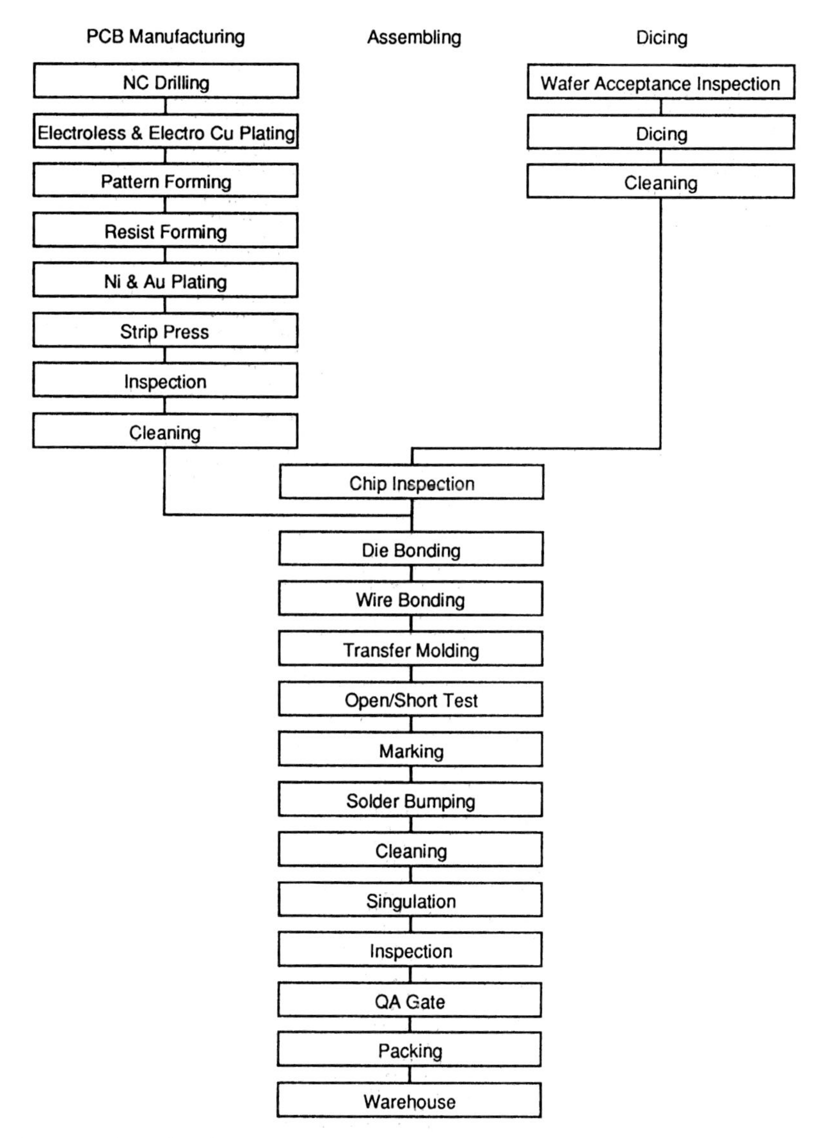

Source: Citizen Watch

Figure 18.5 Citizen BGA flowchart.

Several companies (Bosch Blaupunkt, Kyushu Matsushita, and Valtronic) use screen printing to form the solder balls on the packages, but, many companies use a ball transfer process. The ball transfer process required the development of special equipment. A robotic solder ball placement cell, featuring a Seiko Instruments robot, has been developed by Motorola.[2] Several equipment vendors are introducing new equipment or processes for ball placement. Raychem Corporation has introduced a new process for ball placement that can also be used for replacement of solder balls. In the Raychem process, a water-soluble paper holds solder balls or columns in the desired pattern. After flux is placed on the BGA substrate, the paper holder containing solder balls is aligned and held with a U-shaped fixture. The balls or columns reflow in the desired arrangment after heating. A small amount of water dissolves the laminated paper holder, and allows it to be peeled and removed in a few seconds.[3]

18.5.1 Substrate suppliers

Organic materials for BGA substrates include glass epoxy, polyimide, or bismaleimide-triazine (BT) resin. The majority of the plastic BGA packages today are fabricated with BT resin substrates. There is only one large supplier of BT resin, Mitsubishi Gas Chemical. Allied Signal, through a patent arrangement with Mitsubishi Gas Chemical can also manufacture BT resin. Printed circuit board companies that supply substrates used inside the plastic BGA include many Japanese companies such as Canon, Eastern, Ibiden, Japanese Circuits, Inc. (JCI), and Matsushita Electric Works (MEW). Taiwanese companies supplying substrates for plastic BGAs today include Compeq, while in the United States, Acsist Associates and Continental Circuits are suppliers.

18.5.2 Materials for overmolding or encapsulation

Glob-top materials, molding compounds, and encapsulation materials are widely available from worldwide sources. In the overmolded plastic BGA packages, Citizen Watch uses a special molding compound that was developed jointly with Shin-Etsu Chemical Company, Ltd. (U.S. Patent 4,859,722). Motorola has also developed a transfer molding compound (U.S. Patent 5,132,778). Several additional companies offer molding compounds including Ciba Geigy (OCG), Shin-Etsu Chemical, Sumitomo, and Plaskon. Ciba Geigy's molding compound, ARATRONIC 2210-9, is said to have equivalent performance to the Shin-Etsu Chemical compound used by Citizen Watch. Ciba Geigy has also developed a new molding compound with improvements including lower internal stress and reduced thermal shrinkage. Additional compounds with higher thermal conductivity are expected to be introduced.

Many vendors offer plastic BGA fabrication equipment. Molding systems are available from ASM Fico, KRAS Corporation, and New

Dynamics. Many pick-and-place equipment vendors offer both automated and manual equipment. Full in-line systems are offered by Kyushu Matsushita Electric (KME) of Japan, a pick-and-place system that can be integrated into an existing surface mount line is supplied by Universal Instruments of the United States. Microelectronic Systems makes a vision-aided manual pick-and-place system.

18.5.3 Test sockets

While companies providing sockets to the plastic and ceramic pin grid array markets are common, test sockets for the BGA market are relatively new. Today Enplas, 3M (Textool), Plastronics, Yamaichi, and Wells Electronics supply test sockets. Cinch, a division of the French company Labinal Systems and Components, provides custom sockets for 225 I/O and 300 I/O parts and is developing open-tooled, burn-in, and test sockets. 3M (Textool) has more than 80 parts tooled for tape, plastic, and ceramic BGAs. The development of test sockets will be greatly enhanced by the propagation of standards for BGA packages.

18.5.4 Inspection equipment

BGAs differ from conventional surface mount packages in that the leads cannot be inspected by visual means. This difference calls for the application of equipment for inspection such as x-ray equipment. X-ray equipment is used primarily for inspection during the evaluation and production characterization phase. Typically, x-ray equipment is not utilized in production because it decreases line throughput time. X-ray inspection systems for use in the BGA industry include equipment provided by Fein Focus USA, Inc.; Four Pi Systems; IRT Corporation; Lixi, Inc.; and Nicolet. Inspection systems are supplied by Hitachi, Sonoscan, and Sonix.

18.6 BGA Standards

The plastic and tape ball grid array package has been registered as a JEDEC standard family of packages. The JEDEC standard was proposed under Letter Ballot JC-11.11.92-163, Item 11-354, and is registered in JEDEC publication 95. While not all BGA packages follow the JEDEC standards, the increased use of standard packages will aid infrastructure development by providing guidelines for production and test equipment suppliers. The ceramic BGA packages have not been registered at the time of this writing.

The plastic BGA standard consists of the following configurations:

1. Body size ranging from 7 to 35 mm per side is standard, with increasing increments of 2 mm. After 35 mm, body size increases by 2.5 mm, up to 50 mm maximum. A tolerance of +0.2 mm on the outer dimensions of the package is allowed.
2. Lead pitch variations of 1.0, 1.27, and 1.5 mm are standard. The standard allows the package to be depopulated in a staggered pattern resulting in effective pitches of 1.41, 1.80, and 2.12 mm, respectively.
3. Maximum possible lead counts for staggered matrices for lead counts greater than 200 are provided. Staggered matrices are anticipated primarily for higher lead count applications in the 1.0- and 1.27-mm lead pitch families. Odd array sizes are used for maximum lead counts greater than 400; staggered arrays are symmetrical for odd array sizes only.
4. An overhang standard (the distance from the package edge to center of the outer ball row) has also been provide. For both 1.27- and 1.5-mm lead pitches, minimum overhangs of 1.25 and 2.0 mm are used for maximum lead counts of 100 balls or less and greater than 100, respectively. The 1.0-mm lead pitch assumes advances in package technology and uses minimum overhangs of 1.0 (<400 solder balls) and 1.25 mm (>400 solder balls). The lid of the package may extend to the periphery and may consist of molding compound, epoxy, metal, ceramic, or other material. The lid may extend above and/or below the package body surface, or it may be incorporated within the package body.
5. Solder ball dimensions and package coplanarity guidelines are provided in the standards.

The tape ball grid array JEDEC registration also provides guidelines. Standard body sizes range from 20.32 to 40.64 mm in increments of 2.54 mm, with a constant distance of 1.27 mm from the edge of the package to the first row of solder balls. Solder balls around the periphery may be removed to allow for a larger overhang if necessary. Solder ball array sizes range from 15 × 15 to 31 × 31 in odd configurations only. The array may be depopulated and a staggered matrix constructed by depopulating every other solder ball in an interstitial pattern. The standard calls for lead pitch of 1.27 mm and a package thickness of 1.45 mm.

18.7 BGA Users Today

For the plastic BGA package, little modification to existing backend equipment is required for conversion to BGA chip assembly. Assembly to the printed circuit board does not require specialized equipment or high accuracy placement, and specially handling is not required because there are no leads. Test of the package is done with a test socket. There are several companies using BGA packages today.

Many companies are currently evaluating BGA packages. It is simpler to list companies that are not investigating BGAs than it is to list companies with evaluation programs. Companies investigating the use or supply of BGAs (organic and inorganic) include many well-known companies.

18.8 Summary

The BGA family of packages offers an interesting new option for surface mount. Building on the existing infrastructure of plastic and ceramic pin grid array packages, the industry infrastructure should develop rapidly as standards are introduced and adopted.

Company	Package type
AT&T	Plastic (internally produced)
Bosch	Plastic (plastic cap)
Bull	Ceramic
Motorola	Ceramic, plastic (OMPAC and glob-top versions)
Compaq	Plastic (OMPAC version)
Ericsson Telecommunications	Ceramic
IBM	Ceramic, T-BGA
Hitachi	Ceramic
NEC	Ceramic

SOURCE: TechSearch International, Inc.

Company	Company
Apple Computer	Altera
AMD	AT&T/NCR
Bosch	DEC
Delco	Ericsson
Goldstar	Hewlett-Packard
Hitachi	Hyundai
Intel	LSI Logic
Matsushita	National Semiconductor
Northern Telecom	Samsung
Siliconware	SGS-Thomson
Sun Microsystems	Texas Instruments
Toshiba	Valtronic
Xilinx	

SOURCE: TechSearch International, Inc.

18.9 References

1. Andros, Frank E. and Richard B. Hammer, "Area Array TAB Package Technology," *Proceedings Fifth International TAB/Advanced Packaging Symposium,* February 1993, p. 163A.
2. Martin, Fonzell D. J., "C-5 Solder Sphere Robotic Placement Cell for Overmolded Pad Array Carrier," *Proceedings International Electronics Packaging Society Conference,* September 1993, pp. 740–748.
3. Westwater, Jim, Mike Vrcelj, Peter Dutton, and Mark McCaul, "Devices for Attaching Solder Balls and Solder Columns to Grid Array Packages," *Electrecon '94 Proceedings,* May 1994 p. 17.

Chapter

19

Packaging Glossary

Ronald W. Gedney

19.1 Introduction

The idea for a packaging glossary actually arose from a conversation with Prof. Sillo, chairman of the mechanical engineering department at Binghamton University. We were discussing electronic packaging course material and he said he wished a glossary of packaging terms were available. As I had edited a packaging applications guide for IBM in 1990, I knew that it contained a partial glossary for both packaging acronyms and terminology. So I foolishly said I thought it would not be too difficult to compile such a document, and would see if I could provide it for him.

I sat down with my computer, pulled together some books along with my applications guide and set to work. Some 4 weeks later, I realized this was going to be a much lengthier task than I had planned. By that time, I was up to the "Rs," had compiled over 30 pages, and felt it was too late to quit. So I dug in, kept going, and finally, finally, arrived at the "Zs."

My goal was to include a clear, but brief, definition of as many package acronyms, types, and terminologies as possible. The glossary very thoroughly covers first-level packages (chip carriers) and second-level packages (printed circuit boards). This includes not only package types, but thermal management, process technology—including chemical terms, reliability and quality, contacts and connectors, and electrical terms. The original intent was to provide a reference document for students majoring in electronic packaging (either through the ME or EE curriculum) at universities. It should also be useful to people who are new to—or do not work every day with—the field of electronic packaging. However, electronic packaging is such a complex, interdisciplinary field, that even the experienced practitioner may find it useful to have all the terminology in one reference document.

I would like to thank Dr. Donald P. Seraphim for taking the time to do a thorough review of the glossary and providing me with excellent feedback. I would also like to thank Richard J. Prosman of the Integrated Electronics Engineering Center at Binghamton University for his interest and encouragement, without which this might not have been completed at all.

19.2 Terms

accelerated (stress) testing A test meant to produce failures caused by the same wearout failure mechanism as expected in field operation but in significantly shorter time.

accelerator A chemical additive that hastens a chemical reaction under specific conditions.

acceptable quality level (1) The lowest quality level a supplier is permitted to present continually for acceptance. (2) The maximum percentage of defects or number of defective parts considered to be an acceptable average for a given process or technique.

activating A treatment that renders nonconductive material receptive to electroless deposition. Also called *seeding, catalyzing,* or *sensitizing.*

active components Electronic components, such as transistors, diodes, etc., which can operate on an applied electrical signal so as to change its basic characteristics (i.e., rectification, amplification, switching, etc.).

active trimming A method of adjusting a circuit function under operating conditions.

additive process (plating) A chemical reduction process for obtaining conductive patterns by the selective deposition of conductive material on clad or unclad base material.

adhesion The ability of a conductor or insulator material to withstand a force attempting to separate it from the substrate.

adhesion promotion The chemical process of preparing a surface to provide for a uniform, well-bonded interface.

Advanced Statistical Analysis Program (ASTAP) An IBM circuit analysis and simulation program. It performs DC, time domain, and frequency domain simulations. Statistics can be applied to all simulations to predict operating tolerances. (Also see **SPICE.**)

alloy (1) Any of a large number of substances having metallic properties and consisting of two or more elements, which are usually metallic. (2) To make or melt an alloy.

alpha particle Decay product of some radioactive isotopes. It is a high-energy (mv range) helium nucleus capable of generating electron/hole pairs in microelectronic devices and switching cells, causing soft errors in some devices.

alumina Aluminum oxide, Al_2O_3. Alumina substrates are made of formulations that are primarily alumina. Nearly all ceramic modules are manufactured with alumina ceramics, as alumina is easy to work with, has fairly good thermal conductivity, and, for flip chip, has a TCE of 6.5 versus 3.0 for silicon.

aluminum nitride Chemical symbol is AlN. This material has a thermal conductivity of 200 m^2/K versus 20 for alumina. Also its thermal coefficient of expansion is very close to that of silicon (3.0). This is a relatively new ceramic, which is already in use for high-power components and has potential to replace BeO as well as Al_2O_3.

aluminum oxide See **alumina.**

ambient The environment that surrounds and contacts a system or component.

anchoring spurs Extensions of the lands on printed wiring beneath a cover layer to assist in holding the lands to the base material.

angle of attack The angle between the squeegee face of a thick-film printer and the plane of the screen.

anisotropic Exhibiting different properties when tested along axes in different directions. In magnetics, capable of being magnetized more readily in one direction than in a transverse direction.

anisotropic conductive adhesive An adhesive that can be made conductive in the vertical, or z axis, while remaining an insulator in the horizontal or x and y axes.

annular ring The portion of conductive material that completely surrounds a hole.

antihalation backing The coating on the rear surface of a film or glass plate that reduces the reflection of light into the areas of the emulsion layer that are to be unexposed.

Application-Specific Integrated Circuit (ASIC) An integrated circuit chip with personalization customized for a specific product. Personalization refers to wiring on the integrated circuit chip.

aqueous In printed board manufacturing operations, water-soluble processes such as a mild sodium or potassium carbonate solution used for image development.

Area Array Tape Automated Bonding (ATAB) Tape-automated bonding in which edge-located pads and additional pads on the inner surface area of a chip are addressed in the bonding scheme.

array A group of elements (pads, pins) or circuits arranged in rows and columns on one substrate.

artwork An accurately scaled configuration that is used to produce an artwork master or production master.

artwork master An accurately scaled (usually 1:1) pattern that is used to produce the production master.

aspect ratio A ratio of the length or depth of a hole to its preplated diameter.

aspot The active areas in an electrical contact interface where asperities are highly deformed and break through any insulating film material.

assembly A number of parts or subassemblies or any combination thereof joined together.

assembly/rework Terms denoting joining and replacement processes of microelectronic components. Assembly refers to the initial attachment of device and interconnections to the package. Rework refers to the removal of a device including interconnections, preparation of the joining site for a new device, and rejoining of the new device.

A-wire Term coined by IBM for area wire bond package for memory devices. Refers to particular technique of wire bonding from the middle of the semiconductor device to a lead frame resting on an insulator over the device. (See **lead on chip.**)

axial leads Leads that come out of the ends of a discrete component or device along the central axis, rather than out of the sides.

backbonding Bonding active chips to the substrate using the back of the chip, leaving the face, with its circuitry facing up. The opposite of backbonding is face-down bonding.

Back-End-of-the-Line (BEOL) That portion of the integrated circuit fabrication where the active components (transistors, resistors, etc.) are interconnected with wiring on the wafer. It includes contacts, insulator, metal levels, and bonding sites for chip-to-package connections. Dicing the wafer into individual integrated circuit chips is also a BEOL process. The front-end-of-the-line (FEOL) denotes the first portion of the fabrication where the individual devices are patterned in the semiconductor.

back panel or backplane A planar printed circuit board package holding plugged-in lower-level package components (e.g., cards) as well as discrete wires and cables interconnecting these components. (See also **mother board.**)

Backside Metallurgy (BSM) A metallization pad electrically connected to internal conductors within a multilayered ceramic package to which pins are brazed.

ball bond A bond formed when an interconnection wire with a ball-shape end is deformed by thermocompression against a metallized pad. Also called *nail head bond* from the appearance of the flattened ball.

ball grid array package For very high speed integrated circuits, a package embodying a rather novel technology in which solder-contact pads are not just around the package periphery (as with chip carriers), but cover the entire bottom surface in checkerboard fashion. Also known as *pad array carrier* (PAC), *pad array package, land grid array,* or *pad-grid array package.*

Ball Limiting Metallurgy (BLM) The solder wettable terminal metallurgy which defines the size and area of a soldered connection, such as a C-4 pad on a chip. The BLM limits the flow of the solder to the desired area, and provides adhesion and contact to the chip wiring.

ball mill A cylindrical jar filled with ceramic or metal balls that is rotated to mix and grind ceramic powders.

bare chip (die) An unpackaged silicon device. For multichip modules, manufacturers purchase bare die, tested and ready for assembly to a package.

barrel-cracking One of the major reliability problems with printed boards is barrel-cracking, or the cracking of the plated through-hole copper during temperature cycling.

BAT Acronym for Bond, Assembly and Test. Usually refers to device assembly into a finished module or package.

bayonet coupling A quick-coupling device for plug and receptacle connectors, accomplished by the rotation of the two parts under pressure.

beam lead chip A chip employing electrical terminations in the form of tabs extending beyond the edge of the chip for direct bonding to a mounting substrate.

bellows contact A connector contact that is a flat spring folded to provide a uniform spring rate over the full tolerance range of the mating unit.

beryllia Beryllium oxide, BeO. A substrate material used where extremely good thermal conductivity is desired.

bifet An integrated circuit that has both bipolar and FET transistors integrated together on the same piece of silicon.

bifurcated contact A connector contact (usually a flat spring) that is slotted lengthwise to provide independently operated points of contact, or redundant or tandem operation.

bifurcated squeegee A two-part squeegee used in a screen printer so that the thick-film composition fluid can be pumped through the squeegee to provide a continuous supply of fluid. This arrangement is especially useful for a highly pseudoplastic or thixotropic fluid used for high resolution printing.

bimetal mask A mask formed by different metals combined by electroforming or cladding. Apertures are selectively etched through one metal to form an image in a second metal.

binders Materials added to thick-film compositions, unfired substrates, and so forth to give sufficient strength for prefire handling.

bipolar transistor A transistor that uses charge carriers of both polarities.

BIST See **built-in self-test.**

blade contact A flat male contact, used in multiple-contact connectors, designed to mate with a tuning fork or a flat formed female contact.

bleed The loss of acuity at the edge of a deposit due to fluidity of the deposited material.

blister (1) A localized swelling and separation between any of the layers of a base material or dielectric, or between base material/dielectric and conductive foil or fired conductor. A form of delamination. (2) Air or entrapped volatiles that rise under the solder mask during thermal shocks.

block copolymer A copolymer compound resulting from the chemical reaction between *n* number of molecules, which are a block of one monomer, and "n" number of molecules, which are a block of another monomer. Example: stearine (rigid) with silicone (elastic).

blow hole A void in a solder connection caused by outgassing or a void in a fired dielectric.

board This package element can best be defined as an organic printed-circuit card or board on which smaller cards or modules can be mounted. Its connections to the next higher level involve discrete wire or cables either soldered or plugged into connector systems.

body (1) Main, or largest, portion of a connector to which other portions are attached or inserted. (2) Structural portion of a ceramic particle or the material or mixture from which it is made.

boiling Phase change and formation of bubbles in a superheated liquid.

bond strength The force per unit area required to separate two adjacent layers of a package (substrate, board, etc.) by a force perpendicular to the package surface.

bondability Those surface characteristics and conditions of cleanliness of a bonding area which must exist in order to provide a capability for successfully bonding an interconnection material by one of several methods such as ultrasonic or thermocompression wire bonding, or soldering.

bonding The joining together of two materials; for example, the attachment of wires to an integrated circuit or the mounting of an integrated circuit to a substrate.

bonding layer An adhesive layer used in bonding together other discrete layers of a multilayer printed board during lamination.

bonding pads Areas of metallization on the integrated circuit die that permit connection to the die of fine wires or circuit elements.

boot A form placed around the wire termination of a multiple-contact connector to contain a liquid potting compound. Also, a protective housing usually made from a resilient material to prevent entry of moisture into a connector.

boundary-scan A method of testing a chip whereby scannable test cells are placed at each input/output pin to allow control of the chip boundaries. The test circuitry provides a virtual electronic bed of nails by way of only a 4-wire test bus. This same test bus can also provide access to initiate or evaluate on-chip built-in self-test.

bow The deviation from flatness of a board, characterized by a roughly cylindrical or spherical curvature such that, if the board is rectangular, its four corners are in the same plane.

braze A joint formed between two different materials by formation of liquid at the interface.

brazing The joining of metals by melting a nonferrous, filler brazing metal, such as eutectic gold-tin alloy, having a melting point lower than that of the base metals. Also known as *hard soldering*.

breakaway The screen-to-substrate separation required during off-contact screen printing. Also known as the *snap-off distance*.

brown oxide An oxide layer on copper, formed intentionally by chemicals, to enhance adhesion to polymers.

B-stage resin In a thermosetting reaction, a resin that is in an intermediate state of cross-link which is capable of further flow upon heating. The cure is normally completed during the laminating cycle. (See **C-stage resin** and **prepreg.**)

bt resins Bis-maleimide-triazine-type resins are mixed with epoxies to achieve the desired laminate properties for use in printed boards. Advantages include high heat resistance, low dielectric constant and excellent electrical insulation resistance even after moisture absorption.

BTAB The acronym for Bumped Tape Automated Bonding where the raised bump for each bond site is prepared on the tape material as opposed to the bump being on the chip.

Built-In Self-Test (BIST) An electrical test scheme whereby hardware is added on the chip to allow the integrated circuit (IC) to test itself with minimal external test equipment. By embedding the test circuitry within the same silicon as the IC's normal application circuitry, BIST is not prone to the loading effects or signal limitations of an external test system. BIST also provides at-speed testing of the chip.

bulk resistance The resistance that is due to the length, cross section and material of a conductive element.

bump A means of providing connections to the terminal areas of a device. A small mound is formed on the device (or substrate) pads and is used as a contact for face-down bonding.

bump contacts Contacting pads that rise substantially above the surface level of the chip. Also, raised pads on the substrate that contact the flat land areas of the chip. Also called *ball contacts, raised pads,* or *pedestals.*

Bumped Tape (BTAB) A tape used in the tape automated bonding process where the inner-lead bond sites have been formed into raised metal bumps on the tape rather than on the chip. This ensures mechanical and electrical separation of the inner-lead bonds and the nonpad areas of the chip (die) being bonded.

burned plating Rough, dull electrodeposit caused by excessive plating current density.

burn-in The process of exposing a component part to elevated temperatures with a voltage stress applied for the purpose of screening out marginal parts.

burnishing Smoothing the conductor materials, after printing and firing, with a fiberglass brush, eraser, or other abrasive material.

burn-off Removal of unwanted materials—typically organic from greensheet or organic contaminants from substrates.

bus bar A conduit, such as a conductor on a printed board, for distributing electrical energy. (See **plating bar,** a subcategory of this term.)

butt Placing two conductors together end to end (but not overlapping) with their axes in line.

camber A term that describes the amount of overall warpage present in a substrate.

capacitance The electrostatic element that stores charge. In packaging systems, it is used: (1) in lumped equivalent circuits to represent part of a line discontinuity; (2) in a distributed system to represent the electrostatic storage property of a transmission line; (3) to filter powering systems, because it delivers current in response to a change in voltage.

capacitive coupling The electrical interaction between two conductors caused by the capacitance between them.

card A printed-circuit panel (usually multilayer) that provides the interconnection and power distribution to the electronics on the panel, and provides interconnect capability to the next level package. It is also known as a daughter board. It plugs into a mother (printed circuit) board.

Card-on-Board (COB) Package technology in which multiple printed-circuit panels (cards) are connected to a (large) printed-circuit panel (board) at 90° angles.

case temperature The temperature measured at a specified location on the (external) case of a device.

catalyzing See **activating.**

CDIP (CERDIP) Acronym for Ceramic Dual In-line Package. The package is composed of a ceramic header and lid, a stamped-metal lead frame, and a frit glass that is used to secure the structure.

ceramic Inorganic, nonmetallic material, such as alumina, beryllia, or glass-ceramic, whose final characteristics are produced by subjection to high temperatures.

Ceramic Ball Grid Array package (CBGA) A ceramic package designed for surface mount applications. It is similar to a ceramic pin grid array except that, instead of pins, the input/output terminals are composed of solder balls. (See **ball grid array package.**)

cermet A solid homogeneous material usually consisting of a finely divided admixture of a metal and ceramic in intimate contact. Cermet thin films are normally combinations of dielectric materials and metals, usually used as resistor elements in hybrid circuits.

channel For a printed circuit board, a channel is the space between the grid of through holes (vias).

characteristic impedance (Z_0) The ratio of voltage to current in a propagating wave; that is, the impedance that is offered to this wave at any point on the line. In printed wiring, its value depends on the width of the conductor, the distance from the conductor to ground planes, and the dielectric constant of the media between them.

chemical milling The process in which metal is formed into intricate shapes by masking certain portions and then etching away the unwanted material.

Chemical Vapor Deposition (CVD) Depositing circuit elements on a substrate by chemical reduction of the vapor of a volatile chemical in contact with the substrate.

chemorheology The study of the processability or flow (rheology) and the chemistry of the polymer system. Processability parameters include, for instance, heating rates, hold temperatures, injection speeds, and compaction pressures. The chemical aspect, on the other hand, involves the rate of reaction, the mechanisms, the kinetics, and the cessation of the chemical reaction at the end of the polymerization.

chip The individual semiconductor element or integrated circuit after it has been cut or separated out of the processed semiconductor wafer, distinct from a completely packaged or encapsulated integrated circuit with leads attached. Also referred to as a *die*. Sometimes used to refer to uncased, normally leadless, passive components (e.g., chip capacitor).

chip carrier An integrated circuit package, usually square, which may have a cavity for the chip in the center. Its connections are usually on all four sides.

Chip-on-Board (COB) One of many configurations in which a chip is directly bonded to a printed circuit board or substrate. [See also **Direct Chip Attach (DCA).**]

Chip-on-Lead (COL) Refers to a package structure whereby some inner leads (of a lead frame) pass under the chip, and the tips of the leads protrude under the two short sides of the die where the bonding pads are located. This allows the ratio of the die size to package size to be substantially increased, and enhances the package thermally.

circuitization Denoting the circuit fabrication process of imaging, plating, etching, and stripping.

clearance hole A hole in the conductive pattern that is larger than, but coaxial with, a hole in the printed board base material.

clinched lead A component lead that extends through a hole in the printed board and is formed to prevent components from falling out prior to soldering.

coated metal core substrate A substrate consisting of an organic or inorganic insulation coating bonded to metal. Insulated surface or surfaces are used for circuit deposition. The metal core may be used for its thermal properties or to control the coefficient of expansion of the overall package.

coaxial A cable configuration having two cylindrical conductors with coincidental axes, such as a conductor with a tabular shield surrounding the conductor and insulated from it.

COB (1) Acronym for Chip-on-Board, which see. (2) Acronym for Card-on-Board, which see.

Coefficient of Thermal Expansion (CTE) The ratio of the change in dimensions to the change in temperature-per-unit starting length, usually expressed in cm/cm/°C.

Coffin-Manson equation Often used to relate the strain amplitude to the fatigue life. Basically the equation states that the number of cycles to failure is proportional to the reciprocal of the strain squared. In reliability testing it can be used to predict field life based on accelerated temperature cycling in the laboratory.

cofiring (1) A process in which two thick-film compositions are printed and dried one after the other and then fired at the same time. (2) Processing thick-film conductors and dielectrics through the firing cycle at the same time to form multilayer structures (e.g., multilayer ceramic substrates).

coined lead A cylindrical lead that has been formed to have parallel surfaces approximating a ribbon lead configuration.

cold flow Permanent deformation of the insulation due to mechanical forces, without the aid of heat softening of the insulating material.

cold solder joint A soldered connection exhibiting poor wetting and a dull, porous appearance due to insufficient heat or inadequate cleaning prior to soldering or because of excessive impurities in the solder.

cold weld Forming a hermetic seal in a metal package by welding the lid to the frame using pressure alone.

collimated light In photoprinting, the state of all light rays traveling in parallel at the exposure plane. This is in comparison to a point source, which generates light rays at varying angles.

complementary metal-oxide semiconductor (cmos) This refers to logic in which cascaded field effect transistors (FET) of opposite polarity are used to minimize power consumption.

compliant bond A bond which uses an elastically and/or plastically deformable member to impart the required energy to the lead.

component An individual functional element in a physically independent body that cannot be further reduced or divided without destroying its stated function, for example, a resistor, capacitor, diode, or transistor.

component lead The formed conductor that extends from a component and serves as a mechanical and/or electrical connection for that component.

component side That side of the printed wiring board on which most of the components are mounted.

compression seal A seal made between an electronic package and its leads. The seal is formed as the heated metal case, upon cooling, shrinks around the glass insulator, thereby forming a tight joint.

conditioning Time-limited exposure of a test specimen to specified environment(s) prior to testing.

conductance A measure of the ability of any material to conduct an electrical charge. Conductance is the ratio of current flow to the potential difference. The reciprocal of electrical resistance.

conduction Thermal transmission of heat energy from a hotter region to a cooler region in a conducting medium.

conductive adhesive An adhesive material, usually epoxy, that has metal powder added to increase electrical conductivity. The usual conductor material added is silver.

conductive epoxy An epoxy material (polymer resin) that has been made conductive by the addition of a metal powder, usually gold or silver.

conductive foil A thin sheet of metal that may cover one or both sides of the base material and is used for forming the conductive pattern.

conductive pattern The configuration or design of the conductive material on the base material. For an organic printed board, the pattern includes conductors, lands, and through connections (vias), when these connections are an integral part of the manufacturing process. When applied to flat ceramic packages, a conductive pattern is accomplished by means of screened-on metallization, forming the internal portions of leads and other conductive area.

conductivity, electrical The capability of a material to carry an electrical current; that is, conductances of a unit cube of any material. The reciprocal of resistivity.

conductor, electrical A class of materials that conduct electricity easily. They have very low resistivity which is usually expressed in micro-ohm-cm. The best conductors include silver, copper, gold, and superconducting-ceramics.

conductor spacing The observable distance between adjacent edges (not centerline to centerline) of isolated conductive patterns in a conductor layer.

conductor, thermal A class of materials, such as copper, aluminum, and beryllia, that conduct heat.

conformal coating A thin, nonconductive coating, either plastic or inorganic, applied to a circuit for environmental and/or mechanical protection.

connector Generally, all devices used to provide rapid connect/disconnect service for electrical cable and wire terminations, board to board.

connector assembly A mated plug and receptacle.

contact The element in a connector that makes the actual electrical contact. Also, the point of joining in an electrical connection.

contact alignment The overall side play of contacts within the insert cavity to permit self-alignment of mated contacts. Also called *amount of contact float.*

contact angle The angle enclosed within the fillet, between a plane tangent to the solder/base-metal surface and a plane tangent to the solder/air interface.

contact arc The electrical (current) discharge that occurs between mating contacts when a circuit is being made or broken.

contact area The common area between a conductor and a connector across which the flow of electricity takes place.

contact length Length of travel made by one contact in contact with another during assembly or disassembly of the connector. (See also **wiping action.**)

contact printing A method of screen printing in which the screen is almost (within a few mils) in contact with the substrate. Used for precision printing or with nonflexible screens. Also a method of photoprinting in which the mask is in direct contact with the photoresist on the part.

contact rating The Aerospace Industries Association of America, Inc. defines contact rating as "The maximum current for a given type of load (that is, voltage, frequency, and nature of impedance) which the relay (contacts) will make, carry, and break (unless otherwise specified) for its rated life."

contact resistance Excess electrical resistance in series with the bulk conductor resistance of two contacting electrical conductors arising from the nature of the contact geometry and surface properties of the contacting surfaces.

contact retention The minimum axial load in either direction that a contact must withstand while remaining fixed in its normal position within an insert.

contact spacing The distance between centerlines of adjacent contact areas.

contact spring (1) A current-carrying spring, usually a flat-leaf type or wire member, to which the contacts are fastened. Contact-spring bearing contacts are assembled between insulators in the contact assembly, to form contact combinations. (2) A non-current-carrying spring that positions or tensions a current-carrying member.

contact (spring) tension The contact pressure developed, usually resulting from the specified adjustment of movable contacts against mating stationary springs, when the relay is unenergized.

contact weld (1) The point of attachment of a contact to its support when accomplished by resistance welding. (2) A contacting failure due to the fusing of contacting surfaces under load conditions to the point that the contacts fail to separate when expected.

contact wipe The scrubbing action between mating contacts resulting from contact overtravel or follow.

continuity An uninterrupted path for the flow of electrical current in a circuit.

Controlled Collapse Chip Connection (C4) A solder joint connecting a substrate and a flip chip, where the surface tensions force of the liquid solder supports the weight of the chip and controls the height (collapse) of the joint (the height being a function of the two contacting areas and the total volume of solder in the joint).

controlled impedance For high-speed circuits, the impedance (Z_0) is fixed by design; i.e., the line thickness, spacing, distance to a ground plane, and the dielectric constant of the base material are part of the design. The actual tolerance on the nominal design depends on manufacturing control of these parameters.

convection cooling (self-convection) Generally refers to a system component in still air. The heat actually radiates away from the component without the benefit of forced air in the system environment generating air currents.

conversion efficiency In regard to photoemissive devices, the ratio of maximum available luminous- or radiant-flux output to total input power.

coplanar leads Ribbon-type leads (flat) extending from the sides of the circuit package, all of which lie in the same plane.

copyboard The part of a process camera used to hold the artwork in place.

cordierite A crystalline ceramic material of composition 2 magnesium oxide, 2 aluminum oxide, and 5 silicon oxide that can be crystallized from glass of the same composition or sintered from powders.

core material The fully cured, inner-layer segments, with circuitizing on one or both sides, that form the multilayer circuit.

corner mark The mark at the corners of printed circuit board artwork, the inside edges of which usually locate the borders and establish the contour of the board. Also called *crop mark*.

COS Acronym for Chip-on-Substrate, synonymous with Chip-on-Board, which see.

COT Acronym for Chip-on-Tape, referring to the practice of testing chips after mounted on a continuous reel of TAB tape.

coupled noise The electromagnetic and electrostatic linkages between two nearby conductors that allow one line to induce a signal on the other. (See also **crosstalk.**)

coupler A chemical agent, frequently an organosilane, used to enhance the bond between a resin and a glass reinforcement.

cover lay Outer layer(s) of insulating material applied over the conductive pattern on the surface of the printed board. Also called *cover layer* or *cover coat.*

cover seal The seal at the perimeter of the cover or lid, when joined to the package body. In hybrid fabrication, the cover-sealing operation itself. The seal my be accomplished by resistance welding, cold welding (solid-phase bond), brazing, soldering, or other means.

cracking A condition in metallic and/or nonmetallic coatings consisting of breaks that extend through to an underlying surface.

crazing A network of fine cracks on the surface or through the layers of plastic or glass materials, such as conformal coatings or glass encapsulants. This condition can manifest itself in the form of connected white spots or "crosses" below the surface of the base material and is usually related to mechanically or thermally induced stress. (See also **measling.**)

creep The dimensional change with time of a material under load.

crimp termination A connection in which a metal sleeve is secured to a conductor by mechanically crimping the sleeve with pliers, presses, or automated crimping machines. Splices, terminals, and multicontact connectors are typical terminating devices attached by crimping.

crosshatching The breaking up of large conductive areas by the use of a pattern of voids in the conductive material.

cross link Intermolecular bonds produced between long-chain molecules in a material to increase molecular size and weight by chemical or photon bombardment, resulting in a change in physical properties in the material.

crossover The transverse crossing of metallization paths without mutual electrical contact achieved by the deposition of an insulating layer between the conducting paths at the area of crossing.

cross-sectional area The area of the cut surface of an object that has been cut at right angles to the long axis of the object.

crosstalk The undesirable interference caused by the coupling of energy between signal paths.

crystallization Formation of crystalline phaseout of amorphous material during high-temperature processing. Undesirable or uncontrollable crystallization is called devitrification.

C-stage resin A resin that is fully cross-linked at room temperature. (See also **B-stage resin.**)

CTE Coefficient of Thermal Expansion, which see.

CTE mismatch The difference in the coefficients of thermal expansion (CTEs) of two materials or components joined together, producing strains and stresses at the joining interfaces or in the attachment structures (solder joints, leads, and so on).

cure To change the physical properties of a material (usually from a liquid to a solid) by chemical reaction or by the action of heat and catalysts, alone or in combination, with or without pressure.

curing agent An inorganic or organic compound that initiates the polymerization of a resin.

curing cycle For a thermosetting material, commonly for a resin compound such as a bonding adhesive, it is the combination of total time-temperature profile to achieve the desired result; for example, the complete irreversible hardening of the material, resulting in a strong bond.

current-carrying capacity The maximum current that can be carried continuously, under specified conditions, by a conductor without causing objectionable degradation of electrical or mechanical properties of the conductor or the substrate on which it resides.

Current-Mode Logic (CML) Basically equivalent to emitter-coupled logic (ECL), except that it operates in the nonsaturated mode and has greater speed.

current penetration The depth that a current of a given frequency penetrates into the surface of a conductor carrying the current, commonly called *skin depth.*

curtain coating Coating process in which a fluid forced through a narrow slot creates a falling liquid film.

custom design A form of design in which the choice and arrangement of components and wiring on a package may vary arbitrarily within tolerances from a regular array.

cut and strip Method of producing artwork by cutting the pattern and stripping away the unwanted areas of a two-layer system.

cut-through resistance The ability of a material to withstand mechanical pressure, usually a sharp edge of prescribed radius, without separation.

decoupling capacitor A shunt-placed capacitor that is used to filter transients on a power distribution system.

definition The fidelity of reproduction of the printed board relative to the production master.

delamination A separation between plies within the base material, or between the base material and the conductive foil, or both.

delay time The time interval from the point at which the leading edge of the input pulse has reached 10 percent of its maximum amplitude to the point at which the leading edge of the output pulse has reached 90 percent of its maximum amplitude.

delta-I noise See **switching noise.**

dendritic growth The electrolytic transfer of metal from one conductor to another similar to electroplating except that dendritic growth usually, though not always, forms from cathode to anode. The dendrite assumes the form of a stalk with branches resembling a tree. When this dendritic growth touches the opposite conductor, there is an abrupt rise in current. This phenomenon is dependent on the presence of moisture, current and the nature of the conductive surfaces. Also a "tree-like" structure in electrolytic plating. Usually the result of high current density and additives in the plating bath. (See also **whiskers** and **electromigration.**)

deposition The "laying down" of films or metal or insulators on a substrate, by evaporative deposition, sputtering or electroplating.

derating factor A factor used to reduce the current-carrying capacity of a wire when the wire is used in other environments than that for which the value was established.

device An individual electrical element usually in an independent body, that cannot be further reduced without destroying its stated function. Also, an electronic part consisting of one or more active passive elements.

devitrification The undesirable formation of crystals in glass during firing. The desirable process is called crystallization.

devitrify To deprive of glassy luster and transparency or to change from a vitreous to a crystalline condition.

dewetting (1) The development and formation of a nonwetting condition after wetting has already commenced. (2) A condition that results when molten solder has coated a surface and then receded, leaving irregularly shaped mounds of solder separated by areas covered with a thin solder film. The basis metal is not exposed, however.

dicing Separating a semiconductor wafer into individual dies.

die Integrated circuit chip as cut (diced) from finished wafer. (See **chip.**)

die bond Mechanical attachment of the silicon die or chip to the substrate usually by solder, epoxy, or gold-silicon eutectic. The die bond is made to the back (inactive) side of the chip with the circuit side (face) up.

die pick-and-place machine A die-sort system designed specifically to (1) remove (pick) previously tested semiconductor devices automatically out of a scribed, diced, or broken wafer mounted on a taut plastic tape and (2) transfer each individual good die for deposition into a consecutive cavity of a specific storage container.

dielectric Material that does not conduct electricity. Generally, *dielectric* refers to materials that are to be used as capacitors, whereas *insulators* refers to materials that are primarily used as electrical insulators. Both materials can serve as either or both and the terms are often used interchangeably.

dielectric breakdown The complete failure of a dielectric material characterized by a disruptive electrical discharge through the material due to a deterioration of material or an excessive sudden increase in voltage.

dielectric constant The ratio of the capacitance, C_x, of a given configuration of electrodes with a specified dielectric, to the capacitance, C_v, of the same electrode configuration having a vacuum (or air) as the dielectric.

dielectric loss Electric energy transformed into heat in a dielectric subjected to a changing electric field.

dielectric strength The maximum voltage that a dielectric can withstand, under specified conditions, without resulting in a voltage breakdown (usually expressed as volts per unit dimension).

diffusion bonds Processes accomplished by bringing two conductors to be joined into intimate contact and inducing the atoms of one material to diffuse into the structure of the other.

digitizing Any method of reducing feature locations on a flat plane to give digital representation of *x-y* coordinates.

dimensional stability A measure of dimensional change caused by such factors as temperature, humidity, chemical treatment, age, or stress (usually expressed as delta units per unit).

dip soldering A process in which printed boards with attached components are brought in contact with the surface of a static pool of molten solder for the purpose of soldering the entire exposed conductive pattern and the component leads in one operation.

Direct Chip Attach (DCA) A name applied to any of the chip-to-substrate connections used to eliminate the first level of packaging. (See also **Chip-on-Board.**)

direct emulsion A sensitized liquid coating for producing stencil screens. The material is applied to the screen mesh before exposure and development.

direct emulsion screen A screen whose emulsion is applied by painting directly onto the screen, as opposed to the indirect-emulsion type.

discrete component Individual components or elements, such as resistors, capacitors, etc., that are handled discretely, as opposed to those that are made by screen printing or other deposition methods as parts of a hybrid network.

Distance to Neutral Point (DNP) The separation of a joint from the neutral point on a chip. This dimension controls the strain on the joint imposed by the expansion mismatch between chip and substrate. The neutral point is usually the geometric center of an array of pads and defines the point at which there is no relative motion of chip and substrate in the *X-Y* plane during thermal cycling.

distributed network A network in which the parameters of resistance, capacitance, and inductance cannot be considered localized at any one point in the network.

doctor blade A method of casting slurry into a thin sheet by the use of a knife blade placed over a moving carrier to control slurry thickness.

double-sided substrate A substrate carrying active circuitry on both its topside and bottomside, electrically connected by means of metallized through-holes, edge metallization, or both.

drag soldering A process in which supported, moving printed circuit (wiring) assemblies are brought into contact with the surface of a static pool of molten solder.

drift Permanent change in value of a capacitor or resistor over a period of time because of the effects of temperature, aging, humidity, etc.

drill smear While drilling holes in printed boards, a drill's temperature may exceed the glass transition temperature (T_g) of the dielectric material, resulting in melting. The melts are smeared across the hole walls and copper interconnect surfaces by the penetration and extraction of the drill bit. The smear can be several micrometers thick and interfere with the forming of reliable internal plane connections.

driver The off-chip circuit that supplies the signal voltage and current to the package lines. Also called an *output buffer circuit.*

dross Oxide and other contaminants that form on the surface of molten solder.

dry film For printed board imaging operations, a photopolymer that is coated onto a carrier sheet, which in turn is laminated to the surface of the board.

dry pressing Pressing and compacting together of dry powdered materials with additives in rigid die molds under heat and pressure to form a solid mass, usually followed by sintering to form shapes.

Dual-in-Line Package (DIP) A package having two straight, parallel rows of leads extending at right angles from the base and having standard spacings between leads and between rows of leads.

ductility The capability of a material to deform plastically before fracturing.

duty cycle A statement of energized and deenergized time in repetitious operation; for example, x sec on and y sec off.

dynamic contact resistance A change in contact electrical resistance due to a variation in contact pressure on contacts mechanically closed.

dynamic flex A form of flexible circuitry developed for applications where continued flexure is necessary. In contrast, static flex, once installed, remains fixed.

E glass A low-alkali lime-alumina-borosilicate glass, noted for its good electrical and mechanical properties. Substantially used in making fabric for printed circuit boards.

edge connected module A functional package (containing devices) that is connected to the next level of package on one edge, usually by plugging into a connector socket (although it could also be soldered in place).

edge connector A connector designed specifically for making removable and reliable interconnections between the edge of a printed board and the next level package.

edge definition The fidelity of reproduction of a pattern edge relative to the production master.

edge metallization The metallization applied to the edge of a substrate, wraparound fashion, such that it establishes an electrical connection between circuitry on the top and bottom of the substrate.

edge spacing The distance between a pattern, component, or both, from the edges of the printed board.

elastomer Any elastic, rubberlike substance, such as natural or synthetic rubber. Silicon rubber is one of the most commonly used elastomers.

electrode A conductor through which a current enters or leaves an electrolytic cell, arc furnace, vacuum tube, gas-discharge tube, or other nonmetallic conductor.

electrodeposited (ED) copper ED copper is made in one of two ways: (1) a thin layer of copper is deposited by sputtering or electroless plating on a material and then electroplated up to the final desired thickness; (2) copper is continuously electroplated on a smooth, slowly turning stainless steel (cathode) drum. The deposited-foil thickness is controlled by the speed of the drum's rotation; foil is peeled continuously from the surface of the drum as it emerges from the plating solution. The surface closest to the drum is smooth and even, but the side exposed to the electroplating bath has a pronounced roughness.

electrodeposition The deposition of a conductive material from a plating solution by the application of electrical current.

electroformed mask Usually a trimetal structure in which a core material is selectively electroplated on both sides with a dissimilar metal. The core is then etched back to form a mesh pattern on one side and a stencil cavity on the reverse.

electroless deposition The deposition of conductive material from an autocatalytic plating solution without the application of electrical current.

electroluminescence The direct conversion of electrical energy into light.

electrolysis The separation of chemical components by the passage of current through an electrolyte.

electrolyte Current-conducting solution (liquid or solid) between two electrodes or plates of a capacitor, at least one of which is covered by a dielectric film.

electrolytic cleaning Cleaning in which a current is passed through a solution; the part is set up as one of the electrodes.

electrolytic corrosion Corrosion by means of electrochemical action.

electrolytic tough pitch A term describing the method of raw copper preparation to ensure a good physical- and electrical-grade copper-finished product.

Electromagnetic Compatibility (EMC) A package that has built-in electrically conductive shielding to prevent the escape of electromagnetic radiation. Government standards are available that specify allowable radiation losses.

electromigration The electrolytic transfer of metal from one conductor to another that is separated from the first by a dielectric medium through which the ions move under d-c potential.

electron-beam bonding Bonding two conductors by means of heating with a stream of electrons in a vacuum.

electron-beam patterning E-beam, or EB, patterning of resist by evaporation of the resist material from the heat supplied by the energy of a narrowly focused electron beam.

electronic packaging The technology of interconnecting semiconductor and other electronic devices to provide an electronic function. An electronic package supplies the means for interconnecting power and signal terminals, mechanical stability and protection, and a path for thermal management so that it will both survive and perform under a plurality of environmental conditions.

electronic shielding A physical barrier, usually electrically conductive, designed to reduce the interaction of electric or magnetic fields upon devices, circuits, or portions of circuits.

electroplate The application of a metallic coating on a surface by means of electrolytic action.

electrostatic discharge Discharge of static charge on a surface or body through a conductive path to ground.

element A topologically distinguishable part of a microcircuit that contributes directly to its electrical characteristics.

emitter-coupled logic A nonsaturating form of bipolar logic in which the emitters of the input logic transistors are coupled with the emitter of a reference transistor. Also known as *current-switch logic.*

emulsion The organic material used to coat and/or plug the mesh of a screen to form a circuit pattern.

encapsulate Sealing or covering an element or circuit for mechanical and environmental protection. Typical encapsulating materials are potting, glob top, and molding compounds.

encroachment Solder mask on lands or surface-mount device pads, via bleeding, smearing, or misregistration.

End of Life (EOL) The end of the useful operating life of a component or equipment determined by a "wear-out" or life-terminating mechanism measured in units of time. EOL is usually specified as an objective in reliability calculations.

Engineering Change (EC) A change in design. An electrical design change is frequently implanted by cutting out or adding an electrical path to the manufactured hardware, e.g., laser deleting a line or adding a wire on a ceramic substrate.

escape wiring See **fanout wiring.**

etch factor The ratio of the depth of etch (conductor thickness) to the amount of lateral etch (undercut).

etchant A solution used to remove, by chemical reaction, the unwanted portion of material from a printed board.

etchback A process for the controlled removal of nonmetallic materials from the sidewalls of holes to a specified depth. The process is used to remove resin smear and to expose additional internal conductor surfaces.

etched metal mask A mask formed by etching apertures through a solid metal protected by a photoresist.

etched printed board A board having a conductive pattern formed by the chemical removal of unwanted portions of the conductive foil. (See also **subtractive process.**)

etching A process by which a printed pattern is formed by either chemical or chemical and electrolytic removal of the unwanted portion of conductive material bonded to a base.

eutectic The minimum melting point of a combination of two or more metals or ceramics. The eutectic temperature of a system (if one exists) is always lower than the melting point of any of the individual components of the system.

External Thermal Enhancement (ETE) The mechanical member of the package that conducts heat away from the outside case of the package. This is usually a thermally conductive material (e.g., a heat sink) that is attached to the external surface of a module and projects into an air stream.

external thermal resistance A term describing the thermal resistance from a convenient point of the outside surface of an electronic package to an ambient reference point.

exothermic Characterized by the liberation of heat.

exposure The product of light intensity and time; the total amount of light striking a given area of photographic material.

external leads Electronic package conductors for input and output signals, power and ground.

face-down bonding A method of attaching a component or circuit chip to a substrate by inverting the chip and bonding the chip contacts to the mirror image contact points on the substrate. The pads or bumps that are used at the contact points may have originally been on either the chip or the substrate. The actual bonding process may be some type of thermocompression, ultrasonic, or solder technique. Also called *face bonding.*

failure The termination of the capability of an item to perform a required function.

failure criteria The specification limits beyond which the component is defined to fail to perform its required function.

failure mechanism A structural or chemical process, such as corrosion, carbon tracking, or fatigue, that causes failure.

failure mode Failure at the macro level; that is, the observed effect of failure. Examples include a "short," an open-circuit board, and system malfunction.

failure rate The rate at which devices from a given population can be expected (or were found) to fail as a function of time (e.g., percent per 1000 hours of operation).

fatigue Used to describe a failure of any structure caused by repeated application of stress over a period of time.

fatigue life The number of loading cycles of a specified nature that a design element can sustain before failing.

fanout wiring The wiring layer that directly interconnects a device or small package to the next level of packaging. This may be simply a direct interconnection, but generally is used to describe wiring that "fans out" from the device/package to the wiring or via grid of the next level package. (See also **global wiring.**)

feed-through A connector or terminal block, usually having double-ended terminals that permit simple distribution and busing of electrical circuits.

ferroelectric A crystalline dielectric that exhibits dielectric hysteresis; an electrostatic analog to ferromagnetic materials.

Field Effect Transistor (FET) A transistor in which a voltage applied to a thin conductor over a thin insulator controls current flow in a semiconductor region (gate) of one polar type. This component originates and terminates in two regions of the opposite polar type located at either end of the gate region.

Field Replaceable Unit (FRU) A component or subsystem of an electronic assembly which may be replaced at the site of installation.

filler (1) A substance, often inert, added to a compound to improve properties and/or decrease cost. (2) A material used in a cable to fill large interstices where there are no electrical components.

film Single or multiple layers or coating of thin or thick material used to form interconnections and crossovers (conductors, insulators) or various elements (resistors, capacitors). Thin films can be deposited by vacuum evaporation or sputtering and/or plating. Thick films can be deposited by screen printing. Also, a term to describe thin, plastic sheeting.

film stress The compressive or tensile forces appearing in a film. Internal film stress is the intrinsic stress of a film related to its mechanical structure and deposition parameters. Induced film stress is the component of film stress related to an external force such as mismatched mechanical properties of the substrate.

final seal The manufacturing operation that completes the enclosure of the hybrid microcircuits so that further internal processing cannot be performed without delidding or disassembling the package.

fine leak A leak in a sealed package less than 10^{-6} cm^3/sec at 1 atmosphere of differential air pressure.

fine leak test A test to establish the integrity of a given device package by measuring the leak rate of the package under specified conditions. It usually employs a tracer gas as the test medium.

fingers The short lands containing the contact pads on a substrate to which the chip is bonded. In a flip-chip arrangement, such lands will contain a solder stop—a nonwettable organic or inorganic material to which solder will not adhere—to contain the solder and provide adequate height from chip to substrate.

fire The heating of a thick-film circuit or substrate so that the resistors, conductors, etc., or ceramic body are transformed into their final form.

flame-off The procedure in which the wire is severed by passing a flame across the wire, thereby melting it. The procedure is used in gold wire thermocompression bonding to form a ball for making a ball bond.

flame retardant An inorganic or organic compound added to a polymer mixture that causes the resulting plastic to be self-extinguishing after a flame is removed.

flare The widened portion of a printed circuit line as it enters a land. Also, the undesirable enlarged, tapered area of a punched hole, typical of the side through which material exits during hole formation.

flash point The temperature to which a material must be heated to give off sufficient vapor to form a flammable mixture.

flat cable A cable with two or more parallel, round, or flat conductors in the same plane encapsulated by an insulating material.

flat conductor A wire manufactured in a flattened form, as opposed to round or square.

flatness The long-range deviation from planarity of a film surface, measured in micrometers/millimeter or mils/inch.

flatpack or **flat pac** An integrated circuit package having its leads extending from all four sides and parallel to the base.

flex life The ability of a design element, such as a conductor wire, cable, or flex circuit to withstand repeated bending. (See also **fatigue life.**)

flexible printed wiring (or flex circuit) A patterned arrangement of printed wiring using a flexible base material with or without flexible cover layers.

flip chip A chip that has bumped terminations spaced around the device and is intended for facedown mounting. (See also **facedown bonding.**)

flip-chip attachment A method of attaching a device to a substrate in which the device is flipped so that the connecting conductor pads on the face of the device are set on mirror-image pads on the substrate and bonded by reflowing the solder.

flood bar A bar or other device on a screen-printing machine that will drag pastes back to the starting point after the squeegee has made a printing stroke. The flood stroke returns the paste without pushing it through the meshes, so that it does not print but returns the paste supply to be ready for the next print.

flow soldering See **wave soldering.**

flush conductor A conductor whose outer surface is in the same plane as the surface of the insulating material adjacent to the conductor.

flux In soldering, a material that chemically attacks surface oxides so that molten solder can wet the surface to be soldered, or an inert liquid which excludes oxygen during the soldering process.

footprint The area needed on a substrate for a component or element. Usually refers to specific geometric pattern of a chip. (See also **land pattern.**)

FR-4 Designation of the Electronic Industries Association for a fire-retardant epoxy resin-glass cloth laminate. By common usage, the resin for such a laminate.

frit Glass composition ground up into a powder form and used in thick-film compositions as the portion of the composition that melts upon firing to give adhesion to the substrate and hold the conductive composition together.

from-to list Written wiring instructions in the form of a list indicating termination points.

functional test An electrical test on a package, performed after all component assembly operations are completed, to ensure that the electronic function of the assembly conforms to desired criteria of operation.

fusing The melting of a metallic coating (usually electrodeposited) by means of a heat-transfer medium, followed by solidification.

gas blanket An atmosphere of forming gas flowing over a heated integrated-circuit chip or a substrate during bonding that keeps the metallization from oxidizing.

gate array A semicustom product, implemented from a fully diffused or ion-implanted semiconductor wafer carrying a matrix of identical primary cells arranged into columns with routing channels between them in the x and y directions.

gate, logic Usually an electric circuit which combines information of its two inputs to form its output signal in accordance to the logic function it performs.

gate, structural Term used to designate a hinged frame which contains a number of boards and can swing out for servicing and/or access to the interior of a fixed frame.

glass A hard, brittle substance, usually transparent, made by fusing silicates with soda, lime, etc.

glass-ceramic Inorganic, nonmetallic material obtained by controlled crystallization of glass into a nonporous and fine microstructure.

glass+ceramic Inorganic, nonmetallic material obtained by admixing crystalline ceramic with glass and sintering the composite.

glass fabric Cloth woven from glass yarns which are made of filaments.

glass transition temperature (T_g) In polymer or glass chemistry, the temperature corresponding to the glass-to-liquid transition, below which the thermal expansion coefficient is low and nearly constant, and above which it is very high.

glaze The glassy coating applied to the surface of a formed article or the material or mixture from which the coating is made.

glazed substrate A glass-coated ceramic substrate that effects a smooth and nonporous surface.

glob top A glob of encapsulant material (usually epoxy or silicone or combination thereof) surrounding a chip in the chip-on-board assembly process.

global wiring Wiring interconnecting components mounted on a package (as opposed to the wiring connecting the components to the package. (See also **fanout wiring.**)

golden device A semiconductor device that has been fully characterized and meets all specification requirements (e.g., a "perfect" device), usually used as a test verification tool or to establish standards for the product.

green ceramic Unsintered ceramic, usually a ceramic-loaded organic system.

green density Density of a ceramic after composite pressing.

green sheet A composite organic-inorganic, flexible sheet ready for metallization, if desired, and lamination to form green ceramic entities which upon removal of organic (by firing or sintering) results in a ceramic substrate.

grid An orthogonal network of two sets of parallel equidistant lines used for locating points on a printed board.

ground A common reference point for circuit returns, shielding or heat sinking.

ground plane A conductor layer or portion thereof used as a common reference point for circuit returns, shielding or heat sinking.

ground plane clearance The etched portion of a ground plane around a through hole that isolates the plane electrically and mechanically from the hole.

gull wing A common lead form used to interconnect surface-mounted packages to the printed-circuit board. The leads, normally 100 to 250 microns thick, are bent outward, downward, then again outward from the package body, providing feet for solder interconnection, and some degree of mechanical compliance.

halation The spreading of light outside the intended area of exposure by reflection from the rear surface of the transparent base supporting the emulsion to be exposed.

haloing Mechanically induced fracturing or delamination on or below the surface of the base material; usually evidenced by a light area ("halo") around holes, other machined areas, or both.

hand cut Artwork that has been prepared without the use of a drafting machine.

hard-drawn copper wire Copper wire that has been drawn to size and not annealed.

hard glass Glasses having a high softening temperature (>700°C), such as the borosilicate glasses used to seal feedthrough leads into metal packages.

harness A term used to describe a group of conductors laid parallel or twisted by hand, usually with many breakouts, laced together or pulled into a rubber or plastic sheath, used to interconnect electric circuits.

HASL Acronym for Hot Air Solder Leveling—a technique for controlling the amount and height of solder on a package. The hot air melts (or keeps the solder molten after emerging from a bath) the solder and the velocity of the air forces excess solder off the package pads.

HAST Acronym for Highly Accelerated Stress Test; usually used to describe a test in which components are placed in a pressure chamber and subjected to steam under one atmosphere or more of pressure at correspondingly high temperatures.

header The bottom portion of a device package to which the chip is attached and from which the external leads extend.

heat sink A thermally conductive material, usually metal, to which electronic components, their substrate, or their package are attached. This material has the ability to rapidly transmit heat from the generating source (component).

heel On a wire bond, the part of the lead adjacent to the bond that has been deformed by the edge of the bonding tool used in making the bond. The back edge of the bond.

hermetic Sealed so that the object is gastight. A plastic encapsulation cannot be hermetic as it allows permeation by gases.

hertz dot An intentionally raised area on a connector to concentrate the force within a contact interface, usually used to assist in breaking through oxides on a mating interface.

hertz stress The force divided by the deformed area in a contact system.

High Density Multichip Interconnect (HDMI) A high-density multichip module approach developed by General Electric. The chips are mounted in cavities in a substrate. A polyimide film is laminated to the face of the chips and a laser is used to etch via holes for contact to the chip bonding pads. A thin-film multilevel interconnect structure is built on the polyimide overlay to interconnect the devices and input/output pads.

high pot test A high voltage applied between two terminals seperated by an insulator. The voltage (typically between 100 and 1000 volts) is designed to be lower than the breakdown voltage of the insulator, but adequate to detect any leakage paths that may ultimately cause failure of the package.

hole breakout A condition in which a hole is not completely surrounded by the land.

hole cleaning The process for cleaning conductive surfaces exposed within a hole (e.g., removal of a resin smear).

hole density The quantity of holes in a printed board per unit area.

hole location The dimensional location of the center of the hole.

hole pattern The arrangement of all holes in a printed board with respect to a reference point.

hole pull strength The force necessary to rupture a plated-through hole or its surface terminal pads when loaded or pulled in the direction of the axis of the hole.

hole void A void in the metallic deposit of a plated-through hole that exposes the base material.

horn A cone-shaped member that transmits ultrasonic energy from a transducer to a bonding tool.

hot dip Covering a surface by dipping the surface to be coated into a molten bath of the coating material.

hot-gas reflow The technique in which a heated gas, including air, is impinged on a site to be solder-reflowed, usually to form a solder interconnection.

hot-knife soldering The technique in which a heated blade (electrically or conductively) is used as a heat source for melting solder during package joining. The blade may be used to force mechanical contact throughout the joining process.

hot zone The part of a continuous furnace or kiln that is held at maximum temperature. Other zones are the preheat zone and the cooling zone.

hybrid circuit A circuit that uses two or more fabrication techniques to form the circuit, such as integrated circuit chips attached to a substrate having thin-film devices and conductors.

Hybrid Integrated Circuit (HIC) In general, any integrated circuit that is not monolithic. The term *hybrid* indicates that the circuit elements are made by two or more different technologies. A typical HIC consists of semiconductor chips and capacitors attached to a ceramic substrate carrying printed resistors and interconnections that are vacuum-evaporated or screen printed/fired.

hybrid microcircuit A microcircuit consisting of elements that are a combination of the film circuit type and the semiconductor type, or a combination of one or both of the types with discrete parts.

icicle See **solder projection.**

I-lead A surface-mounted device lead that is formed such that the end of the lead contacts the board pattern at a 90° angle. Also called a butt joint.

ILB Acronym for Inner Lead Bonding, which see.

immersion plating In galvanic displacement, the chemical deposition of a thin metallic coating over certain basis metals by a partial displacement of the basis metal.

impedance The ratio of the effective value of the potential difference between two terminals to the effective value of the current flow produced by that potential difference.

impingement cooling A method of cooling in which the air is mechanically directed from the cooling fan to strike (impinge) directly on the component to be cooled. Sometimes called *jet impingement.*

impregnant A substance, usually liquid, used to saturate an organic dielectric and to replace the air between its fibers and in pinholes. Impregnation usually increases the dielectric strength and the dielectric constant of the assembly.

impregnation In printed circuit boards, the process of coating the substrate (glass cloth) with a resin solution and drying. The dried product is called *prepreg.*

inclusion A foreign particle, metallic or nonmetallic, in a conductive layer, plating or base material.

indirect-direct emulsion A combination film and sensitized-liquid system for producing stencil screens. Materials are applied to the screen mesh before exposure and development.

indirect emulsion A sensitized-film coating used for producing stencil screens. Materials are exposed and developed before application to the screen mesh.

indirect-emulsion screen A screen whose emulsion is a separate sheet or film of material, attached by being pressed into the mesh of the screen (as opposed to the direct-emulsion type).

inductance The property of an electric circuit whereby it resists any change of current during the building up or decaying of a self-induced magnetic field and hence produces a delay in current change, with resulting operational delay. In packaging systems, it is: (1) used in lumped equivalent circuits to represent part of a line discontinuity; (2) used in a distributed system to represent the electromagnetic storage property of a transmission line; and (3) the cause of delta I noise because it induces an opposing voltage in response to a change in current.

inert atmosphere A gas atmosphere such as helium or nitrogen that is nonoxidizing or nonreducing to metals.

infrared light-emitting diode An optoelectronic device containing a semiconductor pn junction that emits radiant energy in the 0.78- to 100-micrometer wavelength region when forward-biased.

Infrared Reflow (IR) The technique in which primarily long wavelength light is used to heat solder joints to the melting temperature. Normally, a circuit board having prepositioned packages is transported through an IR reflow furnace.

injection laser A solid-state semiconductor device consisting of at least one pn junction capable of emitting coherent or stimulated radiation under specified conditions. The device will incorporate a resonant optical cavity.

injection molded Molding by injecting liquefied plastic into a mold of desired shape.

Injection Molded Card (IMC) Card for electronic packages made by injection molding of plastics into a molded cavity of desired shape.

ink Synonymous with composition and paste when relating to screenable thick-film materials and with solder mask when referring to screen printing.

Inner Lead Bond (ILB) Refers to the electrically interconnecting wiring from the device to the next level package. (See also **outer lead bonding.**)

in process test A set of tests (usually nonfunctional) designed to assure integrity of each individual process involved (in constructing a package) to locate defects in construction at the process step where they are created.

input/output (I/O) terminal A chip or package connector (terminal) acting to interconnect the chip to the package or one package level to the physically adjacent level in the hierarchy. Usually refers to the number of contacts necessary to wire to or interconnect an assembly. Pin out, connections, and terminals are other common words to describe the same. Care must be taken to differentiate between the total number of I/Os between levels, signal I/Os, I/Os used to distribute power and reference I/Os.

insert (1) To assemble components, manually or automatically, into a printed board. (2) A part that holds the contacts in the proper arrangement and electrically insulates them from each other and from the shell. (See also **body.**)

insertion force For a contact system, the maximum force encountered while mating a printed board to a connector.

insulation A nonconductive material usually surrounding or separating two conductive materials.

Insulation Resistance (IR) The resistance to current flow when a potential is applied. IR is measured in megohms. Usually intended to evaluate the leakage integrity between interconnections that are supposed to be electrically isolated.

Insulator Metal Substrate Technology (IMST) A substrate, such as one made of porcelainized steel, which is not subject to size limitations and may have superior thermal dissipation characteristics. IMST refers to insulated-metal substrate technology of Sanyo. It is a single-sided aluminum core with epoxy coating and etched copper wiring.

insulators A class of materials with high resistivity. Materials that do not conduct electricity.

integrated circuit A microcircuit consisting of interconnected elements inseparably associated and formed in situ on or within a single substrate to perform an electronic circuit function.

interchip wiring The conducting wiring path connecting circuits on one chip with those on another chip to perform a function.

interconnection The joining of one individual device with another.

interface The borderline region between two different materials, for example, the region where a thick-film resistor composition and its connecting conductor composition meet, intermingle, react, and so on.

interlayer connection An electrical connection (i.e., plated through holes or via holes) between conductive patterns in different layers of a multilayer printed board. (See also **through connection.**)

intermetallic compound (1) An intermediate phase (a homogeneous phase whose composition range does not include any pure metal) in an alloy system that has a narrow range of composition but has an atomic bonding that can be of several types. (2) A stoichiometric compound whose properties differ considerably from those of the metals that make up the compound in terms of strength, brittleness, and hardness.

internal layer A conductive pattern that is contained entirely within a multilayer printed board.

internal resistance A term used to represent thermal resistance from the junction of a device, inside an electronic package, to a convenient point on the outside surface of the package.

Internal Thermal Enhancement (ITE) A direct thermal path from the device to the package case, usually a thermal grease or metal "finger" between the backside of a flip-chip device that provides thermal enhancement inside the first-level package.

interposer To interpose is to place or come between two objects. In packaging, an interposer usually means an interconnection scheme that allows a set of fine-pitch connections (say on a chip) to connect to a second set of connections that is further apart (say on a lead frame). TAB tape is often used in this manner. In connectors, an interposer can provide a 1:1 interconnection from say, a leadless module, to a printed board (for example, an elastomeric connector).

interstitial via hole (1) A plated-through hole connecting two or more conductor layers of a multilayer printed board but not extending fully through all the layers of base material composing the board. (2) A metallized hole connecting two or more conductor layers in a package that is placed within the normal x-y via grid of the package.

invar A trade name (International Nickel Company, Inc.) for a very low thermal expansion alloy of nickel and iron.

ion migration The movement of free ions within a material or across the boundary between two materials under the influence of an applied electric field.

ionizable (ionic) contaminants Process residues, flux activators, fingerprints, and etching and plating salts, all of which exist as ions and, when dissolved, increase electrical conductivity.

irradiation The exposure of a material to high-energy emissions. In insulations, used to alter the molecular structure favorably.

isolation A technique for electrically separating circuit elements. In dielectric isolation, components are isolated by means of insulating layers. In diode isolation, components are isolated by means of reverse-biased pn junctions.

isopak A unique pin-grid array package (trademark of General Dynamics) consisting of kovar pins sealed in a glass-to-kovar plate flush for chip bonding.

JEDEC Acronym for Joint Electron Device Engineering Council. An Electronic Industries Association body that has published numerous standards for first-level packages.

J-lead A surface-mount integrated circuit package whose leads are formed into a J pattern, folding under the device body.

job A term used to indicate a batch of componentry started through a process as a single group having the same history.

jumper wire An electrical connection that is not part of the original design, added between two points on a printed board after the intended conductive pattern has been formed.

junction (1) A contact between two dissimilar metals or materials. (2) A connection between two or more conductors or two or more sections of a transmission line. (3) In solid-state materials, a region of transition between p- and n-type semiconductor materials as in a transistor or diode.

key A device designed to ensure that the coupling of two components can occur in only one position.

keying slot A slot in a printed board that permits the board to be plugged into its mating receptacle, but prevents it from being plugged into any other receptacle. Also, in manufacturing, often used for locating the printed circuit board into a desired position (x, y and z) for photoprinting or drilling operations.

kiln High-temperature furnace used for firing ceramics.

Kirkendall voids The formation of voids by diffusion across the interface between two different materials, in the material having the greater diffusion rate into the other.

kovar An alloy of 53 percent iron, 17 percent cobalt, 29 percent nickel, and trace elements, with thermal expansion matching alumina substrates and certain sealing glasses. Most commonly used lead frame and pin material. Conforms to the ASTM designation F15.

laminate A product made by bonding together two or more layers of material under heat and pressure to form a single structure.

laminate void The absence of resin in an area that normally contains resin.

lamination The process of consolidating sheets of prepreg under heat and pressure to form a solid product. Applied also to the consolidation of prepregs and precircuitized subcomposites to form a composite.

land A portion of a conductive pattern that is usually, but not exclusively, used for the connection or attachment (or both) of components.

Land Grid Array package (LGA) See **Ball Grid Array Package.**

land pattern A combination of lands intended for the mounting of a particular component.

landless hole A plated-through hole or via without a land.

Large Scale Integration (LSI) Term used to describe semiconductor chips with more than 1000 circuits.

laser bonding Effecting a metal-to-metal bond of two conductors by welding the two materials together using a laser beam for a heat source.

laser soldering The technique in which heat to reflow a solder interconnection is provided by a laser.

lay-up The process of registering and stacking layers of a multilayer structure (e.g., PC board or multilayer ceramic module) in preparation for the laminating cycle.

leaching The migration of components of a substrate into working solution.

lead (1) That portion of an electrical component used to connect it to the outside world. (2) A conductive path, usually self-supporting. (3) A soft, heavy metal (chemical symbol Pb) that is used in solder compositions and other alloys.

lead frame The metallic portion of the device package that completes the electrical connection path from the die or dice and from ancillary hybrid circuit elements to the outside world.

Lead on Chip (LOC) Refers to a package structure in which the inner leads (of a lead frame) are located on the chip so that the wires can be bonded inside the chip area. Package density and performance are greatly improved as wire bond sites do not have to be fixed around the edge of the chip, and power distribution can be optimized.

leaded chip carrier A plastic or ceramic chip carrier with compliant leads for terminations.

Leadless Chip Carrier (LCC) A chip carrier with integral metallized terminations and no compliant external leads.

leaf spring contact A leaf spring contact is a connector element consisting of a conductive member configured as a cantilever beam which applies force through the displacement of the beam.

lift-off Patterning of metal by removal or "lift-off" of surrounding materials usually in a solvent.

line See **conductor.**

line discontinuity A load point, consisting of a lumped equivalent circuit or resistance, capacitance, and inductance anywhere on a transmission line that produces spurious reflections.

line loading Externally connected resistance, capacitance and inductance, or combination thereof on a transmission line.

line resistance Resistance of conductor lines in a package, measured in ohms per unit length or for a given cross section, ohms per square.

lines per channel The number of conductive lines between vias in an organic board or ceramic substrate.

Liquid Crystal Display (LCD) Display technology based on liquid crystal materials whose light transmission is changeable by the application of an electrical field. LCD devices are used in numeric readouts and for flat-screen television receivers.

Logic Service Terminal (LST) A terminal (on a package or package component) carrying logic signals as opposed to one used only for electrical power.

loss factor A factor of an insulating material that is equal to the product of its dissipation and dielectric constant. The tangent of the angle between the real and imaginary current.

lot In manufacturing, this term is similar to a "job," but more than one job may be combined to make up a "lot."

low-loss dielectric An insulating material, such as polyethylene, that has relatively low dielectric loss, making it suitable for transmission of radio frequency energy.

lossy signal line Distributed energy dissipative elements in a transmission line, such as series resistance, skin effect, and dielectric conduction.

lug A term commonly used to describe a termination, usually crimped or soldered to the conductor, with provision for screwing down to a terminal.

mask The photographic negative that serves as the master for making thick-film screens and thin-film patterns.

mass spectrometer An instrument used to determine the leak rate of a hermetically sealed package by ionizing the gas outflow, permitting an analysis of the flow rate in cm^3/sec at 1 atmosphere differential pressure.

master drawing A document that shows the dimensional limits or grid locations applicable to any or all parts of a printed board (rigid or flexible), including the arrangement of conductive and nonconductive patterns or elements; the size, type, and location of holes; and any other information necessary to describe the product to be fabricated.

Mean Time to Failure (MTTF) Applicable to individual parts or devices in reliability technology. It is the arithmetic average of the lengths of time-to-failure registered for parts or devices of the same type, operated as a group under identical conditions.

measling An internal condition occurring in laminated base material in which the glass fibers are separated from the resin at the weave intersections. This condition manifests itself in the form of discrete white spots or "crosses" below the surface of the base material and is usually related to thermally induced stress or chemical attack at the interface.

melt extrude To heat a material above its crystalline melt point and extrude it through an orifice.

meniscus In a solder joint, the minimum angle at which the solder tapers from the joint to the flat area of the conductor.

mesh In a mask, number of openings per lineal inch, measured from the center of any wire.

mesh porosity The amount of open area in a mesh versus the amount of closed area. Also expressed as *percentage of open area.*

mesh size The number of openings per inch in a screen. A 200-mesh screen has 200 openings per linear inch and 40,000 openings per square inch.

metal-clad base material (metal clad laminate) Base material covered with foil on one or both of its sides.

metal migration An undesirable phenomenon whereby metal ions, notably silver, are transmitted through another metal or across an insulated surface, in the presence of moisture and an electrical potential.

Metal-Oxide Semiconductor (MOS) A method of construction distinguishes this type of integrated circuit from bipolar integrated circuits. MOS integrated circuits are slower than bipolar integrated circuits, but have the advantage of high circuit density and low cost. The metal acts as a capacitor plate which, when charged across the oxide, draws the opposite charges to the semiconductor surface (i.e. electrons or holes) and bends the conduction band into the conductive state.

Metal-Oxide Semiconductor Field Effect Transistors (MOSFET) The basic element of MOS integrated circuits. An FET consists of diffused source and drain regions on either side of a p- or n-channel region, and a gate electrode insulated from the channel by silicon oxide.

metallization The process of applying a conductive metal film to the surface on an integrated circuit chip or a package base material.

Metallized Ceramic (MC) Ceramic (fired) substrate metallized with thick and/or thin metal films. (In IBM, metallized ceramic refers to a thin-film copper metallization on a fired alumina substrate).

MHO An electrical unit of conductivity that is the conductivity of a body with the resistance of one ohm.

microcircuit A small circuit having a high equivalent circuit element density that is considered to be a single part composed of interconnected elements on or within a single substrate and that performs an electronic circuit function. (This excludes printed wiring boards, circuit card assemblies, and modules composed exclusively of discrete electronic parts.)

microelectronics The area of electronic technology associated with or applied to the realization of electronic systems from extremely small electronic parts or elements.

micron An obsolete unit of length equal to a micrometer.

microsectioning The preparation of a specimen for the metallographic examination of the material to be examined (usually by cutting out a cross section, followed by encapsulation, polishing, etching, staining, and so on).

microstrip A type of transmission line configuration that consists of a conductor over a parallel ground plane, separated by a dielectric.

microstructure Structural features, such as crystal or phase boundaries, or defects or inhomogeneities within a solid, usually resolved at high magnification.

microwave A short electrical wave, usually a wavelength of less than 300 mm (12 in) or over 1000 MHz.

microwave integrated circuits Distributed passive elements (resistors, capacitors) and interconnections that are used as transmission line components at microwave frequencies and that use the convenience of the open-face nature of the microstrip structure and build them into a compact unit along with the active devices.

mill (1) A machine for grinding or mixing material, for example, a ball mill and a paint mill. (2) Grinding or mixing a material, for example, milling a thick-film composition.

minimum annular ring The minimum width of metal, at the narrowest point, between the edge of the hole and the outer edge of the land. This measurement is made to the drilled hole on internal layers of multilayer printed boards and to the edge of the plating on outside layers of multilayer boards and double-sided boards.

minimum electrical spacing The minimum allowable distance between adjacent conductors that is sufficient to prevent dielectric breakdown, corona, or both between the conductors at any given voltage and altitude.

mismatch A termination having a different impedance from that for which a circuit or cable is designed.

misregistration The lack of dimensional conformity between successively produced features or patterns.

module A chip carrier on which the chip terminals are fed out by various means to terminals spaced to suit the spacing and dimensions of wires on the next higher level of package (i.e., card or board). It may also contain wiring planes and power planes interconnecting several of its chips, and thus be equivalent to a card. A separable unit in a packaging scheme.

modulus of elasticity The ratio of stress to strain in an elastic material.

mold release An organic compound added to a molding compound or powder that migrates to the mold surface to form a waxy layer between the plastic and mold metal and to allow easy removal of the part from the mold.

monolithic integrated circuit An integrated circuit consisting of elements formed on or with a semiconductor substrate with at least one of the elements formed within the substrate.

Monolithic Systems Technology (MST) A hybrid ceramic package technology manufactured in IBM by screen-printing silver palladium conductors onto alumina substrates with pins attached by swaging techniques and chips bonded with solder connections (C4).

monomer A term denoting a single property or ingredient. A molecule of low molecular weight used as a starting material for polymerization to produce molecules of larger molecular weight, called polymers.

mother board A printed board assembly used for interconnecting arrays of plug-in electronic modules or cards.

M-quad® Trademark of Olin Corp. Particular design of a quad flat pack having a metal case for improved thermal properties.

Multichip Module (MCM) A module or package capable of supporting several chips on a single package. The industry recognizes three types of multichip modules depending on their construction:

MCM-C Modules constructed on cofired ceramic substrates using thick-film (screen-printing) technologies to form the conductor patterns.

MCM-D Modules whose interconnections are formed by the thin-film deposition of metals on deposited dielectrics, which may be polymers or inorganic dielectrics.

MCM-L Modules using advanced forms of printed wiring board (PWB) technologies to form the copper conductors on plastic laminated-based dielectrics.

multichip package An electronic package that carries a number of chips and interconnects them through several layers of conductive patterns. Each one is separated by insulative layers and interconnected via holes.

multifilament mesh Woven material with multiple-strand threads.

Multilayer Ceramic (MLC) Ceramic substrate consisting of multiple layers of metals and ceramics interconnected with vias. All with thick film wiring.

multilayer printed board The general term for completely processed printed circuit or printed wiring configurations consisting of alternate layers of conductive patterns and insulating materials bonded together, with conductive patterns in more than two layers, and with conductive patterns interconnected as required. The term includes both flexible and rigid organic multilayer boards and ceramic multilayer boards used to interconnect arrays of integrated circuits.

multiple-image production master A production master that is used in the process of making two or more printed boards simultaneously.

mutual capacitance Capacitance between two conductors when all other conductors, including ground, are connected together and then regarded as an ignored ground.

nail head bond See **ball bond.**

nail heading Term to describe the ductile deformation of the land pattern in a PCB where the drill bit has punched through the copper plane.

negative Artwork, artwork master, or production master in which the intended conductive pattern is transparent to light, and the areas to be free from conductive material are opaque.

negative-acting resist A resist that is polymerized (hardened) by light and that, after exposure and development, remains on the surface of a laminate in those areas that were under the transparent parts of a production master.

net A group of terminals interconnected to have a common dc electrical potential in a package.

nick A cut or a notch in a wire or conductor.

NIP A PCB with No Inner Plane, but with two surface planes of interconnections.

nonfunctional land A land on an internal or external layer that is not connected to the conductive pattern on its layer. But, it may be used for test purposes by connection through a via to the other surface of a card or NIP allowing all circuits to be tested from just one side.

nonwetting (1) The lack of metallurgical wetting between molten solder and a metallic surface due to the presence of a physical barrier on the metallic surface. (2) A condition in which a surface has contacted molten solder, but the solder has not adhered to all of the surface; basis metal remains exposed.

off-contact In printing, the opposite of contact printing in that the printer is set up with a space between the screen and the substrate.

offset land A land that is intentionally not in physical contact with its associated component hole.

ohm A unit of electrical resistance, the resistance being that of a circuit in which a potential difference of one volt produces a current of one ampere.

ohmic contact A contact between two materials across which the voltage drop is the same regardless of the direction of current flow.

oil canning The movement of entry material in the "z" direction during drilling in concert with the movement of the pressure foot.

OLB See **Outer Lead Bonding.**

opaquer A material that, when added to the resin system, renders the laminate sufficiently opaque that the yarn or weave of the reinforcing material cannot be seen with the unaided eye, using either reflected or transmitted light.

optical interconnection Components and modules composed of optoelectronic devices that are used as circuit building blocks. Interconnections consist of the conversion of electrical energy to light and vice versa through optoelectronic elements.

optoelectronic device A device that detects and/or is responsive to electromagnetic radiation (light) in the visible, infrared, and/or ultraviolet spectra regions; emits or modifies noncoherent or coherent electromagnetic radiation in these same regions; or utilizes such electromagnetic radiation for its internal operation.

Outer Lead Bonding (OLB) The process of joining the outer leads of a package to the next level of assembly.

outgassing Deaeration or other gaseous emission from a printed board assembly (printed board, component, or connector) when exposed to a reduced pressure, or heat, or both.

outgrowth The increase in conductor width at one side of a conductor, caused by plating buildup over that delineated on the production master.

overcoat A thin film of insulating material, either organic or inorganic, applied over circuit elements to provide mechanical protection and/or prevent contamination.

overglaze A glass coating over another component or element, normally used to give physical or electrical protection.

overhang The sum of outgrowth and undercut. If an undercut does not occur, only the overhang is the outgrowth.

overheated solder connection A solder connection characterized by solder surfaces that are dull, chalky, grainy, and porous or pitted.

overlay One material applied over another material.

Overmolded Pad Array Carrier (OMPAC) A new type of solder ball I/O package developed by Motorola. The OMPAC is made from a BT resin substrate, with bottom side solder balls attached, typically on a 0.040- to 0.060-in grid. After device attach, the substrate is overmolded (transfer molding only on the die side of the package).

ozone test Exposure of material to a high concentration of ozone to give an accelerated indication of degradation expected in normal environments.

package In the electronics/microelectronics industry, an enclosure for a single element, an integrated circuit, or a hybrid circuit. It provides hermetic or nonhermetic protection, determines the form factor, and serves as the first-level interconnection externally for the device by means of package terminals. A package generally consists of a bottom part, called the case or header, and a top part, called the cover or lid. These are sealed into one unit. Passive parts may be enclosed in an encapsulant or molded package.

package delay Time delays associated with the interconnections between components in circuits that make up a logical function.

packaging density Quantity of functions (components, interconnection devices, mechanical devices) per unit volume, usually expressed in qualitative terms, such as high, medium, or low.

packaging level A member of a nested interconnected packaging hierarchy (e.g., chip, chip carrier, card, board in order—low to high level).

pad The metallized area on a substrate or on the face of an integrated circuit used for making electrical connections. (See also **land.**)

pad-grid array package See **ball grid array package.**

panel A rectangular or square base material of predetermined size intended for or containing one or more printed boards.

parasitic An inductive, resistive, or capacitive contribution to a circuit that arises from the circuit configuration, as opposed to the design values of deliberately introduced components. Examples are the inductance of conductors, the resistance of interconnections and contacts, and the capacitance between conductors or between regions with a component.

passivation The formation of an insulating layer directly over the semiconductor surface to protect the surface from contaminants, moisture, or particles. Usually an oxide of the semiconductor is used; however, deposition of other materials is also used.

passive chip components Resistor, resistor array, capacitor, and inductor chips, of thick- or thin-film (multilayer) construction, intended for surface mounting on printed boards or hybrid substrates, having top or back electrical contacts, or both.

passive components (elements) An electronic circuit element that displays no gain or control, such as a resistor, inductor, or capacitor.

paste Synonymous with composition and ink when relating to screenable thick-film materials.

pattern The configuration of conductive and nonconductive materials on a panel or printed board. Pattern also denotes the circuit configuration on related tools, drawings, and masters.

pattern plating The selective plating (usually electrolytic) of a conductive layer into a photoresist pattern.

peel bond Similar to lift-off of the bond, with the separation of the lead from the bonding surface proceeding along the interface of the metallization and substrate insulation rather than the bond-metal surface.

peel strength The force per unit width required to peel the conductor or foil from the base material.

permanent mask A resist that is not removed after processing (for example, plating resist used in the fully additive process).

permeability The property of a material, such as solid plastic, that allows penetration by a liquid or gas.

phase diagram State of a metal alloy or ceramic over a wide temperature range. The phase diagram is used to identify phases as a function of composition and temperature.

Phosphosilicate Glass (PSG) Phosphorous-doped silicon dioxide (also known as *P-glass*). It is often used as a dielectric material for insulation between conducting layers, for inhibiting the diffusion of sodium impurities, and for planarization since it softens and flows at 1000 to 1100°C to create a smooth topography for subsequent operations.

photographic reduction dimension A dimension (for example, the distance between lines or between two specified points) on the artwork master to indicate to the photographer the extent to which the artwork master is to be photographically reduced. (The value of the dimension refers to the 1:1 scale and must be specified.)

photolithography The use of light or ultraviolet rays through a mask to define circuit conductor patterns.

photopolymer A polymer that changes characteristics when exposed to light of a given frequency.

photoresist A photosensitive coating that is applied to a laminate and subsequently exposed through a photo tool (film) and developed to create a pattern that can be either plated or etched.

photosensitive Sensitive to light.

phototransistor A transistor (bipolar or field effect) that is used to detect light or that has been optimized for light sensitivity. The base region or gate may or may not be connected to an external terminal.

physical design The mechanical layout of the package including device and circuit element placement.

pick-and-place The mechanical process of selecting and placing chips on the correct substrate site in preparation for joining (interconnecting) the chip to the substrate.

piezoelectric Electricity or electric polarization caused by stress.

pin Electrical terminal and/or mechanical support that has a round cross section.

pin density The quantity of pins on a printed circuit board per unit area.

pin-grid array A package or interconnect scheme, featuring a multiplicity of plug-in type electrical terminals arranged in a prescribed matrix format or array. Used with high input/output-count packages.

pinhole A small hole occurring as an imperfection that penetrates entirely through a layer of material.

pin-in-hole mounting Method of assembly of printed boards whereby the component pins are inserted through holes in the board and mechanically secured before being soldered in place.

pink ring The change in color around a copper pattern where it has lost adhesion to a polymer due to chemical intrusion of the interface oxide layer.

Pin-Through-Hole (PTH) Refers to the class of packages or modules whereby the leads are soldered in plated through holes in the next level package (usually a printed circuit board).

pit A small hole occurring as an imperfection that does not penetrate entirely through the foil.

pitch The nominal distance from center to center of adjacent conductors. Where conductors are of equal size and spacing is uniform, the pitch is usually measured from the reference edge of the adjacent conductor.

placement The actual placing of chip circuits, chips, chip carriers and/or cards in the desired location on corresponding images for a given package level.

plain weave Mesh pattern in which the filaments are woven with one over, one under.

planar package Usually refers to a simple system having one card (sometimes two cabled together) which holds all the modules or electronics for the system. Larger systems usually use Card-on-Board (COB) packaging where several cards are plugged edgewise into a back-panel printed board.

plasma etching The action of an electrically conductive gas, or plasma (composed of ionized gas or molecules), to remove unwanted portions of conductive or insulative pattern.

plastic A polymeric material, either organic, such as epoxy and polyimide, or inorganic, such as silicone, used for conformal coating, encapsulation, or overcoating.

plastic ball grid array package A surface mount package consisting of a plastic substrate whose input/output connections consist of solder balls in an array on the bottom of the package. (See **ball grid array package.**)

plastic encapsulation Environmental protection of a completed circuit by embedding it in a plastic such as epoxy or silicone.

Plastic Dual In-line Package (PDIP) A molded plastic (over lead frame) package with leads on both sides that are all in alignment. Generally used for pin-through-hole type of assembly. (See also gullwing package.)

Plastic Leaded Chip Carrier (PLCC) A molded plastic and lead frame package designed for surface mount applications. The leads, which may be on all four sides of the package are formed in the shape of a "J."

plasticizer A chemical agent added in compounding plastics and/or ceramic "green" bodies to make them softer and more flexible.

plate finish In laminating, the finish present on the metallic surface of metal-clad base material resulting from direct contact with the laminating press plates without modification by any subsequent finishing process.

plated-through hole A hole in which an electrical connection is made between internal and/or external conductive patterns, by the plating of metal on the wall of the hole.

plated-through hole structure test A visual examination of the metallic conductors and plated-through holes of a printed board after the glass-plastic laminate has been dissolved away.

plating (1) Metallic deposit on a surface, formed either chemically or electrochemically. (2) The process of the chemical or electrochemical deposition of metal on a surface.

plating bar The temporary conductive path interconnecting areas of a printed board to be electroplated, usually located on the panel outside the borders of such a board.

plating up The process consisting of the electrochemical deposition of a conductive material on the base materials (surface holes, and so on) after the base material has been made conductive.

plug The part of a connector that is normally removable from, or permanently mounted to, the other part; usually that half of a two-piece connector that contains the pin contacts.

pogo pin A pogo contact is a connector element that consists of a pin assembly containing an integral coiled spring which applies the contact force.

point-to-point wiring An interconnecting technique in which the connections between components are made by wires routed between connection points.

polar ingredient Any ingredient in a material or complex capable of ionization.

polar solvents Solvents with an electrical dipole that are capable of dissolving polar compounds and/or ionics.

polarization A technique of eliminating symmetry within a plane so that parts can be engaged in only one way in order to minimize the possibility of electrical and mechanical damage or malfunction.

polarized slot A slot at the edge of a printed board used to ensure proper insertion and location in a mating connector. (See also **keying slot.**)

polyimides A class of resin compounds containing the NH group which are derived from ammonia and are "imidized" from polyamic acid at temperatures high enough to initiate and complete the imide ring closure. Polyimides are often used as organic dielectric interlevel layers.

polymer A material having molecules of high molecular weight, formed by the polymerization of lower molecular weight molecules.

polymer thick film A class of materials, used for thick-film inks and polymer-based substrates. Also, a technology in which conductive and resistive inks of a polymer base are screen-printed onto a dielectric to create a circuit.

polymerization A chemical reaction in which the molecules of a monomer are cross-linked to form large molecules whose molecular weight is a multiple of that of the original substance.

porcelain A mixture of borosilicate glass with minor quantities of zirconia and other ingredients. Often referred to as *enamel.*

positive An artwork, artwork master, or production master in which the intended conductive pattern is opaque to light and the areas intended to be free from conductive material are transparent.

positive-acting resist A resist that is decomposed (softened) by light and that, after exposure and development, is removed from those areas that were under the transparent parts of a production master.

pot life The length of time a two-part epoxy system remains useful, usually measured in hours; it can be extended by refrigeration.

potting See **encapsulate.**

power cycling Used for reliability testing of finished assemblies by the application of cyclic power to a heat-generating component in the assembly.

power distribution The network of conductors throughout the package which supplies the operating voltages and currents to the circuits.

prefire To fire one thick-film composition, for example, a conductor, before printing a second thick-film composition, for example, a resistor.

preform For soldering or adhesion functions, a form that is circular or square-shaped and is punched out of thin sheets of solder, epoxy, or eutectic alloy. The form is placed on the spot of attachment by soldering or by bonding prior to placing the object there to be heated.

prepreg Sheet material, such as glass fabric impregnated with a resin cured to an intermediate stage (B-stage resin).

pressed alumina Aluminum oxide ceramic formed by applying pressure to the ceramic powder and a binder prior to firing in a kiln.

press-fit contact An electrical contact that can be pressed into a hole in an insulator, printed board (with or without plated-through holes), or a metal plate.

pressure contact Method of interconnection whereby the electrical path is maintained by means of a continually applied force (such as a spring) to the contact points.

print-through An undesirable condition that is typical with surface-mount designs on double-sided, noninner plane, rigid printed boards or cores, in which the image on one side is printed through the substrate to the other side.

printed board A general term for completely processed printed circuit or printed wiring configurations. It includes rigid or flexible boards (organic or ceramic) and single, double, and multilayer printed boards. (This definition and the definitions below are based on ANSI/IPC-T-50C, Ref. 1.)

printed wiring board A board with printed-on point-to-point connections (only).

printed circuit board A board with printed-on components as well as point-to-point connections.

printed board assembly A general term used to denote either a printed wiring assembly or a printed circuit assembly.

printed circuit A conductive pattern comprising printed components, printed wiring, or a combination thereof, all formed in a predetermined design and intended to be attached to a common base.

printed component A part, such as an inductor, resistor, capacitor, or transmission line, that is formed as part of the conductive pattern of the printed board.

printed wiring The conductive pattern intended to be formed on a common base, to provide point-to-point connection of discrete components, but not to contain printed components.

printed wiring assembly A printed wiring board on which separately manufactured components and parts have been added.

probe A pointed conductor used in making electrical contact to a circuit pad for testing.

production master A 1:1 scale pattern that is used to produce one or more printed boards (rigid or flexible) within the accuracy specified on the master drawing. See also **single-image production master** and **multiple-image production master.**

profile In reference to firing, a graph of time versus temperature or, in a continuous thick-film furnace, of position versus temperature.

profilometer An instrument for measuring surface roughness.

propagation delay The time it takes a signal to leave the output terminal of one chip, travel across the package, and enter the input terminal of another chip.

pseudoplastic A characteristic of a fluid whereby its viscosity decreases as the shear rate is increased. This does not imply a change in behavior with time. This type of behavior is often obtained when solids are dispersed in organic vehicles (for example, thick-film compositions).

pulse soldering Soldering a connection by melting the solder in the joint area by pulsing current through a high-resistance point applied to the joint area and the solder.

purple plague One of several gold-aluminum compounds formed when bonding gold to aluminum and activated by reexposure to high temperature. Compound growth is highly enhanced by the presence of silicon to form ternary compounds. The compounds, which are purplish in color, are very brittle, potentially leading to time-dependent failure of the bonds.

pyrolyzed Of a material, characterized by gaining its final form by the action of heat (burning).

q The inverse ratio of the frequency band between half-power points (bandwidth) to the resonant frequency of the oscillating system. Refers to the electromechanical system of an ultrasonic bonder, or the sensitivity of the mechanical resonance to changes in driving frequency.

Quad Flat Pack (QFP) Ceramic or plastic chip carrier with leads projecting down and away on all four sides of a square package. Commonly used to describe chip carriers with gull wing leads.

Quad In-line Package (QUIP) A diplike plastic package with leads coming out in two rows. Half of the leads are bent close to the body and the other half projected out for additional 1.27 mm before being bent down.

quick disconnect A type of connector that permits rapid locking and unlocking of two connector halves.

radiation (1) The combined process of emission, transmission, and absorption of thermal energy between bodies separated by empty space. (2) The emission of electromagnetic energy from an electrical signal in a conductor.

Radio Frequency Interference (RFI) Undesired conducted or radiated electrical disturbances, including transients, that may interfere with the operation of electrical or electronic communications equipment or other electronic equipment.

random mesh registration A mask or screen in which the mesh holes are not positioned in any specific location relative to the cavity pattern.

rated voltage That voltage at which a component can operate for extended periods without undue degradation or safety hazard.

reactance That part of an impedance of an alternating current circuit that is due to capacitance or inductance.

Reaction Injection Molding (RIM) A molding process where two (or more) streams of reactants are metered into a small mixing chamber where turbulent mixing breaks up the fluids into finely interspersed striations for faster reaction. The mixture is then delivered to a mold to complete the polymerization.

real estate The surface area of an integrated circuit or substrate. The surface area required for a component or element.

receptacle Usually, the fixed or stationary half of a two-piece multiple-contact connector. Also, the connector half, usually mounted on a panel and containing socket contacts.

receiver The off-chip circuit that accepts the signal voltages and currents from the package lines. Also called an *input buffer circuit.*

reflow soldering A process for joining parts by tinning the mating surfaces, placing them together, heating until the solder fuses, and allowing them to cool in the joined position.

registration The accuracy of concentricity or relative position of all patterns on any mask with the corresponding patterns of any other mask of a given device series when properly superimposed.

registration marks The marks on a wafer or substrate that are used for aligning successive processing masks. Also known as *alignment marks* or *fiducial marks.*

reliability A collective name for those measures of quality that reflect the effect of time in storage or use of a product, as distinct from those measures that show the state of the product at the time of delivery. Generally, it is the capability of an item to perform a required function under stated conditions for a stated period of time.

Rent's rule An empirical relation, first recorded by E. Rent of IBM, which states that the number of used input/output terminals on a logic package is proportional to a fractional power of the number of subpackages interconnected in the package.

repairing The act of restoring the functional capability of a defective part without necessarily restoring appearance, interchangeability, or uniformity.

resin An organic polymer that cross-links to form a thermosetting plastic when mixed with a curing agent.

resin recession The presence of voids between the barrel of a plated-through hole and the wall of the hole, seen in microsections of plated-through holes in boards that have been exposed to high temperatures.

resin-rich area A significant thickness of a nonreinforced resin layer of the same composition as that within the base material.

resin smear Resin transferred from the base material onto the wall of a drilled hole, covering the exposed edge of the conductive pattern, normally caused by drilling.

resin-starved area An area in a printed board that has an insufficient amount of resin to wet out the reinforcement completely; evidenced by low gloss, dry spots, or exposed fibers.

resist Coating material used to mask or protect selected areas of a pattern from the action of an etchant, solder, or plating.

resistance The property of a conductor that opposes the flow of current by dissipating energy as heat. In packages, it causes voltage and current losses in signal and power distribution systems.

resistance soldering A method of soldering in which a current is passed through and heats the soldering area by contact with one or more electrodes.

resistivity The ability of a material to resist the passage of electric current either through its bulk or on a surface. The unit of bulk resistivity is the ohm-centimeter. The unit of surface resistivity or line resistance is the ohm. A convenient way to express the resistivity of film conductors is in sheet resistivity, which is the electrical resistance measured across the opposite sides of a square of deposited film material. Sheet resistivity is expressed in ohms per square.

resistor A device that offers resistance to the flow of electric current in accordance with Ohm's law: $R = E/I$, where R is resistance, E is voltage, and I is current.

resistor termination A thick-film conductor pad that overlaps and makes contact with a thick-film resistor area.

resolution The fineness of detail of a screen-printed pattern. The ability of photographic materials to reproduce fine detail.

reverse image The resist pattern on a printed board that allows the exposure of conductive areas for subsequent plating.

reversion A chemical reaction in which a polymerized material degenerates at least partially to a polymeric state lower than that of the original monomer.

It is usually accompanied by significant changes in physical and mechanical properties.

rheology The science dealing with the deformation and flow of matter.

rifling Spiral groove or ridge in the substrate due to filling.

right-angle edge connector A connector that terminates conductors at the edge of a printed board, while bringing the terminations out at right angles to the plane of the board conductors.

rise time The time during which the leading edge of a pulse increases from 10 to 90 percent of its maximum amplitude.

roadmap A printed pattern of nonconductive material by which the circuitry and components are delineated on a board to aid in service and repair of the board.

roller coating Coating process in which liquid is transferred to a soft, grooved roller and then to the surface of a printed board.

rosin flux A flux having a rosin base that becomes interactive after being subjected to the soldering temperature.

roughness The microscopic peak-to-valley distances of film-surface protuberances and depressions, measured in angstroms.

routing program An automatic program containing an algorithm which "routs" or places prescribed wiring within a package. Usually used with printed circuit board layout.

rupture In breaking-strength or creep tests, the point at which a material physically comes apart.

sawtooth Small projections on the edge of a screen-printed pattern caused by fillets of emulsion left in the screen mesh and by the mesh itself.

screen A network of metal or fabric strands mounted snugly on a frame and upon which the film circuit patterns and configurations are superimposed by photographic means.

screen printing A process for transferring an image to a surface by forcing suitable media through a stencil screen with a squeegee. The synonym *silk screening* is a nonpreferential term.

screen tension The tautness of a mounted woven mesh, expressed in mils per unit of weight.

sealing Joining the package carrier base with its cover or lid into a sealed unit. For hybrids, sealing connotes an important finishing operation in fabricating a hybrid microcircuit, signaling the stage when the assembly, in the form of a populated package, becomes a bona fide hermetic (or nonhermetic) entity.

seeding The use of a colloid metal or compound to form metallic nuclei, which become the activated sites for initiation of plating (either electrolytic or electroless).

semiadditive process An additive process for obtaining conductive patterns that combines an electroless metal deposition on an unclad or thin-foil substrate with electroplating, etching, or both. (See also **additive process** and **subtractive process.**)

semiaqueous Aqueous chemistry with the addition of an organic solvent.

semiconductor device A device whose essential characteristics are governed by the flow of charge carriers within a semiconductor.

sensitizer Material used to activate a photographic resist to create chemical changes in it.

sequentially laminated multilayer printed board A multilayer board that is formed by laminating through-hole plated double-sided or multilayer boards together. The circuitry layers are interconnected with interstitial via holes and through connections.

Seraphim's theory States that the wiring capacity of a printed board is approximated by the following equation $W_c = 2.25N_tP$, where N_t is the number of terminals per device (module) mounted on the board and P is the pitch (spacing) between them.

shear rate With regard to viscous fluids, the relative rate of flow or movement.

sheath The material, usually an extruded plastic or elastomer, applied outermost to a wire or cable. Also called a *jacket.*

sheet resistance The electrical resistance of a thin sheet of a material with a uniform thickness as measured across opposite sides of a unit square pattern. Expressed in ohms per square.

shelf life The length of time a one- or two-part epoxy system can be stored before use, usually measured in months.

shell An outside case, usually metallic, into which the insert (body) and contacts are assembled. Shells of mating connector halves usually allow proper alignment and polarization and provide protection for projecting contacts.

signal An electrical impulse of predetermined voltage, current, polarity, and pulse width.

signal conductor An individual conductor used to transmit an impressed signal.

signal distribution The network of package conductors that interconnects the drivers and receivers.

signal plane A conductor layer intended to carry signals, rather than serve as a ground or in some other fixed-voltage function.

silk screening See **screen printing.**

silver migration The growth of silver (Ag) crystals between a silver bearing or coated conductor anode and a cathode a few mils apart in a circuit when a dc voltage is applied over a long period of time under conditions of high humidity. Alloying the Ag with a small quantity of a higher melting temperature metal greatly reduces this tendency. (See also **electromigration.**)

simultaneous switch Two or more driver circuits on the same chip or nearby chips changing state in unison thereby creating switching noise on package lines and drawing current simultaneously. When there are many such circuits switching, the large in-rush of current may drag down the voltage at the device terminals creating so-called delta I noise (which see).

single chip module (SCM) A package supporting one chip, as opposed to a multichip which supports several.

single-image production master A production master used in the process of making a single printed board.

Single-In-line (SIP) A plastic molded lead frame package whereby all the leads come off one edge of the package in one uniform pattern (line).

Single In-line Memory Module (SIMM) Usually a small printed circuit board with eight or nine small outline memory modules with contacts on one long edge designed to plug into the next-level package.

Single-Layer Metallization package (SLAM) Ceramic leadless package without cavity, sealed by ceramic or glass to a ceramic cap.

single-sided board A printed board with a conductive pattern on one side only.

singulated A term used to describe the separation of packages produced in multiple or reel format into a single unit to aid in further processing.

sinter To heat without melting, to cause a refractory dielectric material to become a rigid body free of binders, contaminants, and so on.

skin effect At very high frequencies, current travels only on the surface layer of a conductor, not uniformly throughout; this region of current flow is called the *skin*. The skin depth is inversely proportional to the square root of the frequency.

skived tape Tape shaved in a thin layer from a cylindrical block of material.

slip casting See **tape casting.**

slivers Portions of plating overhang (solder, tin, gold, and so on) on conductor edges that are partially or completely detached.

slump A spreading of printed thick film or solder paste after screen printing but before drying. Too much slumping results in a loss of definition.

slurry A thick mixture of liquid and solids, the solids being in suspension in the liquid.

Small Outline J lead (SOJ) Small outline package with "J" leads for surface mount.

Small Outline Package (SOP) Also called *SOIC* (small outline integrated circuit package). A plastic molded lead frame package with leads on two sides similar to a dual-in-line package but smaller with leads on 1.27-, 1.0-, or 0.85-mm spacing. It is meant for surface mounting.

smear removal See **hole cleaning.**

snap-off distance The screen printer distance setting between the bottom of the screen and the top of the substrate.

soak time The length of time a ceramic material, for example, a substrate or thick-film composition, is held at the peak temperature of the firing cycle.

socket contact A female contact designed to receive and mate with a male contact. It is normally connected to the live side of a circuit.

soft error A memory state error induced by a process which produces no permanent alteration of the physical condition of the memory device. Soft errors can be created by alpha particles passing through the device.

soft glass Glasses, typical high-lead content glasses, having a low softening points that could be used to seal ceramic or metal lids to packages below about 450C. Also called solder glasses because of their ability to wet most metal surfaces.

softening point Refers to the temperature at which the log viscosity of glass is 7.6 poises, as defined and measured to ASTM specification.

solder A low melting point alloy, usually of lead(Pb)-tin(Sn), that can wet copper, conduct current, and mechanically join conductors and so on.

solder balls Small spheres of solder adhering to laminate, mask, or conductor surfaces (generally after wave or reflow soldering).

solder bumps The round solder balls bonded to a transistor contact area and used to make connection to a conductor by face-down bonding techniques.

solder coat A layer of solder applied directly to the printed board conductive path from a molten solder bath.

solder column package Devised by IBM, this first level package looks identical to a ceramic pin grid array package. However, instead of hard metal pins, the input/output terminals consist of columns of solder up to 0.150″ long. The package is surface mounted to a printed circuit board by reflowing a selected amount of solder around the base of the columns. The long solder columns provide stress relief between the ceramic package and the board. It is designed for large ceramic chip carriers (e.g., 35 mm to 64 mm on a side).

solder connection An electrical/mechanical connection that employs solder for the joining of two or more metal parts.

solder cup terminal A metallic termination device that has a hollow, cylindrical feature, open on one end, to accommodate the soldering of one or more leads or wires.

solder dam A dielectric composition screened across a conductor to keep molten solder from spreading further onto solderable conductors.

solder eye A solder-type terminal provided with a hole at its end through which a wire can be inserted prior to being soldered.

solder fillet A blended or meniscoid (rounded) configuration of solder around a component or wire lead and land.

solder hierarchy In a complex package assembly, some components have to be assembled (soldered) before others, so these usually utilize a higher melting point solder than later assembled components. Ideally, if the assembly is soldered to a board, the board connection is made with the lowest melting point solder so that previous solder joints are not reflowed or negatively affected by the last joining process.

solder leveling A solder coating process in which heated gas or other media level and remove excess solder after the substrate is dipped in molten solder.

Solid Logic Technology (SLT) Ceramic package technology practiced by IBM in the 1960s by firing AgPd conductors onto dry-pressed and fired alumina substrate.

solder mask coating See **resist.**

solder plugs Cores of solder in the plated-through holes of a printed board.

solder projection An undesirable protrusion of solder from a solidified solder joint or coating. Also called *icicle.*

solder side On boards with components on one side only, the side of a printed board that is opposite the component side.

solder webbing A continuous film or curtain of solder that is parallel to, but does not necessarily adhere to, a surface pattern, or that is between separate conductive patterns that should be free of solder.

solder wicking The capillary rise of solder between individual strands of stranded wire.

solderability The ability of a conductor to be wetted by solder and to form a strong bond with the solder.

soldering A process of joining metallic surfaces with solder, without melting the base material.

soldering flux See **flux.**

soldering oil (blanket) Liquid formulations used in oil-intermix wave soldering equipment. Also used as pot coverings on still and wave solder pots to eliminate dross and reduce solder surface tension.

solderless wrap See **wire wrap.**

solid state Pertaining to circuits and components using semiconductors as substrates.

space transformer A package transforming a spatially dense set of chip connections to a less dense set of connection points on the next level package.

specific mesh registration A mask or screen in which the mesh holes are carefully aligned to correspond with the apertures of the stencil cavity.

spice A simulation program for integrated circuit analysis that is the industry standard for circuit simulation.

spinning A process for coating a smooth surface with a uniform film. Usually used to coat a semiconductor wafer with a photosensitive emulsion by placing the wafer on a rotating chuck and dropping the emulsion on the surface. The combination of centrifugal acceleration and adhesion of the liquid forms a uniform film of emulsion on the surface. Also used to provide thin coatings (for example, of polyimide dielectric) on package elements.

splay The tendency of a rotating drill bit to drill off-center, out-of-round, nonperpendicular holes.

sputter cleaning Bombardment of a surface with energetic argon or other noble gas ions to clean the surface of oxide films and residues that could interfere with subsequent electrical or mechanical contact layers. The bombardment knocks off (or sputters) surface atoms to render the surface clean.

sputtering The ejection of particles from the surface of a material resulting from bombardment by ions and atoms. The material may be used as a source for deposition.

squeegee The block of a screen printer that pushes the liquid composition across the screen and transfers the images through the mesh onto the substrate.

stair stepping Irregular edge definition of a screen-printed pattern caused by interference between the mesh openings and the stencil cavity.

stamped printed wiring Wiring that is produced by die stamping and is bonded to an insulating base.

static flex Flexible wiring circuit carrier, which once installed, remains fixed.

stencil The emulsion or metal layer in which the image apertures are reproduced.

step exposure A technique in which a series of exposures is made to determine the optimum amount of time and distance required for exposing a photosensitive material with any given light source.

step soldering The technique for sequentially soldering connections using solder alloys with different melting temperatures.

stitch bond A thermocompression bond in which a capillary tube is used for feeding the wire and forming the bond.

storage temperature The temperature at which a device, without any power applied, is stored. Usually a maximum storage temperature is specified.

straight-through lead A component lead that extends through a hole in a printed board without subsequent forming of the lead.

strand A single, uninsulated wire.

strand lay The distance of advance of one strand of spirally stranded conductor, in one turn, measured axially.

stress corrosion Refers to the degradation of mechanical properties of brittle materials by crack propagation due to the acceleration of applied stress in the presence of corroding atmospheres such as water. The phenomenon exists in polymers as well as metals.

stress relief The formed portion of a component lead or wire lead, providing sufficient compliancy to minimize stress between terminations. Encapsulation with polymers can provide stress relief.

strip line A type of transmission line configuration that consists of a single, narrow conductor parallel and equidistant to two parallel ground planes. Also, a microwave conductor on a substrate.

substrate In the hybrid industry, that which is used as a base material, usually aluminum oxide (alumina).

subtractive process A process for obtaining conductive patterns by the selective removal of unwanted portions of a conductive foil, usually by chemical etching.

supported hole A hole in a printed board, the inside surface of which is plated or otherwise reinforced.

Surface Mount Technology (SMT) A method of electrically and mechanically connecting components to the surface of a conductive pattern (e.g., as on a printed circuit board) without using through holes.

swaged leads (1) Component lead wires that extend through the printed board and are flattened, or swaged, to secure the component to the board during manufacturing operations. (2) Leads that are inserted through a ceramic substrate and then formed (swaged) to create a lead bulge on both the top and bottom of the substrate, providing a good mechanical joint to the substrate.

swimming Lateral shifting of a thick-film conductor pattern on molten-glass cross-over patterns. Also, lateral movement of etched inner-layer features during lamination of multilayer boards as a result of excess temperature or pressure or both.

switching noise An induced voltage on the power distribution system at the circuit terminals caused by the rapidly changing current involved in the simultaneous switching of many drivers.

Tape Automated Bonding (TAB) The process where silicon chips are joined to patterned metal on polymer tape (e.g., copper on polyimide) using thermocompression bonding, and subsequently attached to a substrate or board by outer lead bonding. Intermediate processing may be carried out in strip form through operations such as testing, encapsulation, burn-in, and excising the individual packages from the tape.

tape bonding The utilization of a metal or plastic tape material as a support to and carrier of a microelectronic component in a gang bonding process.

tape cable A form of multiple conductor consisting of parallel metal strips embedded in insulating material. Also called *flat, flexible cable.*

tape casting In the hybrid industry, the thin, tapelike appearance of ceramic slurry when poured, usually through a flattening device called a "doctor blade," after drying by evaporation of the suspension medium. This method results in long (can be hundreds of feet), thin (typically 5 to 25 mils), ceramic tape that can be cut to size for optimum processing, metallized, via holes punched and filled with metal, and laminated into multilayer structures for final firing. Also called *slip casting.*

taped components Components attached to continuous tape to facilitate automatic component incoming inspection, lead forming, assembling, and testing.

Tape Pak Trademark of National Semiconductor. Particular design of a quad flat pack.

taper pin A pin-type terminal having a tapered end designed to be impacted into a tapered hole to form a connection.

taper tab A flat terminal having tapered sides designed to receive a mating tapered female terminal.

tarnish A discoloration or staining of a conductor or shield wire, caused by exposure to the atmosphere.

temperature coefficient of resistance The amount of resistance change of a material per degree of temperature rise.

tensile strength The greatest longitudinal tensile stress a substance can bear without tearing apart or rupturing.

tension set The condition in which a plastic material shows permanent deformation caused by a stress, after the stress is removed.

tenting A printed board fabrication method of covering over plated-through holes and the surrounding conductive pattern with resist, usually dry film.

terminal A metallic termination device used for making electrical connections.

terminal pad See **land.**

test board A printed board suitable for determining the acceptability of the board or of a batch of boards produced with the same process.

test coupon A portion of the quality conformance test circuitry used for a specific acceptance test or group of related tests.

test pattern A pattern used for inspection or testing purposes.

thermal aging Exposure to a given thermal condition or a programmed series of conditions for prescribed periods of time.

Thermal Conduction Module (TCM) An IBM multichip (100 chips or more) module that is cooled by thermal conduction of pistons in contact with the chips.

thermal conductivity The rate with which a material is capable of transferring a given amount of heat through itself.

thermal cycling A method whereby a cyclic stress is imposed on an assembly of microelectronic components by alternately heating and cooling in a chamber. It is used to accelerated reliability testing of components and assemblies.

thermal mismatch Difference in thermal coefficients of expansion of materials which are bonded together.

thermal resistance The thermal resistance of a unit cube of material.

thermocompression bonding (TC bonding) A process involving the use of pressure and temperature to join two materials by interdiffusion across the boundary.

Thermogravimetric Analysis (TGA) A technique that measures material weight change as a function of increasing temperature.

Thermomechanical Analysis (TMA) A technique that measures the linear expansion or other deformations of a material with respect to changes in temperature.

thermoplastic A resin that is set into its final shape by forcing the melted base polymer into a cooled mold or through a die, after which it is cooled. The hardened polymer can be remelted and reprocessed several times.

thermoset A resin that is cured, set, or hardened, usually by heating, into a permanent shape. The polymerization reaction is an irreversible reaction known as *cross linking*. Once set, a thermosetting plastic cannot be remelted, although most soften with the application of heat.

thermosonic bonding (T/S) A bonding process which uses a combination of thermocompression (TC) bonding and ultrasonic bonding. It is done on what amounts to a gold-wire TC bonder with ultrasonic power applied to the capillary.

thick film A film deposited by screen printing processes having between 5 and 20 micrometer thickness, and fired at high temperature to fuse into its final form.

thin film Refers to coatings that generally range from a few (2–3) atomic layers to a few (1–5) micrometers in thickness. The key distinguishing feature from thick films is that thin films are usually deposited by evaporation, sputtering or chemical vapor deposition.

thin-film packaging An electronic package in which the conductors and/or insulators are fabricated using deposition and patterning techniques similar to those used for integrated circuit chips.

Thin Small Outline Package (TSOP) A small outline package (which see) whose thickness is much less than a standard package. TSOPs can range from 1.5 down to only 0.5 mm in thickness.

thixotropic Pertaining to the tendency of a fluid to decrease in viscosity as the time to exposure to a given shear rate increases. The shear stress versus shear strain rate curve of a thixotropic material should show a hysteresis loop. A purely pseudoplastic material will not give a hysteresis loop because this property is not time-dependent. Most thick-film pastes or inks are thixotropic.

three-layer tape An interconnection medium used in tape automated bonding (TAB), where the tape is comprised of three layers: metallization (usually copper), polymer (usually polyimide), and adhesive (usually epoxy) in between.

through connection An electrical connection between conductive patterns on opposite sides of an insulating base, such as a plated-through hole or clinched jumper wire.

through-hole mounting See **pin-in-hole mounting.**

tie bar See **plating bar.**

tinned Literally, coated with tin, but commonly used to indicate coating with solder.

tinning A process for coating metallic surfaces with a thin coat of solder.

to can See **transistor outline metal can package.**

tombstoning Term used to describe the lifting of one end of a component during surface mount solder reflow.

tooling feature A specified physical feature on a printed board or panel such as a marking, hole, cutout, notch, slot or edge used exclusively to position the board or panel or mount components accurately.

tooling holes A general term for holes placed in a printed board or panel and used to aid in the manufacturing, assembly, or testing process.

Top Side Metallurgy (TSM) Refers to the metallization on the top side of a substrate to which a chip is joined.

Top Surface Metallurgy (TSM) The layer of conductive metal on the top of a substrate or package.

topography The surface condition of a film; bumps, craters, etc.

torque test A test designed to ascertain the stiffness of a material under given environmental conditions.

transfer molding An automated type of compression molding in which a preform of plastic (usually an epoxy-based resin) is poured from a pot into a hot mold cavity.

transfer soldering A soldering process using a hand soldering iron and a measured amount of solder in the form of a ball, chip, or disk.

transient mismatch Thermal mismatch between elements of a structure which, because of thermal lag, varies with time until reaching a steady state value.

transistor An active semiconductor device capable of providing power amplification and having threc or more terminals.

Transistor Outline (TO) package An industry standard package designation established by JEDEC of EIA.

transistor outline metal can package In integrated circuits, a type of package resembling a transistor can but generally larger and having more leads. The pins are arranged in a circle in the base.

transmission cable Two or more transmission lines. If the structure is flat, it is called flat transmission cable to differentiate it from a round structure, such as a jacketed group of coaxial cables.

transmission line A signal-carrying circuit composed of conductors and dielectric material with controlled electrical characteristics used for the transmission of high-frequency or narrow-pulse signals.

trim (1) Of resistors, to change from the as-fired value to the final desired value, usually by removing parts of the body of the resistor (with abrasive particles—"sand blasting"—or with a laser). (2) To make adjustments in resistors, coils, capacitors (often by adding or replacing) to bring electrical performance into exact agreement with specifications.

trim lines Lines that define the borders of a printed board.

triplate structure In a printed board, a signal line may be sandwiched between two reference planes. Such a design is referred to as a *shielded stripline* or a *triplate structure*.

true position The theoretically exact location of a feature or hole established by basic dimensions.

tuning fork contact A U-shaped female contact, either stamped or formed, so named because it resembles a tuning fork.

twill weave Mesh pattern in which the filaments are woven with one over, two under.

two-layer tape A form of tape fabrication for tape automated bonding (TAB), whereby the copper metallurgy is directly deposited on the Kapton carrier tape (i.e., no adhesive is used in bonding the copper).

Ultra Large Scale Integration (ULSI) So far, an extreme in circuit integration, used to indicate presence of one hundred million transistors (or more) on a single semiconductor chip.

ultrasonic bonding A bond formed when a wire is pressed against a bonding pad and the pressing mechanism is ultrasonically vibrated at a frequency above 10 kHz. These high-frequency vibrations break down and disperse the oxide films that are present on the conductor surfaces. As these surface films are removed, diffusion of the conductor materials occurs at the interface. The joints formed are metallurgically sound diffusion bonds.

undercut (1) In process, the distance on one edge of a conductor measured parallel to the board surface from the outer edge of the conductor, including etch resists, to the maximum point of indentation on the copper edge. (2) After fabrication, the distance on one edge of the conductor measured parallel to the board surface from the outer edge of the conductor, excluding overplating and coatings, to the maximum point of indentation on the same edge.

unsupported hole A hole containing neither conductive material nor any other type of reinforcement.

vacuum deposition Deposition of a metal film onto a substrate in a vacuum by metal evaporation techniques.

vapor-phase reflow A technique for solder reflow to form package interconnections. The solder joint is heated by the heat of condensation of an inert vapor.

varnish The mixed and aged catalyzed resin solution used to impregnate glass cloth in the process of producing printed boards.

vehicle The organic system in a composition.

Very High Speed Integrated Circuit (VHSIC) Originally referring to devices with 1.0 micrometer ground rules.

very large scale integration (VLSI) Referring to a single semiconductor chip with more than 10,000 transistors. The upper boundary is not well-defined.

via (hole) An opening in the dielectric layer(s), of a multilayer structure, whose walls are made conductive to enable interconnections between conductive layers.

via, blind A via that does not go all the way through a multilayer structure. It may interconnect the top or bottom surfaces to one or more internal planes.

via, buried A via that interconnects internal planes in a multilayer structure, but does not come out to the top or bottom surfaces.

via, fixed A via built into a package on a predetermined grid.

via, programmable A via whose location does not correspond to the same grid locations as fixed vias. The via and its location is "programmed" into a design only as needed.

via, segmented A fixed via interconnection. A predetermined subset of all wiring planes.

via, through A fixed via passing through all wiring planes.

viscosity The intrinsic property of a fluid that resists internal flow by offering counteracting forces.

vitreous Glassy. As used in ceramic technology, indicating fired characteristics approaching a glassy state but not necessarily totally glassy.

vitrification The formation of a glassy or noncrystalline material.

void The absence of substances in a localized area.

volt A unit of electromotive force.

voltage drop The amount of voltage loss from original input in a conductor of given size and length.

voltage plane A conductor or portion of a conductor layer on or in a printed board that is maintained at other than ground potential. It can also be used as a common voltage source for heat-sinking or for shielding.

voltage plane clearance The etched portion of a voltage plane around a plated-through or non-plated-through hole that isolates the voltage plane from the hole.

voltage regulation The drop in voltage due to load current, usually expressed as a percentage.

VSOP Acronym for Very Small Outline Package.

wafer Usually, a slice of semiconductor crystalline ingot used for substrate material when modified by the addition, as applicable, of impurity diffusion (doping), ion implantation, epitaxy, etc., and whose active surface has been processed into arrays of discrete devices or ICs by metallization and passivation.

warp See **bow** and **flatness.**

water absorption test A method used to determine the water penetration through an insulating material after a given water immersion period.

watt A unit of electrical power; the power of one ampere of current pushed by one volt of electromotive force.

wave soldering A process in which printed boards are brought in contact with the surface of continuously flowing and circulating solder. Also called *mass soldering.*

wavelength The distance, measured in the direction of propagation, of a repetitive electrical pulse or waveform between two successive points that are characterized by the same phase of vibration.

wear track In connectors, the imprint of one contact surface on its mate caused by insertion and/or withdrawal.

weave exposure A surface condition of base material in which the unbroken fibers of woven glass cloth are not completely covered by resin.

weave texture A surface condition of base material in which a weave pattern of glass cloth is apparent although the unbroken fibers of the woven cloth are completely covered with resin.

wedge bond A thermocompression bond in which a wedge-shaped tool is used to apply pressure to the wire being attached.

welding Connections that are made by fusing two conductors, using heat, pressure, or both.

wetting The formation of a relatively uniform, smooth, unbroken, and adherent film of solder to a basis metal.

whiskers Single-crystal growths resembling fine wire, which may extend to 0.64 mm (0.025 in) high. They most frequently occur on boards or components that have been electroplated with tin. The process may be described as an electroless one in that growth requires no voltage. Growth can occur in inert-gas or vacuum environments and is reportedly induced by compressive stress and high temperatures, accelerated by—though not dependent upon—humidity. (See also **electromigration.**)

wicking Capillary adsorption of liquid along the fibers of a base material. Also, the flow of solder away from the desired area by wetting or capillary action.

wiping action The action of two electrical contacts that come in contact by sliding against each other.

wire bond A completed wire connection whose constituents provide electrical continuity between the semiconductor die (pad) and a terminal. These constituents are (1) the fine wire; (2) metal bonding surfaces like die pad and package land; and (3) metallurgical interfaces between wire, and metals on both the chip and substrate.

wire bonding The method used to attach very fine wire to semiconductor components in order to interconnect these components with each other or with package leads.

wire wrap A method of connecting a solid wire to a square, rectangular, or V-shape terminal by tightly wrapping the wire around the terminal with a special tool.

wireability The capability of a package to permit the interconnections of subpackages or chips mounted on it and terminals attached to it measured as the probability of wiring success. Wiring capacity (length), via availability, and terminal access are all considerations in determining package wireability.

wiring (also **routing**) The manual or automatic prescription of routes for wires interconnecting package components.

wiring assignment The manual or automatic prescription of particular pads, pins, connectors, or terminals to which corresponding wires are to be attached.

wiring capacity The total available length of wiring tracks in a package.

wiring channel A linear region on a package wiring plane containing space for at least one wiring track.

wiring demand The product of wiring connection count and average connection length required to interconnect a given set of chips or subpackages.

yellow wire Discrete wires that are yellow in color interconnecting terminals on a package. Originally used in reference to all back panel wiring on early electronic assemblies.

yield The ratio of the number of acceptable items produced in a production run to the total number that were attempted to be produced (i.e., started in the production run).

yield strength The stress at which a material exhibits a specified strain deviation from purely elastic stress-strain behavior (often chosen at 1 percent offset).

zero-insertion-force connector A connector in which the contact surfaces do not mechanically touch or very lightly touch, thus requiring no insertion force. After mating, the contacts are actuated in some fashion to make intimate electrical contact.

19.3 Author's Note

Since putting this glossary together in August, 1993, I have already had to make several revisions to it. Some changes were terms, provided by reviewers, that I had not run across before. Each company tends to define acronyms and terms that best serve them in describing package technology. It is quite doubtful if all the of this terminology has been included in the preceding glossary. Also, several new terms had been coined since the initial compilation, another reminder of the dynamic environment in which we live and work.

19.4 References

1. *Electronic Materials Handbook: Volume 1 Packaging,* published by ASM International, 1989.
2. R. R. Tummala and E. J. Rymaszewski (eds.), *Microelectronics Packaging Handbook,* Van Nostrand Reinhold, 1989.
3. D. A. Doane and P. F. Franzon (eds.), *Multichip Module Technologies and Alternatives: The Basics,* Van Nostrand Reinhold, 1993.
4. John H. Lau (ed.), *Handbook of Tape Automated Bonding,* Van Nostrand Reinhold, 1992.
5. G. Messner, I. Turlik, J. W. Balde, and P. E. Garrou (eds.), *Thin Film Multichip Modules,* International Society for Hybrid Microelectronics, 1992.
6. D. P. Seraphim, R. Lasky, and C.-Y. Li (eds.), *Principles of Electronic Packaging,* McGraw-Hill, 1989.
7. *Low End Packaging Applications Guide,* edited by R. W. Gedney, March, 1990; 160-page guide to package technology used internal to IBM.

Author Biographies

John H. Lau is a senior engineer of Hewlett-Packard Company located in Palo Alto, California. His research activities cover a broad range of electronics packaging and manufacturing technology.

He received the B.E. degree in civil engineering from National Taiwan University, the M.A.Sc. degree in structural engineering from University of British Columbia, the M.S. degree in engineering mechanics from University of Wisconsin, the M.S. degree in management science from Fairleigh Dickinson University, and the Ph.D. degree in theoretical and applied mechanics from University of Illinois.

Prior to joining Hewlett-Packard Laboratories in 1984, he worked for Exxon Production and Research Company and Sandia National Laboratory. He has more than 24 years of research and development experience in applying the principles of engineering and science to the electronic, petroleum, nuclear, and defense industries. He has authored and coauthored over 100 technical publications in these areas and is the author and editor of the books, *Solder Joint Reliability: Theory and Applications, Handbook of Tape Automated Bonding, Thermal Stress and Strain in Microelectronics Packaging, The Mechanics of Solder Alloy Interconnects, Handbook of Fine Pitch Surface Mount Technology,* and *Chip On Board Technologies for Multichip Modules.*

He is an associate editor of the *ASME Transactions, Journal of Electronic Packaging,* and the *IEEE Transactions on Components, Packaging, and Manufacturing Technology.* He has also served as session chairman, program chairman, general chairman, and invited speaker of several IEEE, ASME, ASM, ISHM, and NEPCON conferences and symposiums. He is an IEEE fellow, a registered professional engineer, and is listed in the American Men and Women of Science.

Richard E. Sigliano received his B.S. in Biological Sciences/Chemistry from Fullerton State. He worked for 5 years at Beckman Instruments as a Material Development Engineer in the areas of liquid crystal materials and displays, thin film microcircuits, and thick film materials and microcircuits.

He has been with KYOCERA AMERICA for the past 15 years, and is currently the New Product Development and Marketing Manager.

He is a member of IEPS, IEEE, ISHM, and is currently the company representative at the JEDEC JC-11 Committee for package standardization.

Osamu Fujikawa is Director, Research and Development Group of Ibiden, Ogaki, Japan. He received his B.S. degree in chemistry from Shizuoka University, Japan in 1970 and joined Ibiden in the same year. Since then he has played a key role in developing PCB, COB, and plastic package material and process technology. Currently, his R&D management assignment covers the developing DCA, MCM, plastic package technologies, and other nonelectronics products at Ibiden.

Motoji Kato is Sales and Marketing Manager of Ibiden U.S.A., Santa Clara, California. He received his B.S. degree in nuclear engineering from Nagoya University, Japan in 1983. He joined Ibiden in the same year and as an engineer of material and process development. He has been involved in technical marketing activities at the U.S. subsidiary of Ibiden since 1988.

Patrick Hession is Component Engineer, Compaq Computer Corporation, Houston, Texas. Mr. Hession received his B.S. degree in chemical engineering from the University of Rochester, New York, in 1982. He has over 12 years of experience in the electronics industry dealing with electronic packaging. His experience includes over 6 years at Texas Instruments in Dallas holding various engineering and engineering management positions in their PCB fabrication organization. While at TI, Mr. Hession worked on development of high-density, sequential multilayer PCBs using laser drilled buried and blind vias. Mr. Hession worked for 5 years at Apple Computer in Cupertino, California, managing their packaging and process development organization. His responsibilities included TAB and BGA package and process development, as well as responsibility for Apple's PCB design technology. For the past year, Mr. Hession has been with Compaq Computer Corporation in Houston. He is currently responsible for Compaq's PCB and IC packaging technology, working in Compaq's PCA-Components Engineering organization. His activity includes managing Compaq's PCB technology roadmap, and design and development of IC packaging, including BGAs.

Thomas Caulfield, a senior engineering manager in MLC Packaging Applications at IBM Microelectronics, has an Eng.Sc.D. degree in Metallurgy from Columbia University. Since 1990, he has been both a team leader and engineering manager during the qualification and implementation of CBGA technology within IBM. His current responsibilities include managing New Business Development and New Product Application in the Ceramic Chip Carrier Business Unit.

Marie Cole, an advisory engineer in MLC Packaging Applications at IBM Microelectronics, has a B.S.Ch.E. from Rensselaer Polytechnic Institute and an M.S. in Materials Science from Columbia University. She has worked on the development of CBGA and CCGA packaging since 1988. Her current role is in New Product Applications in the Ceramic Chip Carrier Business Unit.

Frank Cappo has nearly 30 years of experience in packaging process, tooling, and product development. These activities include support for System 360, 370, and 390. His current assignment is in the MLC Electrical Design and Analysis area of the Ceramic Chip Carrier Business Unit of IBM Microelectronics.

Jeffrey Zitz is the lead thermal engineer for the Advanced Thermal Technology department in the Ceramic Chip Carrier Business Unit of IBM Microelectronics. He received his B.S.M.E. and M.S.M.E. from Rensselaer Polytechnic Institute, concentrating in air and immersion cooling, and has more than 8 years experience in the module and system-level packaging of personal, workstation, mini- and mainframe computers. Jeff is a licensed Professional Engineer in the state of New York.

Joseph Benenati has over 30 years of experience in application, design, and fabrication of electronic modules. He currently is a senior technical staff member in the IBM Microelectronics Division, playing a major role in packaging menu selection to fill product application needs for both single and multichip modules. He is also the program manager for ball grid array modules.

Donald R. Banks received a B.S. in Metallurgical Engineering from the Montana College of Mineral Science and Technology in 1984 and an M.S. in Materials Science and Metallurgical Engineering from the University of Notre Dame in 1986. Mr. Banks has published on intermetallic compound growth, solder corrosion, assembly processes, reliability, and array packaging. He has 5 years experience with ball grid array and column grid array technologies. The author worked in IBM's Assembly Process Design Group from 1988 to 1993 in Austin, Texas and Endicott, New York. Mr. Banks is currently employed by Motorola, Inc., in the Advanced Packaging and Assembly Manufacturing group in Austin, concentrating on both plastic and ceramic BGA and CGA packaging.

Karl Hoebener is an Advisory Scientist in the IBM PC Company Austin. He is at present involved in the Advanced Laminate Technology Development in the panel manufacturing facility. Prior to joining the ALT development group Karl was the project leader of the Solder Ball Connect process development and qualification effort. He was a member of the Austin second-level packaging development laboratory and was associated with several projects that included surface mount technology, SMT/PIH hybrid processes. Prior to joining the Austin facility he was involved in first-level packaging projects such as direct chip attachment on ceramic substrates, and chip dicing and picking.

Karl Hoebener is the recipient of five patents and has authored six technical disclosure bulletins. He received the First and Second Invention Achievement Awards, and an Outstanding Technical Achievement Award for Solder Ball Connect Development and Qualification. He has authored a number of internal and external technical papers.

Puligandla Viswanadham is an Advisory Scientist in the IBM PC Company, Austin. He is currently involved in the Advanced Laminate Technology Development in the Panel Manufacturing Facility. Prior to joining the panel plant he was involved in the process development and reliability assessments of fine pitch assemblies that included fine pitch Quad Flat packs and Thin Small Outline Packages in IBM Systems Technology Division in Endicott, New York. He has an M.Sc., degree from Saugor University, India, and a Ph.D. degree in chemistry from the University of Toledo, Toledo, Ohio. Dr. Viswanadham joined IBM, Rochester, Minnesota in 1979. As a member of the Materials and Process Engineering group he was involved in corrosion studies, analytical methods development, plating, and contamination control. He joined IBM, Austin development laboratories in 1986 and has been involved with solderable surface evaluations, Tape Automated Bonding (TAB), and Fine Pitch Quad Flat Pack second-level assembly process development and reliability evaluations. He was Site Materials Engineering/Analytical Labs Manager at Austin during 1989 to 1990. Prior to Joining IBM, his research activities included high-temperature chemistry and thermodynamics of binary and ternary chalcogenides, atomic absorption, slag seed equilibria in coal-fired magnetohydrodynamic energy generation, and astrophysics. He has authored or coauthored over 65 publications in various technical journals, symposium proceedings, and trade magazines and books. He is a Fellow of the American Institute of Chemists and a member of American Chemical Society, National Association of Corrosion Engineers, New York Academy of Sciences, and Scientific Research Society of North America. He received the IBM First and Second Invention Achievement Awards, and Fourth-level Technical Author Recognition Award. Dr. Viswanadham is the author of two patents and 14 Invention Disclosures.

Yung-Cheng Lee received his B.S.M.E. degree from the National Taiwan University in 1978, and the M.S. and Ph.D. degrees from the University of Minnesota, in 1982 and 1984, respectively. He is an Associate Professor of Mechanical Engineering and the Associate Director of the Center for Advanced Manufacturing and Packaging for Microwave, Optical and Digital Electronics, University of Colorado—Boulder. Prior to joining the University in 1989, he was a Member of Technical Staff at AT&T Bell Laboratories, Murray Hill, New Jersey.

His research interests include low-cost prototyping and thermal management of multichip modules, 3-D packaging, self-aligning soldering, thermosonic bonding, optoelectronics packaging, and fuzzy logic modeling and control. He received the Presidential Young Investigator Award from National Science Foundation in 1990, and the Outstanding Young Manufacturing Engineer Award from SME in 1992.

Jay Jui-Hsiang Liu received the B.S. degree in Physics from National Taiwan University in 1977, the M.S. degree in Materials Science from National Tsing Hua University in 1979, and the Ph.D. degree in Materials Science from Cornell University in 1986.

Currently, he is a Senior Staff Engineer with the Advanced Packaging Development Center of Motorola at Chandler, Arizona, where he works on all issues related to high-performance single-chip and MCM package designs. Prior to joining Motorola, Jay was a member of technical staff with AT&T—Bell Labs, Princeton and worked on material and assembly issues associated with surface mount technologies and MCMs.

Chi-Taou Tsai received the B.S. degree in Physics from National Central University in ChungLi, Taiwan, in 1976 and the M.S. and the Ph.D. degrees in Electrical Engineering from University of Utah, Salt Lake City, Utah in 1982 and 1984, respectively.

He joined Motorola at Phoenix, AZ after graduation to work on electronic packaging R/D. Currently, he is with the Advanced Packaging Design and Automation Center of Motorola at Chandler, AZ, where he works on all issues related to the IC packaging electrical performance and signal integrity. His research interests include electrical modeling, simulation, measurement, design, and development of high-speed and high-performance packages, MCMs, and systems.

Yifan Guo is currently with IBM Microelectronic Division. Dr. Guo received his Ph.D. degree in Engineering Mechanics from Virginia Polytechnic Institute and State University in 1989. After graduation, he was appointed as Assistant Professor in the Engineering Science and Mechanics Department at VPI&SU. He taught courses in mechanics and continued his research in the area of experimental mechanics and mechanics of composite materials. He jointed IBM Corporation in 1990 and began his research in the field of microelectronic packaging. He was heavily involved in the development of the ceramic BGA package in IBM. He has received an IBM Division Award for his contributions in the development and applications of moiré interferometry in electronic packaging.

Dr. Guo has published more than 20 technical papers in journals and conference proceedings, and received a best-paper award for his paper published in the *Journal of Experimental Techniques.*

John S. Corbin, Senior Engineer, IBM RISC Systems 6000 Division, Austin, Texas received his B.S. and M.S. degrees in mechanical engineering from the University of Texas at Austin in 1972 and 1974, respectively. He joined IBM in 1974 and spent 10 years in printer development, working primarily in the area of motion control. This was followed by 6 years in the Systems Technology Division Packaging Laboratory in Austin, applying finite element modeling techniques in the area of mechanical packaging reliability. He is currently a Senior

Engineer in the IBM RISC System 6000 Division where his interests relate to system packaging reliability. Mr. Corbin is a registered professional engineer in the state of Texas.

Robert C. Marrs is Vice President & General Manager, MicroSystems Division, Amkor Electronics, Inc., Chandler, Arizona. His responsibilities in this position include strategic and tactical marketing, design, development, and profit/loss management of newly developed technologies. He directs all advanced-technology product groups and design teams. Mr. Marrs has been with Amkor Electronics since 1989 and founded the advanced packaging division for Amkor Electronics in 1991. He is credited with the design and development of numerous new packaging technologies including PowerQuad, low-cost flip chip, ICR, MCM BGAs, and SuperBGAs. He has more than 20 years of experience in IC packaging and assembly and holds numerous packaging technology patents. Prior to his position with Amkor, Mr. Marrs was founder and president of another subcontract manufacturer, and has held a variety of other packaging positions during his career in the IC industry. Prior affiliations include VLSI Technology, National Semiconductor, Siliconix, and TRW. Mr. Marrs received his Bachelor of Science degree cum laude in 1974 from California State University at Long Beach in Industrial Technology.

William B. Mullen is Senior Member of the Technical Staff and Manager of the Advanced Manufacturing Technology Interconnect Technologies Laboratories in the Land Mobile Products Sector of Motorola, Inc., Fort Lauderdale, Florida. The Interconnect Technology laboratories are responsible for the development and characterization of strategic assembly and integrated circuit packaging technologies for their products. During the last 7 years Mr. Mullen has taken the initiative to introduce both CBGA and PBGA technologies into the portable products produced by Motorola. His expertise in interconnect and process design has provided the foundation for the introduction of surface mount technology to Motorola's portable products. Mr. Mullen received his B.S. in Physics from Florida Institute of Technology in 1970 and has since remained focused in the field of electronics manufacturing processes and electronic materials. Included among his 17 issued patents are several related to OMPAC™ and BGA technologies. He is credited with numerous internal publications on interconnect technology and is currently exploring the attributes of DCA.

Robert Darveaux is Senior Staff Engineer in the Land Mobile Products Sector of Motorola in Plantation, Florida. Dr. Darveaux received his B.S. in Nuclear Engineering from Iowa State University and his Ph.D. in Materials Science and Engineering from North Carolina State University. His research and development activities have been in the areas of thermal performance of multichip modules, mechanical characterization of solders, finite element modeling, power device packaging, failure analysis, and reliability prediction. His modeling and fatigue analysis work have been integral to Motorola's BGA implementation process. He has authored over 20 technical articles and has three patents.

Barry M. Miles is Senior Staff Engineer and Manager of the Advanced Manufacturing Technology IC Packaging Lab, Motorola, Plantation, Florida. Barry is responsible for semiconductor package development and the implementation of these packages into the manufacturing environment, specific to portable and mobile electronic communications products. He was one of the original members of the PBGA package development team at Motorola in Florida, with responsibility for package-level reliability testing and moisture-sensitivity studies. Employed by Motorola since 1983, he has also done work in the areas of thin film hybrids, flex and printed circuit board fabrication, and surface mount assembly. Barry received his B.S. degree in Chemical Engineering from the University of South Florida, Tampa, Florida. He has written numerous papers for publishing in technical conference proceedings, and currently holds 10 issued patents.

Allen Hertz is Staff Process Engineer of Paging Products Group of Motorola, Boynton Beach, Florida. Mr. Hertz is responsible for the development and introduction of production processes and overseeing designs for manufacturability of new paging products. Mr. Hertz is best known for his work on plastic ball grid array packages and 1005 (20 × 40) chip components; he holds patents that apply to both of these areas. In 1989 Mr. Hertz was the first to implement PBGA technology into a commercial product and has been instrumental in the success of this technology in Motorola. Mr. Hertz received his B.S. in Ocean Engineering from Florida Atlantic University; he has authored over 10 technical publications and has four issued patents.

Philip E. Rogren is Vice President of Marketing and Sales, Hestia Technologies Inc., Sunnyvale, California. Mr. Rogren has been involved in semiconductor packaging and interface technology since 1974. His involvement in the industry has included both engineering and marketing and sales responsibilities. He has published numerous technical papers in refereed journals and participated as speaker and panelist in several industry workshops. He received a B.S. degree in Ceramic Engineering from the University of Washington in 1974.

Seyed Hassan Hashemi is a Member of the Technical Staff and a project manager in MCC's High Value Electronics (HVE) Division. He has worked at MCC since January 1985. His work includes high-frequency electrical characterization and thermomechanical modeling of single and multichip packaging and interconnect problems. He has also been a technical lead for the power distribution project, where he worked on experimental and simulation of switching-noise problems in high-speed packaging and interconnect applications. He is currently managing two multiclient projects; the Single and Few Chip Packaging (SFCP) project and the Ball Grid Array (BGA) Package Reliability project.

Mr. Hashemi received a B.S. in Electrical Engineering from the University of Houston in 1982, and an M.S. in Electrical Engineering from the University of Texas at Austin in 1984. He holds four U.S. patents, and is the principle author or co-author of over 30 MCC internal reports, 3 book chapters, and over 25 journal and technical publications in the areas of single and few chip packaging, power distribution, electrical characterization, and modeling of packaging and interconnect problems. He is a member of Eta Kappa Nu, Tau Beta Pi, IEEE, ISHM, and IEPS.

David B. Walshak, Jr., is a member of the technical staff in MCC's High Value Electronics Division. He joined MCC in April 1984. He obtained his B.S. in Mechanical Engineering from the University of Texas at Austin in 1982 and an M.S. in Mechanical Engineering from UT in 1988. His primary area of concentration at MCC has been in the fine-pitch assembly process development. Over the last several years he has developed tape automated bonding (TAB) assembly processes for both inner- and outer-lead bonding using thermocompression, thermosonic, and laser energy techniques.

Mr. Walshak was the project leader for the first sets of multichip modules that were assembled at MCC. He developed cost models for various assembly techniques for use in prototype-sized assembly requests. Recently, he has used finite element modeling to conduct parametric studies of various package components to optimize thermal performance for various package types, including BGAs.

He has written and presented several papers relating to fine-pitch assembly and thermal modeling, including a chapter on inner-lead bonding for John Lau's *Handbook of Tape Automated Bonding*. Mr. Walshak is a member of Pi Tau Sigma, Tau Beta Pi, Phi Kappa Phi, IEPS, and SMTA.

Kingshuk Banerji is Senior Staff Engineer in the Advanced Manufacturing Technology Labs of the Land Mobile Product Sector of Motorola in Plantation, Florida. He received his B.S. and M.S. degrees from Indian Institute of Technology and his Ph.D. in materials engineering from Georgia Tech. He works in the areas of process development and interconnect reliability for board assembly and flip chip packaging.

Andrew J. Mawer is a packaging engineer with Motorola Semiconductor Products Sector's Advanced Packaging Technology Department in Austin, Texas. He received his B.S. in Mechanical Engineering from the University of Texas at Austin in 1987 and his Master of Mechanical Engineering from Rice University in 1989. His primary area of interest is Plastic Ball Grid Array (PBGA) solder joint reliability characterization, improvement, and application lifetime prediction. He has also been involved with PBGA assembly process development (concentrating on solder bump attachment), package design, and all aspects of surface mount assembly, including inspection and rework. Prior to joining Motorola, Andrew spent 4 years at Compaq Computer Corporation in the Component Reliability, Materials Analysis, and Advanced Manufacturing Technology Departments working on various levels of packaging and interconnect, such as surface mount devices, PBGA, TAB, PGA, printed circuit boards, keyboards, laptop/desktop PC enclosures, sockets, connectors, and cables. His main areas of responsibility there were component/interconnect reliability, surface mount assembly process analysis, and thermal/mechanical finite element modeling. He was part of the team that implemented PBGA technology to Compaq. Andrew has authored or coauthored several technical articles on PBGA and has also presented at several industry symposiums.

Glenn Dody is Manager, Surface Mount Application Lab and Reliability Test Center in the Semiconductor Products Sector of Motorola in Austin, Texas. He has 20 years experience with the U.S. Air Force in Aviation Electronics, 7 years as Printed Circuit Board Assembly Manager, 2 years in surface mount package development, and 9 years in surface mount technology research and development. Glenn is a member of ISHM, and the Testing subcommittee of the IEEE Compliant Lead Task Force.

Chin-Ching Huang is Manager, Package Characterization, at VLSI Technology, Inc. Since joining the company in 1989, he has been responsible for the development of measurement techniques, modeling, and simulation of thermal, mechanical stress, and electrical characteristics of packages, TAB and MCM. He has five U.S. patents related to package technology at VLSI Technology, Inc. Dr. Huang received a B.S. (Physics) from Soochow University, Taiwan, M.S. (Physics) and M.S.E.E. from Florida Institute of Technology, and a Ph.D. degree in Materials Science from the University of Virginia. Prior to joining VLSI Technology, Inc., he held the position of Principal Engineer at Digital Equipment Corporation, where he was responsible for conducting research on heat dissipation for MCM. Dr. Huang started his career in the industry as a Senior Engineer at National Semiconductor/Fairchild Camera & Instruments Gate Array Division, where he was involved in package characterization with emphasis on ceramic and plastic package development for ECL and CMOS gate arrays.

Ahmad Hamzehdoost is Senior Packaging Engineer, VLSI Technology, Inc., San Jose. His fields of expertise include thermal and stress analysis of microelectronic packages. He received a B.S. degree in Electrical Engineering from University of Arizona and M.S. degree in Mechanical Engineering from Manhattan College. He has authored several papers on the topic of heat transfer in IC packaging.

John A. Adams has over 19 years experience in the development of Automated Inspection and Measurements systems, using both vision and discrete gauging technologies. He has extensive experience in developing vision inspection computer algorithms. He has patented x-ray and x-ray laminography vision systems. He has expertise in system software design, automated systems hardware design, project planning, equipment acceptance test methodology, calibration methodology, statistical process control, capability studies of automated instrumentation, specification generation, technical documentation, manuals, and trade journal papers. He has presented papers at many national and international technical conferences. He has a B.S. degree in Physics from the University of San Diego and Ph.D. degree in Experimental Nuclear Physics from Arizona State University. He is a member of IEPS. He has received IR-100 award in 1984, is named in *Who's Who in Finance and Industry 1994,* and *Who's Who Worldwide 1995.*

Tom C. Chung is Tandem Computers' Visiting Representative at Microelectronics and Computer Technology Corporation (MCC) in

Austin, Texas. Dr. Chung has more than 14 years of experience in the field of advanced computer packaging and interconnect technologies, including surface mount, tape automated bonding, flip-chip, advanced single-chip packaging, multichip module technologies, etc. He worked on semiconductor assembly, test, packaging, and interconnect-related technology developments for United Technologies Corporation and MCC for 5 years, respectively, before joining Tandem Computers, Inc., in 1989 as a Development Engineering Manager in the advanced computer packaging and interconnect department. Since early 1993, he has taken the responsibility as Tandem's technical representative at MCC to lead a group to develop flip-chip-based MCM technology for the next generation of computer workstations. He has a B.S.E. from NCKU in Taiwan, an M.S.M.E. from Texas Tech University, and an M.S. in engineering management and a Ph.D. in engineering from Southern Methodist University, Dallas, Texas. He owns four U.S. patents and has published more than 20 technical papers. He is also a coauthor of two technical books, *Handbook of Tape Automated Bonding* (published in 1992) and *Chip on Board Technologies for Multichip Modules* (published in 1994).

Paul Mescher is a process development engineer in IBM's Assembly Process Design group. Mr. Mescher received his B.S. degree from Arizona State University in Materials Science Engineering and joined IBM Endicott in 1988. He has been working exclusively on BGA process development issues for IBM since 1991, and continues to be a key member of the TBGA package development team. Prior to 1991, he worked in materials support, splitting time between TAB product and PTFE card process development issues.

Christopher A. Schmolze is Senior Technical Service Engineer, 3M Company, Semiconductor Products Business, Austin, Texas. Mr. Schmolze received his B.S.M.E. from the University of Minnesota. Employed by 3M since 1984, he has held positions in sustaining engineering, product development, and technical service. His areas of interest are static dissipative thermoplastics, IC packages, IC sockets, and IC trays and carriers. This work has led to two patents, one for a "Semiconducting Polymeric Composite" and the other for a "Contactless Multipurpose TAB Socket." He currently represents 3M on industry standards committees for IC sockets and IC package mechanical outlines and chairs the JEDEC JC11.13 Quality Subcommittee. Professional affiliations have included ASME and currently include ISHM and IEPS and CTEA.

E. Jan Vardaman, President, TechSearch International, Inc. received her B.A. in Economics and Business in 1979 from Mercer University, and her M.S. in Economics from the University of Texas in 1981. After working as a government computer industry analyst, she joined Microelectronics and Computer Technology Corporation in 1984, where she analyzed developments in semiconductor packaging/interconnect. In 1987, she founded TechSearch International, Inc., a company involved in licensing electronics technology and providing technical information on international developments in the electronics industry. She is a member of the IEEE, CHMT, ISHM, and the International Electronics Packaging Society. She has made presentations on developments in tape automated bonding and multichip modules in Japan, the United States, and Europe and has authored publications in numerous countries. She is also the editor of the recently released book published by IEEE titled, *Surface Mount Technology—Recent Japanese Developments.*

Ronald W. Gedney is a consultant in electronic packaging since retiring from IBM in 1992. Mr. Gedney received the B.S.E.E. degree from Tufts University, and subsequently joined IBM while undertaking part-time postgraduate studies at Syracuse University. At IBM he first developed reliability test techniques and specifications for passive components and ferrite memory cores. With the formation of IBM's Components Division, he become involved with the development of single- and multichip electronic packages. In 1974 he became Program Manager of IBM's Metallized Ceramic program, a thin film, single- and multichip package technology based in Endicott, New York. He has been manager of Packaging Development in IBM's General Technology Division, Burlington, Vermont, and a consultant at the IBM plant in Havant, England. In 1984 he returned to Endicott, where he has held a number of management and engineering assignments in the packaging development laboratory, including the development of TAB tape. He has published a number of papers and holds several patents in electronic packaging. A past president of the IEEE Components, Packaging and Manufacturing Technology Society, he was elected a Fellow of the IEEE in recognition for contributions to the field of electronic packaging.

Index

ABOUT THE EDITOR

John H. Lau is a senior engineer of Hewlett-Packard Company in Palo Alto, California. His research activities cover a broad range of electronics packaging and manufacturing technology. He has more than 24 years of research and development experience in applying the principles of engineering and science to the electronic, petroleum, nuclear, and defense industries. He has authored and coauthored over 100 technical publications in these areas and is the author and editor of the books *Solder Joint Reliability: Theory and Applications, Handbook of Tape Automated Bonding, Thermal Stress and Strain in Microelectronics Packaging, Handbook of Fine Pitch Surface Mount Technology, The Mechanics of Solder Alloy Interconnects,* and *Chip On Board Technologies for Multichip Modules.* He is an associate editor of the *ASME Transactions,* the *Journal of Electronic Packaging,* and the *IEEE Transactions on Components, Packaging, and Manufacturing Technology.*